T0250701

O. Nurmi   E. Ukkonen (Eds.)

Algorithm Theory – SWAT '92

Third Scandinavian Workshop on Algorithm Theory
Helsinki, Finland, July 8-10, 1992
Proceedings

Springer-Verlag

Berlin Heidelberg New York
London Paris Tokyo
Hong Kong Barcelona
Budapest

Series Editors

Gerhard Goos
Universität Karlsruhe
Postfach 69 80
Vincenz-Priessnitz-Straße 1
W-7500 Karlsruhe, FRG

Juris Hartmanis
Department of Computer Science
Cornell University
5149 Upson Hall
Ithaca, NY 14853, USA

Volume Editors

Otto Nurmi
Esko Ukkonen
Department of Computer Science, University of Helsinki
Teollisuuskatu 23, SF-00510 Helsinki, Finland

CR Subject Classification (1991): F.1-2, E.1-2, G.2-3

ISBN 3-540-55706-7 Springer-Verlag Berlin Heidelberg New York
ISBN 0-387-55706-7 Springer-Verlag New York Berlin Heidelberg

Typesetting: Camera ready by author/editor
Printing and binding: Druckhaus Beltz, Hemsbach/Bergstr.
45/3140-543210 - Printed on acid-free paper

Preface

The papers in this volume were presented at SWAT '92, the Third Scandinavian Workshop on Algorithm Theory. The workshop, which continues the tradition of SWAT '88, SWAT '90, and the Workshop on Algorithms and Data Structures (WADS '89, WADS '91), is intended as an international forum for researchers in the area of design and analysis of algorithms. The call for papers sought contributions in algorithms and data structures, in all areas, including combinatorics, computational geometry, data bases, parallel and distributed computing, and graphics. There were 120 papers submitted, of which the program committee selected 34 for presentation. In addition, invited lectures were presented by Leslie G. Valiant (Direct bulk-synchronous parallel algorithms), Alexander A. Razborov (On small depth threshold circuits), Gaston Gonnet (Efficient two-dimensional searching), and Emo Welzl (New results on linear programming and related problems).

SWAT '92 was held in Helsinki, July 8–10, 1992, and was organized in cooperation with the Department of Computer Science of the University of Helsinki. The organizing committee consisted of B. Aspvall (University of Bergen), H. Hafsteinsson (University of Iceland), R. Karlsson (Lund University), E. M. Schmidt (Aarhus University), O. Nurmi, J. Tarhio, and E. Ukkonen (University of Helsinki).

The program committee wishes to thank all referees who aided in evaluating the papers. The organizers would like to thank J. Kivinen, S. Palander, and J. Vilo for excellent service in all organizational matters related to the conference. Finally, we are very grateful to Nordiska Forskarutbildningsakademin, Ministry of Education (Finland), the Academy of Finland, University of Helsinki, EATCS, the Finnish Society for Computer Science, SUN Microsystems Oy, and ICL Data Oy for sponsoring the workshop.

Helsinki, May 1992

Otto Nurmi
Esko Ukkonen

Program Committee

S. Arnborg (Royal Institute of Technology)
B. Aspvall (University of Bergen)
J. Gilbert (Xerox PARC and University of Bergen)
T. Hagerup (Max–Planck–Institut)
T. Leighton (MIT)
C. Levcopoulos (Lund University)
A. Lingas (Lund University)
Th. Ottmann (University of Freiburg)
M. Overmars (University of Utrecht)
M. Penttonen (University of Joensuu)
W. Rytter (University of Warsaw)
N. Santoro (Carleton University)
S. Skyum (Aarhus University)
E. Ukkonen (chairman, University of Helsinki)

List of Referees for SWAT '92

Alt, H.
Anderson, R.
Bang-Jensen, J.
Bodlaender, H. L.
Boyar, J.
Brüggemann-Klein, A.
Carlsson, S.
Casas, R.
Chen, J.
Chlebus, B. S.
Cole, R.
Cowen, L. J.
de Berg, M.
Dietzfelbinger, M.
Diks, K.
Eckerle, J.
Fleischer, R.
Floréen, P.
Forsell, M.
Frandsen, G. S.
Garrido, O.
Gąsieniec, L.
Goldmann, M.
Golin, M. J.
Håstad, J.
Heinz, A.
Huber-Wäschle, F.
Ihler, E.
Jennings, E.
Jonssons, H.
Kahan, S. H.
Kant, G.
Karabeg, D.

Kari, J.
Kari, L.
Karlsson, R.
Karpinski, M.
Katajainen, J.
Kaufmann, M.
Kivinen, J.
Klasing, R.
Klein, R.
Kloks, T.
Kløve, T.
Klugerman, M.
Knuutila, T.
Kowaluk, M.
Kozen, D. C.
Kravets, D.
Krogdahl, S.
Kunde, M.
Kutyłowski, M.
Lagergren, S.
Langemyr, L.
Langholm, T.
Laver, T.
Lenhof, H.-P.
Lisper, B.
Luccio, F.
Mannila, H.
Mattsson, C.
Mehlhorn, K.
Meiser, S.
Miltersen, P. B.
Monien, B.
Müller, N.

Munthe-Kaas, E.
Nevalainen, O.
Newman, I.
Nielsen, M.
Nilsson, B.
Nurmi, O.
Orponen, P.
Pennings, M.
Raita, T.
Raman, R.
Reich, G.
Rossmanith, P.
Schirra, S.
Schmidt, E.
Schuierer, S.
Schwartzbach, M.
Schwarzkopf, O.
Singh, M.
Smid, M.
Syslo, M.
Szymacha, T.
Tarhio, J.
Teia, B.
Teuhola, J.
Uhrig, C.
van Kreveld, M.
Veldhorst, M.
Vilo, J.
Voutilainen, P.
Wanka, R.
Welzl, E.
Williamson, D. P.
Winter, P.

Table of Contents

VIII

Direct Bulk-Synchronous Parallel Algorithms *

Alexandros V. Gerbessiotis and Leslie G. Valiant
Aiken Computation Laboratory
Harvard University
Cambridge, MA 02138, USA

Abstract

We describe a methodology for constructing parallel algorithms that are transportable among parallel computers having different numbers of processors, different bandwidths of interprocessor communication and different periodicity of global synchronisation. We do this for the bulk-synchronous parallel (BSP) model, which abstracts the characteristics of a parallel machine into three numerical parameters p, g, and L, corresponding to processors, bandwidth, and periodicity respectively. The model differentiates memory that is local to a processor from that which is not, but, for the sake of universality, does not differentiate network proximity. The advantages of this model in supporting shared memory or PRAM style programming have been treated elsewhere. Here we emphasise the viability of an alternative direct style of programming where, for the sake of efficiency the programmer retains control of memory allocation. We show that optimality to within a multiplicative factor close to one can be achieved for the problems of Gauss-Jordan elimination and sorting, by transportable algorithms that can be applied for a wide range of values of the parameters p, g, and L. We also give some simulation results for PRAMs on the BSP to identify the level of slack at which corresponding efficiencies can be approached by shared memory simulations, provided the bandwidth parameter g is good enough.

1 The Model

The bulk-synchronous parallel or BSP model as described in [17] consists of three parts:

(i) a number of processor/memory components. Here we will assume that each consists of a sequential processor with a block of local memory.

(ii) a router that can deliver messages (typically to implement read and write operations) point to point among the components, and

(iii) facilities for globally synchronizing, in barrier style, all or a subset of the components.

The parameters of such a machine are p, the number of processor/memory components, L, the minimal time (measured in terms of basic local computation steps)

---

*This research was supported in part by the National Science Foundation under grant CCR-89-02500 and by the Office of Naval Research under grant N00014-85-K-0445.

between successive synchronization operations, and g the ratio of the total throughput of the whole system in terms of basic computational operations, to the throughput of the router in terms of words of information delivered.

Computation on this model proceeds in a succession of supersteps. In each superstep each processor is given a task that can be executed using data already available there locally before the start of the superstep. The task may include local computation steps, message transmissions and message receipts. To simplify the description of the performance of our algorithms, in this paper we shall assume that each superstep is either purely computation or purely communication (and not a mixture of both). In this way the two kinds of costs are easily separated.

The model has two variants, depending on how synchronization is treated, that affect performance to multiplicative constant factors. In one the synchronizer checks at every L time units whether the tasks that are synchronized in the current superstep have all finished, and only when it discerns that they have does the machine proceed to the next superstep. In this paper, again for simplifying the analysis, we will use the alternative model where a superstep can complete any time after the first L units. More specifically, each superstep is charged $\max\{L, x + hg\}$ basic time units, where x is the maximum number of local computational steps executed by any processor, and h is the maximum number of packets that any processor sends or receives in that superstep. [N.B. If h_1 is the maximum number sent and h_2 the maximum number received we are taking $h = \max\{h_1, h_2\}$ here, although $h = h_1 + h_2$ is possible too [16]. Also how the expression for charging combines the x with the h is not crucial if, as we do here, we insist that each superstep is pure computation or pure communication.] We note that any particular machine may impose a lower bound on L for two reasons: There will be a minimum time requirement for doing synchronizations. Also the communication throughput specified by the chosen g may be realisable only for large enough L because of startup or latency considerations in routing. We note that although the model emphasizes global barrier style synchronization, pairs of processors are always free to synchronize pairwise by, for example, sending messages to and from an agreed memory location. These message transmissions, however, would have to respect the superstep rules.

2 Background

The BSP model was originally suggested as a possible "bridging" model to serve as a standard interface between the language and architecture levels in parallel computation. Two modes of programming were envisaged [17]: *automatic mode* where programs are written in a high level language that hides memory distribution from the user (e.g. PRAM style), and *direct mode* where the programmer retains control of memory allocation.

This paper is concerned with the direct mode. Since it is known that, under appropriate conditions, automatic mode achieves optimality to within constant factors, the primary question to ask is whether there are any circumstances at all in which the effort of programming in direct mode is warranted. We can identify four such circumstances:

(a) where small multiplicative constant factors in runtime are important,

(b) where smaller problem instances can be run efficiently in direct mode (i.e. less "slack" is sufficient) than in automatic mode,

(c) where the available machine has a high g factor (since automatic mode generally requires g to be close to unity), and

(d) where the available machine has an L that is sufficiently high for direct, but not for automatic mode, for the problem instance in hand.

In light of the first of these conditions we emphasize multiplicative constant factors and in this paper we restrict ourselves to *one-optimal* algorithms. Although it is generally problematic to measure complexity to this accuracy, since the operations that are counted have to be carefully defined, it can be made meaningful, if, as we do here, we measure ratios between runtimes on pairs of models that have the same set of local instructions. Thus we shall say that a BSP algorithm is *one-optimal in computation* if it performs the same number of operations, to a multiplicative factor of $1 + o(1)$ as $n \rightarrow \infty$, as a specified corresponding sequential or PRAM algorithm. Insistence on one-optimality leads to algorithms that only work in restricted ranges of n and L in terms of p. These ranges can be extended in most cases, however, if the requirements are relaxed to, say, "two-optimality".

We note that the role of a standard bridging model is as important in the direct programming context as in the automatic one. It will soon be inexpensive to augment workstations and personal computers with several processors. As long as efficient transportable software packages can be written for them, substantial reductions in runtime can be gained. We hypothesise that the BSP model, parametrised by p, L, and g and with direct control of memory distribution among the processors, is a simple enough model to serve this bridging role. One can foresee software packages to be written that can be run efficiently on a variety of machines for a wide range of values of these parameters. In these packages the algorithms would be parametrised by p, L, g, and the problem size n, in the same manner as we do in this paper. An efficient instance of the algorithm can be selected by a combination of compile-time and run-time decisions based on these parameters.

The development and analysis of direct BSP algorithms has a secondary motivation, one relating to computer architecture. A currently unresolved question is that of determining whether, if BSP computers are constructed in currently foreseeable technologies, it is efficacious to augment the global operations supported in hardware beyond the router, to, for example, broadcasting or the parallel prefix. To resolve this the complexities of some of the most frequently computed problems with respect to these various options need to be understood.

3 Communication Optimal Simulations

In order to evaluate the usefulness of direct BSP algorithms we have to consider the best PRAM simulations with which they might compete. This competition is fairly stiff since, given sufficient slack, we can find simulations that are one-optimal in communication operations and optimal to within small multiplicative constant factors in computational operations. By a PRAM simulation being *one-optimal in communication* we mean in particular that $T(n)$ read/write operations of a PRAM can be simulated in $(2gT(n)/p)(1 + o(1))$ time units on the BSP. The factor of $2g$ is necessary since each read operation is implemented as a message to a distant memory block followed by a return message, and time g is charged per message (in terms of throughput, not necessarily latency). Asymptotics are in terms of the problem size n as $n \to \infty$. The parameters g and p may or may not depend on n.

The results concern the simulation of CRCW and EREW PRAMs (which we call CPRAMs and EPRAMs for short) on either the BSP or the S*PRAM. The latter model is defined as follows: It consists of p processor/memory components each of which can in each period of g time units direct a message to any other component. Any component that has just one message directed at it receives it successfully and sends an acknowledgement to the sender in the same period. Any component that has more than one message directed at it fails to receive any of them and sends no acknowledgement. This model was suggested by Anderson and Miller [2] and also discussed by Hartmann and Redfield [7]. It models a variety of schemes for optical interconnects (eg [7], [11], [12]) that have been proposed for efficient all-to-all dynamic communication (although at present none of them is competitive with electronic methods).

Since the S*PRAM seems an elegant and potentially important model the question of simulating the BSP on it is of interest. In [16] it was shown how this could be done to within constant factors. Here we consider at what point one-optimality can be reached where the PRAMs are implemented directly on the S*PRAM (by-passing the BSP).

The time bounds hold with high probability assuming the availability of easily computed hash functions, that behave as perfect random functions when distributing addresses to memory blocks. We can use functions as in [4], or use pipelining to evaluate several applications at a time of a possibly less easily computed hash function.

Let $\log x$, $\lg x$ be the logarithms of x to the base e and two respectively.

Theorem 1 *Let w_p be any function of p such that $w_p \to \infty$ as $p \to \infty$. Then the following amounts of slack are sufficient for simulating any one step of the EPRAM or CPRAM on the BSP or S*PRAM in one-optimal expected time for communication (and constant factor optimal time in local operations if $g = O(1)$):*

$$
\begin{aligned}
&(i) \quad (w_p p \, \log p)\text{-}EPRAM \text{ on } p\text{-}BSP, \\
&(ii) \quad (w_p p^2 \, \log p)\text{-}CPRAM \text{ on } p\text{-}BSP, \\
&(iii) \quad (w_p p^2)\text{-}EPRAM \text{ on } p\text{-}S^* PRAM, \\
&(iv) \quad (w_p p^2 \, \log p)\text{-}CPRAM \text{ on } p\text{-}S^* PRAM,
\end{aligned}
$$

where the expressions before the hyphens denote the numbers of processors.

Proof: In each of the four simulations we shall randomly choose an appropriate hash function h that will randomise memory and distribute memory requests evenly among the p memory modules of the p-BSP or p-$S^* PRAM$. Thus, the contents of memory address m will be mapped by this hash function to a module $h(m)$, which we can assume from the previous discussion is random.

Each processor, at every step of a simulation will perform a number of tasks. For example, in case (i), each processor of the simulating BSP will perform $w_p \log p$

tasks of the simulated $EPRAM$, tasks that may need memory accesses distributed arbitrarily among the processors of the BSP.

We note that time g is charged for each message delivery (when performed in bulk). When we discuss "time", in this context, as the number of such operations, this extra factor of g is implied as far as actual time.

i) The simulation is the one given in [16]. After the distribution of $w_p \log p$ tasks to each processor of the BSP, as described above, these tasks are completed in one superstep. The duration of the superstep will be chosen large enough to accommodate the routing of an $(1 + \epsilon)w_p \log p$-relation, thus allowing the choice of an L as large as $(1 + \epsilon) w_p \log p$, for any appropriate $\epsilon > 0$ that depends on the choice of w_p.

Under the perfect hash function assumption, all references to memory are mapped to memory modules independently of each other, and thus the probability that a memory location maps to some module (say that of processor i), is $1/p$. We then use the following bound for the right tail of the binomial distribution [3]. The probability that the number of successes X in n independent Bernoulli trials, each with probability of success P, is greater than $(1 + \epsilon)nP$, for some ϵ $(0 < \epsilon < 1)$ is at most

$$Prob(X > (1 + \epsilon) \, n \, P) \leq e^{-\frac{1}{3}\epsilon^2 \, n \, P}.$$

Thus, the probability that among $w_p \, p \, \log p$ requests to memory, we have more than $(1 + \epsilon)w_p \log p$ requests to a specific memory module is at most $p^{-w_p^{1/3}}$, for, say, $\epsilon = w_p^{-1/3}$. Since there are p such modules, the probability that more than $(1 + \epsilon)w_p \log p$ requests are directed to any of these, is at most $p^{1 - w_p^{1/3}} = p^{-\Omega(w_p^{1/3})}$. It follows that reads or writes can be implemented as desired, in two or one supersteps respectively.

ii) Each processor of the BSP will simulate $w_p \, p \, \log p$ processors of the CPRAM. We shall show how, if one read request is made by each simulated processor, to an arbitrary other simulated processor, then they can be realised in two phases each of $(1 + o(1)) \, w_p \, p \, \log p$ steps of communication, and the same order, to within a constant factor of computation in the BSP. Write requests (with various conventions for concurrent writes) can be simulated in similar fashion, in just one phase.

Suppose we number the BSP processors $0, 1, \ldots, p - 1$. The simulation consists of four parts.

a) Each BSP processor will sort the up to $w_p \, p \, \log p$ read requests it has according to the BSP processors numbers to which they are destined. This can be done in time linear in $w_p \, p \, \log p$ using bucket sort.

b) Each BSP processor hashes the actual addresses in the destination module of each request into a hash table to identify all such address (i.e. not module) collisions. This takes linear expected time. For each destination address one request is considered "unmarked" and the rest are "marked".

c) A round-robin algorithm is used to implement all the "unmarked" requests. In the k-th of the $p - 1$ stages processor j will transmit all its messages destined for memory module $(j + k)(\bmod p)$, and receive back the values read.

d) The "marked" requests (which did not participate in (c)) are now given the values acquired by their "unmarked" representative in phase (c). This takes time linear in the number of requests per processor.

We can implement each of the $p-1$ stages of (c) in two supersteps, where $L = g(1 + \epsilon)w_p \log p$, $\epsilon = w_p^{-1/3}$. This is because the probability that the up to $w_p\ p\ \log p$ read requests from a fixed processor will have more than $(1 + \epsilon)\ w_p\ \log p$ addressed to any fixed module is then at most $\exp\left(-\Omega(w_p^{1/3})\right)$ by a similar estimation to part (i). Since there are p^2 choices of the fixed processor and module, a bound of $p^2\ \exp\left(-\Omega(w_p^{1/3})\right) = \exp\left(-\Omega(w_p^{1/3})\right)$ follows. Finally we note that instead of doing the round robin in $2(p-1)$ supersteps we could do it in just two and allow L as large as $L = g\ (1 + \epsilon)\ w_p\ p\ \log p$.

iii) The simulations is a round robin process similar to (ii) but is asynchronous. The $w_p p^2$ EPRAM requests are distributed evenly among the $p\ S^*PRAM$ processors, which bucket sort on the S^*PRAM destination modules the $w_p\ p^2$ requests they are each given. Then the overall problem can be viewed as one of transmitting p^2 tasks $\{(j, k) \mid 0 < j, k \leq p\}$, where task (j, k) is that of delivering the requests from processor j destined for processor k.

Each processor j will execute the tasks $(j, j+1(\bmod p)), (j, j+2(\bmod p)), \ldots, (j, j+ p - 1(\bmod p))$ in sequence. It will only execute task $(j, j + k)$ however, once it has received a "special" message from processor $j + 1(\bmod p)$ stating that the latter processor has completed its task $(j+1(\bmod p), (j+1)+(k-1)(\bmod p))$ which is destined for the same processor. This ensures that there is never any contention at the destination processors.

We shall analyse the performance of this algorithm using the delay sequence argument introduced in [1], [15].

We construct a directed graph where each task (j, k) is a node. There are directed edges from (j, k) to $(j, k+1(\bmod p))$, and from $(j+1(\bmod p), k+1(\bmod p))$ to $(j, k+1(\bmod p))$ for each $j = 0, \ldots, p - 1$, and $k = j, \ldots, j - 2(\bmod p)$. Hence the nodes (j, j) are the input nodes and nodes $(j, j - 1(\bmod p))$ are the output nodes.

Consider one execution of this algorithm and suppose that the last task finished in T units of time, where one unit is taken for each global message transmission. Suppose the task last finished was (j_p, k_p). When this last task started it must have awaited the completion of either (j_p, k_{p-1}) or of $(j_p + 1(\bmod p), k_p)$. Whichever task it was, we shall label (j_{p-1}, k_{p-1}). In this way we can trace back along a distinct path from an output task (j_p, k_p) along a path $(j_i, k_i) \rightarrow (j_{i-1}, k_{i-1})$ to some input node (j_1, k_1). This sequence of tasks we shall call a delay sequence D.

Let t_j be the number of tasks $(j, *)$ in this delay sequence. Let l_j be the total number of requests issued by these t_j tasks. Then l_j can be considered to be the number of successes in $p\ w_p$ Bernoulli trials with probability of individual success equal to t_j/p. Hence if Δ^D is the total number of requests in the tasks of delay sequence D, then

$$Prob(\Delta^D \geq T) = Prob(\sum_{j=0}^{p-1} l_j \geq T).$$

We can regard the Δ^D as the number of successes in $w_p p^2$ Bernoullis trials with various probabilities but with total expectation $w_p p$. Then the previously used bound on the tail of such distributions gives.

$$Prob(\Delta^D \geq (1+\epsilon)w_p p) \leq e^{-\epsilon^2 w_p p/3}.$$

if $0 < \epsilon < 1$. Since there are at most $p \, 2^{p-1}$ possible delay sequences in the graph (i.e. p destinations, and 2^{p-1} paths to each one) the probability that any one can cause a delay of $(1+\epsilon) \, w_p \, p$ is at most

$$p \, 2^{p-1} \, e^{-\epsilon^2 w_p p/3}.$$

Choosing $\epsilon = w_p^{-1/3}$, for example, will make this probability approach zero as $p \to \infty$.

We note that in addition to the tasks in the delay sequence there may be up to $p - 1$ additional "special" messages on the critical path. Hence the overall time is $(1 + \epsilon) \, w_p \, p + p = (1 + o(1)) \, w_p \, p$.

(iv) The simulation follows that of part (ii), with the elimination of the now redundant step (b). ∎

We note that for simulations such as ours, where the PRAM processors are shared evenly among the processors of the simulating machine, and the memory space is randomised by hashing, parts (i), (ii), and (iv) of the Theorem are optimal in the sense that the removal of w_p in their statement would render them false. This follows from the fact (see [10], page 96, for a proof) that if, for any constant c, $cp \log p$ balls are thrown randomly into p urns then with some constant probability there will be at least $c_1(c) \log p$ balls in at least one urn for some $c_1(c) > c$. For part (i) consider each memory request such a ball. For parts (ii) and (iv) consider the case that for each simulating processor the $cp \log p$ memory requests are to distinct memory locations, but the set of these locations are identical for the p simulating processors. Then the $cp \log p$ requests can be viewed as balls thrown randomly into p urns, but now p copies of each arrive. (We note that randomising the allocation of PRAM processors to the simulating processors does not help here.)

4   The Format

We shall specify the performance of a direct BSP algorithm $\mathcal{A}$ in three parts. First we specify the basic local operations allowed and an algorithm $\mathcal{A}^*$ (typically sequential or PRAM) with which we are comparing it. Second we specify two ratios π and μ. The former, π, is the ratio between the total operation counts of $\mathcal{A}$ and $\mathcal{A}^*$, and may be a function of p, n and L. The latter, μ, is the ratio between the parallel time required by the communication supersteps of $\mathcal{A}$, and the parallel computation time on the p processors if π were equal to one. When specifying μ it is necessary to be explicit about the amount of information that can be conveyed in a single message. Finally we specify the conditions on n, p, and L that are sufficient for the algorithm to make sense and the claimed bounds on π and μ to hold. We state as corollaries some sufficient conditions for the most important optimality criteria, such as $\pi = 1 + o(1)$, $\mu = o(1)$, or $\mu \leq 1$. All asymptotic bounds refer to the problem size as $n \to \infty$. The other parameters may also $\to \infty$ if they are expressed explicitly in terms of n, but otherwise are not assumed to do so. Thus the results in general hold for values of p, g and L that may or may not depend on n.

Fact 1 *Consider the arithmetic straight line model of computation and the standard algorithm $\mathcal{A}^*$ for multiplying an $n \times n$ matrix M by an $n \times 1$ vector x in $2n^2 - n$ steps. This can be implemented by algorithm $\mathcal{A}$ below on a BSP computer in the same arithmetic model with*

$$\pi \leq 1 + p/n, \text{ and } \mu \leq g\,(n+p)^2/(2n^2 - n),$$

provided $L \leq 2n^2/p$ and $p \leq n$.

Corollary 1 *For algorithm $\mathcal{A}$ if $p = o(n)$ then $\pi = 1 + o(1)$ and $\mu = g(\frac{1}{2} + o(1))$.*

Proof: We assume that the elements of both M and x are each initially distributed as evenly as possible among the p processors. Algorithm $\mathcal{A}$ is the following: it divides the rows of M into p groups each of $\lfloor n/p \rfloor$ or $\lceil n/p \rceil$ rows. It routes all elements of the ith row group (up to $n\lceil n/p \rceil$ elements) to processor i. It transmits p copies of each element of x, one copy to every processor. Finally each processor i computes locally the values of the rows of Mx that are in its group.

The actual computation, which is the last step, takes time $(2n-1)\lceil n/p \rceil = ((2n^2 - n)/p)(1 + p/n)$, which gives the claimed bound on π. The two routing operations take time at most

$$g(n\lceil n/p \rceil + p\lceil n/p \rceil) \leq g(n^2/p + n + n + p)$$

which gives the claimed bound on μ. ∎

5 Broadcasting and Parallel Prefix

BSP algorithms for broadcasting and parallel prefix were discussed in [17]. The broadcasting algorithm uses a t-ary tree, for some appropriate t. The initial message is at the root and in each superstep t copies of each existing copy of the message is made and sent to t distinct processors. Hence if there are p processors to broadcast to, and $t = p^{1/k}$ then k supersteps will suffice and the total communication time required will be

$$gk \, \max\{\tfrac{L}{g}, p^{1/k}\} \, .$$

The parallel prefix algorithm can be thought of also as using a tree but proceeds in two phases, first moving from the leaves to the root, and then back to the leaves, making some computations at each step. The computation time for a parallel prefix on a list of p elements, one at each processor, will then be,

$$2k \, \max\{L, p^{1/k}\} \, ,$$

and the communication time

$$2gk \, \max\{\tfrac{L}{g}, p^{1/k}\} \, .$$

For either algorithm we can choose k as large as $\lg p$.

6   Gauss-Jordan Elimination

The Gauss-Jordan elimination algorithm can be used to solve a single system of linear equations, a multiple system of linear equations on the same coefficient matrix, or to invert a non-singular matrix A.

When applied to an $n \times n$ matrix A it consists of n iterations, where at iteration i, $1 \le i \le n$, it eliminates all non-diagonal elements of column i of A. In particular suppose that at the end of iteration $i - 1$, the elements of matrix A are $a_{jl}^{(i-1)}$. Then in the i-th iteration the algorithm leaves row i unchanged but changes each other row $j \ne i$ as follows:

$$a_{jl}^{(i)} = a_{jl}^{(i-1)} - \frac{a_{ji}^{(i-1)}}{a_{ii}^{(i-1)}} \, a_{il}^{(i-1)} \quad \forall l = i, \ldots, n \tag{1}$$

After iteration i, the non-diagonal elements of the first i columns are all zero.

The version we have described does no pivoting and performs $n^3 + O(n^2)$ arithmetic operations. If it is used to solve a system of linear equations $Ax = b$, the row operations specified above for A are also performed on b, but this adds only an $O(n^2)$ term.

Below we show that even if pivoting is added the BSP can execute this algorithm one-optimally within a suitable range of parameters. The algorithm uses a scattered distribution of the data among the processors to even out processor utilisation (c.f. [5]). For simplicity we shall assume here that $\sqrt{p}$ and $n/\sqrt{p}$ are both integers and that p is a power of two.

We are not claiming that the given algorithm yields the best lower order terms, or indeed, that our analysis of the given algorithm yields the best possible bounds for these terms. Our objective is to show that a relatively simple algorithm has a relatively simple expression upper bounding its performance in terms of the parameters n, p, g, and L. Hence if such an algorithm is implemented in a transportable program package, then it can be run on any BSP machine efficiently, if the parameters p, g and L of the machine, and the input size n are taken into account during compilation and runtime. Of course such a transportable algorithm may have general constraints on the parameters ($p \le n^2$ in the case below). If the constraints are not satisfied the algorithm may not be well defined or the analysis applicable.

Theorem 2 *On the arithmetic model of computation, Gauss-Jordan elimination (with or without pivoting), for the solution of a single system of equations, uses $n^3 + O(n^2)$ operations. There is a BSP algorithm for the same problem (with partial row pivoting) that for $p \le n^2$, and any integer k ($1 \le k \le \lg p$) has:*

$$\pi \le 1 + 2 \max\{\tfrac{Lp}{n^2}, \tfrac{\sqrt{p}}{n}\} + k \max\{\tfrac{Lp}{n^2}, \tfrac{p^{1+1/(2k)}}{n^2}\} + \frac{4\sqrt{p}}{n} + \frac{L^2 p^2}{2 n^4},$$

and

$$\mu = g \left(4 k \max\{\tfrac{Lp}{gn^2}, \tfrac{p^{1+1/(2k)}}{n^2}\} + \max\{\tfrac{Lp}{gn^2}, \tfrac{\sqrt{p}}{n}\} + \lg \sqrt{p} \max\{\tfrac{Lp}{gn^2}, \tfrac{2\sqrt{p}}{n}\} \right).$$

Proof: We first split the processors into $\sqrt{p}$ groups P_r, $1 \le r \le \sqrt{p}$, of size $\sqrt{p}$ each. Let the processors of group P_r be numbered

$$P_r = \{(r - 1)\sqrt{p} + m \mid m = 1, \ldots, \sqrt{p}\}.$$

We will sometimes call processor $(r-1)\sqrt{p}+m$ the m-th processor of group P_r.

We assign the elements of column i of A to the group $P_{(i-1)(\bmod\sqrt{p})+1}$ of processors, in such a way that the the j-th element of column i will be assigned to the $[(j-1)(\bmod\sqrt{p})+1]$-st processor of that group. We handle column b as yet another column of A. Each processor therefore gets at most $\frac{n}{\sqrt{p}}\left(\frac{n}{\sqrt{p}}+1\right)$ elements of A or b.

We describe the Algorithm in terms of the operations performed in a succession of stages. We are going to use the following terminology. Let $columnp(i)$ be the set of $\sqrt{p}$ processors holding elements of column i. Similarly, let $rowp(i)$ be the set of $\sqrt{p}$ processors that hold the elements of row i of A (and b). Let $proc(a_{ij})$ be the processor that holds element a_{ij}.

The Algorithm consists of n iterations, where, for $1 \le i \le n$, the i-th iteration executes calls of three procedures, $Pivot(i)$, $Communicate(i)$, and $Compute(i)$ in succession. These procedures are described below.

Procedure $Pivot(i)$
    begin

    Stage 1:
Every processor, among the ones in $columnp(i)$, finds the element with maximum absolute value among the elements that are simultaneously in column i and row j, for some $j \ge i$, that are local to it. It then communicates this maximum and its row number to processor $proc(a_{ii})$.

    [**Analysis of Stage 1:** Two phases are required for this stage, one for computation and one for communication. The computational part requires time $n/\sqrt{p}$, while the communication part requires the routing of a $(\sqrt{p}-1)$-relation, if we regard an (element,row number) pair as one message.]

    Stage 2:
    (i) Processor $proc(a_{ii})$ finds the maximum among the local maxima it received in stage 1, and identifies the original row number of this maximum as, say, l.
    (ii) Processor $proc(a_{ii})$ communicates the value of l to processors $rowp(i)$ and $rowp(l)$.
    [**Analysis of Stage 2:** The single computation phase requires time $\sqrt{p}$ while the communication requires the routing of a $2\sqrt{p}$ relation.]

    Stage 3:
For each column k, the processors of $rowp(i)$ and $rowp(l)$ containing elements from that column exchange the elements of rows i and l they possess.
    [**Analysis of Stage 3:** An $n/\sqrt{p}$-relation needs to be routed in a single superstep here.]
    end

Procedure $Communicate(i)$
    begin

    Stage 1:
Processor $proc(a_{ii})$ broadcasts element a_{ii} to all $\sqrt{p}$ processors of $rowp(i)$.
    [**Analysis of Stage 1:** The routing of a $\sqrt{p}$-relation is required.]

Stage 2:
Every processor of $rowp(i)$ divides the elements of row i that it holds by a_{ii}.

[**Analysis of Stage 2:** Each element of row i must first be divided by a_{ii} before it is broadcast. This takes $n/\sqrt{p}$ time.]

Stages 3 , ..., $\lceil \lg \sqrt{p} \rceil + 2$:
In $\lceil \lg \sqrt{p} \rceil$ stages, every processor $j \in rowp(i)$ broadcasts the elements of row i it computed in the previous stage to all the processors of the processor group P_m it belongs to (i.e. $j \in P_m$). Simultaneously, the j-th processor of $P_t = columnp(i)$ broadcasts the elements of column i it owns to the j-th processors of every other group P_l, $l \neq t$. Since we need to broadcast an element to at most $\sqrt{p}$ processors, $\lceil \lg \sqrt{p} \rceil$ supersteps of binary replication suffice.

[**Analysis of Stage 3 , ..., $\lceil \lg \sqrt{p} \rceil + 2$:** In each of these $\lceil \lg \sqrt{p} \rceil$ supersteps a $2 \, n/\sqrt{p}$-relation needs to be routed.]
    end

Procedure $Compute(i)$
    begin

Stage 1:
Every processor performs the elimination step given by equation (1) on the elements it owns (note that the ratio a_{kl}/a_{kk}, as given by (1), need not be computed separately by each processor, since it has been evaluated in procedure $Communicate(i)$).

[**Analysis of Stage 1:** The elimination of column i requires $T_i = 2 \, \frac{n}{\sqrt{p}} \lceil \frac{n-i}{\sqrt{p}} \rceil + \frac{2\,n}{\sqrt{p}}$ operations. The actual time to be devoted, however, to it will be L, if $L > T_i$.]
    end

Several times in the algorithm we need to broadcast items to $\sqrt{p}$ processors each. If p is large, then it is best to do this in k supersteps each of which does a $p^{1/(2k)}$-relation rather than in one superstep by doing a $\sqrt{p}$-relation. If we follow this philosophy also when we find the maximum in the pivot procedure, we will need to find maxima of sets of size $p^{1/(2k)}$ in each of k successive stages.

We first find the total amount of time spent on computation for the elimination of all columns (pivoting operations included). We can easily get, from our previous discussion, that this time is upper bounded by

$$C(n) \leq 2\,n \max\{L, \tfrac{n}{\sqrt{p}}\} + n\,k \max\{L, p^{1/(2k)}\} + \sum_{i=1}^{n} \max\{L, T_i\} \, .$$

The cost of eliminating column i is less than L, for

$$2\,\frac{n}{\sqrt{p}} \lceil \frac{n-i}{\sqrt{p}} \rceil + \frac{2\,n}{\sqrt{p}} < L.$$

This is true provided that the following condition holds

$$\frac{2\,n}{\sqrt{p}} \left(\frac{n-i}{\sqrt{p}} + 2 \right) < L,$$

or, equivalently,

$$(n - i) \; < \; \frac{L\,p}{2\,n} - 2\,\sqrt{p}.$$

Thus, for each of the last $\frac{L\,p}{2\,n} - 2\,\sqrt{p}$ calls to procedure $Compute(i)$ we charge L time steps.

Overall, we get

$$\sum_{i=1}^{n} \max\{L, T_i\} \; \leq \; \sum_{i=1}^{n} T_i + \sum_{i=n-\frac{L\,p}{2\,n}+2\sqrt{p}+1}^{n} L$$

$$\leq \; \left(\frac{n^3}{p} + \frac{4\,n^2}{\sqrt{p}} - \frac{n^2}{p} \right) + \left(\frac{L^2\,p}{2\,n} - 2\,\sqrt{p}\,L \right).$$

Then, the expression for $C(n)$ can be written in the following form.

$$C(n) \; \leq \; \frac{n^3}{p} \left(2\max\{\tfrac{L\,p}{n^2}, \tfrac{\sqrt{p}}{n}\} + k\max\{\tfrac{L\,p}{n^2}, \tfrac{p^{1+1/(2k)}}{n^2}\} + 1 + \frac{4\,\sqrt{p}}{n} + \frac{L^2\,p^2}{2\,n^4} \right)$$

$$= \; \pi\,\frac{n^3}{p}$$

Similarly, the total communication time can be bounded as follows.

$$M(n) \; \leq \; n\,g \left(4\,k\max\{\tfrac{L}{g}, p^{1/(2k)}\} + \max\{\tfrac{L}{g}, \tfrac{n}{\sqrt{p}}\} + \lg\sqrt{p}\max\{\tfrac{L}{g}, \tfrac{2\,n}{\sqrt{p}}\} \right)$$

$$\leq \; g\,\frac{n^3}{p} \left(4k\max\{\tfrac{L\,p}{g\,n^2}, \tfrac{p^{1+1/(2k)}}{n^2}\} + \max\{\tfrac{L\,p}{g\,n^2}, \tfrac{\sqrt{p}}{n}\} + \lg\sqrt{p}\max\{\tfrac{L\,p}{g\,n^2}, \tfrac{2\,\sqrt{p}}{n}\} \right)$$

$$= \; \mu\,\frac{n^3}{p}$$

■

Several corollaries can be deduced from the result as stated. As we expect, if we have substantial or even moderate slack, then the algorithm is efficient. The case $k = 1$, for example, gives:

Corollary 1 *If $p = 100$, $n = 1000$, $L = 100$, then π is about 7% from optimal, and $\mu \leq 1$ for g up to about 15.*

It can be seen that for $k = 1$, $\pi = 1 + o(1)$ and $\mu = o(1)$ can be achieved under suitable conditions on L and g if $p = o(n^{4/3})$. Beyond that higher values of k are needed. The case $k = \lg p$ yields:

Corollary 2 *If $p\,\log p = o(n^2)$ and $L = o(\frac{n^2}{p\,\lg p})$, then $\pi = 1 + o(1)$. If in addition $g = o(\frac{n}{\sqrt{p}\,\lg p})$, then $\mu = o(1)$ also.*

These results show that even in the absence of combining or parallel prefix operations, Gauss-Jordan elimination can be solved efficiently even when L is about as large as it can be, and communication is moderately slow.

7 Sorting

In this section we present an algorithm for sorting n keys on the p processor BSP computer. The algorithm is one-optimal in computation, when comparison and arithmetic operations are counted, for a range of the parameters p, n and L, and also one-optimal in communication if g is appropriately bounded. The algorithm derives from quicksort [6] and uses the technique of over-sampling previously used in [[14], [9]] in a parallel context.

Let $X = \{x_1, \ldots, x_N\}$ be a set of input keys ordered such that $x_i < x_{i+1}$, for $1 \leq i \leq N - 1$. We are assuming that the keys are distinct since we can always make them so by, for example, appending to them the code for the memory locations in which they are stored. For positive integers k and s, let $Y = \{y_1, \ldots, y_{ks-1}\}$ be a randomly chosen subset of X, also ordered so that $y_i < y_{i+1}$. As soon as we find Y we partition $X - Y$ into k subsets $X_0, \ldots, X_{k-1}$ where

$$X_0 = \{x : x < y_s\},$$

$$X_i = \{x : y_{is} < x < y_{(i+1)s}\} \quad 0 < i < k - 1,$$

$$X_{k-1} = \{x : y_{(k-1)s} < x\}.$$

For $1 \leq i \leq k\,s - 1$ and $1 \leq j \leq N$, let $q_i(j)$ be the probability that $y_i = x_j$. i.e. that y_i is the j-th element of the sorted input. It is straightforward to verify that

$$q_i(j) = \frac{\binom{j-1}{i-1} \binom{N-j}{k\,s-1-i}}{\binom{N}{k\,s-1}}.$$

We follow the standard notation for the binomial coefficients, taking $\binom{i}{j} = 0$ if $j < i$.

For $0 \leq i \leq k - 1$ let $n_i = |X_i|$ and for $0 \leq i \leq k - 1$, $s - 1 \leq j \leq N - k + 1$, let $p_i(j)$ be the probability that $n_i = j$. The following was proved in [6]:

Claim 1 *For all* i,

$$p_i(j) = p(j) = \frac{\binom{j}{s-1} \binom{N-j-1}{(k-1)\,s-1}}{\binom{N}{k\,s-1}}.$$

It can be deduced that for any $0 < \epsilon < 1$ the sum of $p(j)$, for $j > \lceil (1 + \epsilon)(N - k + 1)/k \rceil$ is bounded above by

$$\frac{N}{ks - 1} \sqrt{2\,\pi\,(k\,s - 1)}\, e^{\frac{1}{6\,(k\,s-1)}}\, e^{-\frac{1}{2}\,\frac{\epsilon^2}{(1+\epsilon)}\,s}.$$

Later in this paper $N \leq n$, and k will take values as much as $\sqrt{n}$, while s will be a polylogarithmic function of n. Let $\log_n x = \log x / \log n$.

Claim 2 *Let* $ks < N/2$, $N \leq n$, $0 < \epsilon < 1$, $\rho > 0$, *and*

$$s = \left(2\rho + 2\log_n N + \log_n (k^2/(ks - 1)) + 2\log_n \sqrt{2\pi} + 2\log_n e^{1/6(ks-1)}\right) \frac{1+\epsilon}{\epsilon^2} \log n.$$

Then the probability that any one of the $p_i(j)$ *for* $0 \leq i < k$ *is of size more than* $\lceil (1 + \epsilon) \frac{N-k+1}{k} \rceil$ *is at most* $1/n^\rho$.

The algorithm below for sorting n items performs m iterations, where m may equal 1 or 2 in practical situations. In each iteration $k-1$ "splitters" are chosen , where $k = p^{1/m}$, and the set partitioned into k subsets of approximately equal size by these splitters. The next iteration performs a similar process of subpartitioning each of these k subsets into k further subsets, etc.

At each stage if the subset to be partitioned is of size N the k splitters are selected by first selecting $ks-1$ random elements from the set, sorting this selection and picking out the $k-1$ elements that partition this selection equally. By Claim 2, if we choose s appropriately then we will be guaranteed that these splitters split the whole set of size N almost equally (i.e. to within a factor of $(1+\epsilon)$), with high probability. To specify the size of s we will select a slowly growing function $\omega_n \to \infty$ (e.g. $\sqrt{\log n}$) and choose $\epsilon = (m\,\omega_n)^{-1}$ in Claim 2.

Then the size of every subset produced in the m iterations will be within a factor $(1+\epsilon)^m = (1+1/(m\,\omega_n))^m \le e^{1/\omega_n}$ of the mean with probability at least $1-n^{-\rho}(p-1)/(k-1)$. Since we will want this to be $1-o(1)$ it will be sufficient to choose ρ to exceed $\log(p/k)$ by a constant.

Proposition 1 *On the comparison model of computation sorting requires $n \lg n$ comparisons. For any integers p, k, s, such that p is a power of k, $k \le p$, $k \le \sqrt{n}$, $ks \le n/2$, there is a BSP sorting algorithm that for any integer $l \ge 2$, has, with probability at least $1 - o(1)$,*

$$\pi \le \frac{p \lg p}{n \lg n \lg k} \max\{L, \lceil (1+\frac{1}{\omega_n \frac{\lg p}{\lg k}})^{\frac{\lg p}{\lg k}} \frac{n}{p} \rceil (\lceil \lg k + 1 \rceil)\}$$

$$+ \frac{p}{n \lg n} \max\{L, \lceil \frac{(1+\frac{1}{\omega_n \frac{\lg p}{\lg k}})^{\frac{\lg p}{\lg k}} n}{p} \rceil \lg \lceil \frac{(1+\frac{1}{\omega_n \frac{\lg p}{\lg k}})^{\frac{\lg p}{\lg k}} n}{p} \rceil \}$$

$$+ \frac{p \lg p}{n \lg n \lg k} \max\{L, (k\,s-1) \lg (k\,s-1)\}$$

$$+ \frac{2\,p \lg p}{n \lg n \lg k} l \max\{L, k \lceil p^{1/l} \rceil\} ,$$

and

$$\mu = g\,(\frac{p \lg p}{n \lg n \lg k} \max\{\frac{L}{g}, (k\,s-1)\} + \frac{3\,p \lg p}{n \lg n \lg k} l \max\{\frac{L}{g}, k \lceil p^{1/l} \rceil\}$$

$$+ \frac{p \lg p}{n \lg n \lg k} \max\{\frac{L}{g}, \lceil (1+\frac{1}{\omega_n \frac{\lg p}{\lg k}})^{\frac{\lg p}{\lg k}} n/p \rceil\}).$$

The expression for π includes all arithmetic as well as comparison operations.

Proof: We divide the algorithm into stages.
Initially, the variables *Left*, *Right* have the values $Left = 1$, $Right = p$. In any recursive call of a procedure, these two variables give the range of the processors that will participate. The input is initially evenly distributed among the p processors before the beginning of the execution of the Algorithm.

Program *Sort(Left,Right)*
begin

Stage 0 :

If $Left = Right$ then

begin

Every processor, in parallel with all the other processors, sorts the elements in its possession;

Exit-Program;

end

[**Analysis of Stage 0:** When this step is executed, every processor will hold at most $\lceil (1 + 1/(\omega_n (\lg p/\lg k)))^{\lg p/\lg k} \frac{n}{p} \rceil$ keys. The cost is that of sequential sorting by some algorithm such as [8], [18].]

Stage 1 :

The $P = Right - Left + 1$ processors participating in this call send between them a set of $k\,s$ of their elements, randomly chosen, to processor $Left$.

[**Analysis of Stage 1:** This stage is itself a superstep. We only need to be able to route a $k\,s$-relation here.]

Stage 2 :

(i) Processor $Left$ sorts the sample, and selects the $k - 1$ splitters.

(ii) It then initiates a broadcast of the $k - 1$ splitters to the other $P - 1$ processors of this recursive call as well as, the index of every splitter in the set of splitters, so that the processors receive the splitters effectively sorted.

[**Analysis of Stage 2:** Sequential sorting requires $(ks - 1)\lg(ks - 1)$ comparisons. If a (splitter, index) pair can be sent as one message and we want to broadcast in l supersteps, then in each superstep it is sufficient to route a $k\lceil p^{1/l}\rceil$-relation.]

Stage 3 :

(i) Each processor, in a single superstep, decides the position of every element it holds with respect to the $k - 1$ splitters. It then counts the number of its elements that fall into each of the k intervals.

(ii) A multiple parallel prefix operation is then initiated on the number of elements per processor and per bucket to determine the processor destination of each element within a bucket.

[**Analysis of Stage 3:** For (i) we need time $\lceil (1 + 1/(\omega_n (\lg p/\lg k)))^{\lg p/\lg k} n/p \rceil$ $(\lceil \lg k \rceil + 1)$. The analysis of the parallel prefix operation is similar to that for the broadcasting in the previous stage.]

Stage 4 :

Every processor knows where to route every key in its possession, and performs this routing.

[**Analysis of Stage 4:** We need to route an $\lceil (1 + 1/(\omega_n (\lg p/\lg k)))^{\lg p/\lg k} n/p \rceil$-relation here.]

Stage 5 :

After the routing has been completed, a recursive call has been completed as well.

begin

for $i = 1$ to k do in parallel

$Sort\ \left((i-1)\frac{P}{k} + 1, i\,\frac{P}{k} \right)$

end

[Analysis of Stage 5: The depth of the recursion is $\lg p/\lg k$. **]**
■

By substituting appropriate values for the parameters in the above proposition a variety of bounds on the runtime of the algorithm can be obtained. The best bounds are obtained by choosing m, ω_n with regard to the actual values of n, p and L, and substituting these numerical values. By making more generic choices for m and ω_n and worst case bounds for p and L, more general but less tight results in terms of n alone can be obtained. The following are two such bounds that are nevertheless, still applicable for reasonable values of p, n and L.

Corollary 3 *There exists a BSP sorting algorithm that for any integer p, n and L such that $p \geq 2$, $n \geq 3500$, $p^3 \leq n/\lg^2 n$, $L \leq n/p$, and $g \leq \lg n$ has with probability at least $1 - o(1)$:*

$$\pi \leq 1 + \frac{1}{\sqrt{\log n}} + \frac{2\log 2}{\log n} + \frac{3\,p\,L}{n\,\lg n} + \frac{2.83}{\log^{1.5} n} + \frac{2.2\log^{2/3} n}{n^{1/3}} + \frac{1.22}{n^{2/3}\,\lg^{2/3} n}$$

$$\leq 1 + \frac{1}{\sqrt{\log n}} + \frac{10}{\log n}$$

and

$$\mu \leq \frac{g}{\lg n}\left(1 + \frac{1}{\sqrt{\log n}}\right) + \frac{4\,p\,L}{n\,\lg n} + \frac{3}{\lg^2 n} + \frac{2.5\,\log^{2/3} n}{n^{1/3}} + \frac{p}{n}$$

$$\leq \frac{g}{\lg n}\left(1 + \frac{1}{\sqrt{\log n}}\right) + \frac{8.5}{\log n}.$$

Proof: Let $m = 1$, $k = p$, $l = 1$, $\omega_n = \sqrt{\log n}$ and $s = 4\,\log^2 n$. ■

Corollary 4 *There exists a BSP sorting algorithm that for any integer p, n and L such that $p \geq 2$, $n \geq 25000$, $p^2 \leq n/(6\,\log^3 n)$, $L \leq n/(p\,\lg p)$, and $g \leq \lg n$, has with probability at least $1 - o(1)$:*

$$\pi \leq 1 + \frac{1}{\sqrt{\log n}} + \frac{2}{\lg n} + \frac{2/3}{\log n} + \frac{2.75}{\log^{1.5} n} + \frac{p\,L}{n\,\lg n} + \frac{2\,p\,L\,\lg p}{n\,\lg n}$$

$$\leq 1 + \frac{1}{\sqrt{\log n}} + \frac{5}{\log n}$$

and

$$\mu \leq \frac{g}{\lg n}\left(1 + \frac{1}{\sqrt{\log n}}\right) + \frac{p\,L}{n\,\lg n} + \frac{3\,p\,L\,\lg p}{n\,\lg n} + \frac{\lg p}{\log^3 n} + \frac{2/3}{\log n} + \frac{p}{n}$$

$$\leq \frac{g}{\lg n}\left(1 + \frac{1}{\sqrt{\log n}}\right) + \frac{3.6}{\log n}.$$

The expression for π includes all arithmetic as well as comparison operations.

Proof: Let $m = 1$, $k = p$, $l = \lg p$, $\omega_n = \sqrt{\log n}$ and $s = 4 \log^2 n$. ∎

By allowing more than one iteration the algorithm achieves one-optimality for values of n much closer to p than these. For $p = n^{1-t}$ for any constant t, a constant number m of iterations, that depends on t suffices. If m is allowed to grow with n then one-optimality can be achieved asymptotically with n exceeding p by only a polylogarithmic factor. The former case can be stated as follows.

Corollary 5 *For any constant t $(0 < t < 1)$ there is a BSP sorting algorithm that for all integers p, n and L satisfying $p = n^{1-t}$ and $L = o(n^t \lg n)$ has, with probability at least $1 - o(1)$, $\pi = 1 + o(1)$, and if $g = o(\lg n)$ also, $\mu = o(1)$.*

Proof: Substitute in Proposition 1 any integer $m > t^{-1} - 1$, $k = p^{1/m}$, $l = \lg p / \lg \lg p$, $\omega_n = \sqrt{\log n}$ and an s satisfying Claim 2. ∎

Finally we note that not only can these bounds be expressed in a variety of ways, but also the algorithm itself leaves room for several improvements. As stated, it sorts the samples sequentially at one processor. In appropriate circumstances its performance can be improved by any one of a variety of alternative methods. For example the sorting of the samples could be performed by parallel algorithms either by using our algorithm recursively, or some completely different one (e.g. [13]). Also, the randomly chosen samples need not be assembled at one processor, prior to being distributed for parallel sorting. Their random selection could be performed in parallel also (e.g. [14]).

Acknowledgement

We are grateful to Thanasis Tsantilas for helpful discussions on similar parallel sorting algorithms and for bringing reference [10] to our attention.

References

[1] R. Aleliunas. Randomized parallel computation. In *ACM SIGACT-SIGOPS Symposium on Principles of Distributed Computing*, pages 60–72, August 1982.

[2] Richard J. Anderson and Gary L. Miller. Optical communication for pointer based algorithms. *Technical Report CRI 88-14*, 1988.

[3] D. Anglin and L. G. Valiant. Fast probabilistic algorithms for Hamiltonian circuits and matchings. *J. Computer and System Sciences*, 18:155–193, 1979.

[4] Martin Dietzfelbinger and Friedhelm Meyer auf der Heide. A new universal class of hash functions and dynamic hashing in real time. *International Colloqium on Automata, Languages and Programming*, 443:6–19, 1990.

[5] G. C. Fox, M. A. Johnson, G. A. Lyzenga, S. W. Otto, J. K. Salmon, and D. W. Walker. *Solving Problems on Concurrent Processors, Vol 1: General Techniques and Regular Problems*. Prentice Hall, Englewood Cliffs, New Jersey, 1988.

[6] W. D. Frazer and A. C. McKellar. Samplesort: A sampling approach to minimal storage tree sorting. *Journal of the ACM*, 17:3:496–507, 1970.

[7] Alfred Hartmann and Steve Redfield. Design sketches for optical crossbar switches intended for large scale parallel processing applications. *Optical Engineering*, 29:3:315–327, April 1989.

[8] C. A. R. Hoare. Quicksort. *Computer Journal*, 5:10–15, 1962.

[9] J. S. Huang and Y. C. Chow. Parallel sorting and data partitioning by sampling. *IEEE Computer Society's Seventh International Computer Software and Applications Conference*, pages 627–631, November 1983.

[10] V. F. Kolchin, B. A. Sevastyanov, and V. P. Chistyakov. *Random Allocations*. Translation ed. A.V. Balakrsishnan, Halsted Press, 1978.

[11] Richard A. Linke. Power distribution in a planar-waveguide based broadcast star network. *IEEE Photonics Technology Letters*, 3:9:850–852, September 1991.

[12] Eric S. Maniloff, Kristina M. Johnson, and J. H. Reif. Holographic routing network for parallel processing machines. In *Holographic Optics II: Principles and Applications, Proc. SPIE* (G. M. Morris, ed.), 1136:283–289, 1989.

[13] D. Nassimi and S. Sahni. Parallel permutation and sorting algorithms and a new generalized connection network. *Journal of the ACM*, 29:3:642:667, July 1982.

[14] J. H. Reif and L. G. Valiant. A logarithmic time sort for linear size networks. *Journal of the ACM*, 34:60–76, January 1987.

[15] E. Upfal. Efficient schemes for parallel communication. In *ACM SIGACT-SIGOPS Symposium on Principles of Distributed Computing*, pages 55–59, August 1982.

[16] L. G. Valiant. General purpose parallel architectures. In *Handbook of Theoretical Computer Science* (J. van Leeuwen, ed.), North Holland, pages 945-971, 1990.

[17] L. G. Valiant. A bridging model for parallel computation. *Communications of the ACM*, 33:103–111, August 1990.

[18] M. H. van Emden. Increasing the efficiency of quicksort. *Communications of the ACM*, 13:9:563–567, 1970.

Memory Limited Inductive Inference Machines

Rūsiņš Freivalds
Institute of Mathematics and Computer Science
University of Latvia
Raiņa bulvāris 29
226250, Riga, Latvia

and

Carl H. Smith †
Department of Computer Science and
Institute for Advanced Computer Studies
The University of Maryland
College Park, MD 20742, USA

Abstract. The traditional model of learning in the limit is restricted so as to allow the learning machines only a fixed, finite amount of memory to store input and other data. A class of recursive functions is presented that cannot be learned deterministically by any such machine, but can be learned by a memory limited probabilistic leaning machine with probability 1.

1 Introduction

Various aspects of machine learning have been under empirical investigation for quite some time [17,26]. More recently, theoretical studies have become popular [11,13,24,28]. The research described in this paper contributes toward the goal of understanding how a computer can be programmed to learn by isolating features of incremental learning algorithms that theoretically enhance their learning potential.

In this work, we consider machines that learn programs for recursive (effectively computable) functions. Several authors have argued that such studies are general enough to include a wide array of learning situations [3-5,7,12,20]. For example, a behavior to be learned can be modeled as a set of stimulus and response pairs. Assuming that any behavior associates only one response to each possible stimulus, behaviors can be viewed as *functions* from stimuli to responses. It is possible to encode every string of ascii symbols in the natural numbers. These strings include arbitrarily long texts and are certainly sufficient to express both stimuli and responses. By using suitable encodings, the learning of functions represents several, ostensibly more robust, learning paradigms.

Hence, for the purposes of a mathematical treatment of learning, it suffices to consider only the learning of functions from natural numbers to natural numbers. A variety of models for learning recursive functions have been considered, each representing some different aspect of learning. The result of the learning will be a program

---

† Supported in part by NSF Grant CCR-9020079.

that computes the function that the machine is trying to learn. Historically, these models are motivated by various aspects of human learning [12] and perspectives on the scientific method [23].

We say that learning has taken place because the machines we consider must produce the resultant program after having ascertained only finitely much information about the behavior of the function. The models we use are all based on the model of Gold [12] that was cast recursion theoretically in [5]. First, we briefly review the basics of the Gold model and then proceed to define the memory limited version of the basic model that will be investigated in this paper.

2 The Gold Model

People often hold steadfast beliefs which later they discover to be false. At various points in time the scientific community was convinced that the earth was flat, the earth was the center of the universe, time is absolute, etc. Hence, one can never be absolutely sure that they have *finished* learning all there is to learn about some concept. We must always be prepared to embrace a better explanation of some phenomenon that we thought had been learned. Gold, in a seminal paper [12], defined the notion called *identification in the limit*. This definition concerned learning by algorithmic devices now called *inductive inference machines* (IIMs). An IIM inputs the range of a recursive function, an ordered pair at a time, and, while doing so, outputs computer programs. Since we will only discuss the inference of (total) recursive functions, we may assume, without loss of generality, that the input is received by an IIM in its natural domain increasing order, $f(0)$, $f(1)$, $\cdots$. An IIM, on input from a function f will output a potentially infinite sequence of programs p_0, p_1, $\cdots$. The IIM *converges* if either the sequence is finite, say of length $n + 1$, or there is program p such that for all but finitely many i, $p_i = p$. In the former case we say the IIM converges to p_n, and in the latter case, to p. In general, there is no effective way to tell when, and if, an IIM has converged.

Following Gold, we say that an IIM M *identifies* a function f (written: $f \in EX(M)$), if, when M is given the range of f as input, it converges to a program p that computes f. If an IIM identifies some function f, then some form of learning must have taken place, since, by the properties of convergence, only finitely much of the range of f was known by the IIM at the (unknown) point of convergence. The terms *infer* and *learn* will be used as synonyms for identify. Each IIM will learn some set of recursive functions. The collection of all such sets, over the universe of effective algorithms viewed as IIMs, serves as a characterization of the learning power inherent in the Gold model. This collection is symbolically denoted by EX (for explanation) and is defined rigorously by $EX = \{U \mid \exists M(U \subseteq EX(M))\}$. Mathematically, this collection is set-theoretically compared with the collections that arise from the other models we discuss below. Many intuitions about machine learning have been gained by working with Gold's model and its derivatives. In the next section, we describe the variant of Gold's model that we examine in this paper.

3 Limited Memory Learning

Connectionist networks are a popular mechanism for learning that, at first glance, appear to be unrelated to inductive inference [2,8]. However, it is possible to constrain inductive inference machines so as to functionally appear more like standard connectionist networks. We will not be able to model connectionist networks precisely, but the model proposed below is close enough for difficulty and impossibility results in our model to also hold for the connectionist networks.

One difference is that connectionist networks do not remember verbatim all the data they are trained on, while inductive inference machines are typically assumed to have perfect memory. A study of inference machines with limited memories (current guess and next or selected data only) was initiated in [29] and pursued in [18-20]. The conclusion reached in that work was that restricting the data available to the inference machine also reduces its learning potential. Memory limited learning was also investigated in [14], but that work assumes that the data is presented in *sorted* order and that several *passes* through the data are allowed. We make neither assumption. The model described below constrains the amount of what can be retained by an IIM, without placing any provisions on the content of the remembered data.

To insure an accurate accounting of the memory used by an IIM, we will henceforth assume that each IIM receives its input in such a way that it is impossible to back up and reread some input after another has been read. To circumvent the use of coding techniques, the memory used will be measured in bits, as opposed to integers. Under these conventions, we say that a set $U \subseteq LEX(M)$ iff there is a constant c such that for any $f \in U$, M uses no more than c bits of memory, exclusive of the input, and $f \in EX(M)$. One formalization of this notion considers the memory limited IIMs as Turing machines with and input tape and a work tape. The input tape is read only once (one way) and the work tape has only c bits of storage capacity. An equivalent formalization is to view memory limited IIMs as finite automaton. The collection of all sets of functions inferrible by limited memory inference machines is denoted by LEX, where $LEX = \{U \mid \exists M(U \subseteq LEX(M)\}$.

A few more technical definitions are needed. Natural numbers ($\mathbb{N}$) will serve as names for programs. The function computed by program i will be denoted by φ_i. It is assumed that $\varphi_0, \varphi_1, \cdots$ forms an acceptable programming system [16,25]. Sometimes, it will be convenient to represent a function by a sequence of values from its range. Such a representation is called a *string* representation. So, for example, the sequence $01^2 0^4 3^\infty$ represents the (total) function:

$$f(x) = \begin{cases} 0 & \text{if } x = 0 \text{ or } 3 \leq x \leq 6, \\ 1 & \text{if } 1 \leq x \leq 2 \\ 3 & \text{otherwise.} \end{cases}$$

This example function has two *blocks* of consecutive 0's, one of length 1 and the other of length 4.

4 Preliminary Results

In order to get a rough idea of the relative learning power of LEX type inference, we will employ the set of functions of finite support and the set of self describing functions. These sets were introduced in [5] and used in [7,10] to separate various classes of learnable sets of functions. Let $U_0 = \{f \mid f \text{ is recursive and } \overset{\infty}{\forall} x(f(x) = 0))\}$ and $U_1 = \{f \mid f \text{ is recursive and } \varphi_{f(0)} = f\}$.

PROPOSITION 1. $U_1 \in LEX$.

Proof: The IIM that, upon reciept of input pair $(x, f(x))$, compares x with 0. If there is a match, the value $f(x)$ is copied from the input register to the output register. The memory is needed only to store the constant 0. Clearly, $U_1 \in LEX$. ☒

PROPOSITION 2. $U_0 \notin LEX$.

Proof: Suppose by way of contradiction that M is an IIM such that $U_0 \in LEX(M)$. A pumping lemma type argument is used. Thge string representation of functions is used in this proof. Since M has a finite memory, there are strings σ and τ such that

1.) σ and τ contain only 0's and 1's, and
2.) τ contains at least one 1, and
3.) M, on input $\sigma\tau$, outputs a conjecture while reading τ, and
4.) The contents of M's memory (including current state) is the same after reading $\sigma\tau$ as it was when it had just finished reading σ.

Such a σ and τ must exists, as otherwise M would not be able to learn all the functions in U_0. By the choice of σ and τ, M's internal state, memory contents and most recent conjecture are identical after reading $\sigma\tau$ and $\sigma\tau^2$. Consequently, M cannot distinguish the functions $\sigma\tau 0^\infty$ and $\sigma\tau^2 0^\infty$, both of which are in U_0. ☒

COROLLARY 3. $LEX \subset EX$.

Proof: By definition, $LEX \subseteq EX$. The inclusion is proper by Proposition 2. ☒

Another type if inference that may be relevant to neural netwrks is the class PEX defined in [7] and studied in [6]. A set of functions U is in PEX just in case there is an IIM that outputs only programs for total recursive functions and $U \subseteq EX(M)$. The collection of sets PEX is defined analogously. In this case the witnessing IIM M is called popperian.

After every training episode on a neural net, the weights at the nodes may or may not be adjusted. When the weights at the nodes of a neural net are held fixed, then the network can be viewed as computing a function as it maps inputs to outputs. In the case of feed forward neural network, the function computed by the network will neccessarily converge on all outputs, i.e. it computes a total recursive function. Consequently, each feed forward neural network can be transformed into a popperian IIM as follows. Let N be some feed forward neural network. An IIM, on input σ, a segment of some function, uses σ as a training set for N. After the training is complete (for the sample σ, the setting of the weights of N determine the function that the network is now computing. The IIM outputs a program for that function. Hence, popperial inductive inference machines share some features with feed forward neural networks.

COROLLARY 4. LEX and PEX are incomparable.

Proof: U_0 is in PEX [7], therefore, by Proposition 2, $PEX - LEX \neq \emptyset$. $U_1 \notin PEX$, hence, by Proposition 1, $LEX - PEX \neq \emptyset$. ☒

5 Probabilistic Limited Memory Machines

Probabilistic inductive inference machines were introduced in [21] and studied further in [22]. A *probabilistic* inductive inference machine is an IIM that makes use of a fair coin in it deliberations. We say that $f \in EX(M)\langle p \rangle$ if M learns f with probability p, $0 \leq p \leq 1$. The collection $EX\langle p \rangle$ is defined to be $\{U \mid \exists M(U \subseteq EX(M)\langle p \rangle\}$. Pitt showed that for $p > \frac{1}{2}$, $EX\langle p \rangle = EX$ [21]. Limiting the memory available to a probabilistic IIM, according to the conventions of this paper, gives rise to the class $LEX\langle p \rangle$. This class is the subject of our next theorem.

THEOREM 5. There is a class U of total recursive functions such that $U \in LEX\langle 1 \rangle - LEX$.

Proof: First we define U. Every function in U will take on only four values, 0, 1 and two self referential indices. Members of U will be constructed via suitable recursion theorems. Every function $f \in U$ will have several (perhaps infinitely many) blocks of 0's. Let $\tau_1, \tau_2, \ldots$ denote the length of the first block of 0's, the second block, etc. Similarly, $\sigma_1, \sigma_2, \ldots$ denotes the lengths of the blocks of 1's, in their order of appearance in the range. For a function f to be in U, one of the following two conditions must be met:

$$\sum_{n=1}^{\infty} \frac{1}{2^{\tau_n}} \text{ converges and } \sum_{n=1}^{\infty} \frac{1}{2^{\sigma_n}} \text{ diverges, or} \tag{1}$$

$$\sum_{n=1}^{\infty} \frac{1}{2^{\tau_n}} \text{ diverges and } \sum_{n=1}^{\infty} \frac{1}{2^{\sigma_n}} \text{ converges.} \tag{2}$$

Furthermore, in case (1) occurs, $f \in U$ iff the sequence of values $f(x_1)$, $f(x_2)$, ..., for points x_i immediately following a block of 0's or 1's, converges to a program for f. Similarly, for (2) to qualify a function f for membership in U, the sequence of values $f(y_1)$, $f(y_2)$, ... converges to a program for f, where $y_i = x_i + 1$, e.g. the y_i's are points immediatetly following a point that immediately follows a block of 0's or 1's.

The proof proceeds by showing that $U \in LEX\langle 1 \rangle$. To achieve this subgoal, two probabilistic ω-automata [1,27] are constructed.

These ω-automata will process the string of values representing the range of functions from U. Consequently, they will only have to recognize symbols as being either 0 or 1 or *other*. The state transition graph of automata A_1 and A_2 is given by the schema below. The arc from q_1 to q_2 labeled $1, \frac{1}{2}$ indicates, the when in state q_1, if the automaton sees a 1, it enters state q_2 with probability $\frac{1}{2}$.

24

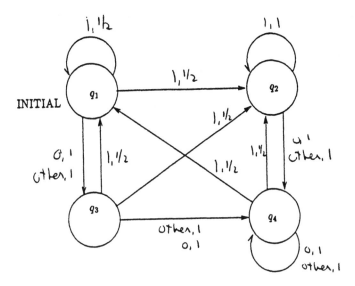

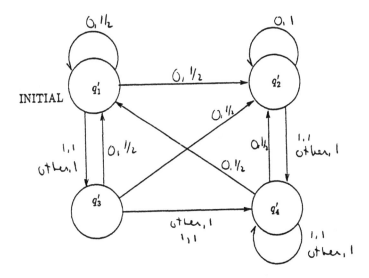

Let some function f from U be given. Consider what happens when the string representation of f is given as input to A_1 and A_2. What will turn out to be important is the number of times A_1 enters state q_3 and how often A_2 enters state

q_3'. Our discussion will be in terms of A_1, q_3 and 1's. The same dialogue will hold for A_2 with q_3 replaced by q_3' and all references to 1 replaced by 0.

State q_3 can only be entered from state q_1. State q_1 is accessible only from states q_3 and q_4. Observation of a 1 is required for a change of state to q_1. Once in state q_1, observing more 1's can keep A_1 in the same state. When A_1 is any state other than q_2, observation of a 1 will move the automaton to state q_1 or state q_2 with equal probability. If A_1 observes n consecutive 1's, then the probability that it will be in state q_1 after seeing them all is 2^{-n}.

Recall that σ_i denotes the length of the the ith block of 1's in the string representation of f. By the above discussion, if

$$\sum_{n=1}^{\infty} \frac{1}{2^{\sigma_n}} \tag{3}$$

diverges, then by the Borel–Cantello lemma, [9] state q_3 will be entered infinitely often with probability 1, and finitely often with probability 0. On the other hand, if (3) converges, then, with probability 1, state q_3 will be entered only finitely often (and infinitely often with probability 0).

Similarly, in A_2, state q_3' will, with probability 1, be entered infinitely often when

$$\sum_{n=1}^{\infty} \frac{1}{2^{\tau_n}} \tag{4}$$

diverges (finitely often with probability 0). Since U was defined to include only functions f such that (3) converges iff (4) diverges, the following statement holds with probability 1 for any member of U:

either state q_3 is entered infinitely often and q_3' only finitely often

or state q_3' is entered infinitely often and q_3 only finitely often.

A probabilistic IIM, M, that infers all the functions in U simulates the behavior of A_1 and A_2, tossing a fair coin every time a probabilistic state transition is made. Suppose that $f \in U$. If observing the value $f(x)$ causes A_2 to enter state q_3', then M outputs the value $f(x+1)$ when it is received as input. If observing the value $f(x)$ causes A_1 to enter state q_3, then M outputs $f(x+2)$ when it is received as input. Under no other circumstances will the IIM produce an output.

With probability 1, M converges to a program that computes f. The memory used by M is bounded by a constant. All that is needed is to remember the transitions of A_1 and A_2, their current states, two bits to remember if q_3 or q_3' was entered, and one final bit to be able to count 2 more inputs before transfering an input to the output tape.

To complete the proof, it remains to show that U can not be identified by any deterministic IIM with a limited memory. Suppose by way of contradiction that M is a IIM such that $U \subseteq LEX(M)$. Choose c, a constant exceeding the number of distinct possible contents of M's memory (including the internal state). Let $a_n = c + \lfloor \log_2 n \rfloor$. Then, there is a c' such that

$$\sum_{n=1}^{\infty} \frac{1}{2^{a_n}} = c' \cdot \sum_{n=1}^{\infty} \frac{1}{2^{\lfloor \log_2 n \rfloor}} > \sum_{n=1}^{\infty} \frac{c'}{n} \quad \text{which diverges.}$$

Define $b_n = a_n + n!$. Consequently,

$$\sum_{n=1}^{\infty} \frac{1}{2^{a_n}} \text{ diverges and } \sum_{n=1}^{\infty} \frac{1}{2^{b_n}} \text{ converges.}$$

The sequences of a_i's and b_i's will be used to determine the sizes of the blocks of consecutive 0's and 1's in the functions that we will construct below. Before defining the function from U that M will fail to identify, we must first describe a transformation on programs. Recall that functions in U have string representations that look like:

$$0^{a_1} x \, y1^{b_1} x \, y0^{a_2} x \, y1^{b_2} x \, y \cdots.$$

If φ_i has a string representation that conforms to the above schema, then $\varphi_{g(i)}$ will appear as:

$$0^{a_1} x \, y1^{b_1} x \, y \cdots x \, y0^{a_c} x \, y1^{b_c} x \, y0^{b_c+1} x \, y1^{a_c+1} x \, y0^{b_c+2} x \, y1^{a_c+2} x \, y \cdots.$$

The transformation g is specified by the construction of $\varphi_{g(i)}$, uniformly in i in effective stages of finite extension below. Consequently, g will be a total recursive function, even if some of the programs in its range compute functions with finite (or empty) domains. $\varphi_{g(i)}^s$ denotes the finite amount of $\varphi_{g(i)}$ determined prior to stage s. $\varphi_{g(i)}^0 = \emptyset$. x^s denotes the least number not in the domain of $\varphi_{g(i)}^s$. Consequently, $x^0 = 0$.

Begin stage 0. Look for the least z such that $\varphi_i(z) = 0$ and there are $2 \cdot c$ numbers $y < z$ such that $\varphi_i(y)$ is defined to some number other than 0 or 1. Set $\varphi_{g(i)}^1 = \{(x, \varphi_i(x)) \mid x < z\}$ and go to stage 1.
End stage 0.

Begin stage $s > 0$. Look for the least $z \geq x^s$ such that $\varphi_i(z)$ is defined and $\varphi_i(z) \neq 0$. If $\varphi_i(z)$ is undefined for some value, then $\varphi_{g(i)}$ has finite domain. Look for the least $w > z + 1$ such that $\varphi_i(w)$ is defined and $\varphi_i(w) \neq 1$. Again, if φ_i is undefined for some value involved in the search, then $\varphi_{g(i)}$ has finite domain. Define:

$$\varphi_{g(i)}^{s+1}(x) = \begin{cases} \varphi_{g(i)}^s(x) & \text{if } x < x^s \\ 0 & \text{if } x^s \leq x < x^s + w - (z+2) \\ \varphi_i(z) & \text{if } x = x^s + w - (z+2) \\ \varphi_i(z+1) & \text{if } x = x^s + w - (z+2) + 1 \\ 1 & \text{if } x^s + w - (z+2) + 1 < x < w \\ \varphi_i(w) & \text{if } x = w \\ \varphi_i(w+1) & \text{if } x = w + 1 \end{cases}$$

Go to stage $s + 1$.
End stage s.

We are now ready to define the function $f \in U$ that cannot be identified by M. By implicit use of the recursion theorem [15], define a function φ_e with string representation:

$$0^{a_1} e \, g(e) 1^{b_1} e \, g(e) 0^{a_2} e \, g(e) 1^{b_2} e \, g(e) \cdots.$$

Clearly, φ_e is a total function. By the choice of the a_i's and b_i's, $\varphi_e \in U$. Let $f = \varphi_e$. To complete the proof, we define a recursive function f' that is different from f but that M cannot distinguish from f. The argument is similar to a pumping lemma argument. Each block of 0's (length a_i) and block of 1's (length b_i) is longer than c in length. The constant c was chosen to exceed the number of possible memory configurations that M can express in its finite memory. So, for each n there is a block of $d_n \leq c$ 0's such that M's memory (and internal state) are in the same state when those 0's are just about to be read, and just after they have all been read. Similarly, there is a block of $j_n \leq c$ 1's. Redrawing the string representation of f, given the above observation yields:

$$\cdots \underbrace{000\cdots0000^{d_n}00\cdots000}_{a_n}e\ g(e)\underbrace{111\cdots1111^{j_n}111\cdots1111}_{b_n}\cdots$$

Since $b_n - a_n = n!$, both d_n and j_n divide $b_n - a_n$, for all $n \geq c$. Consequently, for $n \geq c$, it is possible to expand the nth block of 0's by a multiple of d_n to make the block have exactly b_n 0's. Similarly, it is possible to remove a multiple of j_n 1's from the nth block of 1's, leaving the block with exactly a_n 1's. Performing this transformation on all but the first c blocks of 0's and 1's results in a function that we will call f'. Notice that the second value following each block of 0's or 1's ($g(e)$) is an index for f', hence $f' \in U$. M, on input from f' will be in the same state when it leaves the nth block of 0's (or 1's) as it will be when using f as input. Consequently, M will produce the same outputs on input from f and f'. Hence, M cannot infer both f and f', a contradiction to the assumption that $U \subseteq FEX(M)$. ⊠

6 Conclusions

The traditional model of learning in the limit EX was restricted so as to allow the learning machines only a fixed, finite amount of memory to store input and other data. The resulting class was called LEX. Not surprisingly, LEX was shown to be a proper subset of EX. LEX was shown to be incomparable with the class PEX. Probabilistic learning was also considered. Our main result was to show that $LEX\langle 1\rangle - LEX$ is not empty.

7 Acknowledgements

The results in this paper were obtained while both authors attended ICALP91 in Madrid. We gratefully acknowledge The Universities of Latvia and Maryland, the National Science Foundation and the organizers of ICALP91 for making it possible for the authors to convene in a setting condusive to obtaining the results described in this paper. Our colleague, Bill Gasarch, made valuable comments on an early draft of this paper.

References

1. ABLAEV, F. M. AND FREIVALDS, R. Why sometimes probabilistic algorithms can be more effective. In *Lecture Notes in Computer Science*, 233, 1986

2. ADDANKI, S. Connectionism. In *The Encyclopedia of Artificial Intelligence*, S. Shapiro, Ed., John Wiley and Sons Inc., New York, NY, 1987

3. ANGLUIN, D. AND SMITH, C. H. Inductive inference: theory and methods. *Computing Surveys 15*, 237–269 (1983)

4. ANGLUIN, D. AND SMITH, C. H. Inductive inference. In *Encyclopedia of Artificial Intelligence*, S. Shapiro, Ed., John Wiley and Sons Inc., 1987

5. BLUM, L. AND BLUM, M. Toward a mathematical theory of inductive inference. *Information and Control 28*, 125–155 (1975)

6. CASE, J. AND NGOMANGUELLE, S. Refinements of inductive inference by popperian machines. *Kybernetika* (19??) to appear.

7. CASE, J. AND SMITH, C. Comparison of identification criteria for machine inductive inference. *Theoretical Computer Science 25, 2*, 193–220 (1983)

8. FELDMAN, J. A. AND BALLARD, D. H. Connectionist models and their properties. *Cognitive Science 6, 3*, 205–254 (1982)

9. FELLER, W. *An Introduction to Probability Theory and its Applications*. Wiley, New York, 1968 Third Edition.

10. FREIVALDS, R. AND SMITH, C. *On the power of procrastination for machine learning*. Manuscript.

11. FULK, M. AND CASE, J. *Proceedings of the Third Annual Workshop on Computational Learning Theory*. Margan Kaufmann Publishers, Palo Alto, CA., 1990

12. GOLD, E. M. Language identification in the limit. *Information and Control 10*, 447–474 (1967)

13. HAUSSLER, D. AND PITT, L. *Proceedings of the 1988 Workshop on Computational Learning Theory*. Margan Kaufmann Publishers, Palo Alto, CA., 1988

14. HEATH, D., KASIF, S., KOSARAJU, R., SALZBERG, S., AND SULLIVAN, G. *Learning nested concept classes with limited storage*. Computer Science Department, Johns Hopkins University, Baltimore MD., 1990

15. KLEENE, S. On notation for ordinal numbers. *Journal of Symbolic Logic 3*, 150–155 (1938)

16. MACHTEY, M. AND YOUNG, P. *An Introduction to the General Theory of Algorithms*. North-Holland, New York, 1978

17. MICHALSKI, R., CARBONELL, J., AND MITCHELL, T. *Machine Learning*. Tioga Publishing Co., Palo Alto, CA, 1983

18. MIYAHARA, T. Inductive inference by iteratively working and consistent strategies with anomalies. *Bulletin of Informatics and Cybernetics 22*, 171–177 (1987)

19. MIYAHARA, T. A note on iteratively working strategies in inductive inference. In *Proceedings of the Fujitsu IIAS-SIS Workshop on Computational Learning Theory*, Numazu, Japan, 1989

20. OSHERSON, D., STOB, M., AND WEINSTEIN, S. *Systems that Learn*. MIT Press, Cambridge, Mass., 1986

21. PITT, L. A Characterization of Probabilistic Inference. *Journal of the ACM 36*, 2, 383–433 (1989)

22. PITT, L. AND SMITH, C. Probability and plurality for aggregations of learning machines. *Information and Computation 77*, 77–92 (1988)

23. POPPER, K. *The Logic of Scientific Discovery*. Harper Torch Books, N.Y., 1968 2nd Edition.

24. RIVEST, R., HAUSSLER, D., AND WARMUTH, M. *Proceedings of the Second Annual Workshop on Computational Learning Theory*. Margan Kaufmann Publishers, Palo Alto, CA., 1989

25. ROGERS, H. JR. Gödel numberings of partial recursive functions. *Journal of Symbolic Logic 23*, 331–341 (1958)

26. SHAPIRO, S. *Encyclopedia of Artificial Intelligence*. John Wiley and Sons Inc., New York, NY, 1987

27. TAIMINA, D. YA. AND FREIVALDS, R. On complexity of probabilistic finite automata recognizing superlanguages. In *Methods of logic in construction of effective algorithms*, Kalinin State University, 1966 In Russian.

28. WARMUTH, M. AND VALIANT, L. *Proceedings of the 1991 Workshop on Computational Learning Theory*. Margan Kaufmann Publishers, Palo Alto, CA., 1991

29. WIEHAGEN, R. Limes-erkennung rekursiver funktionen durch spezielle strategien. *Elektronische Informationsverarbeitung und Kybernetik 12*, 93–99 (1976)

Retrieval of scattered information by EREW, CREW and CRCW PRAMs

Faith Fich[*]

University of Toronto, Toronto, Ontario, Canada M5S 1A4
MIT, Cambridge, Massachusetts, USA 02139
fich@theory.toronto.edu

Mirosław Kowaluk[†] and Krzysztof Loryś[‡]

Universität Würzburg, D-W-8700-Würzburg, Germany,
kowaluk@informatik.uni-wuerzburg.dbp.de,
lorys@informatik.uni-wuerzburg.dbp.de

Mirosław Kutyłowski[§]

FB Informatik and Heinz-Nixdorf-Institut, Universität–GH Paderborn,
D-W-4790 Paderborn, Germany, mirekk@uni-paderborn.de

Prabhakar Ragde [¶]

University of Waterloo, Waterloo, Ontario, Canada N2L 3G1
plragde@maytag.waterloo.edu

Abstract. The k-compaction problem arises when k out of n cells in an array are non-empty and the contents of these cells must be moved to the first k locations in the array. Parallel algorithms for k-compaction have obvious applications in processor allocation and load balancing; k-compaction is also an important subroutine in many recently developed parallel algorithms. We show that any EREW PRAM that solves the k-compaction problem requires $\Omega(\sqrt{\log n})$ time, even if the number of processors is arbitrarily large and $k = 2$. On the CREW PRAM, we show that every n-processor algorithm for k-compaction problem requires $\Omega(\log \log n)$ time, even if $k = 2$. Finally, we show that $O(\log k)$ time can be achieved on the ROBUST PRAM, a very weak CRCW PRAM model.

[*]supported by NSERC and the Information Technology Research Centre of Ontario

[†]supported by Deutsche Forschungsgemeinschaft under grant Wa 549/1; present address: Uniwersytet Warszawski, ul. Banacha 2, PL-02-097 Warszawa, Poland

[‡]supported by the Alexander von Humboldt-Stiftung; present address: Uniwersytet Wrocławski, ul. Przesmyckiego 20, PL-51-151, Wrocław, Poland

[§]supported by Deutsche Forschungsgemeinschaft under grant ME 872/1-4; present address: Uniwersytet Wrocławski, ul. Przesmyckiego 20, PL-51-151, Wrocław, Poland

[¶]supported by NSERC and the Information Technology Research Centre of Ontario

1   Introduction

The *k-compaction problem* is defined as follows: given values in array $A[1..n]$, of which at most k are nonzero, rearrange the values so that the nonzero values occupy the initial fragment of the array. This problem has been considered in many contexts; for instance, this is a natural subproblem to consider in the context of processor or task reallocation. Gil and Rudolph [11] gave a deterministic algorithm to solve this problem in time $O(\log k)$ on a concurrent-read concurrent-write parallel random access machine (CRCW PRAM) using n processors. Rudolph and Steiger [21] gave a probabilistic algorithm running in time $O(\log k)$ (with high probability) using k processors. Hagerup and Nowak [16] gave a probabilistic algorithm running in time $O(\log k/\log\log k)$ using n processors. All of these algorithms use an ARBITRARY CRCW PRAM, with the property that when several processors simultaneously write into the same cell, an arbitrary processor succeeds. (The terminology is from [8]).

A weaker concurrent-read concurrent-write model is the COMMON PRAM, in which simultaneous writes must be of the same value. Ragde [20] gave a *k*-compaction algorithm and matching lower bound of $\Theta(\log k/\log\log n)$ for the COMMON PRAM using n processors. The immediate utility of this result was in the development of very fast randomized CRCW PRAM algorithms. Intuitively, if many independent instances of problems are worked on simultaneously by a fast algorithm with a low probability of failure, the failures can be gathered up by compaction and dealt with quickly. Papers such as [1, 9, 10, 12, 14, 15, 18] used compaction or approximate compaction as an essential subroutine in algorithms for space allocation, estimation, sorting, PRAM simulation, generation of random permutations, and computational geometry.

In this paper, we look at the problem of k-compaction on weaker models of computation. Hagerup and Nowak [16] considered k-compaction on the concurrent-read exclusive-write (CREW) PRAM. They gave an algorithm running in $O(k)$ time with n^k processors and gave another running in $O(k^2\log\log n)$ time using n processors. In Section 2, we show that the latter algorithm is optimal for small k by demonstrating an $\Omega(\log\log n)$ lower bound for n processors, even when $k = 2$.

We consider the exclusive-read exclusive-write (EREW) PRAM in Section 3 and show that, for this model, the k-compaction problem requires $\Omega(\sqrt{\log n})$ time, even when $k = 2$. There are very few lower bounds on the EREW PRAM for problems defined on Boolean inputs: Beame, Kik and Kutyłowski [2] show that the problem of broadcasting one value to n processors requires $\Theta(\log n)$ time. Snir's lower bound on searching a sorted table on an EREW PRAM [22] requires a large input domain due to the use of Ramsey theory.

The ROBUST PRAM is a very weak CRCW PRAM [17], in which the result of a simultaneous write might be any value. An algorithm for this model must be correct even if an adversary decides the outcome of every simultaneous write. Every other CRCW PRAM that has been studied can compute the logical OR of n bits in constant time. However, it is not known how to do OR in less than $O(\log n)$ time on the ROBUST PRAM and with $n^{O(1)}$ processors, it requires $\Omega(\sqrt{\log n})$ time [7]. In Section 4 of this paper, we demonstrate an algorithm for k-compaction running in $O(\log k)$ time on the ROBUST PRAM using n processors.

Combined, these results show that an important problem in the context of parallel computation has different complexity depending on whether concurrent reads and concurrent writes are allowed by the model.

2 The bounds for CREW PRAMs

Hagerup and Nowak [16] show that k-compaction of an array of size n can be performed by a CREW PRAM with n^k processors in $O(k)$ time. For $k = 2$, this yields a constant time algorithm with n^2 processors. They also show that, for $n \leq p \leq n^2$, the 2-compaction problem can be solved by a p-processor CREW PRAM in $O(\log \log n - \log \log(p/n))$ time. We will prove a matching lower bound.

Let the k-OR problem be the problem of computing the logical OR of n Boolean variables at most k of which have value one. Note that the k-OR problem can be reduced to k-compaction. Indeed, if we solve the k-compaction problem given a Boolean input that contains at most k ones, then it suffices to inspect the first output cell to determine the value of the OR. Hence any lower bound for computing 2-OR is also a lower bound for the 2-compaction problem. Furthermore, since 2-compaction is a subproblem of k-compaction for $k > 2$, any such lower bound also applies to these problems.

Consider any PRAM and suppose its input variables are $x_1, \ldots, x_n$. For $V \subseteq \{x_1, \ldots, x_n\}$, let I_V denote the inputs in which at most two of the variables in V have value 1 and the rest of the variables have value 0. We say that a processor *depends on at most one variable from the set V at time t* if, for the inputs in I_V, the state of the processor at time t is either fixed or can be expressed as a function of the value of one variable $x_i \in V$. In other words, for each value of x_i, changing the values of the other variables in V does not change the processor's state. Similarly, a memory cell *depends on at most one variable from V at time t* if its contents at time t are either fixed or can be expressed as a function of the value of one variable in V. In particular, if a processor's state or a memory cell's contents can be expressed as a nonconstant function of $x_i \in V$ for all the inputs in I_V, we say that the processor or memory cell depends on x_i.

Lemma 2.1 *Suppose that at the end of its computation, each memory cell of a PRAM depends on at most one variable from a set $V \subseteq \{x_1, \ldots, x_n\}$. If $|V| \geq 2$, then the PRAM does not compute 2-OR.*

Proof. At the end of the computation, the output cell depends on at most one variable in V. Since $|V| \geq 2$, there is a variable $x_i \in V$ that the answer cell does not depend on. Set all other variables to 0. Then the value in the answer cell is fixed, but the OR of the input bits is either 0 or 1, depending on the value of x_i. This means that there is an input in I_V on which the PRAM does not compute OR. □

Theorem 2.2 *A CREW PRAM with $p \geq n$ processors requires more than $(\log \log n - \log \log(5p/n))/2$ steps to solve the 2-OR problem.*

Proof. Consider a p-processor CREW PRAM that computes the OR of the Boolean variables $x_1, \ldots, x_n$, at most two of which have value 1. We prove, by induction on t, that there is a set of variables V_t such that $|V_t| \geq n^{2^t}/(5p)^{2^t-1}$ and, at the end of step t, each processor and each memory cell depends on at most one variable from V_t.

This condition is true at the beginning of the computation, since the state of every processor is fixed and the contents of each memory cell depends on at most one variable from $V_0 = \{x_1, \ldots, x_n\}$. Therefore, assume that $t \geq 0$ and that the claim is true at the end of step t.

Consider the situation just before the read phase in step $t+1$. Suppose all variables not in V_t have been set to 0. Let $G_{t+1} = (V_t, E)$ be the undirected graph with $\{x_i, x_j\} \in E$ if there exists a processor P whose state depends on x_i and a cell C whose contents depends on x_j such that P reads from C for some value of x_i.

Note that $|E| \leq 2p$. Turan's theorem [3, page 282] says that any graph $G = (V, E)$ has an independent set of size at least $|V|^2/(|V| + 2|E|)$. In particular, there is an independent set V'_{t+1} in G_{t+1} such that

$$|V'_{t+1}| \geq |V_t|^2/(|V_t| + 4p) \geq |V_t|^2/5p \geq \left(n^{2^{2t}}/(5p)^{2^{2t}-1}\right)^2/5p \geq n^{2^{2t+1}}/(5p)^{2^{2t+1}-1}.$$

Since the read phase does not cause the contents of memory cells to change and $V'_{t+1} \subseteq V_t$, each memory cell depends on at most one variable from V'_{t+1}. Furthermore, each processor depends on at most one variable from V'_{t+1}. To see this, suppose P depends on x_i and P reads from C and C' when $x_i = 0$ and $x_i = 1$, respectively. Also suppose that C and C' depend only on the variables x_j and x_k, respectively. If $x_i \notin V'_{t+1}$, then x_i is set to 0 and P depends only on x_j. Otherwise, $x_j, x_k \notin V'_{t+1}$, so x_j, x_k are set to 0 and P depends only on x_i.

Now consider the write phase in step $t+1$. Suppose all variables not in V'_{t+1} have been set to 0. Let $G'_{t+1} = (V'_{t+1}, E')$ be the undirected graph with $\{x_i, x_j\} \in E'$ if there exists a processor P whose state depends on x_i and a memory cell C whose contents depends on x_j such that P writes to C for exactly one value of x_i.

Again $|E'| \leq 2p$ and there is an independent set V_{t+1} in G'_{t+1} such that

$$\begin{aligned} |V_{t+1}| &\geq |V'_{t+1}|^2/(|V'_{t+1}| + 4p) \geq |V'_{t+1}|^2/5p \\ &\geq \left(n^{2^{2t+1}}/(5p)^{2^{2t+1}-1}\right)^2/5p \geq n^{2^{2(t+1)}}/(5p)^{2^{2(t+1)}-1}. \end{aligned}$$

Since the write phase does not cause processors' states to change and $V_{t+1} \subseteq V'_{t+1}$, each processor depends on at most one variable from V_{t+1}.

Let C be a memory cell and suppose there are two processors P and P' that depend on different variables x_i and x_j in V_{t+1} and such that P writes to C only when $x_i = a$ and P' writes to C only when $x_j = a'$. Setting $x_i = a$ and $x_j = a'$ causes a write conflict; therefore this situation cannot arise. Thus, which processor (if any) writes to C during step $t+1$ depends on the value of at most one variable.

Suppose that the contents of C at the end of step t depend on the variable x_i and which processor writes to C during step $t+1$ depends on the value of another variable x_j. If $x_i \notin V'_{t+1}$, then x_i is set to 0 and C depends only on x_j. Otherwise, $x_j \notin V'_{t+1}$, so x_j is set to 0. Then the contents of C at the end of step $t+1$ is either fixed or depends only on x_i, depending on whether or not C is written to when $x_j = 0$. Thus at the end of step $t+1$, each memory cell depends on at most one variable from V_{t+1}.

It follows by induction that, after the last step, T, of the computation, there is a set $V_T \subseteq \{x_1, \ldots, x_n\}$ such that $|V_T| \geq n^{2^{2T}}/(5p)^{2^{2T}-1}$ and every memory cell depends on at most one variable from V_T. Furthermore, it follows from Lemma 2.1 that $|V_T| \leq 1$. Therefore $T > (\log \log n - \log \log(5p/n))/2$; otherwise, $(5p/n)^{2^{2T}-1} < n$, which implies that $|V_T| > 1$. $\qquad\square$

Corollary 2.3 *For $k \geq 2$, a CREW PRAM with $p \geq n$ processors requires more than $(\log \log n - \log \log(5p/n))/2$ steps to solve the k-compaction problem.*

3   The bounds for EREW PRAMs

In this section, we prove that an EREW PRAM with arbitrarily many processors requires $\Omega(\sqrt{\log n})$ steps to solve the k-compaction problem for any $k \geq 2$. We also present a new algorithm for 2-compaction that runs in approximately $0.36 \log n$ steps.

The lower bound uses an approach that is similar to the one used to obtain the lower bound for CREW PRAMs.

Theorem 3.1 *An EREW PRAM requires more than $\sqrt{\log n}/2 - 2$ steps to solve the 2-OR problem.*

Proof. Consider an EREW PRAM that solves the 2-OR problem. We prove, by induction on t, that at the end of step t, there is a set of variables $V_t \subseteq \{x_1, \ldots, x_n\}$ such that $|V_t| \geq n/15^t \cdot 4^{t(t-1)}$ and each processor and each memory cell depends on at most one variable from V_t. This condition holds at the beginning of the computation, with $V_0 = \{x_1, \ldots, x_n\}$.

We assume that the property holds for t. In particular, at the end of step t, each processor and memory cell depends on at most one variable from V_t.

Beame, Kik, and Kutyłowski proved that in any EREW PRAM, for any two Boolean inputs that differ on one bit, there are fewer than 4^t processors and 4^t memory cells that can distinguish between them after t steps [2]. Let $z \in I_{V_t}$ denote the input in which all variables are 0 and, for each $x_i \in V_t$, let $z^{\{i\}} \in I_{V_t}$ denote the input in which $x_i = 1$, but all other variables are 0. Then, there are fewer than 4^t processors and 4^t memory cells that distinguish between z and $z^{\{i\}}$ after t steps of the computation. Hence, for each $x_i \in V_t$, there are fewer than 4^t processors and 4^t memory cells that depend on x_i at the end of step t.

Let $G_{t+1} = (V_t, E)$ be the directed graph with $(x_i, x_j) \in E$ if, at the beginning of step $t + 1$, there exists a processor P that depends on x_i and a memory cell C that depends on x_j and, for some input in I_{V_t}, P reads from C during step $t + 1$.

Two processors that depend on different variables cannot both read from the same cell during step $t + 1$. Otherwise, we can set the values of those two variables so that a write conflict arises. Since there are fewer than 4^t cells that depend on each $x_j \in V_t$, the indegree of each node is less than 4^t.

The outdegree of any node $x_i \in V_t$ is less than $2 \cdot 4^t$. This is because there are fewer than 4^t processors that depend on x_i. Each such processor is in one state for those inputs in I_{V_t} in which $x_i = 0$ and in another state for those inputs in I_{V_t} in which $x_i = 1$. Hence each can read from at most 2 different cells during step $t + 1$.

Zarankiewicz's theorem [3, page 284] says that a graph $G = (V, E)$ of degree at most k has an independent set of cardinality at least $|V|/(k+1)$. Since G_{t+1} has degree less than $3 \cdot 4^t$, it has an independent set $V'_{t+1} \subseteq V_t$ such that $|V'_{t+1}| \geq |V_t|/(3 \cdot 4^t)$. As in the proof of Theorem 2.2, it is easy to prove that after the read phase of step $t+1$, each processor and each cell depends on at most one variable from V'_{t+1}.

Now consider the write phase of step $t+1$. Let $G'_{t+1} = (V'_{t+1}, E')$ be the directed graph with $(x_i, x_j) \in E'$ if, immediately before the write phase of step $t+1$, there exists a processor P that depends on x_i and a memory cell C that depends on x_j such that P writes into C for exactly one value of x_i.

As in the proof of Theorem 2.2, if two different processors may write into the same cell at step $t+1$ for inputs in $I_{V'_{t+1}}$, then they depend on the same variable in V'_{t+1}. There are less than 4^t cells that depend on each variable in V'_{t+1} after step t. Thus the indegree of each node is less than 4^t.

To show that the outdegree of each node is less than $4 \cdot 4^t$, consider any $x_i \in V'_{t+1}$. There are less than 4^t processors that depend on x_i at the beginning of step $t+1$. A processor may also depend on x_i by reading a cell during step $t+1$ that depends on x_i. For inputs in $I_{V'_{t+1}}$, no processor that depends on $x_j \in V'_{t+1}$ can read a cell that depends on x_i at step $t+1$. Hence a cell that depends on x_i can either be read by a fixed processor or not be read at all. Since there are less than 4^t such cells, there are less than $2 \cdot 4^t$ processors that depend on x_i after the read phase of step $t+1$. Each of these processors have two possible states they can be in. Thus there are less than $2 \cdot 2 \cdot 4^t$ cells that are written to during step $t+1$ by any of these processors for any input in $I_{V'_{t+1}}$. This implies that the outdegree of x_i is less than $4 \cdot 4^t$.

Since G'_{t+1} has degree less than $5 \cdot 4^t$, it has an independent set $V_{t+1} \subseteq V'_{t+1}$ of cardinality at least $|V'_{t+1}|/(5 \cdot 4^t)$. As in the proof of Theorem 2.2, one can see that after the write phase of step $t+1$ each processor and each cell depends on at most one variable from V_{t+1}. Also,

$$|V_{t+1}| \geq \frac{|V'_{t+1}|}{5 \cdot 4^t} \geq \frac{|V_t|}{3 \cdot 4^t \cdot 5 \cdot 4^t} \geq \frac{n/(15^t \cdot 4^{t(t-1)})}{15 \cdot 4^{2t}} = \frac{n}{15^{t+1} \cdot 4^{(t+1) \cdot t}}.$$

Since $|V_t| \geq n/15^t \cdot 4^{t(t-1)}$, it follows that $|V_t| \geq 2$, for $t \leq \sqrt{\log n/2} - 2$. Hence, by Lemma 2.1, the computation requires more than $t \leq \sqrt{\log n/2} - 2$ steps. $\quad\square$

Corollary 3.2 *For $k \geq 2$, a EREW PRAM with $p \geq n$ processors requires more than $\sqrt{\log n/2} - 2$ steps to solve the k-compaction problem.*

Next, we turn attention to upper bounds.

Lemma 3.3 *The k-compaction problem can be solved by an EREW PRAM with n processors in $\log n + O(1)$ steps.*

The idea of this algorithm is to replace each nonzero value by 1 and then compute the prefix sum to determine the number of nonzero values in each initial segment of the input. Using this information, the original nonzero elements can be moved (in order) to the beginning of the array.

We conjecture that, even for $k = 2$, it is not possible to improve k-compaction by more than a constant factor on an EREW PRAM. However, the following algorithm does solve 2-compaction almost a factor of 3 faster.

Algorithm 3.4 *The 2-compaction problem can be solved on an EREW PRAM using n processors in $0.5 \log_b n + 0(1) \approx 0.36 \log n$ steps, where $b = (3 + \sqrt{5})/2$.*

Proof. Partition the input into $\sqrt{n}$ groups of $\sqrt{n}$ input variables and associate $\sqrt{n}$ processors with each group. In parallel, the processors $P_{i,1}, \ldots, P_{i,\sqrt{n}}$ associated with each group $C_{i,1}, \ldots, C_{i,\sqrt{n}}$ will determine whether there are 0, 1 or 2 nonzero values in their group, using the subroutine described below. Furthermore, if the group contains a single nonzero value, then all the variables in the group are assigned that value. If the variables in the group are all initially zero, they do not change value.

From this point, the solution to the 2-compaction problem can be obtained in at most 2 more steps, as follows. Processors in groups that are entirely zero do nothing. In particular, if the input was initially all zero, the output will be all zero.

If there is a group that originally had 2 nonzero values, 2 processors in the group write these values to the first 2 answer cells. Note that, in this case, all other groups are entirely zero, so no write conflicts will arise.

Now suppose the ith group originally had 1 nonzero value. Then each processor $P_{i,j}$, $j \neq i$, reads $C_{j,i}$, the ith variable in group j. Since there is at most one other group that contains a nonzero value, at most one of these processors can read a nonzero value. If $C_{j,i}$ is nonzero, then processor $P_{i,j}$ writes 0 into $C_{i,1}$ and during the second step, writes the nonzero value originally in the ith group to one of the answer cells. To avoid a conflict with $P_{j,i}$, processor $P_{i,j}$ writes its value to the first answer cell if $i < j$ and to the second answer cell if $i > j$. During the second step, $P_{i,1}$ reads $C_{i,1}$. If it is nonzero, then there was no other group with a nonzero value and $P_{i,1}$ writes this value to the first answer cell.

It remains to describe the subroutine. To be concrete, let $m = \sqrt{n}$ and suppose that cells $C_{i,1}, \ldots, C_{i,m}$ initially contain the values $a_1, \ldots, a_m$. The subroutine is based on the algorithm of Cook, Dwork and Reischuk [5] for computing the OR of m Boolean variables.

We define a function Last that identifies the last occurrence of a nonzero value in a sequence, if one exists. More precisely,

$$\text{Last}([x_1, \ldots, x_v]) = 0 \quad \text{if} \quad x_1 = \ldots = x_v = 0 \text{ and}$$
$$\text{Last}([x_1, \ldots, x_v]) = (u, x_u) \quad \text{if} \quad x_u \neq 0 \text{ and } x_{u+1} = \ldots = x_v = 0.$$

One can easily see that

$$\text{Last}([x_1, \ldots, x_v, x_{v+1}, \ldots, x_w]) = \begin{cases} \text{Last}([x_{v+1}, \ldots, x_w]), & \text{if } \text{Last}([x_{v+1}, \ldots, x_w]) \neq 0, \\ \text{Last}([x_1, \ldots, x_v]), & \text{otherwise.} \end{cases}$$

For $t \geq 0$, let $c(t) = F_{2t+1}$ and $p(t) = F_{2t}$, where F_i is the ith Fibonacci number. Then $c(0) = 1$, $p(0) = 0$, $p(t + 1) = p(t) + c(t)$, and $c(t + 1) = c(t) + p(t + 1)$. Inductively, we will ensure that immediately after step t, each cell $C_{i,j}$ stores the value of $\text{Last}([a_j, \ldots, a_{j+c(t)-1}])$, and each processor $P_{i,j}$ knows the value of $\text{Last}([a_j, \ldots, a_{j+p(t)-1}])$, Here a_k, $k > m$, denotes the value a_j such that $j = k \bmod m$ and $1 \leq j \leq m$.

When $t = 0$, these conditions are true. During step $t + 1$, processor $P_{i,j}$ reads from the cell $C_{i,j'}$ where $j' = j + p(t) \bmod m$. Since $C_{i,j'}$ stores the value of $\text{Last}([a_{j'}, \ldots, a_{j'+c(t)-1}])$ and $j' + c(t) - 1 = j + p(t+1) - 1 \bmod m$, $P_{i,j}$ can easily compute $\text{Last}([a_j, \ldots, a_{j+p(t+1)-1}])$. During step $t + 1$, if $\text{Last}([a_j, \ldots, a_{j+p(t+1)-1}])$ is different from 0, processor $P_{i,j}$ writes this value to cell $C_{i,j''}$, where $j'' = j - c(t) \bmod m$. Otherwise, $P_{i,j}$ does not write. Note that $j+p(t+1)-1 = j''+c(t+1)-1 \bmod m$ so that, after step $t+1$, $C_{i,j''}$ contains the value $\text{Last}([a_{j''}, \ldots, a_{j''+c(t+1)-1}])$, whether or not $P_{i,j}$ wrote.

If $T \geq \log_b m + 1.34$, $F_{2T+1} \geq m$. (See, for example, [6].) Thus, after T steps, each cell $C_{i,j}$ contains (k, a_k), where a_k is the last element in the sequence $[a_j, a_{j+1}, \ldots, a_m, a_1, \ldots, a_{j-1}]$ that is different from 0, if such an element exists. Otherwise, $C_{i,j}$ contains the value 0 for every $j \in \{1, \ldots, m\}$. Note that if both a_k and $a_{k'}$ are nonzero, where $1 \leq k < k' \leq m$, then $C_{i,k} = (k', a_{k'})$ and $C_{i,k'} = (k, a_k)$. In this case, processors $P_{i,k}$ and $P_{i,k'}$ can write the values a_k and $a_{k'}$ to the first and the second answer cells, respectively. If a_k is the only nonzero element among $a_1, \ldots, a_m$, then $C_{i,j} = (k, a_k)$ for all $j \in \{1, \ldots, m\}$. In particular, $C_{i,k} = (k, a_k)$. Using these observations, it is easy for the processors in the ith group to distinguish between these two cases. $\qquad\square$

4 ROBUST PRAM algorithms for the compaction problem

The k-compaction problem can be solved by an n-processor COMMON PRAM in $O(\log k / \log \log n)$ time [20]. If an upper bound k on the number of ones in the input is not known, the algorithm still works correctly using $O(\log m)$ steps, where m is the exact number of ones in the input. We present similar algorithms that work on the ROBUST PRAM.

We begin by describing a simple n-processor ROBUST PRAM that solves the 2-compaction problem in $O(1)$ steps. Suppose the input is located in cells $C_{i,j}$, $i, j \in \{1, \ldots, \sqrt{n}\}$. Each cell $C_{i,j}$ is read by one processor. If the cell stores a nonzero value, the processor attempts to write that value into cells R_i and S_j.

Note that if the input is entirely 0, then $R_1, R_2, \ldots, R_{\sqrt{n}}$ and $S_1, S_2, \ldots, S_{\sqrt{n}}$ will all contain their initial value, 0. If the input contains exactly one nonzero value then that value will occur exactly once among $R_1, R_2, \ldots, R_{\sqrt{n}}$ and exactly once among $S_1, S_2, \ldots, S_{\sqrt{n}}$. Now suppose that there are two input cells $C_{i,j}$ and $C_{i',j'}$ that contain nonzero values. Then either $i \neq i'$ or $j \neq j'$, so either $R_1, R_2, \ldots, R_{\sqrt{n}}$ or $S_1, S_2, \ldots, S_{\sqrt{n}}$ will contain both these values.

Next, the 2-compaction problems for $R_1, R_2, \ldots, R_{\sqrt{n}}$ and $S_1, S_2, \ldots, S_{\sqrt{n}}$ are each solved by the $O(1)$ time CREW PRAM algorithm that uses $(\sqrt{n})^2 = n$ processors [17]. A constant number of additional steps suffice to combine the solutions to these two problems into the solution of the original problem.

With sufficiently many processors, a ROBUST PRAM can perform k-compaction in $O(\log k)$ time. The idea is to assign one processor to each subset of at most k input variables. In $O(\log k)$ steps, the processors determine which of these sets contain only nonzero values. Call these sets *full*. Every processor whose set is full writes its processor number to a cell associated with its set's size. There is only one full set of maximum size, so a unique processor writes to the corresponding cell. This cell can be identified in $O(\log k)$ steps, since all cells corresponding to larger sets are not written to. Once the full set of maximum size is identified, it is easy to produce the output for the k-compaction problem.

Lemma 4.1 *[13]. The k-compaction problem can be solved in $O(\log k)$ time by a ROBUST PRAM with $\sum_{i=1}^{k} \binom{n}{k}$ processors.*

Using this algorithm as a subroutine, we are able to obtain an algorithm that uses significantly fewer processors.

Theorem 4.2 *The k-compaction problem of size n can be solved in $O(\log k)$ time by an n-processor ROBUST PRAM.*

Proof. If $k \geq n^{1/5}$, the result follows from Lemma 3.3. Therefore, we assume that $k \leq n^{1/5}$.

First, the input cells are partitioned into groups of l cells, where

$$
l = \begin{cases}
2k(k-1) & \text{if } k \leq \frac{\log n}{4 \log \log n}, \\
\frac{(k-1)\log n}{3 \log \log n - 1} & \text{if } \frac{\log n}{4 \log \log n} < k \leq \log n, \text{ and} \\
\frac{(k-1)\log n}{3 \log k - 1} & \text{if } \log n \leq k \leq n^{1/5}.
\end{cases}
$$

We apply the EREW PRAM algorithm from Lemma 3.3 to solve the k-compaction problem within each group in $O(\log l) = O(\log k)$ steps.

Let $y_j = j$ if the jth group contains a cell with a nonzero value and let $y_j = 0$ otherwise. Given a solution of the k-compaction problem for the values $y_1, \ldots, y_{n/l}$, replace each nonzero value j in the solution by the number of nonzero values in the jth group. Then compute the prefix sum of these numbers in $O(\log k)$ steps and use this information to move the nonzero values in each group to their appropriate places in the answer. Thus it suffices to solve this smaller instance of k-compaction.

There are more than $0.6q/\ln q$ prime numbers in the interval $[q, 2q]$ for $q \geq 20.5$ (see [19]). The number l has been chosen so that $0.6n^{(k-1)/l}/\ln\left(n^{(k-1)/l}\right) \geq l$, for n sufficiently large. Therefore there are at least l prime numbers between $n^{(k-1)/l}$ and $2n^{(k-1)/l}$. Let $p_1, \ldots, p_l$ be distinct prime numbers in this range.

Claim *If $a_1, \ldots, a_k \in [1, n]$ are distinct integers, then for each a_j, there is some prime p_i such that*

$$a_r \neq a_j \mod p_i \quad \text{for all } r \neq j.$$

To prove the claim, consider any number $j \leq k$. Let $Z_{j,r} = \{p_i : i \leq l \text{ and } a_j = a_r \mod p_i\}$. It suffices to show that there exists $p_i \notin \bigcup_{r \neq j} Z_{j,r}$. Since $|a_j - a_r| < n$, there are less than $l/(k-1)$ indices i such that $p_i \mid (a_j - a_r)$; otherwise $|a_j - a_r| \geq (n^{(k-1)/l})^{l/(k-1)} = n$. Hence $|\bigcup_{r \neq j} Z_{j,r}| < (k-1) \cdot l/(k-1) = l$ and there exists $p_i \notin \bigcup_{r \neq j} Z_{j,r}$.

For each prime p_i, our algorithm uses p_i auxiliary cells $S_i(0), \ldots, S_i(p_i - 1)$. In parallel, for each $y_j \neq 0$ and each prime p_i, one processor attempts to write the value j into cell $S_i(j \bmod p_i)$. Of course, write conflicts may occur producing unpredictable results. However, by the claim, for each $y_j \neq 0$, there is some prime p_i such that exactly one processor writes into $S_i(j \bmod p_i)$.

To eliminate garbage caused by write conflicts, we check the correctness of the data stored in these cells. Namely, if a cell $S_i(q)$ contains a nonzero value j, but $y_j = 0$, then 0 is written into $S_i(q)$. Since there are at most $l2n^{(k-1)/l} \in O(n)$ of these cells, there are sufficiently many processors to accomplish this in constant time.

If $k > \log n/4 \log \log n$, then $l2n^{(k-1)/l} \in k^{O(1)}$. Otherwise, we reduce the number of cells to $lk \in k^{O(1)}$ by performing k-compaction on $S_i(0), \ldots, S_i(p_i - 1)$, for each $i \leq l$. Since $k \leq \log n/4 \log \log n$ implies that $l(2n^{(k-1)/l})^k \in O(n)$, these l different k-compaction problems can be solved simultaneously in $O(\log k)$ steps using the algorithm of Lemma 4.1.

Next, all but one copy of each nonzero value in these $k^{O(1)}$ cells are set to 0. This can be done in $O(\log k)$ steps using Cole's CREW PRAM sorting algorithm [4]. Since the number of cells is also in $O(n)$, there is a sufficient number of processors available.

Finally, the k-compaction algorithm of Lemma 3.3, is applied to these cells to obtain the solution of the k-compaction problem for the values $y_1, \ldots, y_{n/l}$. □

Note that the algorithm presented has the property that during a computation no memory cell has to store a word that is longer than $\max\{\lceil \log n \rceil + 1, w\}$, where w is the maximal length of the words being the initial contents of the input cells.

When the number of cells with nonzero values is small, but an upper bound on this number is not known, it is still possible to produce the answer quickly.

Theorem 4.3 *A ROBUST PRAM with n processors can solve the n-compaction problem so that, if the input contains exactly k cells with nonzero values, then, after $O(\log k)$ steps, these values are in the first k output cells and none of the output cells are subsequently modified.*

Proof. It suffices to solve the compaction problem when, for each i, the ith input variable has value either 0 or i. This is because, when a processor writes the value i to an output cell, it could write the initial value of the ith input cell instead.

The k-compaction algorithm in Theorem 4.2 can also be applied when there are more then k cells with nonzero values. In this case, the algorithm yields some result, but not all of the nonzero values can be contained in the first k output cells. However, even in this case, each output value is correct and cannot occur more than once in the output cells.

Our algorithm works by repeatedly finding nonzero values in the input, appending them to the output, and removing them from the input (by setting them to 0). The k-compaction algorithm in Theorem 4.2 is applied for successively larger values of k. Specifically, the jth stage performs k_j-compaction, where $k_1 = 2$ and $k_{j+1} = k_j^2$, for $j \geq 1$. In total, $1 + \lceil \log \log n \rceil$ stages are performed, and, in particular, the last stage performs n-compaction. During stage $j + 1$, if there are more than k_{j+1} nonzero values in the input, not all of them (or, perhaps, none of them) are detected by this procedure, but the values that are output are all correct.

Now suppose the input initially contains k nonzero values, where $k_{j-1} < k \leq k_j$. Since the number of nonzero values in the input only decreases as the algorithm

proceeds, there are no more than k nonzero values in the input at the beginning of stage j. Then, during stage j, all remaining nonzero values are identified. During later stages, no nonzero values remain in the input, so the contents of the output cells will not be modified. The number of steps performed during the stages 1 through j is $O(\log(k_1) + \cdots + \log(k_j)) = O(\log(k_j^{1/2^{j-1}}) + \cdots + \log(k_j) = O(\log k_j) = O(\log k)$.
$\square$

For the inputs that contain at most k cells with nonzero values, the PRAM described in Theorem 4.3 produces the correct output within $O(\log k)$ steps, but does not halt until $\Omega(\log n)$ steps have been performed. The problem is that, without a bound on the number of cells with nonzero values in the input, it is not known whether any additional output cells will be written into. For example, when the input contains many cells with nonzero values, it is possible that, due to write conflicts, all the output cells may remain 0 for a long time. This behavior is not the fault of our algorithm, but rather is inherent in any solution, due to the fact that any n-processor ROBUST PRAM requires $\Omega(\sqrt{\log n})$ steps to compute the OR of n Boolean variables [7].

References

[1] H. Bast and T. Hagerup: *Fast and Reliable Parallel Hashing.* In: Proc. 3rd Annual ACM Symposium on Parallel Algorithms and Architectures, 1991, pp. 50–61.

[2] P. Beame, M. Kik and M. Kutyłowski: *Information broadcasting by Exclusive Read PRAMs.* Manuscript.

[3] C. Berge: *Graphs and Hypergraphs.* North–Holland, Amsterdam, 1976.

[4] R. Cole: *Parallel Merge Sort.* SIAM J. Comput. 17, 1988, pp. 770–785.

[5] S. Cook, C. Dwork and R. Reischuk: *Upper and Lower Time Bounds for Parallel Random Access Machines Without Simultaneous Writes.* SIAM J. Comput. 15, 1986, pp. 87–97.

[6] M. Dietzfelbinger, M. Kutyłowski and R. Reischuk: *Exact Time Bounds for Computing Boolean Functions on PRAMs Without Simultaneous Writes.* To appear in J. Computer and System Sciences.

[7] F. E. Fich, R. Impagliazzo, B. Kapron, V. King and M. Kutyłowski: *Limits on the Power of Parallel Random Access Machines with Weak Forms of Write Conflict Resolution.* Manuscript.

[8] F. E. Fich, P. Ragde, A. Wigderson: *Relations Between Concurrent-Write Models of Parallel Computation.* SIAM J. Comput. 17, 1988, pp. 606–627.

[9] J. Gil and Y. Matias: *Fast Hashing on a PRAM – Designing by Expectation.* In: Proc. 2nd Annual ACM Symposium on Discrete Algorithms, 1991, pp. 271–280.

[10] J. Gil, Y. Matias, and U. Vishkin: *Towards a Theory of Nearly Constant Parallel Time Algorithms.* In: Proc. 32nd Annual IEEE Symposium on Foundations of Computer Science, 1991, pp. 698–710.

[11] J. Gil and L. Rudolph: *Counting and Packing in Parallel.* In: Proc. 1986 International Conference on Parallel Processing, pp. 1000–1002.

[12] M. Goodrich: *Using Approximation Algorithms to Design Parallel Algorithms that may Ignore Processor Allocation.* In: Proc. 32nd Annual IEEE Symposium on Foundations of Computer Science, 1991, pp. 711–722.

[13] T. Hagerup. Personal communication

[14] T. Hagerup: *Fast and Optimal Simulations between CRCW PRAMs.* In: Proc. 9th Symposium on Theoretical Aspects of Computer Science, 1992, pp. 45–56.

[15] T. Hagerup: *The Log-Star Revolution.* In: Proc. 9th Symposium on Theoretical Aspects of Computer Science, 1992, pp. 259–280.

[16] T. Hagerup and M. Nowak: *Parallel retrieval of scattered information.* In: Proc. 16th International Colloquium on Automata, Languages, and Programming, 1989, pp. 439–450.

[17] T. Hagerup and T. Radzik: *Every ROBUST CRCW PRAM can Efficiently Simulate a PRIORITY PRAM.* In: Proc. 2nd ACM Symposium on Parallel Algorithms and Architectures, 1990, pp. 125–135.

[18] Y. Matias and U. Vishkin: *On Parallel Hashing and Integer Sorting.* In: Proc. 18th International Colloquium on Automata, Languages, and Programming, 1991, pp. 729–743.

[19] J. B. Rosser and L. Schoenfeld: *Approximate Formulas for Some Functions of Prime Numbers.* Illinois J. Math. 6, 1962, pp. 64–94.

[20] P. Ragde: *The Parallel Simplicity of Compaction and Chaining.* In: Proc. 17th International Colloquium on Automata, Languages, and Programming, 1990, pp. 744–751.

[21] L. Rudolph and W. Steiger: *Subset Selection in Parallel.* In: Proc. 1985 International Conference on Parallel Processing, pp. 11–14.

[22] M. Snir: *On parallel searching.* SIAM J. Comput. 14, 1985, pp. 688–708.

On Small Depth Threshold Circuits

Alexander A. Razborov*

Steklov Mathematical Institute

Vavilova 42, 117966, GSP–1, Moscow, RUSSIA

Abstract

In this talk we will consider various classes defined by small depth polynomial size circuits which contain threshold gates and parity gates. We will describe various inclusions between many classes defined in this way and also classes whose definitions rely upon spectral properties of Boolean functions.

1. Introduction

The main goal of the computational complexity theory is to be able to classify computational problems accordingly to their inherent complexity. At the first stage the problems are combined into large collections called complexity classes, each class consisting of problems which can be efficiently solved by an algorithm from a certain family. This allows one to unify many heterogeneous questions into only a few major problems about possible inclusions of one complexity class into another. Unfortunately, we are not even nearly close to solving the most important problems of this kind like the P vs. NP question or the NC vs. P question.

This talk will be devoted to a fragment of the complexity hierarchy (lying well below the class NC^1) where the existing machinery does allow us to answer questions on possible inclusions between complexity classes and in fact the result known at the moment give more or less complete picture of the fine structure within that fragment.

More precisely, we will be mostly interested in small depth circuits which contain threshold gates. There are two reasons for studying them.

The first reason is that threshold circuits are very closely connected to neural nets which is one of the most active areas in computer science. The basic element of a neural net is close to a threshold gate.

Another reason is that the complexity classes defined by small depth threshold circuits contain many interesting Boolean functions and are closely related to other complexity classes defined both in terms of small depth circuits and the spectral behavior of the function in question.

The paper is organized as follows. In Section 2 we introduce the necessary notation. Section 3 is devoted to (linear) threshold circuits of bounded depth. In Section

---

*This paper was prepared while the author was visiting Department of Mathematics at MIT partially supported by the Sloan foundation.

4 we consider complexity classes defined both in terms of the spectral representation and also in terms of polynomial thresholds. In Section 5 we merge together the two hierarchies of complexity classes considered in previous sections. The concluding section 6 contains some applications of the general theory to computing very concrete Boolean functions.

2. Notation

We will consider Boolean functions but for notational simplicity we will be working over $\{-1, 1\}$ rather than $\{0, 1\}$ where we let -1 correspond to 1 and 1 to 0. Thus variables will take the values $\{-1, 1\}$ and a typical function will be from $\{-1, 1\}^n$ to $\{-1, 1\}$. In this notation the parity of a set of variables will be equal to their product and thus we will speak of a monomial rather than of the parity of a set of variables. If we have a vector x of variables (indexed like x_i or x_{ij}) then a monomial will be written in the form x^α where α is a 0, 1-vector of the same type.

A *threshold gate* with n inputs is determined by n integer *weights* $(w_1, w_2, \ldots, w_n)$ and a *threshold* T. On an input $x = (x_1, \ldots, x_n) \in \{-1, 1\}^n$ it takes the value $\operatorname{sign}(x_1 w_1 + \cdots + x_n w_n + T)$ (we will always assume w.l.o.g. that the linear form $x_1 w_1 + \cdots + x_n w_n + T$ never evaluates to 0). The parameter $\sum_{i=1}^n |w_i| + |T|$ is called the *total weight* of the corresponding threshold gate.

Circuits considered in this paper will be mostly assembled from threshold gates and gates which compute monomials (= parity gates in the $\{0, 1\}$-terminology). We define the *size* of a circuit to be the number of gates.

3. Linear Threshold Circuits

In this section we will consider (linear) *threshold circuits* that is circuits consisting entirely of threshold gates. Let LT_d denote the class of functions computable by polynomial size depth d threshold circuits. Note that it is not quite clear a priori that functions computable even by, say, a single threshold gate have polynomial size circuits (of an arbitrary depth). The following well-known result (see e.g. [HKP91]) takes care of this.

Theorem 3.1. *For each threshold gate with n inputs there exists a threshold gate which computes the same function and has the total weight at most $\exp(O(n \log n))$.*

A model which is more natural from the "polynomial" point of view is to have the restriction that absolute values of all (integer!) weights are bounded by a polynomial in the length of the input. We will refer to this restriction as the *small weights* restriction and let $\widehat{LT}_d$ denote the class of Boolean functions computable by polynomial size depth d small weights threshold circuits. It can be easily seen that $\widehat{LT}_d$-circuits can be further simplified to consist of MAJORITY gates and of negations which appear on input variables only.

Now we review lower bounds known for linear threshold circuits.

It is easy to see that LT_1 does not contain all Boolean functions in n variables for any $n \geq 2$. In fact, even such a simple function in just two variables as $x_1 x_2$ is outside of LT_1.

An example of a function in $LT_1 \setminus \widehat{LT}_1$ was first presented in [MK61]:

Theorem 3.2 (Myhill, Kautz). *Any linear threshold gate computing the LT_1-function*

$$\text{sign} \left(\sum_{i=1}^{q+1} 2^{i-1} x_i + \sum_{j=1}^{q} \left(2^q - 2^{j-1}\right) y_j - 2^q \right)$$

must have a coefficient which is at least as large as 2^q.

In fact, Myhill and Kautz gave also an example for which the better bound $\Omega\left(2^n/n\right)$ holds but the proof of this latter result is much harder. The separation between $\widehat{LT}_1$ and LT_1 also follows from more general Theorem 5.4 below.

In depth 2 the first lower bounds were proven in the seminal paper [HMP+87]. Namely, they established the following.

Theorem 3.3 (Hajnal, Maass, Pudlák, Szegedy, Turán). *Any depth 2 small weights threshold circuit computing the function INNER PRODUCT MOD 2 (which is defined as $IP2_n(x_1,\ldots,x_n,y_1,\ldots,y_n) \rightleftharpoons (x_1 \wedge y_1) \oplus \cdots \oplus (x_n \wedge y_n)$ in $\{0,1\}-$notation) must have size $\exp(\Omega(n))$.*

Krause [Kra91] and Krause, Waack [KW91] slightly generalized and extended this result.

Note that $IP2_n \in \widehat{LT}_3$. Hence Theorem 3.3 gives the separation $\widehat{LT}_2 \neq \widehat{LT}_3$ (which, given Theorems 5.1 and 3.7, can be also deduced from Theorem 5.3 below).

No superpolynomial lower bounds are known for $\widehat{LT}_3$-circuits or even for LT_2-circuits. Maass, Schnitger and Sontag [MSS91] proved an $\Omega\left(\frac{\log \log n}{\log \log \log n}\right)$ bound on the size of depth 2 threshold circuits computing an explicitly given Boolean function. The following result was proved in [GHR92]:

Theorem 3.4 (Goldmann, Håstad, Razborov). *Any depth 2 threshold circuit computing INNER PRODUCT MOD 2 has size at least $\Omega(n/\log n)$.*

In fact, this is a direct consequence of Theorem 3.3 and the technique used for proving Theorem 3.7 below.

It seems that the only lower bound known for depth three comes from [HG90]. The *generalized inner product* mod 2, $GIP2_{n,s}$ is the Boolean function in ns variables defined (in $\{0,1\}$-notation) as follows:

$$GIP2_{n,s}(x_{ij}) \rightleftharpoons \bigoplus_{i=1}^{n} \bigwedge_{j=1}^{s} x_{ij}.$$

In particular, $IP2_n \equiv GIP2_{n,2}$.

Theorem 3.5 (Håstad, Goldmann). *Any depth 3 small weights threshold circuit which computes* $GIP2_{n,s}$ *and has fan-in at most* $(s-1)$ *at the bottom level, must have size* $\exp\left(\Omega\left(\frac{n}{s4^s}\right)\right)$.

This result gives an exponential lower bound but only for circuits with fan-in at most $\epsilon \log n$ at the bottom level. It would be extremely interesting to strengthen Theorem 3.5 because of the following simulation discovered in [Yao90]. Let ACC be the class of functions computable by polynomial size bounded depth circuits over $\{\neg, \wedge, \vee, MOD_{m_1}, \ldots, MOD_{m_k}\}$ where $m_1, \ldots, m_k$ are fixed integers and $MOD_m(x_1, \ldots, x_n) = 1$ iff $\sum_{i=1}^{n} x_i$ is divisible by m.

Theorem 3.6 (Yao). *If* $f_n \in ACC$ *then* f_n *is also computable by depth 3 small weights threshold circuits of size* $\exp\left((\log n)^{O(1)}\right)$ *and with fan-in at most* $(\log n)^{O(1)}$ *at the bottom level.*

So far superpolynomial lower bounds for ACC-circuits are known only for the bases $\{\neg, \wedge, \vee, MOD_q\}$ where q is a power of a prime (Razborov [Raz87], Smolensky [Smo87], Barrington [Bar86]).

Let's now see how efficiently general threshold circuits can be simulated by threshold circuits with small weights. The results of Chandra, Stockmeyer, Vishkin [CSV84] and Pippenger [Pip87] imply that the function ITERATED ADDITION (that is addition of n n-bit numbers) is computable by constant depth polynomial size *small weights* circuits. A direct consequence of this is that $LT_1 \subseteq \widehat{LT}_d$ for some constant d which was estimated as $d = 13$ in [SB91]. A better construction (based in fact on the spectral technique to be discussed in the next section) was given by Siu and Bruck [SB91]. Namely, they showed that the ITERATED ADDITION is in $\widehat{LT}_3$ which implies $LT_1 \subseteq \widehat{LT}_3$ and, moreover, $LT_d \subseteq \widehat{LT}_{2d+1}$ for any d which in general may depend upon the number of variables. For fixed d this was further improved in [GHR92]:

Theorem 3.7 (Goldmann, Håstad, Razborov). $LT_d \subseteq \widehat{LT}_{d+1}$ *for any fixed* $d > 0$.

This implies that the classes defined by general threshold circuits and by small weights threshold circuits form the following alternating hierarchy:

$$\widehat{LT}_1 \subseteq LT_1 \subseteq \widehat{LT}_2 \subseteq LT_2 \subseteq \widehat{LT}_3 \subseteq \ldots \tag{1}$$

Let me recall that the inclusion $\widehat{LT}_1 \subseteq LT_1$ is proper by Theorem 3.2 (or by Theorem 5.4), whereas LT_1 and $\widehat{LT}_2$ are trivially separated by the PARITY function. The inclusion $\widehat{LT}_2 \subseteq LT_2$ was shown to be proper by Goldmann, Håstad and Razborov [GHR92] (it is a consequence of Theorem 5.3 below). The question whether LT_2 is different from higher levels of the hierarchy (1) (or whether it contains NP) is open.

4. Spectral Representation and Polynomial Thresholds

Any Boolean function $f : \{-1,1\}^n \longrightarrow \{-1,1\}$ can be uniquely represented as a multilinear polynomial over reals:

$$f(x_1,\ldots,x_n) = \sum_{\alpha \in \{0,1\}^n} a_\alpha(f) x^\alpha. \tag{2}$$

This representation is called the *spectral representation* of f and its coefficients $\{a_\alpha(f) \mid \alpha \in \{0,1\}^n\}$ are *spectral coefficients* of f. We define

$$L_1(f) \rightleftharpoons \sum_{\alpha \in \{0,1\}^n} |a_\alpha(f)|$$

and

$$L_\infty(f) \rightleftharpoons \max_{\alpha \in \{0,1\}^n} |a_\alpha(f)|.$$

Similarly we might define the Euclidean norm $L_2(f)$ but it turns out that $L_2(f)$ equals 1 for any f. In fact, it implies that

$$L_1(f) \geq 1 \geq L_\infty(f), \ L_1(f) \cdot L_\infty(f) \geq 1. \tag{3}$$

In general, the spectral approach is a very useful tool in the study of Boolean functions (see e.g. [KKN88, LMN89, BOH90, KM91]). But in this survey we are exclusively interested in its applications to threshold circuits.

Along these lines Bruck and Smolensky [BS92] explicitly defined the class PL_1 which consists of all functions f_n with $L_1(f) \leq n^{O(1)}$ and the class $PL_\infty \rightleftharpoons \{f_n \mid L_\infty(f)^{-1} \leq n^{O(1)}\}$. Note that by (3), $PL_1 \subseteq PL_\infty$.

The classes which provide a strong link between threshold circuits and spectral properties of Boolean functions were defined by Bruck in [Bru90]. Namely, the class PT_1 consists of all functions f_n which allow a representation of the form

$$f_n(x_1,\ldots,x_n) = \text{sign}\left(\sum_{\alpha \in A} w_\alpha x^\alpha\right) \tag{4}$$

where $A \subseteq \{0,1\}^n$, $|A| \leq n^{O(1)}$. Note that in the $\{0,1\}$-notation PT_1 equals the class of all functions computable by polynomial size depth 2 circuits with a (general) threshold gate at the top and parity gates at the bottom. On the other hand, the definition of PT_1 bears the obvious similarity with (2).

The class $\widehat{PT}_1$ is defined in the same way, only now we additionally require the weights w_α in (4) to be small ([Bru90]).

Bruck [Bru90] showed a general lower bound for PT_1-circuits which in our notation basically amounts to the following:

Theorem 4.1 (Bruck). $PT_1 \subseteq PL_\infty$.

Bruck and Smolensky [BS92] established the dual result:

Theorem 4.2 (Bruck, Smolensky). $PL_1 \subseteq \widehat{PT}_1$.

So we have the hierarchy

$$PL_1 \subseteq \widehat{PT}_1 \subseteq PT_1 \subseteq PL_\infty. \tag{5}$$

The inclusion $PL_1 \subseteq \widehat{PT}_1$ was shown to be proper in [BS92]:

Theorem 4.3 (Bruck, Smolensky). *The function*

$$EXACT_n(x_1, \ldots, x_n) = 1 \rightleftharpoons \sum_{i=1}^{n} x_i = n/2$$

is in $\widehat{PT}_1 \setminus PL_1$.

$\widehat{PT}_1 \subseteq PT_1$ was shown to be proper by Goldmann, Håstad and Razborov [GHR92] (see more general Theorem 5.4 below). The inclusion $PT_1 \subseteq PL_\infty$ is proper just because the class PL_∞ contains almost all functions; an explicit function separating those two classes was presented in [BS92].

5. The Fine Structure

In this section we combine the two hierarchies (1) and (5) into one powerful picture.

It is clear that $\widehat{LT}_1 \subseteq \widehat{PT}_1$ and $LT_1 \subseteq PT_1$. Less obvious inclusions were established in [Bru90]:

Theorem 5.1 (Bruck). $\widehat{PT}_1 \subseteq \widehat{LT}_2$ *and* $PT_1 \subseteq LT_2$.

At the moment we have the following picture.

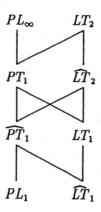

It turns out that this picture reflects *all* possible inclusions between the eight classes shown there. Let's review the reasons.

The following result was proved in [Bru90]:

Theorem 5.2 (Bruck). *The Complete Quadratic Function CQ_n which in the $\{0,1\}$-notation is given by $CQ_n(x_1,\ldots,x_n) = \bigoplus_{1\leq i<j\leq n}(x_i \wedge x_j)$, is in $\widehat{LT}_2 \setminus PL_\infty$.*

The next three easy separations were noticed by Goldmann, Håstad and Razborov in [GHR92].

1. Although we do not know any explicit superlinear lower bounds for LT_2-circuits, we still can claim that $PL_\infty \not\subseteq LT_2$ just because PL_∞ contains almost all functions.

2. If we consider MAJ_n instead of $EXACT_n$ in Theorem 4.3, then it can be improved to give $\widehat{LT}_1 \not\subseteq PL_1$.

3. The PARITY function is in $PL_1 \setminus LT_1$.

The two separations which are still needed to claim that our picture is complete, are $LT_1 \not\subseteq \widehat{PT}_1$ and $PT_1 \not\subseteq \widehat{LT}_2$. In other words, we need lower bounds analogous to those given by Theorem 3.3 but for simpler functions. Such bounds were proven in [GHR92].

Let

$$p_n(x,y) = \text{sign}\left(1 + 2\sum_{i=0}^{n-1}\sum_{j=0}^{2n-1} 2^i y_j(x_{i,2j} + x_{i,2j+1})\right)$$

and

$$U_n(x) = \text{sign}\left(1 + 2\sum_{i=0}^{n-1}\sum_{j=0}^{4n-1} 2^i x_{i,j}\right).$$

Obviously, $p_n(x,y)$ and $U_n(x)$ are in PT_1 and LT_1 respectively.

Theorem 5.3 (Goldmann, Håstad, Razborov). *Any depth 2 small weights threshold circuit computing $p_n(x,y)$ must have size $\exp(\Omega(n))$. Hence $PT_1 \not\subseteq \widehat{LT}_2$.*

Theorem 5.4 (Goldmann, Håstad, Razborov). *For any representation*

$$U_n(x) = \text{sign}\left(\sum_{\alpha\in A} w_\alpha x^\alpha\right)$$

of $U_n(x)$ in the form (4) we have $\sum_{\alpha\in A}|w_\alpha| \geq \exp(\Omega(n))$. Hence $LT_1 \not\subseteq \widehat{PT}_1$.

As we noted before, Theorems 5.3 and 5.4 generalize and strengthen many of previous results.

A few words should be said about the method of proof of Theorems 3.3, 5.3, 5.4. Assume that $f_n(x_1,\ldots,x_n,y_1,\ldots,y_n)$ is a Boolean functions with its variables divided into two groups, x-variables and y-variables. Denote by $C_{1/2-\epsilon}(g;1 \to 2)$ the *probabilistic one-way communication complexity of g with error $1/2 - \epsilon$ i.e. with advantage ϵ* ([Yao79]). We consider the model in which the probability of being correct is at least $1/2+\epsilon$ for every pair of inputs, the random string is shared by both parties and the complexity is measured as the number of bits sent in the *worst* case (not the average). Let $C(g;1 \to 2)$ be the corresponding deterministic measure.

The following lemma which was implicit in [HMP+87] is the key stone to proving Theorems 3.3, 5.3, 5.4:

Lemma 5.5. *Let* $w, d \geq 0$ *and* $f_n(x_1, \ldots, x_n, y_1, \ldots, y_n)$ *be computed by a depth 2 threshold circuit with a threshold gate of the total weight* w *at the top and arbitrary gates* g *satisfying* $C(g; 1 \to 2) \leq d$ *at the bottom. Then*

$$C_{1/2 - 1/(2w)}(f_n; 1 \to 2) \leq d.$$

In fact, the paper [HMP$^+$87] dealt with the two-way communication complexity and also Krause [Kra91] and Krause, Waack [KW91] used similar arguments. It is not clear however whether the proof of Theorems 5.3, 5.4 can be carried over in the context of two-way complexity.

6. Applications to Concrete Functions

In this concluding section we will see that inclusions summarized in our main picture have been extremely useful for designing threshold circuits for very concrete Boolean functions.

We will be interested in such important functions as the ADDITION, MULTIPLICATION, DIVISION, COMPARISON (of two n-bit numbers), POWERING (computing x^n where x is an n-bit number), ITERATED ADDITION, ITERATED MULTIPLICATION, MAXIMUM and SORTING (of n n-bit numbers). In fact, some of these functions allow a naive implementation by constant depth polynomial size small weights circuits and the remaining functions can be implemented so using the reductions of Chandra, Stockmeyer and Vishkin [CSV84] and the results of Beame, Cook and Hoover [BCH86] and of Pippenger [Pip87]. This was observed in [HMP$^+$87] (see also [SB91]). However the depth of resulting circuits is far from optimal. We already noted in Section 3 that the circuits for the ITERATED ADDITION obtained in this way have depth 13; the MULTIPLICATION seems to require depth 10. We will present below more recent results many of which are based on the general theory from previous sections. They lead to much better (and in many cases tight) upper bounds in terms of depth.

Siu and Bruck [SB91] showed that the ADDITION and COMPARISON of two n-bit numbers are both in PL_1 and hence are doable in $\widehat{LT}_2$. A constructive version of Siu and Bruck's result was presented by Alon and Bruck [AB91]. Quite recently Siu and Roychowdhury [SR92] have used Theorem 3.7 to show that even the ITERATED ADDITION is in $\widehat{LT}_2$. All these results are optimal in depth since none of the three functions is in $\widehat{LT}_1$ (this is obvious for the ADDITION and ITERATED ADDITION; for the case of the COMPARISON see [SB91]).

Siu and Bruck [SB91] also showed that the MULTIPLICATION (of two n-bit numbers) is in $\widehat{LT}_4$ (this was later rediscovered in [HHK91] with a better bound on the circuit size). Siu and Roychowdhury [SR92] showed that in fact the MULTIPLICATION is doable in depth 3. This latter result is depth-optimal since [HMP$^+$87] proved before that the MULTIPLICATION is not in $\widehat{LT}_2$.

DIVISION and POWERING were shown to be in $\widehat{LT}_4$ by Siu, Bruck, Kailath and Hofmeister [SBKH91] and in $\widehat{LT}_3$ by Siu and Roychowdhury [SR92]. To the best of my knowledge, it is open whether they are in $\widehat{LT}_2$ or not.

ITERATED MULTIPLICATION was shown to be in $\widehat{LT}_5$ by Siu, Bruck, Kailath and Hofmeister [SBKH91] and in $\widehat{LT}_4$ by Siu and Roychowdhury [SR92]. It is open whether it is doable in depth 3 or not.

Siu and Bruck [SB91] also considered the MAXIMUM and SORTING of n n-bit numbers and, using their method, placed them into $\widehat{LT}_4$ and $\widehat{LT}_3$ respectively. Siu and Roychowdhury [SR92] improved on the second result showing that in fact the SORTING is also in $\widehat{LT}_3$. Whether these functions can be done in depth 2 seems to be open.

We summarize in the following table our knowledge on the depth-optimal constructions for the functions we've been discussing in this section.

Function	upper bound	lower bound
ADDITION	2 [SB91]	2
ITERATED ADDITION	2 [SR92]	2
MULTIPLICATION	3 [SR92]	3 [HMP$^+$87]
ITERATED MULTIPLICATION	4 [SR92]	3 [HMP$^+$87]
DIVISION	3 [SR92]	2
POWERING	3 [SR92]	2
COMPARISON	2 [SB91]	2 [SB91]
MAXIMUM	3 [SB91]	2
SORTING	3 [SBKH91]	2

7. Acknowledgement

I am highly indebted to Jehoshua Bruck and Michael Goldmann for valuable remarks promptly made on an earlier version of this paper.

References

[AB91]   N. Alon and J. Bruck. Explicit constructions of depth-2 majority circuits for comparison and addition. Technical Report RJ 8300 (75661), IBM Research Division, August 1991.

[Bar86]   D. A. Barrington. A note on a theorem of Razborov. Technical report, University of Massachusetts, 1986.

[BCH86]   P. Beame, S. Cook, and H. Hoover. Log depth circuits for division and related problems. *SIAM Journal on Computing*, 15:994–1003, 1986.

[BOH90]   Y. Brandman, A. Orlitsky, and J. Hennesy. A spectral lower bound technique for the size of decision trees and two-level AND/OR circuits. *IEEE Transactions on Computers*, 39(2):282–287, February 1990.

[Bru90]   J. Bruck. Harmonic analysis of polynomial threshold functions. *SIAM Journal on Discrete Mathematics*, 3(2):168–177, May 1990.

[BS92]   J. Bruck and R. Smolensky. Polynomial threshold functions, AC^0 functions and spectral norms. *SIAM Journal on Computing*, 21(1):33–42, February 1992.

[CSV84]   A. K. Chandra, L. Stockmeyer, and U. Vishkin. Constant depth reducibility. *SIAM Journal on Computing*, 13:423–439, 1984.

[GHR92]   M. Goldmann, J. Håstad, and A. Razborov. Majority gates vs. general weighted threshold gates. In *Proceedings of the 7th Structure in Complexity Theory Annual Conference*, 1992.

[HG90]   J. Håstad and M. Goldmann. On the power of small-depth threshold circuits. In *Proceedings of the 31st IEEE FOCS*, pages 610–618, 1990.

[HHK91]   T. Hofmeister, W. Honberg, and S. Köling. Some notes on threshold circuits and multiplication in depth 4. *Information Processing Letters*, 39:219–225, 1991.

[HKP91]   J. Hertz, R. Krogh, and A. Palmer. *An Introduction to the Theory of Neural Computation*. Addison-Wesley, 1991.

[HMP+87] A. Hajnal, W. Maass, P. Pudlák, M. Szegedy, and G. Turán. Threshold circuits of bounded depth. In *Proceedings of 28th IEEE FOCS*, pages 99–110, 1987.

[KKN88]   J. Kahn, G. Kalai, and Linial N. The influence of variables on Boolean functions. In *Proceedings of the 29th IEEE Symposium on Foundations of Computer Science*, pages 68–80, 1988.

[KM91]   E. Kushilevitz and Y. Mansour. Learning decision trees using the Fourier spectrum. In *Proceedings of the 23rd ACM STOC*, pages 455–464, 1991.

[Kra91]   M. Krause. Geometric arguments yield better bounds for threshold circuits and distributed computing. In *6-th Structure in Complexity Theory Conference*, pages 314–322, 1991.

[KW91]   M. Krause and S. Waack. Variation ranks of communication matrices and lower bounds for depth two circuits having symmetric gates with unbounded fan-in. In *Proceedings of the 32th IEEE Symposium on Foundations of Computer Science*, pages 777–782, 1991.

[LMN89]   N. Linial, Y. Mansour, and N. Nisan. Constant depth circuits, Fourier transforms and learnability. In *Proceedings of the 30th IEEE Symposium on Foundations of Computer Science*, pages 574–579, 1989.

[MK61]   J. Myhill and W.H. Kautz. On the size of weights required for linear-input switching functions. *IRE Trans. on Electronic Computers*, EC10(2):288–290, June 1961.

[MSS91]   W. Maass, G. Schnitger, and E. Sontag. On the computational power of sigmoid versus boolean threshold circuits. In *Proceedings of the 32nd IEEE Symposium on Foundations of Computer Science*, pages 767–776, 1991.

[Pip87] N. Pippenger. The complexity of computations by networks. *IBM J. Res. Develop.*, 31:235–243, 1987.

[Raz87] A. Razborov. Lower bounds on the size of bounded-depth networks over a complete basis with logical addition. *Mathematical Notes of the Academy of Sciences of the USSR*, 41(4):598–607, 1987. English translation in 41:4, pages 333-338.

[SB91] K.-I. Siu and J. Bruck. On the power of threshold circuits with small weights. *SIAM Journal on Discrete Mathematics*, 4(3):423–435, 1991.

[SBKH91] K.-I. Siu, J. Bruck, T. Kailath, and T. Hofmeister. Depth-efficient neural networks for division and related problems. Technical Report RJ 7946, IBM Research, January 1991. To appear in IEEE Trans. Information Theory.

[Smo87] R. Smolensky. Algebraic methods in the theory of lower bounds for Boolean circuit complexity. In *Proceedings of the 19th ACM Symposium on Theory of Computing*, pages 77–82, 1987.

[SR92] K.-Y. Siu and V. Roychowdhury. On optimal depth threshold circuits for multiplication and related problems. Manuscript, 1992.

[Yao79] A. Yao. Some complexity questions related to distributive computing. In *Proceedings of the 11th ACM STOC*, pages 209–213, 1979.

[Yao90] A. Yao. On *ACC* and threshold circuits. In *Proceedings of the 31th IEEE FOCS*, pages 619–627, 1990.

An Elementary Approach to Some Analytic Asymptotics

Nicholas Pippenger

Department of Computer Science, The University of British Columbia
Vancouver, BC V6T 1Z2, Canada

Abstract. Fredman and Knuth have treated certain recurrences, such as $M(0) = 1$ and

$$M(n + 1) = \min_{0 \leq k \leq n} \big(\alpha M(k) + \beta M(n - k)\big),$$

where $\min(\alpha, \beta) > 1$. Their treatment depends on certain auxiliary recurrences, such as

$$h(x) = \begin{cases} 0, & \text{if } 0 \leq x < 1; \\ 1 + h(x/\alpha) + h(x/\beta), & \text{if } 1 \leq x < \infty. \end{cases}$$

The asymptotic behavior of $h(x)$ as $x \to \infty$ with α and β fixed depends on whether $\log \alpha / \log \beta$ is rational or irrational. The solution of Fredman and Knuth used analytic methods in both cases, and used in particular the Wiener-Ikehara Tauberian theorem in the irrational case. We show that a more explicit solutions to these recurrences can be obtained by entirely elementary methods, based on a geometric interpretation of $h(x)$ as a sum of binomial coefficients over a triangular subregion of Pascal's triangle. Apart from Stirling's formula, we need in the irrational case only the Kronecker-Weyl theorem (which can itself be proved by elementary methods), to the effect that if ϑ is irrational, the fractional parts of the sequence $\vartheta, 2\vartheta, 3\vartheta, \ldots$ are uniformly distributed in the unit interval.

1 Introduction

The analysis of algorithms and data structures, as well as of constructions for systems such as sorting and switching networks, often leads to recurrences. Because recursive algorithms, data structures and constructions often involve choices that should be made in an optimal way, the recurrences often involve minimization. In their paper "Recurrence Relations Based on Minimization", Fredman and Knuth [1] treat a large number of related recurrences by a combination of combinatorial and analytic methods. The goal of the present paper is to show how in many cases it is possible to replace the analytic component of their solutions with elementary arguments. (Here the terms "analytic" and "elementary" are used in accordance with the practice in number theory: "analytic" refers to methods based on properties of analytic functions of a complex variable, especially residues or integral transforms, while "elementary" refers to the absence of such methods. In particular, "elementary" does not refer to either simplicity or brevity.) As a bonus, we shall see that our analysis leads to a more explicit and informative solution in some cases.

Of the recurrences treated by Fredman and Knuth, the one which best illustrates our contribution is $M(0) = 1$ and

$$M(n+1) = \min_{0 \le k \le n} \left(\alpha M(k) + \beta M(n-k) \right), \tag{1.1}$$

where α and β are fixed parameters with $\min(\alpha, \beta) > 1$. (This is the case "$g(n) = \delta_{n0}$" dealt with in their Section 6.) By straightforward and elementary arguments, Fredman and Knuth reduce the study of (1.1) to that of the function h defined by

$$h(x) = \begin{cases} 0, & \text{if } 0 \le x < 1, \\ 1 + h(x/\alpha) + h(x/\beta), & \text{if } 1 \le x < \infty. \end{cases} \tag{1.2}$$

The analysis of Fredman and Knuth proceeds by considering the integral transform

$$G(s) = \int_1^\infty \frac{h(t)\, dt}{t^{s+1}}$$

of h, which with the aid of (1.2) can be shown to be $G(s) = 1/s(1 - \alpha^{-s} - \beta^{-s})$. This function is analytic in the open half-plane $\mathrm{Re}(s) > \gamma$, where γ is the unique real solution to

$$\alpha^{-\gamma} + \beta^{-\gamma} = 1. \tag{1.3}$$

Furthermore, $G(s)$ has a simple pole at $s = \gamma$ with residue $C = 1/(\alpha^{-\gamma} \log \alpha^\gamma + \beta^{-\gamma} \log \beta^\gamma))$, as is easily calculated. This pole will ultimately give rise to a factor Cx^γ in the asymptotic behavior of $h(x)$.

The behavior of $G(s)$ on the remainder of the critical line $\mathrm{Re}(s) = \gamma$ depends on whether $\log \alpha / \log \beta$ is rational or irrational. If this quotient is irrational, the pole a $s = \gamma$ is the only one on the critical line, and a Tauberian theorem due to Wiener, Ikehara and Landau (Lemma 4.3 in Fredman and Knuth) leads to the conclusion that

$$h(x) \sim Cx^\gamma \tag{1.4}$$

in this case. If the quotient $\log \alpha / \log \beta$ is rational, $G(s)$ has additional poles periodically disposed along the critical line. Application of Cauchy's residue theorem leads to the conclusion that

$$h(x) \sim D(x)x^\gamma \tag{1.5}$$

in this case, where $D(x)$ is a periodic function of $\log x$ whose period is determined by the spacing between poles along the critical line, and whose Fourier coefficients are determined by the residues at those poles.

In this paper we shall derive (1.4) and (1.5) in an elementary fashion. This new derivation has the merit of giving a simple explicit formula for the function $D(x)$ in (1.5). We shall also want the solution to the related recurrence

$$h'(x) = \begin{cases} 0, & \text{if } 0 \le x < 1, \\ 1 + \alpha h(x/\alpha) + \beta h(x/\beta), & \text{if } 1 \le x < \infty. \end{cases} \tag{1.6}$$

By analogous elementary methods, we shall show that

$$h'(x) \sim C' x^{\gamma+1}, \qquad (1.7)$$

where $C' = 1/(\alpha^{-\gamma} \log \alpha^{\gamma+1} + \beta^{-\gamma} \log \beta^{\gamma+1}))$ in the irrational case, and

$$h'(x) \sim D'(x) x^{\gamma+1}, \qquad (1.8)$$

where $D'(x)$ is periodic function of $\log x$ which will be determined explitily in the rational case.

Fredman and Knuth showed that (1.4) implies that

$$M(n) \sim A x^{1+1/\gamma}, \qquad (1.9)$$

where A is an explicitly determined constant in the irrational case. We shall show that (1.5) and (1.8) together imply

$$M(n) \sim B(n) n^{1+1/\gamma}, \qquad (1.10)$$

where $B(n)$ is an explicitly determined periodic function of $\log n$ in the rational case. (Once the form of the functions $D(x)$ and $D'(x)$ are explicitly know, it is possible to go back and derive these results by extending the analysis of Fredman and Knuth. This would involve showing that certain Fourier series converge to certain periodic functions. But since there is no general procedure for identifying a function from its Fourier series, it does not appear to be possible to extend the analysis of Fredman and Knuth without knowing what $D(x)$ and $D'(x)$ are by some other method.)

2 The Rational Case

Our analysis begins with the observation that $h(x)$ is the number of words over the alphabet $\{\alpha, \beta\}$ for which the product of the letters is at most x. Indeed, if $0 \leq x < 1$, then $h(x) = 0$ and there are no such words (provided that, as usual, we interpret the product with no factors as unity). If $1 \leq x < \infty$, then $h(x) = 1 + h(x/\alpha) + h(x/\beta)$, and any word for which the product of the letters is at most x must be either be empty (and there is 1 such word) or consist of an α followed by a word for which the product of the letters is at most x/α (and there are $h(x/\alpha)$ such words), or consist of an β followed by a word for which the product of the letters is at most x/β (and there are $h(x/\beta)$ such words). Since there are exactly $\binom{i+j}{i}$ words that contain i α's and j β's, we have established the following explicit formula for $h(x)$:

$$h(x) = \sum_{\alpha^i \beta^j \leq x} \binom{i+j}{i}. \qquad (2.1)$$

Taking logarithms in the constraint of the summation, we see that $h(x)$ may be interpreted as the sum of the binomial coefficients $\binom{i+j}{i}$ in Pascal's triangle over the triangular subregion bounded by the inequalities $i \geq 0$, $j \geq 0$ and

$$i \log \alpha + j \log \beta \leq \log x. \qquad (2.2)$$

Suppose that $\log \alpha / \log \beta$ is the rational number p/q, where p and q are positive integers such that $\gcd(p, q) = 1$. Then $\log_{\alpha\beta} \alpha = p/(p+q)$, $\log_{\alpha\beta} \beta = q/(p+q)$, and if we set

$$\varrho = (\alpha\beta)^{1/(p+q)},$$

then (2.2) becomes

$$pi + qj \le \log_\varrho x.$$

Since p, q, i and j are integers, we see that $h(x)$ remains constant as x increases except when $\log_\varrho x$ passes through an integer k, when it jumps by

$$S(k) = \sum_{pi+qj=k} \binom{i+j}{i}. \tag{2.3}$$

We shall see below that $S(k)$ has the asymptotic formula

$$S(k) \sim (C \log \sigma) \sigma^k, \tag{2.4}$$

where

$$\sigma = \varrho^\gamma = (\alpha\beta)^{\gamma/(p+q)}.$$

If we set $S^*(l) = \sum_{0 \le k \le l} S(k)$, it follows that

$$S^*(l) \sim (C\sigma \log \sigma /(\sigma - 1)) \sigma^l.$$

This formula gives the asymptotic value $S^*(l)$ of $h(x)$ when x is a "magic" number of the form $x = \varrho^l$. The asymptotic formula for arbitrary x follows from this and the fact that $h(x)$ remains constant between magic values of x. If we write $\log_\varrho x = l+\lambda$, where $l = \lfloor \log_\varrho x \rfloor$ (the integral part of $\log_\varrho x$) and $\lambda = \{\log_\varrho x\}$ (the fractional part of $\log_\varrho x$), then

$$h(x) \sim (C\sigma \log \sigma /(\sigma - 1)) \sigma^l$$
$$\sim (C\sigma^{1-\lambda} \log \sigma /(\sigma - 1)) \sigma^{l+\lambda}$$
$$\sim P(\{\log_\varrho x\}) x^\gamma,$$

where

$$P(\lambda) = C\sigma^{1-\lambda} \log \sigma /(\sigma - 1).$$

This establishes (1.5) with $D(x) = P(\{\log_\varrho x\})$, which is periodic in $\log x$ (with period $\log \varrho$), as claimed.

It remains to establish (2.4). We shall just sketch the main steps here; the similar sum $\sum_{2i+3j=k} \binom{i+j}{i} \phi^{i+j} /(i + j)$ (where $\phi = (1 + \sqrt{5})/2$) is estimated in detail by Miller *et al.* ([6], pp. 52–57), and (2.3) is analogous in every respect. First, we approximate the binomial coefficients $\binom{i+j}{i} = (i+j)!/i!j!$ in (2.3) by applying Stirling's formula

$$n! = n^n e^{-n} \sqrt{2\pi n}(1 + O(n^{-1})),$$

to their constituent factorials. If we separate the approximation into algebraically varying factors (arising from the factors $\sqrt{2\pi n}$) and exponentially varying factors (arising from the factors n^n), we see that the exponentially varying factors impart to the summand a peaking reminiscent of the central limit theorem: the greatest contribution to the sum comes when i and j are in the fixed ratio $\alpha^{-\gamma}/\beta^{-\gamma}$. This variation allow the terms of the sum not near the peak to be neglected. The resulting truncated sum is then estimated by an integral; the error in this estimation is at most the total variation of the summand, which is (since the summand is unimodal) is at most twice the largest term. The resulting integral can be transformed into the well known integral $\int_{-\infty}^{+\infty} e^{-y^2}\,dy = \sqrt{\pi}$ by adjoining negligible tails. The result is (2.4).

We should mention here that the special cases $p = q = 1$, where $S(k) = 2^k$, and $p = 1$, $q = 2$, where $S(k) = F_{k+1} \sim \phi^{k+1}/\sqrt{5}$ (in which F_n is the n-th Fibonacci number; see Knuth [3], Section 1.2.8, Equation (15) and Exercise 16) are well known, and the analysis just given can be regarded as a generalization of these cases.

In Section 4 we shall also want the solution to the recurrence (1.6) for $h'(x)$ in the rational case. Let us call the product of the letters in a word over the alphabet $\{\alpha, \beta\}$ the *weight* of the word. Then $h'(x)$ is the sum of the weights of all words whose weight is at most x, and thus we have the explicit formula

$$h'(x) = \sum_{\alpha^i \beta^j \le x} \binom{i+j}{i} \alpha^i \beta^j.$$

The treatment of this sum is completely analgous to that of (2.1); the result is

$$h'(x) \sim P'(\{\log_\varrho x\})\, x^{\gamma+1},$$

where

$$P'(\lambda) = C' \tau^{1-\lambda} \log \tau/(\tau - 1),$$

in which

$$\tau = \varrho^{\gamma+1} = (\alpha\beta)^{(\gamma+1)/(p+q)}.$$

This establishes (1.8) with $D'(x) = P'(\{\log_\varrho x\})$.

3 The Irrational Case

When $\log \alpha / \log \beta$ is irrational the analysis of the preceding section is not applicable, for as x increases new binomial coefficients enter the sum one by one, rather than in the regularly spaced platoons of the rational case. Furthermore, the order of their entry is very irregular, with small coefficients near the axes being interspersed with large ones near the main diagonal. The analysis of this section is based on a regularity of averages amid this irregularity of detail, as expressed by the "ergodicity of an irrational rotation of the circle". We shall use in particular the Kronecker-Weyl theorem in the form stated below. Weyl's orginal proof (and probably still the simplest proof) of this theorem was based on Fourier Series, which by some tastes might not be counted as elementary. A subsequent proof based on continued fractions (see Nivin [7], Chapter 6, Section 3) is incontestably elementary, however.

Let ϑ be a real number. We shall say that the sequence $\{\vartheta\}, \{2\vartheta\}, \{3\vartheta\}, \ldots$ is *uniformly distributed* if, for every $\varepsilon > 0$ and $\delta > 0$, there exists n_0 such that, for all $n \geq n_0$, we have

$$(1 - \delta)\varepsilon n \leq \#\{1 \leq m \leq n : 0 \leq \{m\vartheta\} \leq \varepsilon\} \leq (1 + \delta)\varepsilon n.$$

Theorem: (L. Kronecker, H. Weyl) The sequence $\{\vartheta\}, \{2\vartheta\}, \{3\vartheta\}, \ldots$ is uniformly distributed if and only if ϑ is irrational.

(This theorem as stated is due to Weyl [8]; Kronecker [4] proved that the sequence $\{\vartheta\}, \{2\vartheta\}, \{3\vartheta\}, \ldots$ is dense in the unit interval if and only if ϑ is irrational.)

By straightforward differencing arguments, we can generalize the condition of uniform distribution to ensure that

$$(1 - \delta)\varepsilon n \leq \#\{l + 1 \leq m \leq l + n : 0 \leq \{\lambda + m\vartheta\} \leq \varepsilon\} \leq (1 + \delta)\varepsilon n$$

holds for all integers l and real numbers λ (so that any n consecutive terms of the sequence $\vartheta, 2\vartheta, 3\vartheta, \ldots$ are uniformly distributed with respect to all intervals of length ε).

Let $\varepsilon > 0$ be fixed. Define the function $h_\varepsilon(x)$ by

$$h_\varepsilon(x) = \sum_{xe^{-\varepsilon} < \alpha^i \beta^j \leq x} \binom{i + j}{i}. \tag{3.1}$$

Taking logarithms in the constraint of the summation, we see that $h_\varepsilon(x)$ may be interpreted as the sum of the binomial coefficient over the trapezoidal region bounded by the inequalities $i \geq 0$, $j \geq 0$ and

$$\log x - \varepsilon < i \log \alpha + j \log \beta \leq \log x. \tag{3.2}$$

We shall see below that $h_\varepsilon(x)$ satisfies the asymptotic inequalities

$$C\gamma\varepsilon(1 - \varepsilon)e^{-\gamma\varepsilon} x^\gamma \lesssim h_\varepsilon(x) \lesssim C\gamma\varepsilon(1 + \varepsilon)x^\gamma \tag{3.3}$$

as $x \to \infty$ with ε fixed. (Here $f(x) \lesssim g(x)$ means that $\limsup_{x \to \infty} f(x)/g(x) \leq 1$.) If we set $l = \lfloor \log x/\varepsilon \rfloor + 1$, then $xe^{-l\varepsilon} < 1$, so we have

$$h(x) = \sum_{0 \leq k \leq l} h_\varepsilon(xe^{-k\varepsilon}).$$

Observing that the sum of the terms with $k \geq l/2$ is negligible, applying (3.3) to the remaining terms, and adding a negligible infinite series of terms, we obtain

$$C\gamma\varepsilon(1 - \varepsilon)e^{-\gamma\varepsilon} \sum_{0 \leq k < \infty} (xe^{-k\varepsilon})^\gamma \lesssim h(x) \lesssim C\gamma\varepsilon(1 + \varepsilon) \sum_{0 \leq k < \infty} (xe^{-k\varepsilon})^\gamma,$$

and therefore

$$C\gamma\varepsilon(1-\varepsilon)e^{-\gamma\varepsilon}x^\gamma/(1-e^{-\gamma\varepsilon}) \lesssim h(x) \lesssim C\gamma\varepsilon(1+\varepsilon)x^\gamma/(1-e^{-\gamma\varepsilon}).$$

Since this holds for every $\varepsilon > 0$, we may let ε tend to 0 and obtain (1.4).

It remains to establish (3.3). The proof follows the same general lines as that for (2.4), but is complicated by the fact that the lattice points (i,j) are not equally spaced in the trapezoid (3.2) as they were along the boundary of the triangle (2.2). Our salvation comes from the Kronecker-Weyl theorem, which shows that though they are not "equally spaced", they are "uniformly distributed".

Suppose that $\varepsilon < \log\beta$ and set $\vartheta = \log\alpha/\log\beta$, so that ϑ is irrational. Let us fix i and ask whether there exists a j such that i and j satisfy the inequalities (3.2). Clearly there exists such a j if and only if $\{i\vartheta\}$ falls in the interval $((\log x - \varepsilon)/\log\beta, \log x/\log\beta]$. (Since the length of this interval is $\varepsilon/\log\beta < 1$, it contains either no integers or one integer.) Let $\delta > 0$ be fixed. By the Kronecker-Weyl theorem, we may choose L sufficiently large that among any L consecutive i's, there are $(1 \pm \delta)L\varepsilon/\log\beta$ for which there exists a corresponding j satisfying (3.2). This allows the trapezoid (3.2) to be broken into pieces, each of which is sufficiently large that it contains a number of lattice points approximately proportional to its area, yet sufficiently small that the binomial coefficients associated with these lattice points are approximately equal. The sum over these pieces can then be approximated by an integral as in the proof of (2.4). The result is (3.3). We observe that the same method works to establish the asymptotic formula (1.7) for $h'(x)$ in the irrational case.

Though we have derived (1.4) and (1.5) by parallel arguments, there is an important difference between these derivations. We could have done the analysis in Section 2 to obtain an O-estimate for the error in (1.5); the most straightforward way of doing this yields a factor of $\left(1 + O((\log\log x)^{3/2}(\log x)^{-1/2})\right)$. No such sharpening is possible for (1.4), however, since the Kronecker-Weyl theorem, in the form we have cited, give no estimate for the rate of convergence to the uniform distribution. The same phenomenon arises for the analytic proof using the Wiener-Ikehara theorem, for while convergence follows from the behavior of $G(s)$ on the critical line and the right half-plane it bounds, the rate of convergence depends on how closely the poles in the left half-plane approach the critical line as their imaginary parts grow [9, 2, 5]. With either method, the missing information depends on how well the irrational number $\log\alpha/\log\beta$ can be approximated by rational numbers as the denominators of these rational numbers grow. This is the crux of the difference: all rational numbers are alike, but each irrational number is irrational in its own way.

Since we have not made any quantitative hypothesis concerning the irrationality of $\log\alpha/\log\beta$, we cannot expect to draw any conclusion about the rate of approach in (1.4). If however we assume that $|\log\alpha/\log\beta - p/q|$ is bounded away from 0 by a function of q, the elementary method here (as well as the analytic method used by Fredman and Knuth) can be adapted to yield an explicit O-estimate in (1.4).

4 Conclusion

After deriving (1.4) and (1.5) in a new way, and obtaining explicit descriptions of the functions $D(x)$ and $D'(x)$ appearing in (1.5) and (1.8), we shall exhibit in this section the consequences of these explicit descriptions for the original recurrence (1.1).

Fredman and Knuth show, by elementary arguments, that

$$M(n) = 1 + (\alpha + \beta - 1)W(n), \tag{4.1}$$

where $W(n)$ is the sum of the weights of the n words having the smallest weights. By the definitions of $h(x)$ and $h'(x)$, we have $W(h(x)) = h'(x)$. Let us assume that $\log \alpha / \log \beta$ is rational. Recall that a value of x is "magic" if $x = \varrho^l$ for some natural number l. We have $D(x) = P(0)$ and $P'(x) = P'(0)$ for all magic values of x, and the asymptotic formulas

$$h(x) \sim P(0) \, x^\gamma \tag{4.2}$$

and

$$h'(x) \sim P'(0) \, x^{\gamma+1}, \tag{4.3}$$

valid for magic values of x.

Let us say that a value of n is "magic" if $n = h(x)$ for some magic value of x. Then (4.2) and (4.3) yields the asymptotic formula

$$W(n) \sim P'(0)(n/P(0))^{1+1/\gamma}, \tag{4.4}$$

valid for magic values of n.

To extend (4.4) to arbitrary values of n, we observe that as n increases between magic values, $W(n)$ increases by the addition of equal weights. Thus the formula for arbitrary n is obtained by linearly interpolating between the values given by (4.4) for magic values of n. This gives

$$W(n) \sim P'(0)Q(\{\log_\sigma(n/P(0))\}) \, (n/P(0))^{1+1/\gamma},$$

where $Q(\lambda) = (1 - \lambda + \lambda\tau)\tau^{-\lambda}$, which establishes (1.10) with

$$B(n) = (\alpha + \beta - 1)P'(0)Q(\{\log_\sigma(n/P(0))\}) \, / \, P(0)^{1+1/\gamma},$$

which is periodic in $\log n$ (with period $\log \sigma$), as claimed.

Acknowledgment. The author is indebted to an anonymous referee for pointing out a gaffe in Section 3 of an earlier version of this paper.

5 References

1. M. L. Fredman and D. E. Knuth, "Recurrence Relations Based on Minimization", *J. Math. Anal. Appl.*, 48 (1974) 534–559.

2. S. Ikehara, "An Extension of Landau's Theorem in the Analytical Theory of Numbers", *J. Math. and Phys.*, 10 (1931) 1–12.

3. D. E. Knuth, *The Art of Computer Programming–Volume 1: Fundamental Algorithms*, Addison-Wesley, Reading, MA, 1968.

4. L. Kronecker, "Näherungsweisse Ganzzahlige Auflösung Linearer Gleichungen", *Sitzungsber. Berliner Akad. Wiss.*, 46 (1884) 1071 ff.

5. E. Landau, "Über den Wienerschen neuen Weg zum Primzahlsatz", *Sitzber. Preussische Akad. Wiss.*, (1932) 514–521.

6. R. E. Miller, N. Pippenger, A. L. Rosenberg and L. Snyder, "Optimal 2,3-Trees", *SIAM J. Comput.*, 8 (1979) 42–59.

7. I. Niven, *Irrational Numbers*, Mathematical Association of America, 1956.

8. H. Weyl, "Über die Gleichverteilung von Zahlen modulo Eins", *Math. Annalen*, 77 (1916) 313–352.

9. N. Wiener, "A New Method in Tauberian Theorems", *J. Math. and Phys.*, 7 (1928) 161–184.

An optimal parallel algorithm for computing a near-optimal order of matrix multiplications[*]

Artur Czumaj[†]

Institute of Informatics, Warsaw University

Abstract

This paper considers the computation of matrix chain products of the form $M_1 \times M_2 \times \cdots \times M_{n-1}$. The order in which the matrices are multiplied affects the number of operations. The best sequential algorithm for computing an optimal order of matrix multiplication runs in $O(n \log n)$ time while the best known parallel NC algorithm runs in $O(\log^2 n)$ time using $n^6/\log^6 n$ processors. This paper presents the first approximating optimal parallel algorithm for this problem and for the problem of finding a near-optimal triangulation of a convex polygon. The algorithm runs in $O(\log n)$ time using $n/\log n$ processors on a CREW PRAM, and in $O(\log \log n)$ time using $n/\log \log n$ processors on a weak CRCW PRAM. It produces an order of matrix multiplications and a partition of polygon which differ from the optimal ones at most 0.1547 times.

1 Introduction

The problem of computing an *optimal order of matrix multiplication* (*the matrix chain product problem*) is defined as follows. Consider the evaluation of the product of $n - 1$ matrices

$$M = M_1 \times M_2 \times \cdots \times M_{n-1}$$

where M_i is a $w_{i-1} \times w_i$ ($w_i \geq 1$) matrix. Since matrix multiplication satisfies the associative law, the final result is the same for all orders of multiplying. However, the order of multiplication greatly affects the total number of operations to evaluate M. The problem is to find an optimal order of multiplying the matrices, such that the total number of operations is minimized. Here, we assume that the number of operations to multiply a $p \times q$ matrix by a $q \times r$ matrix is pqr. It was shown in [Cha-75] that an arbitrary order of matrix multiplications may be as bad as $O(T_{opt}^3)$, where T_{opt} is the minimal number of operations required to compute matrix chain products.

One can show that this problem is equivalent to the problem of finding an *optimal triangulation of a convex polygon* (see [HS-80]). Given a convex polygon $(v_0, v_1, \ldots, v_n)$. Divide it into triangles, such that the total cost of partitioning is the smallest possible. By the total cost of triangulation we mean the sum of costs

---

[*]This work was supported by the grant GR-61 from the Polish Ministry of National Education

[†]Warsaw University, Department of Mathematics, Institute of Informatics, Poland, 02-097 Warszawa, ul. Banacha 2, E-mail : czumaj@plearn.bitnet

of all triangles in this partitioning. The cost of a triangle is the product of weights in each vertex of triangle.

Both these problems can be solved sequentially in $O(n \log n)$ time [HS-80]. The best known approach to design parallel algorithms is based on dynamic programming. It gives us NC algorithms which run in $O(\log^2 n)$ time using $O(n^6 / \log^k n)$ processors on a CREW PRAM for some k ([Ryt-88], [HLV-90] and [GP-92]). Algorithms which increase the total time-processor product from $O(n \log n)$ to $O(n^6 / \log^k n)$ are not of much of practical value. This suggests designing approximating algorithms for these problems. They would run fast and use few processors what would fill the gap between the total work in their serial versions and parallel ones.

[Chin-78] described a sequential approximating algorithm for finding a near optimal order of matrix multiplications. This algorithm was later improved, analysed and transformed to the problem of finding a near-optimal triangulation of a convex n-gon in [HS-81]. The algorithm runs in linear time and has an error at most 15.47% (i.e., its cost is at most $1.1547 \times$ the cost of an optimal order of product of matrices). Especially interesting is the fact that this error is decreasing when n is growing.

In this paper we describe the first optimal parallel algorithm which solves these two problems approximatelly. It runs in $O(\log n)$ time using $n / \log n$ processors on a CREW PRAM and in $O(\log \log n)$ time using $n / \log \log n$ processors on a CRCW PRAM. Moreover we can improve these running times if the domain of input is restricted. In the matrix chain product problem there are always integer values of matrix dimensions. We will show that if dimensions are drawn from a domain $[k \ldots k+s]$ we can implement our algorithm to run in $O(\alpha(n) + \log \log \log s)$ time with linear time-processor product on a priority CRCW PRAM[1]. Our algorithm produces the order of matrix multiplications and the triangulation of a convex polygon with an error ratio at most 0.1547 - the same as in Chin-Hu-Shing algorithm.

The models of parallel computation that are used in this paper are the concurrent-read exclusive-write (CREW) and the concurrent-read concurrent-write (CRCW) parallel random access machines (PRAM). A PRAM employs synchronous processors all having access to a common memory. A CREW PRAM allows several processors to read the same entry of memory simultaneously, but forbids multiple concurrent writes to a cell. A CRCW PRAM allows simultaneous access by more than one processor to the same memory cell for both reads and writes. In this paper we mainly focus on the weakest model of a CRCW PRAM - *weak CRCW*, in which the only concurrent writes allowed are of the value 1.

This paper is organized as follows. Section 2 contains some basic definitions and notions concerning the algorithm. Section 3 recalls Chin-Hu-Shing sequential algorithm. Section 4 shows reduction of the initial problem to the one of finding an optimal triangulation in a basic polygon. Section 5 describes the new algorithm which is then implemented on a CREW PRAM and on a CRCW PRAM in section 6. In the last section we give some extensions of our algorithm.

---

[1] $\alpha(n)$ denotes very slowly growing the inverse-Ackerman function, for definition see e.g. [BV-89]

2 Basic notions and definitions

Firstly we introduce some basic notions and facts.

The fact stated below converts the matrix chain product problem to the problem of finding an optimal triangulation of a convex polygon, see [HS-80].

Fact 2.1 *Any order of multiplying $n - 1$ matrices corresponds to a partition of a n-sided convex polygon.*

Corollary 2.2 *The problem of finding an optimal order of multiplying a chain of matrices is equivalent to the problem of finding an optimal triangulation of a convex polygon. Here by the cost of partitioning of a polygon we mean the sum of costs of all triangles in a given partition, and by the cost of a triangle we mean the product of values of three vertices in this triangle.*

This equivalence and transformation are clear, so we give the following observation without a proof.

Observation 2.3 *Any partition of a convex polygon can be transformed to an order of multiplication of chain of matrices in constant time with n processors on a CREW PRAM. Also any order of multiplication of chain of matrices can be transformed to a partition of a convex polygon in constant time with n processors on a CREW PRAM.*

From now on, only the partitioning problem will be discussed.

Throughout this paper we will use $w_0, w_1, \ldots, w_{n-1}$ and $v_0, v_1, \ldots, v_{n-1}$ to denote vertices as well as their weights in a convex polygon. For simplicity we assume that all weights are distinct. If there are some vertices with the same weights then we assume that a particular ordering is chosen and remains fixed. We can choose e.g. lexicographically ordering[2].

Define a vertex v_i to be the *smallest* (minimum) one if for each other vertex v_j we have $v_i < v_j$. Similarly we define the kth *smallest* vertex v_i if there are exactly $k - 1$ vertices smaller than v_i.

We will need the following problems.

The all nearest smaller values problem

Given an array $A = (a_0, a_1, \cdots, a_n)$. For each a_i $(0 \leq i \leq n)$, find the nearest element to its left (and the nearest element to its right) that is less than a_i, if such an element exists. That is, for each i, $1 \leq i \leq n$, find the maximal j $(0 \leq j < i)$ and the minimal k $(i < k \leq n)$ such that $a_j < a_i$ and $a_k < a_i$.

Fact 2.4 ([BSV-88], [BBGSV-89]) *The all nearest smaller values problem can be solved in $O(\log n)$ time using $n/\log n$ processors on a CREW PRAM and in $O(\log \log n)$ time using $n/\log \log n$ processors on a weak CRCW PRAM.*

The prefix minima problem

Given an array $A = (a_0, a_1, \cdots, a_n)$. For each a_i $(0 \leq i \leq n)$, find the minimum among $a_0, \ldots, a_i$.

---

[2] That is, $v_i \prec v_j$ iff $v_i < v_j$, or $v_i = v_j$ and $i < j$

Fact 2.5 *([Sch-87], [BSV-88]) The prefix minima problem can be solved in $O(\log n)$ time using $n/\log n$ processors on a CREW PRAM and in $O(\log \log n)$ time using $n/\log \log n$ processors on a weak CRCW PRAM.*

The all nearest zero problem

Let $A = (a_0, a_1, \ldots, a_n)$ be an array of bits. Find for each bit a_i the nearest zero bit both to its left and right.

Fact 2.6 *The all nearest zero problem can be solved in $O(\log n)$ time using $n/\log n$ processors on a CREW PRAM and in $O(\alpha(n))$ time using $n/\alpha(n)$ processors on a weak CRCW PRAM ([BV-89]).*

Evaluation of logical 'AND' in a tree

Given a binary tree T (with possible chains) of n nodes. Let each vertex has assigned a logical value *false* or *true*. Evaluate for each internal vertex v logical 'AND' of v and all its descendants.

Remark. This problem can be transformed to the problem of evaluation of expression with logical operation 'AND' in each internal vertex.

Lemma 1 .

Suppose we are given an infix order in a tree T. That is, for each node v_i we know its infix rank i. Assume also that for each node there is given the number of all its descendants in a tree. Then we can evaluate logical 'AND' for all vertices in a tree T in $O(\log n)$ time using $n/\log n$ processors on a CREW PRAM and in $O(\alpha(n))$ time using $n/\alpha(n)$ processors on a weak CRCW PRAM.

Proof: We show how to reduce this problem to the all nearest zero problem. Let $A = (a_1, \ldots, a_n)$ be an array of bits such that $a_i = 1$ iff the ith (w.r.t. to an infix order) vertex v_i in a tree T has assigned value *true*. Because of infix order we know that for given vertex v_i (with ith number) all its descendants are in the consecutive subarrays to its left (*descendants of its left son*) and right (*descendants of its right son*). Therefore we start with finding for each a_i (that is, for each vertex v_i) the nearest zero bit to its left and rigth. Then we have to check if these zeros are in a subtree rooted at v_i. Because we know the number of its descendants we can check this in constant time with one processor for each node v. □

Remark. These results can be extended to the case when we are given a prefix or postfix order and the number of all descendants or if we are given in the array the Euler tour (see e.g. [GR-88]) of the input tree.

3 Chin-Hu-Shing approximating algorithm

The Chin-Hu-Shing ([Chin-78] and [HS-81]) algorithm is based on two intuitions:

- If a vertex has a very large weight, then it should be cut in the optimal partition.

- If none of vertices has a very large weight, then we should join the smallest vertex with all other vertices.

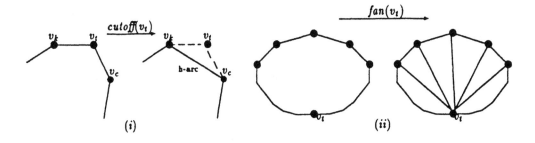

Figure 1: Procedures (i) $cutoff(v_t)$ and (ii) $fan(v_t)$.

Let v_m be the smallest vertex of a convex polygon, and v_t be a vertex with v_k and v_c as its two neighbours. Define vertex v_t to be *large* iff

$$\frac{1}{v_k} + \frac{1}{v_c} > \frac{1}{v_t} + \frac{1}{v_m}.$$

In Chin-Hu-Shing algorithm the following procedures are used.

Procedure *cutoff (v_t)*::
Let v_k and v_c be two neighbours of vertex v_t. Join v_k with v_c (i.e., cut off v_t) and call this arc *h-arc*. Later we consider a polygon without vertex v_t.

Procedure *fan (v_t)*::
Join v_t with all vertices in a polygon. These arcs are called *fan-arcs*.

Additionally there are three procedures which pop a given vertex off the stack $(pop(v_i))$, push onto the stack $(push(v_i))$ and return the top of the stack (top).

Chin, Hu and Shing described the following heuristic linear-time algorithm.

Chin-Hu-Shing Algorithm

Input : - a sequence of weights of vertices in a polygon - $w_0, w_1, \ldots, w_{n-1}$
Find the smallest vertex w_i and shift indices of vertices such that w_i is the first element of a sequence; that is, $v_0 = w_i, v_1 = w_{(i+1) \bmod n}, \cdots, v_j = w_{(i+j) \bmod n}, \cdots$.

```
push(v_0);
i := 1;
while i ≤ n - 1 do
    if top ≠ v_0 and top is large then
        cutoff(top); pop(top);
    else
        push(v_i);
    fi;
od;
fan(v_0);
```

In the sequel this algorithm will be called *algorithm CHS*.

What is especially interesting, in most cases algorithm CHS yields the optimal solution or a solution which takes only a few percent worse than the optimal one[3]. As was mentioned in the introduction the (worst case) error ratio is inversely proportional to n and is maximum when $n = 5$. In general, if value $t = \max_i\{\frac{v_i}{v_0}\}$ has upper bound, then Hu and Shing [HS-81] calculated the maximum error ratio for any given input n. This ratio is given by the following formula:

$$\text{maximum error ratio R} = \frac{t - 1}{t^2 + t + (n - 4)}$$

For example, if $t = 2$, then $R = 1/(n+2)$, and if $t \cong \sqrt{n}$, and n approaches ∞, then $R = \frac{1}{2\sqrt{n}}$.

Fact 3.1 ([HS-80], [HS-81]) *Algorithm CHS finds a near-optimal partitioning of a polygon and a near-optimal order of computing the matrix chain products with the maximum error ratio $\simeq 0.1547$ (exactly it is $\frac{\sqrt{3}}{6+3\sqrt{3}}$).*

4 Reduction to the problem of finding an optimal triangulation in a basic polygon

The first step of our algorithm is a partition of a convex polygon into smaller nonintersecting *basic polygons*.

Fact 4.1 ([HS-80]) *There exists an optimal triangulation of a convex polygon containing arcs or sides between the smallest vertex and the second and the third smallest ones*

Define a *basic polygon* to be a polygon containing sides between the smallest vertex (v_0) and the second (v_1) and the third (v_{n-1}) ones. Fact 4.1 implies a partition of a convex n-gon into smaller nonintersecting basic subpolygons which are in an optimal triangulation. Such a partition can be found in the following way.

First divide a polygon into two parts by joining the smallest vertex with the second smallest one. Consider two obtained subpolygons independently. Assume that v_0, v_{n-1} are two smallest vertices and v_0 is the smallest one.

Observation 4.2 *Vertex v_t is the third smallest one if and only if v_t is a vertex with the greatest index between v_0 and v_{n-1}, such that there is no vertex less than v_t in the sequence $v_1, v_2, \ldots, v_{t-1}$.*

Using the observation above we can easy design parallel algorithm which divides a convex polygon into basic polygons.

Let $v_0, v_1, \ldots v_{n-1}$ be the weights of a polygon with v_0 as the smallest vertex and v_{n-1} as the second smallest one. Solve the prefix minima problem with respect to weights $(v_1, \ldots, v_{n-1})$. Join v_i with v_0 iff this vertex is the minimal in the range $v_1, \ldots, v_i$. Hence we obtain the following fact.

---

[3] less than 1 percent on the average, see simulation results in [Chin-78]

Fact 4.3 *Partitioning of the polygon into nonintersecting basic subpolygons can be done in $O(\log n)$ time using $n/\log n$ processors on a CREW PRAM and in $O(\log \log n)$ time using $n/\log \log n$ processors on a weak CRCW PRAM.*

5 Outline of a new algorithm

In this section we describe an overview of a new algorithm which produces the same partitioning as that of Chin-Hu-Shing. An input of this new algorithm is a basic polygon with v_0 as the smallest vertex, v_1 as the second smallest one, and v_{n-1} as the third smallest one.

First, our algorithm finds the set of candidates for h-arcs. We show a necessary condition for an arc to be an h-arc. Moreover this condition gives us the set of edges in a polygon which do not intersect, what implies that this set of candidates is not too big (it counts exactly n-3 arcs). Then from the set of all candidates for h-arc, we eliminate these edges which are not chosen by algorithm CHS. At the end we find all the fan-arcs.

Let us define a *candidate* to be an arc (v_i, v_j) such that for each k, $i < k < j$, the inequality $v_i < v_k$, $v_j < v_k$ hold. We show that arcs with this property are candidates for h-arcs (i.e., this is a necessary condition to be h-arc).

Observation 5.1 *No candidates intersect.*

Observation 5.2 *Let v_k, v_c be neighbours of vertex v_t. If v_t is large then the following condition holds*

$$v_c < v_t, \text{ and } v_k < v_t.$$

Now we give some observations concerning algorithm CHS. We can see that during an execution of the procedure $cutoff(v_t)$, with v_k and v_c as neighbours of v_t, two invariants hold:

1. $k < t < c$

2. for each i $(k < i < c, i \neq t)$, $v_t < v_i$

Hence we obtain the following crucial lemma.

Lemma 2 .
h-arc is always a candidate.

Let us look at the polygon after finding all candidates. This polygon is completely divided into $n - 2$ triangles. We introduce a tree of candidates. Triangle (v_i, v_j, v_l) (where $i < j < l$) is said to lie lower than arc (v_i, v_l). Let us take candidate (v_k, v_c). It is obvious that each such an arc (if $k < c - 1$) has exactly one lower triangle. We can define the *tree of candidates* as follows.

For given candidate (v_k, v_c) we define its sons to be arcs (except sides of polygon) in the lower triangle for (v_k, v_c). It is easy to see that this tree is a binary tree (i.e., each vertex has at most two sons) with $n - 3$ vertices.

Observation 5.3 *For each candidate (v_k, v_c), all its descendants are of the form (v_i, v_j), where $k \leq i < j \leq c$.*

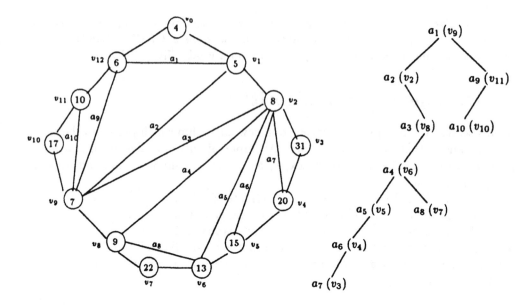

Figure 2: Partitioning of a convex polygon by candidates and the corresponding tree of candidates. In the tree each vertex is represented by the corresponding candidate and in brackets we present vertices which store given candidate.

Given a tree of candidates, we need to verify for all the candidates whether they are h-arcs.

Algorithm CHS begins by creating h-arcs and then non-divided subpolygon is partitioned by fan-arcs. Always when vertex v_t is cut new h-arc is added. Therefore during the exectuting the loop in algorithm CHS there is always only one non-divided convex polygon. Thus we obtain the following

Fact 5.4 *The loop in algorithm CHS finishes with exactly one convex polygon without internal arcs.*

This fact implies the following important observation.

Observation 5.5 *Let (v_k, v_c) be the h-arc obtained by algorithm CHS. Then all the candidates (v_i, v_j), for $k \leq i < j \leq c$, are generated by CHS algorithm as h-arcs.*

Proof: From the above fact, we see that there is no fan-arc in the polygon $(v_k, v_{k+1}, \ldots, v_c)$. Hence this polygon must be triangulated by h-arcs. □

The following is a new algorithm for partitioning of a basic polygon.

Algorithm Par-CHS

{ *Input :* basic polygon with vertices $v_0, v_1, \ldots, v_{n-1}$ }

1. Create the tree of candidates.

2. Verify all candidates - i.e., find all h-arcs.

3. Find all fan arcs in a polygon.

Theorem 3 .
Algorithm Par-CHS produces the same partition of a convex polygon as CIIS algorithm.

Proof: It follows from the previous comments and from the fact that all arcs divided a convex polygon into basic subpolygons are fan-arcs. □

6 An optimal parallel algorithm

In this section we show how to implement efficiently algorithm Par-CHS on CREW PRAM and CRCW PRAM machines.

Begin with building a tree of candidates. From previous observations we can establish condition when procedure $cutoff(v_t)$ is called. Let v_k and v_c $(k < t < c)$ be two neighbours of v_t. Then a vertex v_t is poped off the stack if and only if all vertices between v_k and v_c (in the original convex polygon) are greater than v_t (and $v_k < v_t$ and $v_c < v_t$). That is, for each i, $k < i < c, i \neq t$, $v_t < v_i$.

Therefore, to find such a triple it is enough to find for each vertex v_t two vertices v_k and v_c which lie on the both sides of v_t and satisfy the above condition. It is clear that this is equivalent to the problem of finding all nearest smaller values. For each v_t choose v_k to be the nearest smaller vertex to the left side and v_c to be the nearest smaller vertex than v_t to the right side. And now create the set of pairs - (k, c) which are stored in the array $node(t)$.

Then we need to create the candidate's tree. That is, we have to find fathers for all nodes (i.e., for all pairs (k, c)). The root of the tree is a pair $(1, n-1)$ stored in $node(r)$, where r is this vertex which was cut off by arc (v_1, v_{n-1}). Definition of the father's relation implies that the father of candidate (v_i, v_j) is either $node(j) = (i, r)$, if $v_i < v_j$, or $node(i) = (l, j)$, if $v_j < v_i$, for some r or l. Hence we can find fathers in $O(1)$ time using n processors on a CREW PRAM.

From the previous section we know that a candidate is an h-arc iff

- it cuts off a large vertex (with respect to this arc) and

- all candidates below it (in the tree of candidates) are h-arcs too

One can easily show that if $node(t) = (k, c)$ and (v_k, v_c) is an h-arc, then (v_k, v_c) cuts off v_t as a large vertex. Hence to check whether given candidate is an h-arc we can solve straight-line program which evaluates boolean expressions. For each leaf $node(t) = (k, c)$ of the tree of candidates assign value *true* if vertex v_t is large with v_k and v_c as its two neighbours and *false* otherwise. For every internal vertex $node(t) = (k, c)$ assign the logical operation 'AND' if v_t is large with v_k and v_c as its two neighbours, otherwise set value *false*. Then compute a tree of expression. If $node(t) = (k, c)$ has computed value *true* then candidate (v_k, v_c) is an h-arc. But standard algorithms for expression evaluation (see e.g. [CV-88] or [Ryt-90]) run in at least $O(\log n)$ time on any of PRAM's models. Therefore we will need new ideas to solve this problem.

Let us look at the tree of candidates. Each candidate (v_k, v_c) is held in $node(t)$. For each such a node we know the number of all its descendants (exactly c-k-2) and morever $t - 1$ is its infix number (see e.g. a tree from figure 2). Thus we can use the algorithm from lemma 1. So a node corresponding to a candidate (v_k, v_c) has value *true* if and only if it is an h-arc.

At the end we need only to find all other vertices in the polygon, i.e., all fan-arcs. From the previous step we have in the array $node(i)$ some h-arcs, and all others candidates are not in the partitioning. Hence we can check whether node v_i was cut off during executing algorithm CHS, simply by checking if $node(i)$ is an h-arc. If $node(i)$ is not an h-arc, then join vertex v_i with a vertex v_0 and put this information into $node(i)$. It can be done in constant time with n processors.

This completes the proof of the following result.

Theorem 4 .
The problem of finding a near-optimal order of matrix chain products and the problem of finding a near-optimal partition of a convex polygon can be solved

- *in $O(\log n)$ time using $n/\log n$ processors on a CREW PRAM*
- *in $O(\log \log n)$ time with $n/\log \log n$ processors on a weak CRCW PRAM*

7 Conclusion and extensions

We gave very fast optimal parallel algorithm for finding a near-optimal order of matrix chain product and for finding a near-optimal partitioning of a convex polygon. It runs in $O(\log n)$ time on a CREW PRAM and in $O(\log \log n)$ time on a weak CRCW PRAM. This algorithm has optimal linear time-processor product. These bounds seem to be optimal among linear-work algorithms on both CREW and CRCW PRAM.

One can improve these bounds if the domain of input values is restricted. In the matrix chain products problem there are always integer values of matrix dimension. Therefore if dimensions are drawn from a domain $[k \ldots k + s]$ we can find a near-optimal order of matrix multiplications in $O(\alpha(n) + \log \log \log s)$ time with linear work on a priority CRCW PRAM. This result follows from the algorithm with the same bound for the nearest smaller values problem (and also the prefix minima problem) where the domain is restricted to the set $[0 \ldots s]$ [Ber-92].

We can speed-up our optimal algorithm to $\alpha(n)$ time on a common CRCW PRAM if the difference between two succesive elements (i.e., dimension or weights of vertices of a polygon) is bounded by some constant k. It also follows from the result for the all nearest smaller values problem [BV-91].

As in [Cha-75] the algorithm can be generalized to a larger class of functions by assuming that the multiplication of a $p \times q$ matrix and a $q \times r$ matrix takes $\tau(p, q, r)$ operations. Most of the results (except the error ratio) will stay true as long as $\tau(p, q, r)$ is nonnegative and *reasonable*, i.e., $\tau(p, q, r)$ is monotonically nondecreasing in p, q, r and $\tau(p, q, r) = \tau(r, p, q)$.

Acknowledgment

The author would like to thank B. Chlebus, K. Diks and W. Rytter for carefully reading draft of the paper and many helpful comments.

References

[Ber-92] O. Berkman, personal communication, 1992.

[BBGSV-89] O. Berkman, D. Breslauer, Z. Galil, B. Schieber, U. Vishkin, "Highly parallelizable problems", *Proceedings of the 21st Annual ACM Symposium on Theory of Computing* 1989, pp. 309-319.

[BSV-88] O. Berkman, B. Schieber, U. Vishkin, "Some doubly logarithmic optimal parallel algorithms based on finding all nearest smaller values", UMIACS-TR-88-79, University of Maryland, Institute for Advanced Computer Studies, 1988.

[BV-89] O. Berkman, U. Vishkin, "Recursive *-tree parallel data-structure", *Proceedings of the 30th Annual Symposium on Foundations of Computer Science*, IEEE Computer Society, 1989, pp. 196-202, *also UMIACS-TR-90-40, University of Maryland, Institute for Advanced Computer Studies, 1990*.

[BV-91] O. Berkman, U. Vishkin, "Almost fully-parallel parentheses matching", UMIACS-TR-91-103, University of Maryland, Institute for Advanced Computer Studies, 1991.

[Cha-75] A.K. Chandra, "Computing matrix chain products in near-optimal time", IBM Research Report RC 5625(#24393), IBM T.J. Watson Res. Ctr., Yorktown Heights, N.Y., 1975.

[Chin-78] F.Y. Chin, "An $O(n)$ algorithm for determining a near-optimal computation order of matrix chain products", Communications of the ACM, Vol. 21, No. 7, 1978, pp. 544-549.

[CV-88] R. Cole, U. Vishkin, "Optimal parallel algorithms for expression tree evaluation and list ranking", *Proceedings of the 3rd Aegean Workshop Comput.*, 1988.

[GP-92] Z. Galil, K. Park, "Parallel dynamic programming", manuscript 1992.

[GR-88] A. Gibbons, W. Rytter, "Efficient parallel algorithms", Cambridge University Press, 1988.

[HLV-90] S-H.S. Huang, H. Liu, V. Viswanathan, "Parallel dynamic programming", *Proceedings of the 2nd IEEE Symposium on Parallel and Distibuted Processing*, 1990, pp. 497-500.

[HS-80] T.C. Hu, M.T. Shing, "Some theorems about matrix multiplication", *Proceedings of the 21st Annual Symposium on Foundations of Computer Science*, IEEE Computer Society, 1980, pp. 28-35.

[HS-81] T.C. Hu, M.T. Shing, "An $O(n)$ algorithm to find a near-optimum partition of a convex polygon", Journal of Algorithms, Vol. 2, 1981, pp. 122-138.

[Ryt-88] W. Rytter, "On efficient parallel computations for some dynamic drogramming problems", Theoretical Computer Science, Vol. 59, 1988, pp. 297-307.

[Ryt-90] W. Rytter, "On parallel computation of expression and straight-line programs", Computer and Artificial Intelligence, Vol. 9, No. 5, 1990, pp. 427-429.

[Sch-87] B. Schieber, "Design and analysis of some parallel algorithms", PhD Thesis, Tel Aviv University, Tel Aviv, 1987.

Generating Sparse 2—spanners

Guy Kortsarz [*]   David Peleg[*] [†]

Abstract

A k—spanner of a connected graph $G = (V, E)$ is a subgraph G' consisting of all the vertices of V and a subset of the edges, with the additional property that the distance between any two vertices in G' is larger than that distance in G by no more than a factor of k. This note concerns the problem of finding the sparsest 2-spanner in a given graph, and presents an approximation algorithm for this problem with approximation ratio $\log(|E|/|V|)$.

1   Introduction

The concept of *graph spanners* has been studied in several recent papers, in the context of communication networks, distributed computing, robotics and computational geometry [ADDJ90,Cai91,Che86,DFS87,DJ89,LL89,PS89,PU89]. Consider a connected simple graph $G = (V, E)$, with $|V| = n$ vertices. A subgraph $G' = (V, E')$ of G is a $k - spanner$ if for every $u, v \in V$,

$$\frac{dist(u, v, G')}{dist(u, v, G)} \leq k,$$

where $dist(u, v, G')$ denotes the distance from u to v in G', i.e., the minimal number of edges in a path connecting them in G'. We refer to k as the *stretch factor* of G'.

In the Euclidean setting, spanners were studied in [Cai91,DFS87,DJ89,LL89]. Spanners for general graphs were first introduced in [PU89], where it was shown that for every n—vertex hypercube there exists a 3-spanner with no more than $7n$ edges. Spanners were used in [PU89] to construct a new type of synchronizer for an asynchronous network. For this, and other applications, it is desirable that the spanners be as *sparse* as possible, namely, have few edges. This leads to the following problem. Let $S_k(G)$ denote the minimal number of edges in a k—spanner for the graph G. The *sparsest k-spanner* problem involves constructing a k—spanner with $S_k(G)$ edges for a given graph G.

[*]Department of Applied Mathematics and Computer Science, The Weizmann Institute, Rehovot 76100, Israel.

[†]Supported in part by a Walter and Elise Haas Career Development Award and by a grant from the Basic Research Foundation.

It is shown in [PS89] that the problem of determining, for a given graph $G = (V, E)$ and an integer m, whether $S_2(G) \leq m$ is NP-complete. This indicates that it is unlikely to find an exact solution for the sparsest $k-$spanner problem even in the case $k = 2$. Consequently, two possible remaining courses of action for investigating the problem are establishing global bounds on $S_k(G)$ and devising approximation algorithms for the problem.

In [PS89] it is shown that every $n-$vertex graph G has a polynomial time constructible $(4k + 1)-$spanner with at most $O(n^{1+1/k})$ edges, or in other words, $S_{4k+1}(G) = O(n^{1+1/k})$ for every graph G. Hence in particular, every graph G has an $O(\log n)-$spanner with $O(n)$ edges. These results are close to the best possible in general, as implied by the lower bound given in [PS89]. The construction of [PS89] is based on the concept of *sparse covers* or *partitions* (cf. [AP90]). Consequently, faster algorithms for constructing sparse covers, in either the sequential, parallel or distributed modes [LS91,ABCP91,ABCP,ABCP92], directly translate into faster algorithms for spanner construction as well.

The results of [PS89] were improved and generalized qin [ADDJ90] to the weighted case, in which there are positive weights associated with the edges, and the distance between two vertices is the weighted distance. Specifically, it is shown in [ADDJ90] that given an $n-$vertex graph and an integer $k \geq 1$, there is a polynomially constructible $(2k + 1)-$spanner G' such that $|E(G')| < n \cdot \lceil n^{\frac{1}{k}} \rceil$. Again, this result is shown to be the best possible.

The algorithms of [ADDJ90,PS89] provide us with *global* upper bounds for sparse $k-$spanners, i.e., general bounds that hold *for every graph*. However, it may be that for specific graphs, considerably sparser spanners exist. Furthermore, the upper bounds on sparsity given by these algorithms are small (i.e., close to n) only for large values of k. It is therefore interesting to look for *approximation algorithms*, that yield near-optimal *local* bounds applying to the specific graph at hand, by exploiting its individual properties.

In the sequel we concentrate on the sparsest 2-spanner problem. For this case, the best *global* upper bound is $S_2(G) = O(n^2)$. To see why this cannot be improved in general, consider the complete bipartite graph having $n/2$ vertices on each side. It is not hard to see that the only $2-$spanner for this graph is the graph itself. Thus there are cases where any 2-spanner requires $\Omega(n^2)$ edges. This lends additional motivation to our interest in approximating the sparsest $2-$spanner for speciefic graphs.

The construction of [ADDJ90] can be thought of as an approximation algorithm for the sparsest $k-$spanner problem. However, for the case of $k = 2$ the ratio provided by this algorithm might be as bad as $\Omega(n)$ (which is also the trivial ratio, since every 2-spanner contains at least $n - 1$ vertices, and every subgraph contains at most $n(n - 1)/2$ edges).

In this note we present an approximation algorithm for the sparsest $2-$spanner problem with approximation ratio $\log \frac{|E|}{|V|}$. That is, given a graph $G = (V, E)$, our algorithm generates a $2-$spanner $G' = (V, E')$ with $|E'| = O(S_2(G) \cdot \log \frac{|E|}{|V|})$ edges. In the next three sections we give some preliminary definitions, describe the algorithm and analyze its performance.

2 Preliminaries

We start by introducing some definitions. Let $U \subseteq V$ be a subset of the vertices. The graph induced by U is denoted by $G(U)$. The set of edges in $G(U)$ is denoted by $E(U)$. The *density* of U in G is defined as

$$\rho_G(U) = \frac{|E(U)|}{|U|} .$$

The *maximal density* of the graph G is defined to be

$$\rho(G) = max_{U \subseteq V}\{\rho_G(U)\}.$$

We recall the following fact (cf. Ch. 4 of [Law76]).

Lemma 2.1 [Law76] *Given a graph G, the problem of finding a subset $U \subseteq V$ with density $\rho_G(U) = \rho(G)$, can be solved polynomially using flow techniques.*

We make use of an alternative characterization of $k-$spanners, given in the following lemma of [PS89].

Lemma 2.2 [PS89] *The subgraph $G' = (V, E')$ is a $k - spanner$ of the graph $G = (V, E)$ iff $dist(u, v, G') \leq k$ for every $(v, u) \in E$.*

Next, we introduce the definition of a k-spanner of a subset $E' \subset E$ of the edges.

Definition 2.3 *Let E' be a subset of the edges. An optimal k-spanner for E' is a minimal subset $E'' \subset E$ such that every edge $e \in E' \setminus E''$ lies on a cycle of length $k + 1$ or less with the edges of E''.*

Thus the sparsest $2-$spanner problem can be restated as follows: we look for a minimal subset of edges $E' \subset E$ such that every edge e that does not belong to E' lies on a triangle with two edges that do belong to E'. Since a spanning graph of any set E' is also a spanning graph of any subset $E'' \subset E'$, the following fact holds.

Fact 2.4 *Let E_1 be an arbitrary subset of E_2, and let E_1' and E_2' be the edge sets of an optimal $k-$spanner for E_1 and E_2, respectively. Then $|E_1'| \leq |E_2'|$.* ∎

Given a graph G, we denote by $N(v)$ the set of neighbors of v in G, i.e.,

$$N(v) = \{u \mid (u, v) \in E\} .$$

Let E' be an arbitrary set of edges and U an arbitrary subset of vertices. Denote by $\mathcal{R}(E', U)$ the subset of the edges in the induced graph $G(U)$ *restricted* to E', namely,

$$\mathcal{R}(E', U) = E(U) \cap E' .$$

We denote

$$dom_G(E', v) = |\mathcal{R}(E', N(v))|$$

and say that v *dominates* (or covers) the edges of $\mathcal{R}(E', N(v))$ in G. Note that if all the edges adjacent to v are in the spanner, then all the edges of $\mathcal{R}(E', N(v))$ lie on a triangle with these spanner edges, and thus are taken care of. Denote the graph of neighbors of v restricted to E' by

$$N(E', v) = (N(v), \mathcal{R}(E', N(v))) .$$

Denote the maximal density of this restricted neighborhood graph by

$$\rho(E', v) = \rho(N(E', v)) .$$

3 The approximation algorithm

Let us first explain the idea behind our approximation algorithm for the 2–spanner problem. Throughout the run of the algorithm we maintain a *partition* of the edge set E into three pairwise disjoint sets of edges, denoted H^s, H^c and H^u. The set H^s contains *spanner edges*, i.e., edges that were already added to the constructed spanner. The set H^c consists of *covered edges*, i.e., edges that are not in the spanner, but lie on a triangle with two edges that are included in H^s. That is, at any given moment, for every edge $e \in H^c$ there exist two edges $e_1, e_2 \in H^s$ such that e, e_1 and e_2 form a triangle. Finally, H^u consists of *unspanned edges*, i.e., edges that are still neither in the spanner nor covered by it.

Our algorithm operates by repeatedly performing the following operation. For every vertex v, we consider the graph $N(H^u, v)$, consisting of the set of neighbors of v, with the edge set restricted to the unspanned edges H^u. Let P be the procedure for computing the densest subgraph of a given graph G, whose existence is asserted in Lemma 2.1. We use Procedure P to look for a subset U_v of maximal density $\rho(H^u, v)$ in the graph $N(H^u, v)$. Then we choose the densest such set among all the sets $\{U_v \mid v \in V\}$. Assume that the chosen set is U_w.

After finding U_w, we add the "star" composed of the edges connecting U_w and w, to the edge set of the spanner H^s. In this way we *cover* a "large" set of edges (namely, those in $H^u \cap E(U_w)$), while adding only a "small" number of new edges (specifically, $|U_w|$) to the spanner.

This operation is repeated until all sets U_v are "sufficiently sparse," whence the algorithm halts and $H^s \cup H^u$ is taken to be the edge set of the resulting spanner.

We now state our approximation algorithm more precisely.

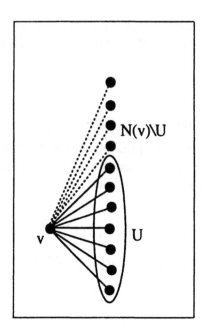

Figure 1: The set U represents a dense subset of $N(v)$. The solid edges are the ones added to the constructed spanner.

Algorithm 3.1 *An approximation algorithm for the 2−spanner problem*

Input: a graph $G = (V, E)$.

1. *Set* $H^u \leftarrow E$; $H^c \leftarrow \emptyset$; $H^s \leftarrow \emptyset$;

2. **While** *there exists some* v *for which* $\rho(H^u, v) \geq 1$ **do**

   (a) *Choose a vertex* v *for which* $\rho(H^u, v)$ *is maximal.*

   (b) *Let* U_v *be the corresponding dense subset of* $N(v)$.
   $$H^s \leftarrow H^s \cup \{(u, v) \mid u \in U_v\}.$$
   $$H^c \leftarrow (H^c \cup \mathcal{R}(H^u, U_v)) \setminus H_s.$$
   $$H^u \leftarrow H^u \setminus (H^s \cup H^c).$$

   End-While.

3. *Return(*$H^s \cup H^u$*)*

4 Analysis

Let us first note that the algorithm is correct, in the sense that the output set of edges indeed forms a 2−spanner of G, since every edge in H^c lies on a triangle with two edges of H^s.

Next, denoting the edge set of an optimal 2–spanner for G by H^*, let us now proceed to bound from above the ratio between the sizes of the sets $H^* \cup H^u$ and H^*, thus establishing the approximation ratio of our algorithm.

Let us break the execution of the main loop of the algorithm into phases as follows. Denote $r = \frac{|E|}{|V|}$ and $f = \lceil \log r \rceil$. Note that since the set H^u decreases in size at every step, $\rho(H^u, v)$ is monotonically decreasing as well.

Definition 4.1 *We define the first phase to include all the iterations during which for every selected vertex v, $\rho(H^u, v) \geq \frac{r}{2}$. For $2 \leq i \leq f$, the i'th phase consists of the iterations during which every selected vertex v satisfies*

$$\frac{r}{2^{i-1}} > \rho(H^u, v) \geq \frac{r}{2^i} .$$

Let H_i^s (respectively, H_i^c) be the set of edges added to H^s (resp., H^c) in the i'th phase, and let H_i^u be the set of edges left in H^u at the *end* of the i'th phase. Let H_i^* be the edge set of an optimal 2-spanner for H_i^u. We denote by X_i the set of vertices selected by the algorithm during step (a) of the iterations of the i'th phase (namely, those vertices for which $\rho(H^u, v)$ was maximal in the iterations of the i'th phase).

Note that a vertex v may be picked more than once during a phase, and in more than one phase. Consider a particular phase i. Each time that the vertex v is picked in the i'th phase, a subset $S_v = \{(w, v) | w \in U_v\}$ of its adjacent edges is added to H^s, namely, those edges connecting it to U_v. Also, there is a corresponding set $C_v = \mathcal{R}(H^u, U_v)$ of edges from H^u that lie on a triangle with the edges of S_v and are thus added to H^c. Since $|S_v| = |U_v|$, by definition of ρ, these sets S_v and C_v satisfy

$$\frac{|C_v|}{|S_v|} = \rho(H_u, v).$$

Denote the cardinality of the union of these sets C_v added during the i'th phase by $h_i^c(v)$, and the cardinality of the union of the sets S_v by $h_i^s(v)$, for every vertex $v \in X_i$. Note that by the definition of the i'th phase, it follows from the above that for every $v \in X_i$,

$$h_i^c(v) \geq \frac{r}{2^i} \cdot h_i^s(v) . \tag{1}$$

Observe that an edge $e = (v, u)$ may belong to two different sets S_v, S_u, hence

$$|H_i^s| \leq \sum_{v \in X_i} h_i^s(v) . \tag{2}$$

On the other hand, edges are included in sets C_v atmost once, hence

$$|H_i^c| = \sum_{v \in X_i} h_i^c(v) . \tag{3}$$

It follows from (1),(2) and (3) that

$$|H_i^c| \geq \frac{r}{2^i}|H_i^s| . \tag{4}$$

We now prove the following claim. Let $G_i^* = (V, H_i^*)$ be an optimal 2-spanner of H_i^u, and for every $v \in V$ let $d_i^*(v)$ be the degree of v in the graph G_i^*. Recall that $dom_{G_i^*}(H_i^u, v)$ is the number of edges of H_i^u dominated by v in G_i^*. Denote

$$\rho_i(v) = \frac{dom_{G_i^*}(H_i^u, v)}{d_i^*(v)} .$$

Lemma 4.2 *For every $v \in V$,*

$$\rho_i(v) < \frac{r}{2^i} .$$

Proof: Let $N^*(v)$ be the set of vertices adjacent to v in G_i^*. Thus $|N^*(v)| = d_i^*(v)$. Also

$$dom_{G_i^*}(H_i^u, v) = |\mathcal{R}(H_i^u, N^*(v))| = |E(N^*(v)) \cap H_i^u| .$$

Thus $\rho_i(v)$ is the density of $N^*(v)$ in the restricted neighborhood graph $N(H_i^u, v)$, i.e., $\rho_i(v) = \rho_{N(H_i^*, v)}(N^*(v))$. This density is no larger than the maximal density of the graph $N(H_i^u, v)$, namely, $\rho_i(v) \leq \rho(H_i^u, v)$. Thus the required claim follows directly from the definition of the i'th phase. ∎

Lemma 4.3 *For every $1 \leq i \leq f$,*

$$\frac{|H_i^u|}{|H^*|} < \frac{r}{2^{i-1}} + 1 .$$

Proof: First let us remark that

$$|H_i^u| \leq |H_i^*| + \sum_{v \in V} dom_{G_i^*}(H_i^u, v), \tag{5}$$

since every edge $e \in H_i^u$ either belongs to H_i^* or is dominated by some vertex in G_i^*. Secondly, by Fact 2.4 we have that

$$|H_i^*| \leq |H^*| . \tag{6}$$

Thirdly, note that

$$|H_i^*| = \frac{1}{2} \sum_{v \in V} d_i^*(v) . \tag{7}$$

Combining Eq. (5), (6) and (7) we conclude that

$$\frac{|H_i^u|}{|H^*|} \leq \frac{|H_i^u|}{|H_i^*|} \leq \frac{|H_i^*| + \sum_{v \in V} dom_{G_i^*}(H_i^u, v)}{|H_i^*|}$$

$$= 1 + \frac{\sum_{v \in V} dom_{G_i^*}(H_i^u, v)}{\frac{1}{2} \sum_{v \in V} d_i^*(v)}$$

$$\leq 1 + 2 \cdot \max_{v \in V} \left\{ \frac{dom_{G_i^*}(H_i^u, v)}{d_i^*(v)} \right\}$$

$$= 1 + 2 \cdot \max_{v \in V} \{\rho_i(v)\}.$$

Thus by Lemma 4.2 we have

$$\frac{|H_i^u|}{|H^*|} < 2 \cdot \frac{r}{2^i} + 1 = \frac{r}{2^{i-1}} + 1 . \quad \blacksquare$$

We now proceed to prove our main lemma.

Lemma 4.4 *For every $1 \le i \le f$,*

$$\frac{|H_i^s|}{|H^*|} < 4 + \frac{2^i}{r} .$$

Proof: We first prove the claim for $i = 1$. We may assume w.l.o.g that $n \ge 2$. In this case by Eq. (4) and by the choice of r

$$\frac{|H_1^s|}{|H^*|} \le \frac{\frac{2}{r} \cdot |H_1^c|}{|H^*|} \le \frac{\frac{2 \cdot n}{|E|} \cdot |E|}{|H^*|} \le \frac{2 \cdot n}{n-1} \le 4 .$$

We now prove the claim for $i > 1$. By Eq. (4) and by the fact that $H_i^c \subseteq H_{i-1}^u$, we have

$$\frac{|H_i^s|}{|H^*|} \le \frac{\frac{2^i}{r} \cdot |H_i^c|}{|H^*|} \le \frac{2^i}{r} \cdot \frac{|H_{i-1}^u|}{|H^*|} .$$

Using Lemma 4.3 we get

$$\frac{|H_i^s|}{|H^*|} < \frac{2^i}{r} \cdot (\frac{r}{2^{i-2}} + 1) = 4 + \frac{2^i}{r} . \quad \blacksquare$$

Corollary 4.5

$$\frac{|H^s|}{|H^*|} = O(\log r) .$$

Proof: By Lemma 4.4 and the choice of f,

$$\begin{aligned}
\frac{|H^s|}{|H^*|} &= \frac{\sum_{i=1}^{f} |H_i^s|}{|H^*|} \\
&\le \sum_{i=1}^{f} (4 + \frac{2^i}{r}) \\
&= 4 \cdot f + \frac{1}{r} \cdot \sum_{i=1}^{f} 2^i = 4 \cdot f + O(1) = O(\log r). \quad \blacksquare
\end{aligned}$$

Furthermore, by Lemma 4.3 we have

Corollary 4.6 $\frac{|H^u|}{|H^*|} < 3 .$ $\blacksquare$

Combining Corollaries 4.5 and 4.6 we conclude our main result.

Theorem 4.7 *Algorithm 3.1 is an $O(\log \frac{|E|}{|V|})$ approximation algorithm for the sparsest 2-spanner problem.* ∎

Finally, let us consider the time complexity of our algorithm. Note that the "maximal density" procedure P asserted by Lemma 2.1 can be implemented using any appropriate maximum-flow algorithm of the numerous algorithms suggested in the literature, and will take time polynomial in the size of its input [Law76]. Let $F(n,m)$ be the time taken by Procedure P. Each iteration of our algorithm makes n invocations of this procedure, for computing $\rho(H^u, v)$ for every vertex v. Each iteration adds at least one edge to H_c. Since every edge is added to H_c at most once, there are at most m iterations. Hence, the complexity of the algorithm is bounded by $n \cdot m \cdot F(n,m)$, which is still polynomial in the input size.

5 Conclusion and open problems

We have shown that there exists an approximation algorithm for the sparsest 2−spanner problem with a worst case approximation ratio of $O(\log \frac{|E|}{|V|})$. Note that while the worst case ratio of the algorithm is $O(\log n)$, it performs better for sparse graphs. The next immediate problem is to approximate the sparsest k−spanner problem for an arbitrary fixed value of k with a similar ratio. Another interesting problem is to give an approximation algorithm for the weighted version of this problem.

Acknowledgment

We are grateful to Noga Alon for his helpful comments, and for directing us to Lemma 2.1 in [Law76].

References

[ABCP]   Baruch Awerbuch, Bonnie Berger, Lenore Cowen, and David Peleg. Low diameter graph decomposition is in NC. These proceedings.

[ABCP91] Baruch Awerbuch, Bonnie Berger, Lenore Cowen, and David Peleg. Fast constructions of sparse neighborhood covers. Unpublished manuscript, 1991.

[ABCP92] Baruch Awerbuch, Bonnie Berger, Lenore Cowen, and David Peleg. Fast network decomposition. In *Proc. 11th ACM Symp. on Principles of Distributed Computing*, 1992. To appear.

[ADDJ90] I. Althöfer, G. Das, D. Dobkin, and D. Joseph. Generating sparse spanners for weighted graphs. *Proc. 2nd Scandinavian Workshop on Algorithm Theory*, pages 26–37, July 1990.

[AP90]   Baruch Awerbuch and David Peleg. Sparse partitions. In 31st *IEEE Symp. on Foundations of Computer Science*, pages 503–513, October 1990.

[Cai91]   L. Cai. Tree 2-spanners. Technical Report No. 91-4, Simon Fraser University, 1991.

[Che86]   L.P. Chew. There is a planar graph almost as good as the complete graph. In *ACM Symposium on Computational Geometry*, pages 169–177, 1986.

[DFS87]   D.P. Dobkin, S.J. Friedman, and K.J. Supowit. Delaunay graphs are almost as good as complete graphs. In *Proc. 91'st IEEE Symp. on Foundations of Computer Science*, pages 20–26, 1987.

[DJ89]   G. Das and D. Joseph. Which triangulation approximates the complete graph? In *International Symposium on Optimal Algorithms*, 1989.

[Law76]   E.L. Lawler. *Combinatorial Optimization: Networks and Matroids*. Holt, Rinehart and Winston, 1976.

[LL89]   Levcopoulos and Lingas. There are planar graphs almost as good as the complete graph and as short as minimum spanning trees. In *International Symposium on Optimal Algorithms*, 1989.

[LS91]   A.L. Liestman and T. Shermer. Grid and hypercube spanners. Technical Report No. 91-1, Simon Fraser University, 1991.

[PS89]   D. Peleg and A. Schäffer. Graph spanners. *J.of Graph Theory*, 13:99–116, 1989.

[PU89]   D. Peleg and J.D. Ullman. An optimal synchronizer for the hypercube. *SIAM J. on Comput.*, 18:740–747, August 1989.

Low-Diameter Graph Decomposition is in NC

Baruch Awerbuch[*][1] Bonnie Berger[**][1] Lenore Cowen[***][1] David Peleg[†][2]

[1] Dept. of Mathematics and Lab. for Computer Science, M.I.T., Cambridge, MA 02139.
[2] Department of Applied Mathematics and Computer Science, The Weizmann Institute, Rehovot 76100, Israel.

Abstract. We obtain the first deterministic NC algorithm for the low-diameter graph decomposition problem on arbitrary graphs. We achieve this through derandomizing an algorithm of Linial and Saks. Our algorithm runs in $O(\log^5(n))$ time and uses $O(n^2)$ processors.

1 Introduction

For an undirected graph $G = (V, E)$, a (χ, d)-*decomposition* is defined to be a χ-coloring of the nodes of the graph that satisfies the following properties:

1. each color class is partitioned into an arbitrary number of disjoint *clusters;*
2. the distance between any pair of nodes in a cluster is at most d, where *distance* is the length of the shortest path connecting the nodes in G,
3. clusters of the same color are at least distance 2 apart.

A (χ, d)-decomposition is said to be *low-diameter* if χ and d are both $O(poly \log n)$.

The graph decomposition problem is a problem in algorithmic graph theory that originated in the theory of distributed computing. It was introduced in [3, 6], as a means of partitioning a network into local regions. For further work on graph decomposition in the distributed model, see [8, 7, 11, 5, 1, 14, 9]. Linial and Saks [11] have given the only algorithm that finds a graph decomposition in polylogarithmic time in the distributed model. In addition, their algorithm is randomized and gives a low-diameter decomposition with $\chi = O(\log n)$ and $d = O(\log n)$. (Linial and Saks also proved that their *low-diameter* decomposition is optimal, i.e. there exist families of graphs for which one cannot achieve better than a $(\log n, \log n)$-decomposition.) It is easy to see that the Linial-Saks algorithm can be run on the PRAM and thus places the low-diameter graph decomposition problem in the class RNC.

In this paper, we achieve the first polylogarithmic-time deterministic parallel algorithm for (χ, d)-decomposition. The algorithm decomposes an arbitrary graph

* Supported by Air Force Contract TNDGAFOSR-86-0078, ARO contract DAAL03-86-K-0171, NSF contract CCR8611442, DARPA contract N00014-89-J-1988, and a special grant from IBM.
** Supported by an NSF Postdoctoral Research Fellowship.
*** Supported in part by DARPA contracts N00014-87-K-0825 and N00014-89-J-1988, Air Force Contract OSR-89-02171, Army Contract DAAL-03-86-K-0171 and Navy-ONR Contract N00014-19-J-1698
† Supported in part by an Allon Fellowship, by a Bantrell Fellowship and by a Walter and Elise Haas Career Development Award.

into $O(\log^2 n)$ colors, with cluster diameter at most $O(\log n)$. Thus we place the low-diameter graph decomposition problem into the class NC.

The algorithm uses a non-trivial scaling technique to remove the randomness from the algorithm of Linial-Saks. In Section 2.1, we review the Linial-Saks algorithm. Section 2.2 gives our new modified RNC algorithm, whose analysis is shown in Section 2.4 to depend only on *pairwise* independence. This is the crux of the argument. Once we have a pairwise independent RNC algorithm, it is well known how to remove the randomness to obtain an NC algorithm. In Section 2.6 we are a bit more careful, however, in order to keep down the blowup in the number of processors. Our (deterministic) NC algorithm runs in $O(\log^5(n))$ time and uses $O(n^2)$ processors.

The (χ, d)-decomposition problem is related to the *sparse t-neighborhood cover* problem [8], which has applications to sequential approximation algorithms for all-pairs shortest paths [4] and finding small edge cuts in planar graphs [15]. We believe the NC algorithm in this paper will also have applications to parallel graph algorithms.

2 The Algorithm

In this section, we construct a deterministic NC algorithm for low-diameter graph decomposition. This is achieved by modifying an RNC algorithm of Linial-Saks to depend only on pairwise independence, and then removing the randomness. To get our newly-devised pairwise independent *benefit function* [10, 13] to work, we have to employ a non-trivial scaling technique. Scaling has been used previously only on the simple measure of node degree in a graph.

2.1 The RNC Algorithm of Linial-Saks

Linial and Saks's randomized algorithm [11] emulates the following simple greedy procedure. Pick a color. Pick an arbitrary node (call it a *center* node) and greedily grow a ball around it of minimum radius r, such that a constant fraction of the nodes in the ball lie in the interior (i.e. are also in the ball of radius $r - 1$ around the center node). It is easy to prove that there always exists an $r \leq \log n$ for which this condition holds. The interior of the ball is put into the color class, and the entire ball is removed from the graph. (The *border* (those nodes whose distance from the center node is exactly r) will not be colored with the current color). Then pick another arbitrary node, and do the same thing, until all nodes in the graph have been processed. Then return all the uncolored nodes (the border nodes) to the graph, and begin again on a new color.

To emulate the greedy algorithm randomly, Linial-Saks still consider each of $O(\log n)$ colors sequentially, but must find a distribution that will allow all center nodes of clusters of the same color to grow out in parallel, while minimizing collisions. If all nodes are allowed to greedily grow out at once, there is no obvious criterion for deciding which nodes should be placed in the color-class in such a way that the resulting coloring is guaranteed both to have small diameter and to contain a substantial fraction of the nodes.

Linial-Saks give a randomized distributed (trivially also an RNC) algorithm where nodes compete to be the center node. In their algorithm, in a given phase they select which nodes will be given color j as follows. Each node flips a candidate radius n-wise independently at random according to a truncated geometric distribution (the radius is never set greater than B, which is set below). It is assumed that each node has a unique ID associated with it. Each node y then broadcasts the triple $(r_y, ID_y, d(y, z))$ to all nodes z within distance r_y of y. For the remainder of this paper $d(y, z)$ will denote the distance between y and z in G. (This is sometimes referred to as the *weak* distance, as opposed to the *strong* distance, which is the distance between y and z in the subgraph induced by a cluster which contains them.) Now each node z elects its center node, $C(z)$, to be the node of highest ID whose broadcast it received. If $r_y > d(z, y)$, then z joins the current color class; if $r_y = d(z, y)$, then z remains uncolored until the next phase.

Linial and Saks show that if two neighboring nodes were both given color i, then they both declared the same node y to be their winning center node. This is because their algorithm emulates a greedy algorithm that sequentially processes nodes from highest to lowest ID in a phase. The diameter of the resulting clusters is therefore bounded by $2B$. Setting $B = O(\log n)$, they can expect to color a constant fraction of the remaining nodes at each phase. So their algorithm uses $O(\log n)$ colors. (See their paper [11] for a discussion of trade-offs between diameter and number of colors; in [11], Linial-Saks also give a family of graphs for which these trade-offs between χ and d are the best possible.)

The analysis of the above algorithm cannot be shown to work with constant-wise independence; in fact, one can construct graphs for which there will be no good sample point in a sample space with only constant-wise independence. It even seems doubtful that the Linial-Saks algorithm above would work with polylogarithmic independence. So if we want to remove randomness, we need to alter the randomized algorithm of Linial-Saks.

2.2 Overview of the Pairwise Independent RNC Algorithm

Surprisingly, we show that there is an alternative RNC algorithm where each node still flips a candidate radius and competes to be the center of a cluster, whose analysis can be shown to depend only on pairwise independence.

The new algorithm will proceed with *iterations* inside each *phase*, where a phase corresponds to a single color of Linial-Saks. In each iteration, nodes will grow their radii according to the same distribution as Linial-Saks, except there will be some probability (possibly large) that a node y does not grow a ball at all. If a node decides to grow a ball, it does so according to the same truncated geometric distribution as Linial-Saks. We get our *scaled* truncated distribution as follows:

$$Pr[r_y = NIL] = 1 - x$$
$$Pr[r_y = j] \quad = xp^j(1-p) \text{ for } 0 \leq j \leq B - 1$$
$$Pr[r_y = B] \quad = xp^B$$

The design of the algorithm proceeds as follows: we devise a new *benefit function* whose expectation will be a lower bound on the probability a node is colored by a given iteration (color) of the algorithm, plus pairwise independence will suffice to compute this benefit function. The pairwise-independent benefit function will serve as a good estimate to the n-wise independent "benefit function" of the Linial-Saks algorithm, *whenever nodes y in the graph would not expect to be reached by many candidate radii z.* This is why it is important that some nodes not grow candidate balls at all.

To maximize the new pairwise-independent benefit function, the probability x that a node grows a ball at all will be scaled according to a measure of local *density* in the graph around it (see the definition of the measure T_y below.) Since dense and sparse regions can appear in the same graph, the scaling factor x, will start small, and double in every iteration of a phase (this is the $O(\log n)$ blowup in the number of colors). We argue that in each iteration, those y's with the density scaled for in that iteration, will have expected benefit lower bounded by a constant fraction. Therefore, in each iteration, we expect to color a constant fraction of these nodes (Lemma 2). At the beginning of a phase x is reset to reflect the maximum density in the remaining graph that is being worked on. In $O(\log n)$ phases of $O(\log n)$ iterations each, we expect to color the entire graph.

2.3 The RNC Algorithm

Define $T_y = \sum_{z \mid d(z,y) < B} p^{d(z,y)}$, and $\Delta = \max_{\forall y \in G} T_y$. Each phase will have $O(\log n)$ iterations, where each iteration i colors a constant fraction of the nodes y with T_y between $\Delta/2^{i+1}$ and $\Delta/2^i$. Note that T_y decreases from iteration to iteration, but Δ is fixed.

The algorithm runs for $O(\log n)$ phases of $O(\log n)$ iterations each. At each iteration, we begin a new color. For each iteration i of a phase, set $x = 2^i/(5\Delta)$.

Each node y selects an integer radius r_y pairwise independently at random according to the truncated geometric distribution scaled by x (defined in Section 2.2). We can assume every node has a unique ID [11]. Each node y broadcasts (r_y, ID_y) to all nodes that are within distance r_y of it. After collecting all such messages from other nodes, each node y selects the node $C(y)$ of highest ID from among the nodes whose broadcast it received in the first round (including itself), and gets the current color if $d(y, C(y)) < r_{C(y)}$. (A NIL node does not broadcast.) At the end of the iteration, all the nodes colored are removed from the graph.

2.4 Analysis of the Algorithm's Performance

We fix a node y and estimate the probability that it is assigned to a color, S. Linial and Saks [11] have lower bounded this probability for their algorithm's phases by summing over all possible winners of y, and essentially calculating the probability that a given winner captures y and no other winners of higher ID capture y. Since the probability that $y \in S$ can be expressed as a union of probabilities, we are able to lower bound this union by the first two terms of the inclusion/exclusion expansion as follows:

$$Pr[y \in S] \geq$$
$$\sum_{z \mid d(z,y) < B} \left(Pr[r_z > d(z,y)] - \sum_{u > z \mid d(u,y) \leq B} Pr[(r_z > d(z,y)) \wedge (r_u \geq d(u,y))] \right)$$

Notice that the above lower bound on the probability that y is colored can be computed using only pairwise independence. This will be the basis of our new benefit function. We will indicate why the Linial and Saks algorithm cannot be shown to work with this weak lower bound.[3] However, we can scale x so that this lower bound suffices for the new algorithm.

More formally, for a given node z, define the following two indicator variables:

$X_{y,z}: r_z \geq d(z,y)$
$Z_{y,z}: r_z > d(z,y)$

Then we can rewrite our lower bound on $Pr[y \in S]$ as

$$\sum_{z \mid d(z,y) < B} E[Z_{y,z}] - \sum_{\substack{u > z \mid d(z,y) < B \\ d(u,y) \leq B}} E[Z_{y,z}X_{y,u}]$$

The *benefit* of a sample point $R = <r_1, \ldots, r_n>$ for a single node y, is now defined as

$$B_y(R) = \sum_{z \mid d(z,y) < B} Z_{y,z} - \sum_{\substack{u > z \mid d(z,y) < B \\ d(u,y) \leq B}} Z_{y,z}X_{y,u}$$

Hence, our lower bound on $Pr[y \in S]$ is, by linearity of expectation, the expected benefit.

Recall that $T_y = \sum_{z \mid d(z,y) < B} p^{d(z,y)}$. This is a local density measure, in the sense that each node close to y will contribute proportionally to T_y according to how close it is to y in the graph. In the proof of the next lemma, we show first that pxT_y will correspond exactly to the expected number of balls which have y in the interior, and $px^2(T_y^2/2)$ is an overestimate of the number of pairs of balls our benefit function subtracts off.

Lemma 1. *If $p \leq 1/2$, then $E[B_y(R)] \geq pxT_y - px^2(T_y^2/2)$.*

Proof Since $E[X_{y,z}] = xp^{d(z,y)}$, and $E[Z_{y,z}] = xp^{d(z,y)+1}$, we can rewrite

$$E[B_y(R)] = pxT_y - px^2 \left(\sum_{\substack{u > z \mid d(z,y) < B \\ d(u,y) \leq B}} p^{d(z,y)+d(u,y)} \right)$$

It thus remains to show that

$$\frac{T_y^2}{2} \geq \sum_{\substack{u > z \mid d(z,y) < B \\ d(u,y) \leq B}} p^{d(z,y)+d(u,y)} \tag{1}$$

---

[3] We can, in fact, construct example graphs on which their algorithm will not perform well using only pairwise independence, but in this paper we just point out where the analysis fails.

We proceed as follows:

$$\frac{T_y^2}{2} = \frac{1}{2}\left[\sum_{z|d(z,\,y)\,<\,B} p^{2d(z,y)} + \left(2\sum_{\substack{z_1\,<\,z_2|\ d(z_1,\,y)\,<\,B \\ d(z_2,\,y)\,<\,B}} p^{d(z_1,y)+d(z_2,y)}\right)\right]$$

$$= \frac{1}{2}\sum_{z|d(z,\,y)\,<\,B} p^{2d(z,y)} + \sum_{\substack{u\,>\,z|\ d(z,\,y)\,<\,B \\ d(u,\,y)\,\le\,B}} p^{d(z,y)+d(u,y)} - \sum_{\substack{u\,>\,z|\ d(z,\,y)\,<\,B \\ d(u,\,y)\,=\,B}} p^{d(z,y)+B}$$

Since the middle term is exactly the right hand side of inequality 1, it remains only to show that the difference of the first and last terms is positive. We have, if $d(z,y) < B$, then $2d(z,y) \le d(z,y) + B - 1$. Therefore

$$\frac{1}{2}\sum_{z|d(z,\,y)\,<\,B} p^{2d(z,y)} \ge \frac{1}{2}\sum_{z|d(z,\,y)\,<\,B} p^{d(z,y)+B-1}$$

$$= \frac{1}{2p}\sum_{z|d(z,\,y)\,<\,B} p^{d(z,y)+B}$$

Since by assumption $p \le 1/2$, and since conditioning a sum further can only decrease the number of terms, we get

$$\frac{1}{2}\sum_{z|d(z,\,y)\,<\,B} p^{2d(z,y)} \ge \sum_{\substack{u|d(u,\,y)\,=\,B, \\ z|d(z,\,y)\,<\,B}} p^{d(z,y)+B}$$

$\square$

To get the desired high value for $E[B_y(R)]$, Lemma 1 says to choose x so that the second (negative) term does not "swamp out" the first. Setting $x = 1$ corresponds to the Linial-Saks algorithm, and notice that this is a bad setting, since the second term will dominate. For each node y, we want to set $x = \theta(1/T_y)$. We cannot just set x once for each phase, since there can be many different T_y values in the same graph. The smallest x required (for the most dense regions) will be $x = \max(T_y) = \Delta$. Then x is doubled at each iteration, until $x = 1$ at the final iteration of a phase.

Define the set D_i at the ith iteration of a phase as follows:

$$D_i = \{y | \Delta/2^{i+1} \le T_y \le \Delta/2^i \wedge (y \notin D_h \text{ for all } h < i)\}$$

It is clear from the definition that each y can be in at most one set D_i. In the proof of Lemma 3 we will argue that every y falls into D_i for some iteration i as well. The sets D_i, dynamically constructed over the iterations of a phase, therefore form a partition of the nodes of the graph.

Given a sample point $R = <r_1, \ldots, r_n>$, define the benefit of the ith iteration of a phase as:

$$B_I(R) = \sum_{D_i} B_y(R). \tag{2}$$

Recall that $\Delta = \max_{\forall y \in G} T_y$, and at the ith iteration of a phase, $x = 2^i/(5\Delta)$. In the analysis that follows, we show that we expect to color a constant fraction of the nodes which have $y \in D_i$ in the ith iteration.

Lemma 2. *In the ith iteration, we expect to color $2p/25$ of those $y \in D_i$.*

Proof

$$E[\text{ \# of } y \in S | D_i] = \sum_{y \in D_i} Pr[y \in S]$$

$$\geq E[B_I(R)]$$

$$\geq \sum_{y \in D_i} p \frac{2^i}{5\Delta} T_y - p \left(\frac{2^{2i}}{25\Delta^2} \right) \frac{T_y^2}{2}$$

by Lemma 1. Since we want a lower bound, we substitute $T_y \leq \frac{\Delta}{2^i}$ in the positive term and $T_y \geq \frac{\Delta}{2^{i+1}}$ in the negative term, giving

$$E[\text{ \# of } y \in S \,|\, y \in D_i] \geq \sum_{y \in D_i} p \frac{1}{10} - p \frac{1}{50}$$

$$= \frac{2}{25} p |D_i|$$

$\square$

The next lemma gives us that the expected number of phases is $O((\log n)/(\log(2p/25))) = O(\log n)$.

Lemma 3. *Suppose $V' \subseteq V$ is the set of nodes present in the graph at the beginning of a phase. After $\log(5\Delta)$ iterations of a phase, the expected number of nodes colored is $(2p/25)|V'|$.*

Proof Since for all y, $T_y \geq 1$, over all iterations, and since $x \to 1$, then there must exist an iteration where $xT_y \geq 1/10$. Since T_y cannot increase (it can only decrease if we color and remove nodes in previous iterations), and $xT_y \leq 1/5$ in the first iteration for all y, we know that for each y there exists an iteration in which $1/5 \geq xT_y \geq 1/10$. If i is the first such iteration for a given vertex y, then by definition, $y \in D_i$, and the sets D_i form a partition of all the vertices in the graph. By Lemma 2, we expect to color $2p/25$ of the vertices in D_i at every iteration i, and every vertex is in exactly one set D_i, so we expect to color a $(2p/25)$ fraction overall. $\square$

It remains to set B so that we can determine the maximum weak diameter of our clusters, which also influences the running time of the algorithm. By Lemma 3, we have that the probability of a node being colored in a phase is $2p/25$. Thus, the probability that there is some node which has not been assigned a color in the first l phases is at most $n(1 - (2p/25))^l$. By selecting B to be $\frac{25 \log n + \omega(1)}{2p}$, it is easily verified that this quantity is $o(1)$.

Theorem 4. *There is a pairwise independent RNC algorithm which given a graph $G = (V, E)$, finds a $(\log^2 n, \log n)$-decomposition in $O(\log^3 n)$ time, using a linear number of processors.*

2.5 The Pairwise Independent Distribution

We have shown that we expect our RNC algorithm to color the entire graph with $O(\log^2 n)$ colors, and the analysis depends on pairwise independence. We now show how to construct a pairwise independent sample space which obeys the truncated geometric distribution. We construct a sample space in which the r_i are pairwise independent and where for $i = 1, \ldots, n$:

$$Pr[r_i = NIL] = 1 - x$$
$$Pr[r_i = j] \quad = xp^j(1-p) \text{ for } 0 \le j \le B - 1$$
$$Pr[r_i = B] \quad = xp^B$$

Without loss of generality, let p and x be powers of 2. Let $r = B \log(1/p) + \log(1/x)$. Note that since $B = O(\log n)$, we have that $r = O(\log n)$. In order to construct the sample space, we choose $W \in Z_2^l$, where $l = r(\log n + 1)$, uniformly at random. Let $W = <\omega^{(1)}, \omega^{(2)}, \ldots, \omega^{(r)}>$, each of $(\log n + 1)$ bits long, and we define $\omega_j^{(i)}$ to be the jth bit of $\omega^{(i)}$.

For $i = 1, \ldots, n$, define random variable $Y_i \in Z_2^r$ such that its kth bit is set as

$$Y_{i,k} = <bin(i), 1> \cdot \omega^{(k)},$$

where $bin(i)$ is the $(\log n)$-bit binary expansion of i.

We now use the Y_i's to set the r_i so that they have the desired property. Let t be the most significant bit position in which Y_i contains a 0. Set

$$r_i = NIL \text{ if } t \in [1, .., \log(1/x)]$$
$$= j \quad \text{ if } t \in (\log(1/x) + j \log(1/p), .., \log(1/x) + (j+1)\log(1/p)], \text{ for } j \ne B-1$$
$$= B \quad \text{ otherwise.}$$

It should be clear that the values of the r_i's have the right probability distribution; however, we do need to argue that the r_i's are pairwise independent. It is easy to see [10, 13] that, for all k, the kth bits of all the Y_i's are pairwise independent if $\omega^{(k)}$ is generated randomly; and thus the Y_i's are pairwise independent. As a consequence, the r_i's are pairwise independent as well.

2.6 The NC Algorithm

We want to search the sample space given in the previous section to remove the randomness from the pairwise independent RNC algorithm; i.e. to find a setting of the r_y's in the ith iteration of a phase for which the benefit, $B_I(R)$, is at least as large as the expected benefit, $E[B_I(R)]$.

Since the sample space is generated from r $(\log n)$-bit strings, it thus is of size $2^{r \log n} \le O(n^{\log n})$, which is clearly too large to search exhaustively. We could however devise a quadratic size sample space which would give us pairwise independent r_y's with the right property (see [10, 12, 2]). Unfortunately, this approach would require $O(n^5)$ processors: the benefit function must be evaluated on $O(n^2)$ different processors simultaneously.

Alternatively, we will use a variant of a method of Luby [13] to binary search a pairwise independent distribution for a good sample point. We can in fact naively apply this method because our benefit function is a sum of terms depending on one or two variables each; i.e.

$$B_{\mathcal{I}}(R) \;=\; \sum_{y \in D_i} B_y(R) \;=\; \sum_{y \in D_i} \left(\sum_{z \mid d(z,y) \,<\, B} Z_{y,z} \;-\; \sum_{\substack{u \,>\, z \mid \, d(z,y)\, \le\, B \\ d(u,y)\, \le\, B}} Z_{y,z} X_{y,u} \right) \qquad (3)$$

where recall $D_i = \{y \mid \Delta/2^{i+1} \le T_y \le \Delta/2^i \wedge (y \notin D_h \text{ for all } h < i)\}$. The binary search is over the bits of W (see Section 2.5): at the qt-th step of the binary search, $\omega_i^{(q)}$ is set to 0 if $E[B_{\mathcal{I}}(R) \mid \omega_1^{(1)} = b_{11}, \omega_2^{(1)} = b_{12}, \ldots, \omega_i^{(q)} = b_{qt}]$, with $b_{qt} = 0$ is greater than with $b_{qt} = 1$; and 1 otherwise. The naive approach would yield an $O(n^3)$ processor NC algorithm, since we require one processor for each term of the benefit function, expanded as a sum of functions depending on one or two variables each.

The reason the benefit function has too many terms is that it includes sums over pairs of random variables. Luby gets around this problem by computing conditional expectations on terms of the form $\sum_{i,j \in S} X_i X_j$ directly, using $O(|S|)$ processors. We are able to put our benefit function into a form where we can apply a similar trick. (In our case, we will also have to deal with a "weighted" version, but Luby's trick easily extends to this case.)

The crucial observation is that, by definition of $Z_{y,z}$ and $X_{y,z}$, we can equivalently write $E[Z_{y,z} X_{y,u}]$ as $pE[X_{y,z} X_{y,u}]$; thus, we can lower bound the expected performance of the algorithm within at least a multiplicative factor of p of its performance in Lemmas 2 and 3, if we upper bound the latter expectation.

It will be essential throughout the discussion below to be familiar with the notation used for the distribution in Section 2.5. Notice that our indicator variables have the following meaning:

$$X_{y,z} \equiv Y_{z,k} = 1 \quad \text{for all } k, \; 1 \le k \le d(z,y)\log(1/p)$$
$$Z_{y,z} \equiv Z_{z,k} = 1 \quad \text{for all } k, \; 1 \le k \le (d(z,y)+1)\log(1/p)$$

If we fix the outer summation of the expected benefit at some y, then the problem now remaining is to show how to compute

$$E[\sum_{(z,u) \in S} X_{y,z} X_{y,u} \mid \omega_1^{(1)} = b_{11}, \omega_2^{(1)} = b_{12}, \ldots, \omega_i^{(q)} = b_{qt}], \qquad (4)$$

in $O(\log n)$ time using $O(|S|)$ processors. For notational convenience, we write (z,u) for $z \ne u$. Below, we assume all expectations are conditioned on $\omega_1^{(1)} = b_{11}, \ldots, \omega_i^{(q)} = b_{qt}$.

Note that we only need be interested in the case where both random variables $X_{y,z}$ and $X_{y,u}$ are undetermined. If $q > d(i,y)\log(1/p)$, then $X_{y,i}$ is determined. So we assume $q \le d(i,y)\log(1/p)$ for $i = z, u$. Also, note that we know the exact value of the first $q - 1$ bits of each Y_z. Thus, we need only consider those indices $z \in S$ in Equation 4 with $Y_{z,j} = 1$ for all $j \le q - 1$; otherwise, the terms zero out. Let $S' \subseteq S$ be this set of indices.

In addition, the remaining bits of each Y_z are independently set. Consequently,

$$E[\sum_{(z,u)\in S'} X_{y,z}X_{y,u}] = E[\sum_{(z,u)\in S'} \gamma(z,y)\gamma(u,y)Y_{z,q}Y_{u,q}]$$

$$= E[(\sum_{z\in S'} \gamma(z,y)Y_{z,q})^2 - \sum_{z\in S'} \gamma(z,y)^2 Y_{z,q}^2],$$

where $\gamma(z,y) = 1/2^{d(z,y)\log(1/p)-q}$

Observe that we have set t bits of $\omega^{(q)}$. If $t = \log n+1$, then we know all the $Y_{z,q}$'s, and we can directly compute the last expectation in the equation above. Otherwise, we partition S' into sets $S_\alpha = \{z \in S' \mid z_{t+1}\cdots z_{\log n} = \alpha\}$. We further partition each S_α into $S_{\alpha,0} = \{z \in S_\alpha \mid \sum_{i=1}^{t} z_i\omega_i^{(q)} = 0 \pmod 2\}$ and $S_{\alpha,1} = S_\alpha - S_{\alpha,0}$. Note that given $\omega_1^{(1)} = b_{11}, \ldots, \omega_t^{(q)} = b_{qt}$,

1. $Pr[Y_{z,q} = 0] = Pr[Y_{z,q} = 1] = 1/2$,
2. if $z \in S_{\alpha,j}$, and $u \in S_{\alpha,j'}$, then $Y_{z,q} = Y_{u,q}$ iff $j = j'$, and
3. if $z \in S_\alpha$ and $z' \in S_{\alpha'}$, where $\alpha \neq \alpha'$, then $Pr[Y_{z,q} = Y_{u,q}] = Pr[Y_{z,q} \neq Y_{u,q}] = 1/2$.

Therefore, conditioned on $\omega_1^{(1)} = b_{11}, \ldots, \omega_t^{(q)} = b_{qt}$,

$$E[\sum_{(z,u)\in S'} X_{y,z}X_{y,u}]$$

$$= E[\sum_{(z,u)\in S'} \gamma(z,y)\gamma(u,y)Y_{z,q}Y_{u,q}]$$

$$= E[\sum_\alpha \sum_{(z,u)\in S_\alpha} \gamma(z,y)\gamma(u,y)Y_{z,q}Y_{u,q} + \sum_{(\alpha,\alpha')}\sum_{z\in S_\alpha}\sum_{u\in S_{\alpha'}} \gamma(z,y)\gamma(u,y)Y_{z,q}Y_{u,q}]$$

$$= \sum_\alpha E[\sum_{(z,u)\in S_{\alpha,0}} \gamma(z,y)\gamma(u,y)Y_{z,q}Y_{u,q} + \sum_{(z,u)\in S_{\alpha,1}} \gamma(z,y)\gamma(u,y)Y_{z,q}Y_{u,q}$$

$$+ 2\sum_{z\in S_{\alpha,0}}\sum_{u\in S_{\alpha,1}} \gamma(z,y)\gamma(u,y)Y_{z,q}Y_{u,q}] + \sum_{(\alpha,\alpha')} E[\sum_{z\in S_\alpha}\sum_{u\in S_{\alpha'}} \gamma(z,y)\gamma(u,y)Y_{z,q}Y_{u,q}]$$

$$= \sum_\alpha \left[\frac{1}{2}\sum_{(z,u)\in S_{\alpha,0}} \gamma(z,y)\gamma(u,y) + \frac{1}{2}\sum_{(z,u)\in S_{\alpha,1}} \gamma(z,y)\gamma(u,y) + 0 \right]$$

$$+ \sum_{(\alpha,\alpha')} \frac{1}{4}\left(\sum_{z\in S_\alpha} \gamma(z,y)\right)\left(\sum_{u\in S_{\alpha'}} \gamma(u,y)\right)$$

$$= \frac{1}{2}\sum_\alpha \left[\left(\sum_{z\in S_{\alpha,0}} \gamma(z,y)\right)^2 - \sum_{z\in S_{\alpha,0}} \gamma(z,y)^2 + \left(\sum_{z\in S_{\alpha,1}} \gamma(z,y)\right)^2 - \sum_{z\in S_{\alpha,1}} \gamma(z,y)^2 \right]$$

$$+ \frac{1}{4}\left[\left(\sum_\alpha \sum_{z\in S_\alpha} \gamma(z,y)\right)^2 - \sum_\alpha \left(\sum_{z\in S_\alpha} \gamma(z,y)\right)^2 \right]$$

Since every node $z \in S'$ is in precisely four sums, we can compute this using $O(|S|)$ processors.

In the above analysis, we fixed the outer sum of the expected benefit at some y. To compute the benefit at iteration i, we need to sum the benefits of all $y \in D_i$. However, we argued in the proof of Lemma 3 that the sets D_i form a partition of the vertices. Therefore we consider each y exactly once over all iterations of a phase, and so our algorithm needs only $O(n^2)$ processors, and we obtain the following theorem.

Theorem 5. *There is an NC algorithm which given a graph $G = (V, E)$, finds a $(\log^2 n, \log n)$-decomposition in $O(\log^5 n)$ time, using $O(n^2)$ processors.*

Acknowledgments

Thanks to John Rompel for helpful discussions.

References

1. Y. Afek and M. Ricklin. Sparser: A paradigm for running distributed algorithms. *J. of Algorithms*, 1991. Accepted for publication.
2. N. Alon, L. Babai, and A. Itai. A fast and simple randomized parallel algorithm for the maximal independent set problem. *J. of Algorithms*, 7:567–583, 1986.
3. B. Awerbuch. Complexity of network synchronization. *J. of the ACM*, 32(4):804–823, Oct. 1985.
4. B. Awerbuch, B. Berger, L. Cowen, and D. Peleg. Fast sequential constructions of sparse neighborhood covers. Unpublished manuscript, Nov. 1991.
5. B. Awerbuch, B. Berger, L. Cowen, and D. Peleg. Fast distributed network decomposition. In *Proc. 11th ACM Symp. on Principles of Distributed Computing*, Aug. 1992.
6. B. Awerbuch, A. Goldberg, M. Luby, and S. Plotkin. Network decomposition and locality in distributed computation. In *Proc. 30th IEEE Symp. on Foundations of Computer Science*, May 1989.
7. B. Awerbuch and D. Peleg. Network synchronization with polylogarithmic overhead. In *Proc. 31st IEEE Symp. on Foundations of Computer Science*, pages 514–522, 1990.
8. B. Awerbuch and D. Peleg. Sparse partitions. In *Proc. 31st IEEE Symp. on Foundations of Computer Science*, pages 503–513, 1990.
9. Y. Bartal, A. Fiat, and Y. Rabani. Competitive algorithms for distributed data management. In *Proc. 24th ACM Symp. on Theory of Computing*, 1992. to appear.
10. R. M. Karp and A. Wigderson. A fast parallel algorithm for the maximal independent set problem. *J. of the ACM*, 32(4):762–773, Oct. 1985.
11. N. Linial and M. Saks. Decomposing graphs into regions of small diameter. In *Proc. 2nd ACM-SIAM Symp. on Discrete Algorithms*, pages 320–330. ACM/SIAM, 1991.
12. M. Luby. A simple parallel algorithm for the maximal independent set problem. *SIAM J. on Comput.*, 15(4):1036–1053, Nov. 1986.
13. M. Luby. Removing randomness in parallel computation without a processor penalty. In *Proc. 29th IEEE Symp. on Foundations of Computer Science*, pages 162–173. IEEE, Oct. 1988.
14. A. Pasconesi and A. Srinivasan. Improved algorithms for network decompositions. In *Proc. 24th ACM Symp. on Theory of Computing*, 1992. to appear.
15. S. Rao. Finding small edge cuts in planar graphs. In *Proc. 24th ACM Symp. on Theory of Computing*, 1992. to appear.

Parallel Algorithm for Cograph Recognition with Applications

Xin He [1]

Department of Computer Science
State University of New York at Buffalo
Buffalo, NY 14260

Abstract. We present a parallel algorithm for recognizing cographs and constructing their cotrees. The algorithm takes $O(\log^2 n)$ time with $O(n + m)$ processors on a CRCW PRAM, where n and m are the number of vertices and edges of the graph. Using cotree representation, we obtain a parallel algorithm for the permutation representation problem for cographs using $O(\log n)$ time with $O(n)$ processors. We also present a parallel algorithm for the depth-first spanning tree problem for permutation graphs (a class properly contains cographs) which takes $O(\log^2 n)$ time with $O(n)$ processors.

1. Introduction

The *complement reducible graphs* (also called *cographs*) are defined as follows: (1) A graph consisting of a single vertex is a cograph; (2) If $G_1, G_2, \ldots, G_k$ are cographs, so is their union $G_1 \cup G_2 \cup \ldots \cup G_k$; (3) If G is a cograph, so is its complement.

Cographs have arisen in disparate areas of Computer Science and have been studied by various researchers [1, 2, 3, 4, 5, 7, 8, 9, 11, 12, 16]. The names synonymous with cographs include: D^*-graphs [11], P_4 restricted graphs [7, 8], Hereditary Dacey Graphs [16] and so on. The most important property of cographs for algorithm design is that G is a cograph iff G has a *cotree* representation (to be defined in section 2) [7]. Most algorithms for solving various cograph problems require the cotree as input. So the following problem is central to cograph algorithms: Determine if a given graph G is a cograph or not, and if so construct its cotree.

A linear time sequential algorithm for recognizing cographs was given in [8]. A parallel algorithm for this problem was presented in [12] which takes $O(\log^2 n)$ time with $O(n^3/\log^2 n)$ processors. Another parallel cograph recognition algorithm was developed in [4] which requires $O(\log^2 n)$ time with $O(nm)$ processors. In [14], a parallel algorithm for the modular decomposition problem was developed using $O(\log n)$ time with $O(n^3)$ processors. This implies a cotree construction algorithm with the same time and processor bounds. A cotree construction algorithm was developed in [13] which uses $O(\log n)$ time with $O(\frac{n^2 + mn}{\log n})$ processors. Using cotree representation, NC algorithms for many cograph problems were presented in [1, 2, 3, 4, 5, 12]: maximum clique, minimum coloring, minimum domination, minimum fill-in and isomorphism testing (using $O(\log n)$ time with a linear number of processors); Hamiltonian path, Hamiltonian cycle, minimum path covering, and maximum matching (using $O(\log^2 n)$ time with $O(n^2)$ processors).

We present a new algorithm for recognizing cographs and constructing cotrees. Our algorithm takes $O(\log^2 n)$ time with $O(n+m)$ processors, a substantial improvement over the previously known algorithms. Since cotree construction dominates the

---

[1] Research supported by NSF grant CCR-9011214.

complexity of the algorithms in [1, 2, 3, 4, 5, 12] for solving various cograph problems, the efficiency of these algorithms is uniformly improved. We also present parallel algorithms for solving two other problems. The first one is the permutation representation problem of cographs. Cographs are a proper subclass of *permutation graphs* [9]. The permutation representation of permutation graphs can be found in $O(\log^2 n)$ time with $O(n^4)$ processors [10]. We show that if the cotree is given, the permutation representation of cographs can be constructed in $O(\log n)$ time with $O(n)$ processors. The second problem is the depth-first spanning tree (DFST) problem. We show that the DFST problem for permutation graphs (hence cographs) can be solved in $O(\log^2 n)$ time with $O(n)$ processors.

The parallel computation model we use is CRCW PRAM. The model consists of a number of processors and a common memory. In each time unit, a processor can read a memory cell, perform an arithmetic or logic operation, and write into a memory cell. Both concurrent read and concurrent write are allowed. If a write conflict occurs, the processor with the lowest processor number succeeds.

This paper is organized as follows. Section 2 introduces the definitions and properties of cographs and cotrees. In sections 3 and 4, several lemmas needed by our algorithm are proved. The recognition algorithm is presented in section 5. The permutation representation algorithm is discussed in section 6. The depth first spanning tree algorithm for permutation graphs is presented in section 7. Most proofs are omitted due to space limitation and will be given in the full paper.

2. Properties of Cographs and Cotrees

Let $G = (V, E)$ be a graph. For $V_1 \subset V$, $G - V_1$ denotes the subgraph obtained from G by deleting the vertices in V_1. For each $v \in V$, define $N(v) = \{u \in V | (u, v) \in E\}$ and $\Gamma(v) = N(v) \cup \{v\}$. If $N(u) = N(v)$, u and v are called a *false twin* and denoted by $u \sim_f v$. If $\Gamma(u) = \Gamma(v)$, u and v are called a *true twin* and denoted by $u \sim_t v$. Note that $u \sim_f v$ implies $(u, v) \notin E$ and $u \sim_t v$ implies $(u, v) \in E$. Clearly both $\sim_f$ and $\sim_t$ are equivalence relations. The equivalence classes of V under the relation $\sim_t$ ($\sim_f$, resp.) are called $\sim_t$ ($\sim_f$, resp.) classes.

Let T be a rooted tree. An internal node is a *tip node* iff all of its children are leaf nodes. $L(T)$ denotes the set of leaf nodes of T. For each node x, $p_T(x)$ denotes the parent of x in T. $T(x)$ denotes the subtree of T rooted at x. *Leaf(x)* denotes the set of the leaf *children* of x. The first common node of the paths from x and y to r is called the *lowest common ancestor* of x and y and denoted by $lca_T(x, y)$.

Definition 1: A *cotree* is a rooted tree T such that: (a) The internal nodes are labeled 0 or 1; the root of T is a 1-node; the children of a 1-node are 0-nodes; the children of a 0-node are 1-nodes. (b) Each internal node has at least two children, except that the root r might have only one child.

A cograph G can be represented by a rooted tree T as follows. The leaf nodes of T represent the vertices of G. Each internal node of T represents either a "union" or a "complement" operator. A "union" node has at least two children. A "complement" node has exactly one child which is a "union" node. We can modify T to be a cotree as follows: Merge each "union" node and its parent (which is a "complement" node) into a single node. Then label the nodes so that Definition 1 (a) holds. The most important property of this representation is as follows [7]:

Property of Cotree: Let $G = (V, E)$ be a cograph and T be the cotree con-

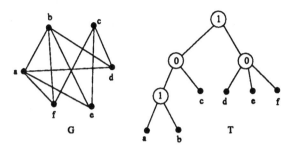

Figure 1: A graph G and its cotree T.

structed as above. For any $x, y \in V$, $(x, y) \in E$ iff $lca_T(x, y)$ is a 1-node.

Fig 1 shows a cograph G and its cotree T. The following theorem is well-known.

Theorem 1 [7, 8]: G is a cograph iff G has a cotree representation T. Moreover, the cotree representation is unique up to the permutation of the children of the internal nodes of T. $\square$

From the property of cotree, it's not hard to prove the following:

Lemma 2 [12]: Let G be a cograph with cotree T.

(1) $u \sim_t v$ ($u \sim_f v$, respectively) iff $p_T(u) = p_T(v)$ is a 1-node (0-node, resp.)

(2) For each 1-node α, $Leaf(\alpha)$ is a $\sim_t$ class. For each 0-node β and each $x \in Leaf(\beta)$, $\{x\}$ is a $\sim_t$ class.

(3) For each 0-node β, $Leaf(\beta)$ is a $\sim_f$ class. For each 1-node α and each $x \in Leaf(\alpha)$, $\{x\}$ is a $\sim_f$ class. $\square$

For example, the $\sim_f$ classes of the graph G shown in Fig 1 are: $\{d, e, f\}$, $\{c\}$, $\{a\}$ and $\{b\}$. The $\sim_t$ classes of G are: $\{a, b\}$, $\{c\}$, $\{d\}$, $\{e\}$ and $\{f\}$.

3. Contracting Cotree

The basic idea of our cograph recognition algorithm is as follows. Given a graph G, the algorithm deletes some vertices from G to form a subgraph G' with the following properties: (a) $|G'| \leq c|G|$ for some constant $c < 1$. (b) G is a cograph iff G' is a cograph. (c) If G' is a cograph with cotree T', the cotree T of G can be constructed from T' easily. The algorithm is recursively applied to G'. If G' is not a cograph, G is rejected. If G' is a cograph, the recursive call returns the cotree T' of G'. Then the cotree T of G is constructed from T'.

The problem is how to construct G'. Our method is developed as follows. Assume G is a cograph with cotree T. We define two operations *Trim-tree* and *Cut-tree*, which can be applied to T to produce a cotree T' such that $|T'| \leq c|T|$. We then define two operations *Trim-graph* and *Cut-graph* that, when applied to G, produce a cograph G' corresponding to T'. We show Trim-graph and Cut-graph can be applied to *any* G to produce a graph G' satisfying the three requirements listed above. In this section we discuss the Trim-tree and Cut-tree operations.

Definition 2: Let T be a cotree. Let α be an internal node and $Q = Leaf(\alpha)$ with $|Q| > 1$. Let $q \in Q$. $Trim(T, q, Q)$ is the cotree obtained from T as follows: If α is a tip node then delete all nodes in Q and relabel α by q. If α is a non-tip node then delete all nodes in $Q - \{q\}$.

Definition 3: Let T be a cotree. *Trim-tree(T)* is the cotree obtained from T

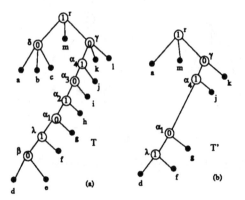

Figure 2: Trim-tree and Cut-tree operations.

as follows. For each internal node α with $|Leaf(\alpha)| > 1$, select $q \in Leaf(\alpha)$ and perform $Trim(T, q, Leaf(\alpha))$.

An internal node α of T is *special* if α has exactly one leaf child and one non-leaf child. A *trail* of T is a sequence $\{\alpha_1, \alpha_2, \dots, \alpha_l\}$ of special nodes such that $\alpha_i = p_T(\alpha_{i-1})$ for $1 < i \leq l$. A *chain* of T is a *maximal* trail $C = \{\alpha_1, \alpha_2, \dots, \alpha_l\}$ such that α_1 is a 0-node and α_l is a 1-node. For each i, let u_i be the leaf child of α_i. We say u_i *belongs to* C. The sequence $L = \{u_1, \dots, u_l\}$ is called a *line* of T.

Definition 4: Let T be a cotree and $L = \{u_1, \dots, u_l\}$ be a line of T with $l \geq 4$. $Cut(T, L)$ is the cotree obtained from T as follows: Delete α_k, u_k for $2 \leq k \leq l - 1$ and make α_1 a child of α_l.

Definition 5: *Cut-tree(T)* is the cotree obtained from T as follows. Let $L_1 \dots L_r$ be all lines of T of length ≥ 4. Perform $Cut(T, L_i)$ for $1 \leq i \leq r$.

Fig 2b shows the cotree T' obtained by performing the Trim-tree and Cut-tree operations on the cotree T shown in Fig 2a. In this figure, $\{\alpha_1, \alpha_2, \alpha_3, \alpha_4\}$ is a chain.

Lemma 3: Let T' be the cotree obtained by performing the Trim-tree and Cut-tree operations on a cotree T. Then $|L(T')| \leq \frac{17}{18}|L(T)|$. $\square$

We next define two operations $UnTrim$ and $UnCut$ which are the "inverses" of the $Trim$ and Cut operations.

Definition 6: Let T be a cotree, q be a leaf node and $\alpha = p_T(q)$. Let Q be a set such that $Q \cap L(T) = \{q\}$. $UnTrim_t(T, q, Q)$ ($UnTrim_f(T, q, Q)$, resp.) is the cotree obtained from T as follows: If α is a 1-node (0-node, resp.) then for each $x \in Q - \{q\}$ add a new leaf child x to α. If α is a 0-node (1-node, resp.) then delete q; add a new 1-node (0-node, resp.) β, make β a child of α and for each $x \in Q$ add a new leaf child x to β.

Definition 7: Let T be a cotree and u_1, u_l be two leaf nodes of T such that $\alpha_1 = p_T(u_1)$ is a 0-node, $\alpha_l = p_T(u_l)$ is a 1-node and $\alpha_l = p_T(\alpha_1)$. Let $L = \{u_1, u_2, \dots, u_l\}$ be a set such that l is even and $L \cap L(T) = \{u_1, u_l\}$. $UnCut(T, L)$ is the cotree obtained from T as follows. Add a sequence of new nodes $\alpha_2, \dots, \alpha_{l-1}$ between α_1 and α_l. Label α_k's ($2 \leq k \leq l - 1$) so that the 0-nodes and 1-nodes alternate on the path $\{\alpha_1 \dots \alpha_l\}$. For each $2 \leq k \leq l - 1$, add a leaf child u_i to α_i.

4. Contracting Graphs

In this section we define the operations on a graph G that correspond to the effects of the Trim-tree and Cut-tree operations on T. By the property of cotree, it's easy to prove the following:

Lemma 4: Let $G = (V, E)$ be a graph. Let Q be a $\sim_t$ ($\sim_f$, resp.) class of V with $|Q| > 1$. Let $q \in Q$ and $G' = G - (Q - \{q\})$. Then G is a cograph with cotree T iff G' is a cograph with cotree T', where $T' = Trim(T, q, Q)$ and $T = UnTrim_t(T', q, Q)$ ($T = UnTrim_f(T', q, Q)$, resp.) $\square$

Let $G = (V, E)$ be a graph. Consider the equivalence classes of V under the relations $\sim_t$ and $\sim_f$. Let $Q_1, \ldots, Q_s$ be the $\sim_t$ classes containing more than one vertex. Let $R_1, \ldots, R_t$ be the $\sim_f$ classes containing more than one vertex. The collection $\Psi(G) = \{Q_1, \ldots, Q_s, R_1, \ldots, R_t\}$ is called the *class partition* of G. If $Q_i \cap R_j = \emptyset$ for $1 \le i \le s$ and $1 \le j \le t$, we say $\Psi(G)$ is *valid*.

Definition 8: Let $\Psi(G) = \{Q_1, \ldots, Q_s, R_1, \ldots, R_t\}$ be the class partition of G. If $\Psi(G)$ is valid, then *Trim-graph(G)* is the graph obtained from G as follows: For each $Q_i \in \Psi(G)$, select $q_i \in Q_i$ and delete the vertices in $Q_i - \{q_i\}$. For each $R_j \in \Psi(G)$, select $r_j \in R_j$ and delete the vertices in $R_j - \{r_j\}$.

Lemma 5: (1) If G is a cograph then the class partition $\Psi(G)$ is valid. (2) If $\Psi(G)$ is valid, then G is a cograph iff $G' = Trim\text{-}graph(G)$ is a cograph.

Proof: (1) Let T be the cotree of G. By Lemma 2, $\Psi(G) = \{Leaf(\alpha) \mid \alpha$ is an internal node of T and $|Leaf(\alpha)| > 1\}$, which is clearly valid.

(2) Suppose G is a cograph with cotree T. The elements of $\Psi(G)$ are disjoint $\sim_t$ or $\sim_f$ classes. By repeated applications of Lemma 4, $T' = Trim\text{-}tree(T)$ is a cotree of G'. Hence G' is a cograph. Suppose G' is a cograph with cotree T'. Let T be the cotree obtained from T' by performing $UnTrim_t(T', q_i, Q_i)$ for each $\sim_t$ class $Q_i \in \Psi(G)$ and $UnTrim_f(T', r_j, R_j)$ for each $\sim_f$ class $R_j \in \Psi(G)$. By repeated applications of Lemma 4, it can be shown T is a cotree of G. So G is a cograph. $\square$

We next investigate the operation on a graph that corresponds to the Cut operation. Let G be a cograph with cotree T. Let $\{\alpha_1, \alpha_2, \alpha_3, \alpha_4\}$ be a chain of T and u_i be the leaf child of α_i ($1 \le i \le 4$). The following definition is motivated by the relationships of the neighborhoods of u_i's in G.

Definition 9: Let $G = (V, E)$ be a graph. Let u_1, u_2, u_3, u_4 be four vertices.

(1) $\{u_1, u_2, u_3\}$ is a *false trio* iff: (1a) $N(u_3) \subset N(u_1)$ and $N(u_1) - N(u_3) = \{u_2\}$; and (1b) both $\{u_1\}$ and $\{u_3\}$ are $\sim_t$ class and $\sim_f$ class.

(2) $\{u_2, u_3, u_4\}$ is a *true trio* iff: (2a) $\Gamma(u_2) \subset \Gamma(u_4)$ and $\Gamma(u_4) - \Gamma(u_2) = \{u_3\}$; and (2b) both $\{u_2\}$ and $\{u_4\}$ are $\sim_t$ class and $\sim_f$ class.

(3) $\{u_1, u_2, u_3, u_4\}$ is a *4-twin* of G iff $\{u_1, u_2, u_3\}$ is a false trio and $\{u_2, u_3, u_4\}$ is a true trio.

Lemma 6: Let G be a cograph with cotree T. Let $u_1, u_2, u_3, u_4 \in V$ and $\alpha_i = p_T(u_i)$ ($1 \le i \le 4$).

(1) $\{u_1, u_2, u_3\}$ is a false trio iff they are positioned in T as shown in Fig 3a. $\{u_2, u_3, u_4\}$ is a true trio iff they are positioned in T as shown in Fig 3b.

(2) $\{u_1, u_2, u_3, u_4\}$ is a 4-twin iff: (a) α_1 and α_3 are 0-nodes, α_2 and α_4 are 1-nodes; (b) $\alpha_{i+1} = p_T(\alpha_i)$ for $1 \le i < 4$; (c) For $1 \le i \le 4$, u_i is the only leaf child of α_i (Fig 3c).

(3) For any $u_2, u_3 \in V$, u_2 and u_3 belongs to at most one true trio $\{u_2, u_3, u_4\}$.

(4) If $\{u_1, u_2, u_3, u_4\}$ is a 4-twin, then $N(u_1) \subset N(u_2)$ and $N(u_3) \subset N(u_2)$.

Proof: (1) was proved in [12]. (2) and (3) immediately follow from (1). (4) immediately follows from (2) and the property of cotree. $\square$

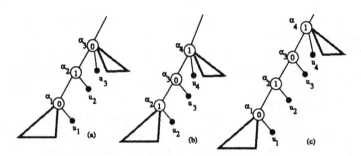

Figure 3: (a) false trio, (b) true trio, (c) 4-twin.

A set $L = \{u_1, \ldots, u_l\} \subset V$, ($l$ is even and ≥ 4), is called a *contractible sequence* iff $\{u_i, u_{i+1}, u_{i+2}, u_{i+3}\}$ is a 4-twin for each odd $i \leq l - 3$. We can prove:

Lemma 7: Let $G = (V, E)$ be a graph and $L = \{u_1, \ldots, u_l\} \subset V$. Let $G_1 = G - (L - \{u_1, u_l\})$. Suppose L is a contractible sequence of G. Then G is a cograph with cotree T iff G_1 is a cograph with cotree T_1 such that $\alpha_1 = p_{T_1}(u_1)$ is a 0-node, $\alpha_l = p_{T_1}(u_l)$ is a 1-node, and $\alpha_l = p_{T_1}(\alpha_1)$, where T and T_1 are related by $T_1 = Cut(T, L)$ and $T = UnCut(T_1, L)$. $\square$

Let F be a directed forest. A vertex v of F is a *branch vertex* if v has at least two incoming arcs. A *walk* of F is a path P such that: (1) the internal vertices of P are non-branch vertices; (2) the starting vertex of P is either a branch vertex or a leaf; and (3) the end vertex of P is either a branch vertex or a root. The *walk decomposition* of F, denoted by $WD(F)$, is the set of all walks of F.

In order to define the Cut-graph operation on a graph $G = (V, E)$, we first construct a directed graph $H(G)$ as follows. The vertex set of $H(G)$ is V. For each 4-twin $\{u_1, u_2, u_3, u_4\}$ of G, there are three arcs $u_1 \to u_2$, $u_2 \to u_3$, $u_3 \to u_4$ in $H(G)$. If $H(G)$ is a forest and each walk in $WD(H(G))$ corresponds to a contractible sequence of G, we say $H(G)$ is a *consistent decomposition* of G.

Definition 10: Suppose $H(G)$ is a consistent decomposition of G. Let $\{L_1 \ldots L_r\}$ be the set of contractible sequences obtained from $WD(H(G))$. For each L_i, suppose $L_i = \{u_1^i, u_2^i, \ldots, u_{l_i}^i\}$ for some l_i. *Cut-graph(G)* is the graph obtained from G as follows: For each $L_i \in WD(H(G))$, delete the vertices u_k^i ($2 \leq k \leq l_i - 1$) from G.

Lemma 8: (1) If G is a cograph then $H(G)$ is a consistent decomposition of G.

(2) Suppose $H(G)$ is a consistent decomposition of G. Let $\{L_1, \ldots, L_r\}$ be the set of the contractible sequences obtained from $WD(H(G))$. Then G is a cograph iff $G' = Cut\text{-}graph(G)$ is a cograph whose cotree T' satisfies the following conditions: For each $L_i = \{u_1^i, \ldots, u_{l_i}^i\}$, $p_{T'}(u_1^i)$ is a 0-node, $p_{T'}(u_{l_i}^i)$ is a 1-node, and $p_{T'}(u_{l_i}^i)$ is the parent of $p_{T'}(u_1^i)$ in T'.

Proof: (1) Let T be the cotree of G. Consider the maximal trails $C_1, \ldots, C_r$ of T. Suppose $C_i = \{\alpha_1^i, \ldots, \alpha_{l_i}^i\}$. Let u_k^i be the leaf child of α_k^i ($1 \leq k \leq l_i$). By Lemma 6 (2), $\{u_j^i, u_{j+1}^i, u_{j+2}^i, u_{j+3}^i\}$ ($1 \leq j \leq l_i - 3$ and j is odd) is a 4-twin and it introduces three arcs $u_j^i \to u_{j+1}^i \to u_{j+2}^i \to u_{j+3}^i$ into $H(G)$. By Lemma 6 (2), these are the only arcs in $H(G)$. Clearly $H(G)$ is a forest and each walk in $WD(H(G))$ is a contractible sequence of G. So $H(G)$ is a consistent decomposition of G.

(2) Since G' is obtained from G by deleting vertices in the disjoint contractible

sequences L_i, this can be shown by repeated applications of Lemma 7. □

5. Recognition Algorithm

We are now ready to present our cograph recognition algorithm **Test**. The algorithm takes two parameters: The first one is the input graph G. The second one is an integer *depth* used to control the recursion depth. On a graph G of n vertices, the algorithm is invoked by Test$(G, \lceil \log_{18/17} n \rceil)$.

Algorithm 1: Test$(G, depth)$

(0) If $depth = 0$ and $|G| > 1$ then G is not a cograph. Return "no".

(1) If G has only one vertex, then G is a cograph. Return its trivial cotree.

(2) Find the class partition $\Psi(G) = \{Q_1, \ldots, Q_s, R_1, \ldots, R_t\}$. If $\Psi(G)$ is not valid, return "no".

(3) Construct the directed graph $H(G)$. If $H(G)$ is not a consistent decomposition of G, return "no". Otherwise let $WD(H(G)) = \{L_i\}$ be the set of contractible sequences obtained from $H(G)$.

(4) Using $\Psi(G)$ and $WD(H(G))$ found in Steps (2) and (3), perform the Trim-graph and Cut-graph operations on G. Let G' be the resulting graph.

(5) Call Test$(G', depth - 1)$.

(6) If G' is not a cograph, then G is not a cograph. Return "no".

(7) Suppose G' is a cograph with cotree T'. For each $L_i \in WD(H(G))$, check if the conditions in Lemma 8 (2) hold. If not, G is not a cograph, return "no".

(8) Construct T as follows. For each $Q_i, R_j \in \Psi(G)$, perform $UnTrim_t(T', q_i, Q_i)$ and $UnTrim_f(T', r_j, R_j)$. For each $L_i \in WD(H(G))$, perform $UnCut(T', L_i)$. Return the resulting tree T.

Theorem 9: The algorithm Test is correct.

Proof: By induction on $|G|$. If $|G| = 1$, the algorithm is correct by Step 1. Suppose the algorithm is correct on graphs with $< n$ vertices. Consider a graph G with n vertices. Suppose G is a cograph. By Lemmas 5 and 8, the valid class partition $\Psi(G)$ and the consistent decomposition $H(G)$ of G can be found in Steps 2 and 3; and the graph G' obtained in Step 4 is a cograph. By the induction hypothesis, the algorithm recognizes G' as a cograph and constructs its cotree T'. By Lemma 8, the tests in Step 7 are satisfied and the tree T constructed in Step 8 is a cotree of G. So the algorithm recognizes G as a cograph and outputs T.

Suppose G is not a cograph. Five cases may occur: (a) $\Psi(G)$ is not valid. (b) $H(G)$ is not a consistent decomposition. (c) G' constructed in Step 4 is not a cograph. (d) G' is a cograph, but at least one test fails in Step 7. (e) After $\lceil \log_{18/17} n \rceil$ recursive calls, the graph still contains more than one vertex. In any case the algorithm correctly rejects G. □

We now turn to the implementation of the algorithm. We show each iteration of Test takes $O(\log n)$ time with $O(n + m)$ processors. We only discuss Steps 2 and 3. (Other steps can be easily implemented within the required time and processor bounds).

Let $\{1, \ldots, n\}$ denote the vertices of G. For each $i \in V$, $deg(i)$ denotes the degree of i. $N[i]$ is an array containing $deg(i)$ entries. Each entry corresponds to a vertex in $N(i)$. $\Gamma[i]$ is an array containing $deg(i) + 1$ entries. Each entry corresponds to a vertex in $\Gamma(i)$. We sort the entries in $N[i]$'s and $\Gamma[i]$'s in the increasing order. This can be done in $O(\log n)$ time with $O(n + m)$ processors [6].

Given two arrays $X[1..a]$ and $Y[1..b]$, define $diff_L(X,Y) = \min\{i \mid X[i] \neq Y[i]\}$. (If $a < b$ and $X[i] = Y[i]$ for all $1 \leq i \leq a$, define $diff_L(X,Y) = a+1$.) Define $X <_L Y$ iff $X[d] < Y[d]$ where $d = diff_L(X,Y)$. (If $a < b$ and $X[i] = Y[i]$ for all $1 \leq i \leq a$, define $X <_L Y$). Define $diff_R(X,Y) = \min\{i \mid X[a+1-i] \neq Y[b+1-i]\}$. The relation $X <_R Y$ is defined analogous to $<_L$.

To implement Step 2, we need to find the $\sim_l$ classes of V. We sort the set of the arrays $\{N[i] \mid 1 \leq i \leq n\}$ by $<_L$, using the sorting algorithm in [6]. Since $i \sim_l j$ iff $N[i] = N[j]$, each $\sim_l$ class corresponds to a contiguous block in the sorted list. An array $N[i]$ of size $k = deg(i)$ is assigned k processors. The comparison between two arrays $N[i]$ and $N[j]$ can be done in $O(1)$ time using the processors assigned to them. The total number of processors needed is $\sum_{i=1}^n deg(i) = 2m$ with $O(\log n)$ time. The $\sim_t$ classes can be similarly computed.

In order to implement Step 3, we need to find all 4-twins of G. This is the most difficult part of the implementation. We break it into several easier problems. Three vertices $\{u, v, w\}$ are called a *triple* of G if $\Gamma(u) \subset \Gamma(w)$ and $\Gamma(w) - \Gamma(u) = \{v\}$. We discuss how to find the triples of G.

For $u, w \in V$ define: $l(u,w) = diff_L(\Gamma[u], \Gamma[w])$ and $r(u,w) = diff_R(\Gamma[u], \Gamma[w])$. Clearly $\{u, v, w\}$ is a triple iff: (1) $deg(u) = deg(w) - 1$; (2) $l(u,w) + r(u,w) = deg(w) + 2$; and (3) $\Gamma[w][l(u,w)] = v$. So the problem is reduced to computing $l(u,w)$'s and $r(u,w)$'s. We only discuss the computation of $l(u,w)$'s.

Lemma 10: (1) If $\{u, v, w\}$ is a triple, then $\Gamma[w] <_L \Gamma[u]$.

(2) Let $w, u_1, u_2 \in V$ with $deg(u_1) = deg(u_2) = deg(w) - 1$. If $\Gamma[w] <_L \Gamma[u_1] <_L \Gamma[u_2]$, then $l(u_1, w) \geq l(u_2, w)$. $\square$

We partition the set of the arrays $\{\Gamma[i] \mid i \in V\}$ into blocks $B_1, \ldots, B_\Delta$ (Δ is the maximum degree of G), where $B_k = \{\Gamma[i] \mid deg(i) = k\}$. The arrays within each block B_k are sorted in non-increasing order under $<_L$. For each $w \in V$, we want to find u (if it exists) such that $\{u, v, w\}$ is a triple for some v. If $\{u, v, w\}$ is a triple, then $\Gamma(u) \subset \Gamma(w)$. This implies $u \in N(w)$. So we only need to search the neighbors of w for u. Lemma 10 suggests a binary-search-like algorithm as follows.

Algorithm 2:

(1) Let $U = \{u_1, \ldots, u_q\}$ be the set of the neighbors of w with degree $deg(w) - 1$. (U is sorted in non-increasing order of $\Gamma(u_i)$ under $<_L$.)

(2) Do a binary search in the set $\{\Gamma[u_i] \mid u_i \in U\}$ to find the largest index l such that $\Gamma(w) <_L \Gamma(u_l)$. For each j with $l < j \leq q$, let $l(u, w_j) = -1$. (By Lemma 10 (1), u_j (for $j > l$) and w cannot form a triple with any v. So $l(u_j, w)$ is assigned a dummy value -1.)

(3) Call $Find(1, l, 1, deg(w) + 1)$.

The steps (1) and (2) clearly take $O(\log n)$ time with $k = deg(w)$ processors. The recursive procedure $Find(a, b, c, d)$ finds $l(u_i, w)$ for each $u_i \in U$ with $a \leq i \leq b$. The procedure call is concerned only with the array entries whose indices are between c and d. This procedure uses a subroutine $Comp(\Gamma[u], \Gamma[w], c, d)$ which, given two arrays $\Gamma[u]$, $\Gamma[w]$ and two integers $c \leq d$, computes $\min\{j \mid c \leq j \leq d, \Gamma[u][j] \neq \Gamma[w][j]\}$. This takes $O(1)$ time with $d - c$ processors.

Procedure Find(a,b,c,d):

If $c = d$ then for all i with $a \leq i \leq b$, let $l(u_i, w) = c$.

Else do:

$\quad e = \lceil (a+b)/2 \rceil; \qquad f = Comp(\Gamma[u_e], \Gamma[w], c, d);$

let $l(u_e, w) = f$;

In parallel, call $Find(a, e - 1, c, f)$ and $Find(e + 1, b, f, d)$;

From Lemma 10 (2), it is easy to show Algorithm 2 correctly computes $l(u_i, w)$ for all u_i with $i \leq l$. Since $|U| \leq k = deg(w)$ and each array $\Gamma[u_i]$ ($u_i \in U$) has size $\leq k$, this computation takes $O(\log n)$ time with k processors. Thus $l(u, w)$'s for all u, w can be computed in $O(\log n)$ time with $O(n+m)$ processors. Similarly $r(u, w)$'s for all u, w can be computed in $O(\log n)$ time with $O(n+m)$ processors. From $l(u, w)$ and $r(u, w)$, all triples of G can be identified in $O(1)$ time with $O(n+m)$ processors.

A triple $\{u, v, w\}$ is a true trio iff both $\{u\}$ and $\{w\}$ are $\sim_f$ and $\sim_t$ classes. This can be checked in $O(1)$ time with $O(n + m)$ processors.

For each true trio $\{u, v, w\}$, we need to determine if there exists a vertex z such that $\{z, u, v, w\}$ is a 4-twin. If $\{z, u, v, w\}$ is a 4-twin, then $u \in N(z) - N(v)$ which implies $z \in N(u)$. Moreover, by Lemma 6 (4), $N(z) \subset N(u)$ and $N(v) \subset N(u)$. Let $p = deg(u)$ and $W = \{j \in N(u) \mid deg(j) \leq p\}$. Thus our task is to search the set of the arrays $\{N[j] \mid j \in W\}$ against the array $N[v]$ to find z such that $N(v) \subset N(z)$ and $N(z) - N(v) = \{u\}$. This task can be carried out by using the similar method for finding the triples. It takes $O(\log n)$ time with p processors. So the total number of processors needed for finding all 4-twins is $O(n + m)$.

If $\{u, v, w, z\}$ is a 4-twin of G, three arcs $u \to v \to w \to z$ are added in to $H(G)$. The walk decomposition $WD(H(S))$ can be computed in $O(\log n)$ time with $O(n)$ processors by using the pointer jump method. In summary we have:

Theorem 11: The cograph recognition and cotree construction problem can be solved in $O(\log^2 n)$ time with $O(n + m)$ processors on a CRCW PRAM.

6. Permutation Representation

A graph $G = (V, E)$ is a *permutation graph* if V can be labeled by $\{1, \ldots, n\}$ and there is a permutation $\pi : \{1, \ldots, n\} \to \{1, \ldots, n\}$ such that $(i, j) \in E$ iff $(i - j)(\pi(i) - \pi(j)) < 0$. π is called a *permutation representation* (PR) of G. Cographs are a proper subset of permutation graphs [9]. In this section, we show a PR of a cograph can be found in $O(\log n)$ time with $O(n)$ processors.

We describe an interpretation of the PR. Consider an $n \times n$ grid. Let π be a PR of a permutation graph $G = (V, E)$. For each i ($1 \leq i \leq n$), plot the vertex i at the grid point $P(i) = (i, \pi(i))$. Then $(i, j) \in E$ iff either $P(i)$ is located to the north-west of $P(j)$ or $P(j)$ is located to the north-west of $P(i)$. If the vertices of G can be plotted on the grid in this way, the PR of G can be easily obtained.

Let G be a cograph with cotree T. We plot the vertices of G on an $n \times n$ grid. For each node α of T, let $G(\alpha)$ denote the subgraph of G induced by the vertices in $L(T(\alpha))$. (Recall that $L(T(\alpha))$ denote the set of the leaf nodes in the subtree $T(\alpha)$ of T rooted at α). The vertices of $G(\alpha)$ will be plotted on an $n(\alpha) \times n(\alpha)$ square region $R(\alpha)$ on the grid, where $n(\alpha)$ is the number of vertices in $G(\alpha)$. We represent $R(\alpha)$ by $[x_1(\alpha), y_1(\alpha), x_2(\alpha), y_2(\alpha)]$ where $(x_1(\alpha), y_1(\alpha))$ is the lower-left corner and $(x_2(\alpha), y_2(\alpha))$ is the upper-right corner of $R(\alpha)$.

Our algorithm is a top-down computation on T. For the root r of T, let $R(r) = [1, 1, n, n]$. Consider an internal node α with children $\beta_1, \ldots, \beta_k$. Suppose $R(\alpha) = [x_1(\alpha), y_1(\alpha), x_2(\alpha), y_2(\alpha)]$ has been computed, we describe how to compute $R(\beta_i) = [x_1(\beta_i), y_1(\beta_i), x_2(\beta_i), y_2(\beta_i)]$ for each $1 \leq i \leq k$.

Case 1: α is a 0-node. Since the subgraphs $G(\beta_i)$'s are pairwise disconnected, the

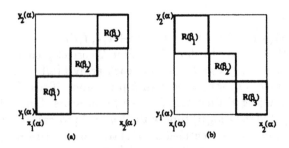

Figure 4: Grid drawing of a graph.

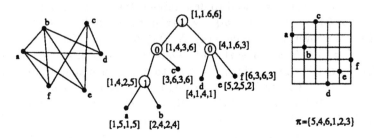

Figure 5: A cograph and its permutation representation.

squares $R(\beta_i)$'s are placed on the grid running from the south-west to the north-east (Fig 4a). Therefore $R(\beta_i)$ can be computed by using the following formulas:

$$x_1(\beta_i) = x_1(\alpha) + \sum_{j=1}^{i-1} n(\beta_j); \quad y_1(\beta_i) = y_1(\alpha) + \sum_{j=1}^{i-1} n(\beta_j);$$
$$x_2(\beta_i) = x_1(\alpha) + \sum_{j=1}^{i} n(\beta_j) - 1; \quad y_2(\beta_i) = y_1(\alpha) + \sum_{j=1}^{i} n(\beta_j) - 1.$$

Case 2: α is a 1-node. Since the subgraphs $G(\beta_i)$'s are completely connected to each other, $R(\beta_i)$'s are placed on the grid running from the north-west to the south-east (Fig 4b). So $R(\beta_i)$ can be computed by using the following formulas:

$$x_1(\beta_i) = x_1(\alpha) + \sum_{j=1}^{i-1} n(\beta_j); \quad y_1(\beta_i) = y_1(\alpha) + \sum_{j=i+1}^{k} n(\beta_j);$$
$$x_2(\beta_i) = x_1(\alpha) + \sum_{j=1}^{i} n(\beta_j) - 1; \quad y_2(\beta_i) = y_1(\alpha) + \sum_{j=i}^{k} n(\beta_j) - 1.$$

This gives the following algorithm for constructing a PR for cographs.

Algorithm 3: Permutation representation

(1) Compute $n(\alpha)$ for each node α of T by using the tree contraction algorithm.

(2) For each internal node α with children $\beta_1, \ldots, \beta_k$, calculate the pre-sum and the post-sum of the sequence $\{n(\beta_1), \ldots, n(\beta_k)\}$.

(3) Assign the root r of T the value $R(r) = [1, 1, n, n]$. Assign the internal nodes of T the operators as defined above. Perform the top-down tree contraction algorithm [1] to T. Each leaf node u of T gets a value $R(u) = [x_1(u), y_1(u), x_2(u), y_2(u)]$. Plot the vertex u at the grid point $(x_1(u), y_1(u))$.

(4) Define a permutation π such that $\pi(x_1(u)) = y_1(u)$ for each $u \in V$.

Fig 5 shows an example of this algorithm. Since pre-sum, post-sum and tree contraction can be done in $O(\log n)$ time with $O(n)$ processors [1], we have:

Theorem 12: Given the cotree as input, the permutation representation of a cograph can be constructed in $O(\log n)$ time with $O(n)$ processors.

7. Depth First Spanning Tree Algorithm

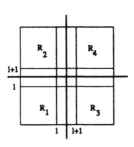

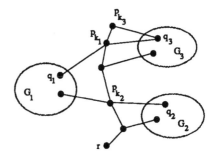

Figure 6: Divide a grid into 4 regions. Figure 7. Constructing a DFST.

Let $G = (V, E)$ be a connected graph and H be a spanning tree of G rooted at $r \in V$. H is a *depth first spanning tree* (DFST) of G if there exist no cross edges. We show the DFST problem can be solved efficiently for permutation graphs.

Consider the grid drawing R of a permutation graph G. Let $l = \lfloor n/2 \rfloor$. Draw a horizontal line with y-coordinate $l + 0.5$ and a vertical line with x-coordinate $l + 0.5$. These two lines divide R into four regions R_1, R_2, R_3, R_4 (Fig 6). Let n_i ($1 \le i \le 4$) be the number of vertices in R_i. Observe that $n_1 + n_2 = n_1 + n_3 = l$. This implies $n_2 = n_3$. Since each vertex in R_2 is adjacent to every vertex in R_3, we can construct a cycle C consisting of all vertices in R_2 and R_3, where the vertices of C alternate between R_2 and R_3. Moreover if C is removed from G, the graph becomes disconnected and each connected component has at most $n/2$ vertices. The following algorithm solves the DFST problem for permutation graphs.

Algorithm 4: Depth First Spanning Tree

(1) Construct the cycle C consisting of the vertices in R_2 and R_3 as above.

(2) Find a path P_1 from the specified root r to a vertex z on C. Let x be a vertex such that (x, z) is an edge of C. Let $P = P_1 \cup (C - \{(x, z)\})$. Number the vertices of the path P as $\{p_1, p_2, \ldots, p_t\}$ starting at $r = p_1$.

(3) Let $G' = G - P$. Compute the connected components $G_1, \ldots, G_s$ of G'.

(4) For each G_i ($1 \le i \le s$), let p_{k_i} be the vertex of P with the largest index such that p_{k_i} is adjacent to a vertex q_i in G_i.

(5) For each i ($1 \le i \le s$), recurse on G_i to find a DFST T_i of G_i rooted at q_i.

(6) Return the tree $T = P \cup (\cup_{i=1}^s T_i) \cup (\cup_{i=1}^s \{(q_i, p_{k_i})\})$.

Fig 7 illustrates the construction of the Algorithm 4. It is easy to verify that the tree T constructed in Step 6 has no cross edges and hence is a DFST of G. It can be shown that each iteration of the algorithm can be implemented in $O(\log n)$ time with $O(n)$ processors. Since the recursion depth is $O(\log n)$, we have:

Theorem 13: Let G be a permutation graph given by the permutation representation. A DFST of G can be constructed in $O(\log^2 n)$ time with $O(n)$ processors.

Acknowledgements: In an earlier version of this paper, the cotree recognition algorithm takes $O(\log^2 n)$ time with $O(n^2)$ processors. The DFST algorithm for permutation graphs takes $O(\log^2 n)$ time with $O(n + m)$ processors. Mark Novick pointed out that the first algorithm can be implemented using $O(n + m)$ processors and the second algorithm can be implemented using $O(n)$ processors.

References

[1] K. Abrahamson, N. Dadoun, D. G. Kirkpatrick and T. Przytycka, A Simple Parallel tree Contraction Algorithm, J. Algorithms 10 (2), 1989, pp. 287-302.

[2] G. S. Adhar and S. Peng, NC Algorithms for Complement Reducible Graphs and Parity Graphs, in Proc. of International Conf. on Comput. and Info., 1989.

[3] G. S. Adhar and S. Peng, Parallel Algorithms for Complement Reducible Graphs and Parity Graphs with Applications, in Proc. Workshop on Algorithms and Data Structures, 1989.

[4] G. S. Adhar and S. Peng, Parallel Algorithms for Cographs and Parity Graphs with Applications, J. Algorithms 11, 1990, pp. 252-284.

[5] G. S. Adhar and S. Peng, Parallel Algorithms for Finding Minimal Path Cover, Hamiltonian Path and Hamiltonian Cycle in Cographs, UMIACS-TR-89-64, Uni. of Maryland, 1989.

[6] R. Cole, Parallel Merge Sort, Proc. 27th IEEE Symposium on Foundation of Computer Science, 1986, pp. 511-517.

[7] D. G. Corneil, H. Lerchs and L. Stewart Burlinham, Complement Reducible Graphs, Disc. Appl. Math. 3, 1981, pp. 163-174.

[8] D. G. Corneil, Y. Perl and L. K. Stewart Burlinham, A Linear Recognition Algorithm for Cographs, SIAM J. on Computing 14 (4), 1985, pp. 926-934.

[9] M. C. Golumbic, Algorithmic Graph Theory and Perfect Graphs, Academic Press, New York, 1980.

[10] D. Helmbold and E. Mayr, Perfect Graphs and Parallel Algorithms, in Proc. 1986 Intl. Conf. on Parallel Processing, pp. 853-860.

[11] H. A. Jung, On a Class of Posets and the Corresponding Comparability Graphs, J. Combinatorial Theory (B) 24, 1978, pp. 125-133.

[12] D. G. Kirkpatrick and T. Przytycka, Parallel Recognition of Complement Reducible Graphs and Cotree Construction, TR 88-1, Department of Computer Science, University of British Columbia, 1988.

[13] Lin and S. Olariu, An NC Algorithm for Cographs, Department of Computer Science, TR-89-32, Old Dominion University, 1989.

[14] M. Novick, Fast Parallel Algorithms for the Modular Decomposition, TR 89-1016, Department of Computer Science, Cornell University, 1989.

[15] D. P. Sumner, Dacey Graphs, J. Australian Math. Soc. 18 (4), 1974, pp. 492-502.

Parallel algorithms for all minimum link paths and link center problems

Subir Kumar Ghosh*   Anil Maheshwari†

Abstract

The link metric, defined on a constrained region R of the plane, sets the distance between a pair of points in R equal the minimum number of segments or links that are needed to construct a path in R between the points. The *minimum link path problem* is to compute a path consisting of minimum number of links between two points in R, when R is the inside of a simple polygon P of size n. Recently Chandru *et al.* [1] proposed a parallel algorithm for computing minimum link path between two points inside P and it runs in $O(\log n \log \log n)$ time using $O(n)$ processors. Here we show that minimum link paths from a point to all vertices of P can be computed in $O(\log^2 n \log \log n)$ time using $O(n)$ processors. Using this result we propose a parallel algorithm for computing the *link center* of P. The link center of P is the set of points x inside P such that the link distance from x to any other point in P is minimized. The algorithm runs in $O(\log^2 n \log \log n)$ time using $O(n^2)$ processors. We also show that a triangle in the *approximate link center* can be computed in $O(\log^3 n \log \log n)$ time using $O(n)$ processors. The complexity results of this paper are with respect to the CREW-PRAM model of computation.

1 Introduction

Consider a point-size robot that moves in straight-line paths and has to navigate in a constrained planar region. In order to reach its destination, it may have to make several turns. Suppose that straight line motion is "cheap", but rotation is "expensive". Minimizing the number of turns is a natural objective to use in planning paths for such motion. This motivates the study of shortest path problems in the *link metric*. For details on applications, see [2, 9, 10].

The link metric, defined with respect to a planar region R, sets the distance between a pair of points (s, t) in R to equal the minimum number of line segments needed to construct a path R that connects s and t. The *minimum link path problem* is to compute a path consisting of the minimum number of links between two points in a region R defined by a simple polygon P. Linear-time sequential algorithms for the minimum link path problem have been proposed by Ghosh [3] and Suri [9].

*Computer Science Group, Tata Institute of Fundamental Research, Homi Bhabha Road, Bombay - 400 005, India, E-Mail : ghosh@tifrvax.bitnet

†Computer Systems and Communication Group, Tata Institute of Fundamental Research, Homi Bhabha Road, Bombay - 400 005, India, E-Mail : manil@tifrvax.bitnet

Recently Chandru *et al.* [1] proposed a parallel algorithm for computing minimum link path between two points of P and it runs in $O(\log n \log \log n)$ time using $O(n)$ processors, where n is the number of vertices of P.

All minimum link paths problem is to compute minimum link paths from a given point s inside P to all vertices of P. Suri [9] proposed a linear-time sequential algorithm for this problem. The problem can be solved in parallel by assigning n-processors to each vertex of P and then computing the minimum link path from s to each vertex of P by the algorithm of Chandru *et al.* [1]. The complexity of this algorithm is $O(\log n \log \log n)$ time using $O(n^2)$ processors. Here we present an efficient algorithm that runs in $O(\log^2 n \log \log n)$ time using $O(n)$ processors. The algorithm is based on the relationship between *complete visibility* and *greedy path* established in Ghosh [3] and uses the algorithms of Chandru *et al.* [1] and Goodrich *et al.* [5] as major steps.

The *link center* of a simple polygon P is the set of points x inside P such that the link distance from x to any other point in P is minimized. The link center problem has applications in locating a transmitter so that the maximum number of retransmissions needed to reach any point in P is minimized, or in choosing the best location for a mobile unit minimizing the minimum number of turns needed to reach any point in a polygonal region P [2].

Lenhart *et al.* [8] proposed an $O(n^2)$ time sequential algorithm for computing the link center. Ke [6] and Djidjev *et al.* [2] improved the algorithm of Lenhart *et al.* to run in $O(n \log n)$ time. To the best of our knowledge no parallel algorithm is known for computing the link center. Here we propose an algorithm for computing the link center by parallelizing the sequential algorithm of Lenhart *et al.* [8] and it runs in $O(\log^2 n \log \log n)$ time using $O(n^2)$ processors. Our parallel algorithm for all minimum link paths problem is used as a major step in computing link center.

A point z of P is said to be in the *approximate link center* if the maximum link distance from z is at most one more than the value attained from the link center. Lenhart *et al.* [8] presented a sequential algorithm for computing a triangle in the approximate link center and it runs in $O(n \log n)$ time. Following the approach of Lenhart *et al.* [8] we propose an algorithm for computing a triangle in the approximate link center and it runs in $O(\log^3 n \log \log n)$ time using only $O(n)$ processors.

Our complexity results in this paper are with respect to the CREW-PRAM (Concurrent Read and Exclusive Write - Parallel Random Access Machine) model of computation. For details on CREW-PRAM model of computation, see [7].

We assume that the simple polygon P is given as a clockwise sequence of vertices with their respective x and y coordinates. The symbol P is used to denote the region of the plane enclosed by P and $bd(P)$ denotes the boundary of P. If u and v are two points on $bd(P)$ then the clockwise $bd(P)$ from u to v is denoted as $bd(u, v)$. Two points of P are said to be visible if the line segment joining them lies totally inside P. A polygon is said to be a *weak visibility polygon* if every point of the polygon is visible from some point of an internal segment. Let C be a convex polygon inside P. For any point z inside P draw two tangents zc_l and zc_r to C, where c_l and c_r are two vertices of C. The segment zc_l (or zc_r) is called *left* (respectively, *right*) *tangent* if C lies to the right (respectively, left) of a ray drawn from z through c_l (respectively, c_r). The point z is said to be *completely visible* from C if zc_l and zc_r lie inside P.

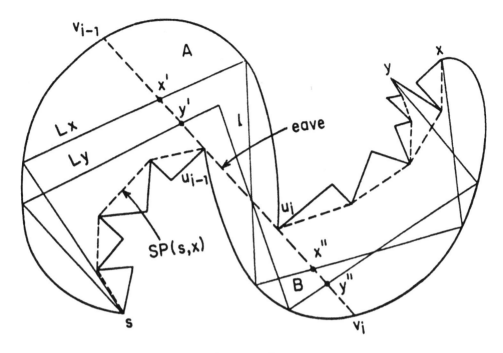

Figure 1: $u_{i-1}u_i$ is an eave

In Section 2 we present an algorithm for all minimum link paths problem. In Section 3 we present an algorithm for computing the link center. In Section 4 we present an algorithm for computing a triangle in the approximate link center. In Section 5 we discuss a few open problems.

2 An algorithm for all minimum link paths problem

In this section we describe an algorithm for computing minimum link paths from a given point s to all vertices of P. Following the approach of Ghosh [3], we design our algorithm to compute the minimum link path from s to all vertices of P. Let $SP(s,t) = (s, u_1, ..., u_k, t)$ be the Euclidean shortest path inside P from s to a vertex t. An edge $u_{i-1}u_i$ of $SP(s,t)$ is called an *eave* if u_{i-2} and u_{i+1} lie on the opposite side of line passing through u_{i-1} and u_i (Figure 1). If an edge $u_{i-1}u_i$ of $SP(s,t)$ is a subsegment of a link, we say that the minimum link path from s to t (denoted as $MLP(s,t)$) contains $u_{i-1}u_i$. Note that the minimum link path, on contrary to the shortest path, between any two vertices is not unique. The following lemma suggests a procedure to partition P into subpolygons such that the link paths can be computed in each subpolygon in parallel.

Lemma 2.1 *If $u_{i-1}u_i$ is an eave common to $SP(s,x)$ and $SP(s,y)$ for two vertices x and y, then there exists $MLP(s,x)$ and $MLP(s,y)$ such that they contain $u_{i-1}u_i$ and their paths from s to u_i are same.*

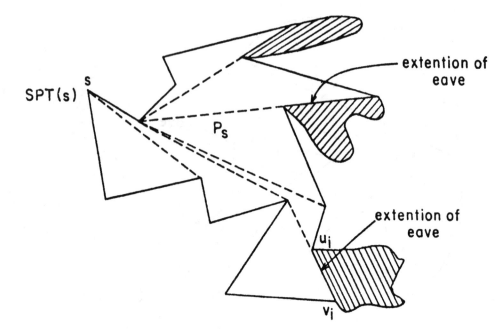

SPT(s)

extention of eave

extention of eave

Figure 2: P_s is not a standard polygon

Proof: Let L_x (or L_y) be a minimum link path from s to x (respectively, y). Extend $u_{i-1}u_i$ from u_{i-1} and u_i to $bd(P)$ and they meet $bd(P)$ at v_{i-1} and v_i respectively (Figure 1). So $v_{i-1}v_i$ partitions P into four disjoint regions. It can be seen that two of these regions do not contain s, x and y; call them A and B. Since P is a closed and bounded region, there exists a link l in L_x that intersects $u_{i-1}u_i$. As any line segment in P that intersects $u_{i-1}u_i$ must have one endpoint in A and the other endpoint in B, one endpoint of l lies in A and the other endpoint lies in B. Let x' (or x'') be the intersection point of $v_{i-1}u_{i-1}$ (respectively, v_iu_i) with L_x. We construct a new link path $MLP(s, x)$ from L_x by removing the portion of L_x between x' and x'' and adding the segment $x'x''$. This alteration has removed the link l totally and therefore, $MLP(s, x)$ is a minimum link path containing the eave $u_{i-1}u_i$. Let y'' be the intersection point of L_y and v_iu_i. We construct a new link path $MLP(s, y)$ containing $u_{i-1}u_i$ from L_y by concatenating $MLP(s, x)$ from s to x', the segment $x'y''$ and L_y from y'' to y. Observe that the paths from s to u_i are same in $MLP(s, x)$ and $MLP(s, y)$. □

Above lemma suggests the following procedure. Compute the *shortest path tree* rooted at s in P by the algorithm of Goodrich *et al.* [5]. The shortest path tree of a polygon rooted at s is the union of Euclidean shortest paths from s to all vertices of P and is denoted as $SPT(s)$. Partition P into subpolygons (called *standard polygons*) by extending each eave $u_{i-1}u_i$ of $SPT(s)$ from u_i to $bd(P)$, where u_{i-1} is the parent of u_i in $SPT(s)$. In each standard polygon P_i compute the minimum link path from the extension of the eave u_iv_i (called *limiting edge*) to all vertices of P_i. Concatenate the link path from s to $u_{i-1}u_i$ to each link path in P_i as stated in the proof of Lemma 2.1.

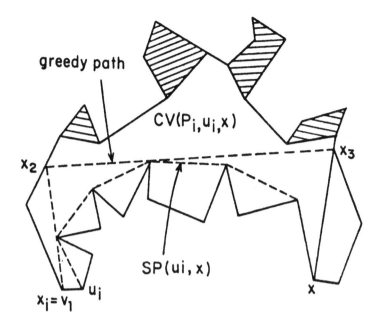

Figure 3: complete visibility and greedy path

Observe that the shortest paths from s to all vertices of P_i passes through u_i. The shortest paths from u_i to all vertices of P_i are either only left turning at each vertex in the path or only right turning at each vertex in the path. Note that the subpolygon P_s containing s is not a standard polygon (Figure 2). Compute the visibility polygon VP from s in P_s, and add all non-polygonal edges (called *constructed edges*) on the boundary VP to decompose P_s into standard polygons. All vertices visible from s can be connected to s by a link.

Now computing minimum link paths from s to all vertices of P reduces to computing minimum link paths from the limiting edge to each vertex in every standard polygon of P. Since the problem is identical in each standard polygon, from now on we restrict our attention to one standard polygon P_i.

We now recall a lemma from Ghosh [3] on the notions of *complete visibility* and *greedy path*, which is used by Chandru *et al.* [1] in computing the minimum link path from the limiting edge $u_i v_i$ to a vertex x of P_i. Let R be the region of P_i bounded by $SP(u_i, x)$ and $bd(x, u_i)$. The *complete visibility polygon* of P_i from $SP(u_i, x)$, denoted as $CV(P_i, u_i, x)$, is the set of points z of R such that both tangents from z to $SP(u_i, x)$ are totally inside R (Figure 3). Let x_1 be the point of intersection of segment $u_i v_i$ and the boundary of $CV(P_i, u_i, x)$. Draw the right tangent of x_1 to $SP(u_i, x)$ and extend it to the boundary of $CV(P_i, u_i, x)$ meeting at a point x_2. Similarly, draw the right tangent of x_2 to $SP(u_i, x)$ and extend it to the boundary of $CV(P_i, u_i, x)$ meeting at a point x_3, and so on until a point x_r is found such that x and x_r are visible. The path $x_1 x_2$, $x_2 x_3, ..., x_{r-1} x_r, x_r x$ is called the *greedy path* from $u_i v_i$ to x. We state the lemma from Ghosh [3].

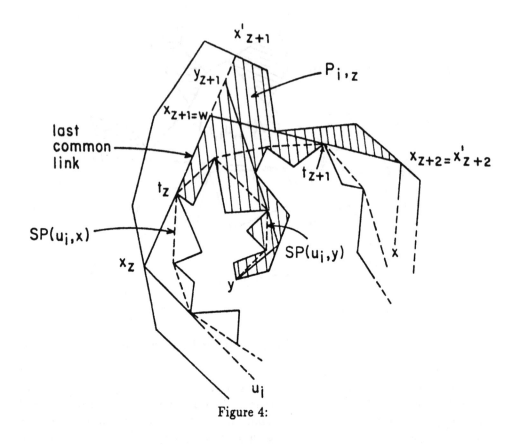

Figure 4:

Lemma 2.2 *(Ghosh [9]) The greedy path* $x_1x_2, x_2x_3, ..., x_{r-1}x_r, x_rx$ *is a minimum link path from the limiting edge* u_iv_i *to* x *in* P_i.

Consider the problem of computing greedy paths from u_iv_i to two vertices x and y of P_i. Observe that two greedy paths are same from u_iv_i upto a point w and then they bifurcate to x and y (Figure 4). Let x_z be the previous turning point of w in the greedy path from u_iv_i to x. The link x_zx_{z+1} is called the *last common link* of x and y. Instead of computing the greedy path to x and y separately, compute the greedy path from u_iv_i to x. Locate the last common link x_zx_{z+1} and compute the greedy path from x_z to y. In order to locate the last common link, we need the following lemma.

Lemma 2.3 *If* x_zx_{z+1} *is the last link in the greedy path from* u_iv_i *to* x *such that* x_zx_{z+1} *is tangential to* $SP(u_i, y)$ *then* x_zx_{z+1} *is the last common link of* x *and* y.

Proof : Let x_zx_{z+1} is tangential at a vertex t_z of $SP(u_i, y)$ (Figure 4). Partition $CV(P_i, u_i, x)$ by the segment x_zt_z into two subpolygons P_1 and P_2. Let u_i belong to P_1. It follows from the definition of complete visibility that $bd(u_i, x_z)$ is same for $CV(P_i, u_i, x)$ and $CV(P_i, u_i, y)$. Therefore, the greedy paths from u_iv_i to x and y are same upto x_z. Since the subsequent links of x_zx_{z+1} are not tangential to $SP(u_i, y)$, they cannot be in the greedy path from u_iv_i to y. Therefore x_zx_{z+1} is the last common link of x and y. $\square$

Corollary 2.1 *Extend $x_z x_{z+1}$ from x_{z+1} and it meets $bd(P_i)$ at x'_{z+1}. The next turning point of x_z in the greedy path from $u_i v_i$ to y lies on $x_z x'_{z+1}$.*

For computing minimum link paths from the limiting edge $u_i v_i$ to all vertices of P_i, our algorithm chooses a vertex x of P_i and computes the minimum link path from $u_i v_i$ to x by the algorithm of Chandru *et al.* [1]. It follows from Lemma 2.3 that for every vertex y of P_i, a link in the greedy path to x is the last common link. Extend each link $x_z x_{z+1}$ to x'_{z+1} as stated in Corollary 2.1. Let $x_z x_{z+1}$ is tangential to $SP(u_i, x)$ at a vertex t_z. In order to compute the next turning point of x_z in the greedy path to y, compute the weak visibility polygon from $t_z x'_{z+1}$ for all z by the algorithm of Goodrich *et al.* [5]. Each visibility polygon is computed in the disjoint regions of P_i and they further partition P_i into standard subpolygons. Our algorithm repeats the process of computation as stated above for each standard subpolygons in parallel till the minimum link path to each vertex of P_i has been computed. In the following we formally state our procedure for computing minimum link paths from the limiting chain $u_i v_i$ to all vertices of standard polygon P_i.

Procedure Minimum_Link_Path(P_i);

Step 1 Locate the leaf x in $SPT(u_i)$ such that $bd(u_i, x)$ and $bd(x, u_i)$ have same number of leaves in $SPT(u_i)$.

Step 2 Compute the minimum link path $MLP(u_i v_i, x)$ from $u_i v_i$ to x in P_i by the algorithm of Chandru *et al.* [1]. Let $MLP(u_i v_i, x) = x_1 x_2, x_2 x_3, ..., x_{r-1} x_r$ where $x_1 \in u_i v_i$ and $x_r = x$.

Step 3 Let $x_0 = u_i$. For each link $x_z x_{z+1}$ $(0 \le z < r)$ *do in parallel*

    3.1 Extend $x_z x_{z+1}$ from x_{z+1} and it meets $bd(P_i)$ at x'_{z+1}.

    3.2 Let $x_z x_{z+1}$ be the tangent to $SP(u_i, x)$ at t_z. Add $t_1 x'_2, t_2 x'_3, ..., t_{r-2} x'_{r-1}$ to P_i and it partitions P_i into disjoint subpolygons. Let $P_{i,z}$ denote the subpolygon containing $t_z x'_{z+1}$ and $t_{z+1} x'_{z+2}$ on its boundary (Figure 4).

    3.3 Compute the weak visibility polygon VP_z from $t_z x'_{z+1}$ in $P_{i,z}$ by the algorithm of Goodrich *et al.* [5] (Figure 4).

    3.4 For all vertices p_α weakly visible from $t_z x'_{z+1}$ in $P_{i,z}$, connect p_α to $t_z x'_{z+1}$ by a link. Construct links by extending each constructed edge of VP_z to $t_z x'_{z+1}$.

    3.5 Add all constructed edges on the boundary of VP_z to partition $P_{i,z}$ into standard subpolygons $P^1_{i,z}, P^2_{i,z}, ..., P^l_{i,z}$, where constructed edges are the limiting edges of standard subpolygons.

Step 4. For all j and w in parallel do Minimum_Link_Path($P^j_{i,w}$).

Step 5. Stop.

Now we analyze the complexity of the algorithm. We assign a processor to each vertex of P. The shortest path tree $SPT(s)$ can be computed in $O(\log n)$ time using $O(n)$ processors by the algorithm of Goodrich *et al.* [5]. Eaves in $SPT(s)$

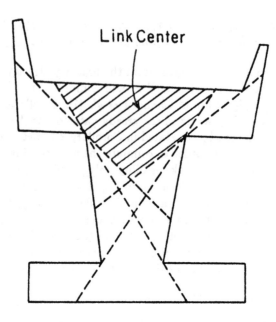

Figure 5: $r_L = 2$, $D_L = 4$

can be identified in $O(1)$ time and they can be extended to $bd(P)$ in $O(\log n)$ time using $O(n)$ processors. Step 1 can be performed by using Euler tour technique of Tarjan and Vishkin [11] and it takes $O(\log n)$ time and $O(n)$ processors. Step 2 can be performed by the algorithm of Chandru et al. [1] in $O(\log n \log \log n)$ time using $O(n)$ processors. Step 3.1 requires $O(1)$ time using $O(n)$ processors. Step 3.2 requires $O(\log n)$ time using $O(n)$ processors for the array representation of the boundary of each subpolygon $P_{i,s}$. Step 3.3 can be performed by the algorithm of Goodrich et al. [5] in $O(\log n)$ time using $O(n)$ processors. Step 3.4 requires $O(1)$ time using $O(n)$ processors. Step 3.5 requires $O(\log n)$ time using $O(n)$ processors for the array representation of each standard polygon. Since Step 2 computes the minimum link path in disjoint standard polygons, the overall complexity of this step is $O(\log n \log \log n)$ time using $O(n)$ processors. The procedure can be executed at most $O(\log n)$ times due to the choice of the leaf x in Step 1. Hence the overall complexity is $O(\log^2 n \log \log n)$ time using $O(n)$ processors. We summarize the result in the following theorem.

Theorem 2.1 *Minimum link paths from a given point to all vertices of an n-sided polygon can be computed in $O(\log^2 n \log \log n)$ time using $O(n)$ processors in CREW PRAM model of computation.*

3 An algorithm for computing link center

Our algorithm for computing the link center of a polygon is the parallelization of the sequential algorithm of Lenhart et al. [8]. We start with a few definitions and notations. The *link distance* $d_L(x, y)$ denotes the number of links in the minimum link path inside P from a point x to another point y. A vertex y is said to be the *furthest link neighbor* of x if all points of P can be reached from x in $d_L(x, y)$ links. The *link diameter* D_L of P is the maximum link distance between any two vertices of P. The *maximum link distance* of x is the link distance from x to its furthest link neighbor and is denoted as $M_L(x)$. The *link center* C_L of P is the set of points x inside P such that the maximum link distance from x is minimized (Figure 5). The link radius r_L of P is the maximum link distance from a point x in the link center i.e. $M_L(x) = r_L$. So, every point of the link center can be reached from any point of P by r_L links. Let $N_r(x)$ denote the region of P such that any point in the region can be reached from x in r links. The algorithm of Lenhart *et al.* [8] is based on the following lemmas.

Lemma 3.1 *(Lenhart et al. [8]) For any simple polygon P, $\left\lceil \frac{D_L}{2} \right\rceil \le r_L \le \left\lceil \frac{D_L}{2} \right\rceil + 1$*

Lemma 3.2 *(Lenhart et al. [8]) The link center of a polygon P is the intersection of the sets $N_r(v)$ over all convex vertices v of P, where $r = r_L$.*

An algorithm for computing the link center

Step 1: For each vertex v of P, compute the link path from v to all vertices of P by the algorithm of Section 2.

Step 2: For each vertex v of P, compute $M_L(v)$.

Step 3: Compute $D_L = max_v\{M_L(v)\}$. Let $r = \left\lceil \frac{D_L}{2} \right\rceil$.

Step 4: For each convex vertex v of P, compute $N_r(v)$.

Step 5: Compute $\cap_v N_r(v)$.

Step 6: If $\cap_v N_r(v)$ is empty then r:=r+1 and goto Step 4.

Step 7: Stop.

We now discuss the computational aspects of the algorithm. We define the *link tree* of v as follows. Every link in the link path from v corresponds to a *node* in the link tree and two nodes are adjacent in the link tree if and only if the corresponding links are consecutive in the link path from v. Make v as the root of the link tree by connecting all nodes of link tree which correspond to the links starting from v.

Consider Step 2. $M_L(v)$ can be computed by computing the depth of each node in the link tree from v. Construct an *Euler tour* on the link tree and then break the tour at v (see [11]). This reduces the tour to a list and each edge of the link tree appears in the forward and reverse direction in the list. Assign a weight of +1 to each edge in the forward direction and a weight of -1 to each edge in the reverse direction. Perform a prefix computation over this list. The value associated with each vertex is its depth from v.

Consider Step 4. Observe that the region $N_r(v)$ is a subpolygon of P such that every point of $N_r(v)$ can be reached from v in r links. Since Step 1 computes the link path from v to all vertices, r^{th} link in all link paths from v can be identified. For every r^{th} link $x_{r-1}x'_r$, cut P by $t_x x'_r$ to decompose P into subpolygons, where

$x_{r-1}x'_r$ is tangent at a vertex t_s. Observe that the subpolygon containing v is $N_r(v)$. If $x_{r-1}x'_r$ contains an eave then take the appropriate vertex for cutting P. An edge of $N_r(v)$ is called a *constructed edge* if it decomposes P into two regions.

Consider Step 5. We know that the boundary of $N_r(v)$ consists of constructed edges and the polygonal boundary. One of the endpoints of any constructed edge is a reflex vertex of P. The rth link from two different vertices can be tangent at the same reflex vertex. So a reflex vertex can be endpoint of several constructed edges. For each reflex vertex it is enough to consider two extreme constructed edges in order to compute $\cap N_r(v)$. The extreme constructed edges can be located as follows. Each constructed edge has removed a region of P; call it *forbidden region*. Locate two extreme constructed edges such that the forbidden region of all other constructed edges are contained inside the forbidden region of one of the extreme constructed edges. From now on we assume that each reflex vertex can be endpoint of atmost two constructed edges. The remaining task is to compute the appropriate portion of each constructed edge on the boundary of $\cap N_r(v)$. If two constructed edges e and e' intersects then we can identify the portion of e and e' lying in the forbidden region of e' and e respectively. The boundary of $\cap N_r(v)$ can be computed by removing the portion of each constructed edge lying in the forbidden region of some constructed edge.

We analyze the complexity of the algorithm. Step 1 takes $O(log^2 n \log \log n)$ time using $O(n^2)$ processors. Link tree from a vertex v can be constructed in $O(1)$ time using $O(n)$ processors. The computational complexity of Step 2 is that of constructing Euler tour and computing prefix sums. Step 2 can be performed in $O(\log n)$ time using $O(n^2)$ processors ([11, 7]). The region $N_r(v)$ of P can be computed in $O(\log n)$ time using $O(n)$ processors. So, Step 4 requires $O(\log n)$ time using $O(n^2)$ processors. In Step 5, the number of intersection points can be of $O(n^2)$. Since there are $O(n^2)$ processors, all the intersection points can be computed in $O(1)$ time. The portion of a constructed edge lying on the boundary of $\cap N_r(v)$ can be identified in $O(\log n)$ time using $O(n)$ processors. Hence Step 5 requires $O(\log n)$ time using $O(n^2)$ processors. We summarize the result in the following theorem.

Theorem 3.1 *The link center of an n-sided simple polygon can be computed in $O(\log^2 n \log \log n)$ time using $O(n^2)$ processors in the CREW PRAM model of computation.*

4 Computing a triangle in the approximate link center

In the previous section we have presented an algorithm for computing the link center of a simple polygon P which requires quadratic number of processors. In this section we present an algorithm for computing a triangle in the *approximate link center* and it runs in $O(\log^3 n \log \log n)$ time using $O(n)$ processors. A point z is said to be in the approximate link center if the maximum link distance of z is at most one more than the value attained from the link center. Our algorithm is based on the following lemmas of Lenhart *et al.* [8].

Lemma 4.1 *(Lenhart et al. [8]) There is a triangle in the triangulation of P such that every point in the triangle is in the approximate link center of P.*

Lemma 4.2 *(Lenhart et al. [8]) Let e be a diagonal of a triangulation of P that partitions P into two simple polygon P_1 and P_2. Let c_1 (or c_2) is the maximum link distance from e in P_1 (respectively, P_2).*
   (i) If $c_1 = c_2$, then both triangles containing e are in the approximate link center.
   (ii) If $c_1 < c_2$, then P_2 contains a triangle in the approximate link center.

We now present our divide and conquer algorithm for computing a triangle in the approximate link center. Triangulate P in $O(\log n)$ time using $O(n)$ processors by the algorithm of Goodrich [4]. Compute the dual tree of the triangulation, where each triangle is represented as a node in the tree and two nodes are connected in the tree if and only if the corresponding triangles are adjacent in the triangulation. We know that the degree of each node in the tree is atmost 3. Remove an edge t in the tree such that number of nodes in each subtree is no more than two-thirds of the original tree. Let t correspond to a diagonal e of P. So, e partitions P into two subpolygons P_1 and P_2. Compute the maximum link distance from e in P_1 and P_2 as follows. Compute the weak visibility polygon from e by the algorithm of Goodrich et al. [5] and it decomposes P_1 and P_2 into disjoint subpolygons. The vertices in the weak visibility polygon can be reached by one link from e. In each subpolygon compute link paths from the constructed edge to all vertices of the subpolygon by the algorithm of Section 2. Hence we can compute the maximum link distance from e in P_1 and P_2 in $O(\log^2 n \log\log n)$ time using $O(n)$ processors. Once the maximum link distance from e in P_1 and P_2 are known, it can be determined whether the desired triangle in the approximate link center is in P_1 or P_2. Apply Lemma 4.2 recursively to compute a triangle in the approximate link center. Note that at any stage of the recursion, if the subpolygon is just a triangle, the triangle is in the approximate link center. Since we partition the dual tree of the triangulation atmost $O(\log n)$ times, the complexity of the algorithm is $O(\log^3 n \log\log n)$ time using $O(n)$ processors. We summarize the result in the following theorem.

Theorem 4.1 *A triangle in the approximate link center of an n-sided polygon can be computed in $O(\log^3 n \log\log n)$ time using $O(n)$ processors in CREW PRAM model of computation.*

5 Concluding Remarks

In Section 2 we have presented a parallel algorithm for computing minimum link path from a given point to all vertices of the simple polygon P. There is a linear time sequential algorithm for this problem [10]. The complexity of the parallel algorithm for this problem needs to be improved.

In Section 3 we have presented a parallel algorithm for computing link center using quadratic number of processors. Since the sequential algorithm for this problem [2, 6] runs in $O(n \log n)$ time, the processor complexity needs to be improved. Following special case of the link center problem may be of interest. Can we compute a point inside the link center of a simple polygon by using only linear number of processors ?

We can compute a *link diameter* of P using the algorithm in Section 2 and it runs in $O(\log^2 n \log \log n)$ time using $O(n^2)$ processors. Can we compute a link diameter using only linear number of processors ? Using the algorithm in Section 2 we can compute a *furthest link neighbor* of the given vertex, i.e. a vertex which is at the maximum link distance from the given vertex. The algorithm runs in $O(\log^2 n \log \log n)$ time using $O(n)$ processors. We can also compute a furthest link neighbor of each vertex using $O(n^2)$ processors. Can we compute a furthest link neighbor of each vertex by using only linear number of processors ?

References

[1] Vijay Chandru, Subir Ghosh, Anil Maheshwari, V T Rajan and Sanjeev Saluja, *NC-Algorithms for minimum link path and related problems*, Technical Report No. CS-90/3, Computer Science Group, Tata Institute of Fundamental Research, Bombay, 1991.

[2] H.N. Djidjev, A. Lingas and J-R Sack, *An $O(n \log n)$ algorithm for computing a link center of a simple polygon*, STACS 89, LNCS, 349 (1989), Springer Verlag.

[3] S.K. Ghosh, *Computing the visibility polygon from a convex set and related problems*, Journal of Algorithms, 12(1991), pp. 75-95.

[4] M.T. Goodrich, *Triangulating a polygon in parallel*, Journal of Algorithms, 10(1989), pp. 327-351.

[5] M.T. Goodrich, B. Shauck, S. Guha, *Parallel algorithms in shortest path and visibility problems*, Proceedings of the 6th ACM Symposium on Computation Geometry, 1990, pp. 73-82.

[6] Y. Ke, *Efficient algorithms for weak visibility and link distance problems in polygons*, Ph.D. Thesis, The John Hopkins University, 1989.

[7] R. M. Karp and R. Vijaya Ramachandran, *Parallel Algorithms for Shared-Memory Machines*, Handbook of Theoretical Computer Science, Edited by J. van Leeuwen, Volume 1, Elsevier Science Publishers B.V., 1990.

[8] W. Lenhart, R. Pollack, J. Sack, R. Seidel, M. Sharir, S. Suri, G. Toussaint, S. Whitesides and C. Yap, *Computing the link center of a simple polygon*, Discrete and Computational Geometry, 3 (1988), pp. 281-293.

[9] S. Suri, *A linear time algorithm for minimum link path inside a simple polygon*, Computer Vision, Graphics and Image Processing, 35(1986), pp. 99-110.

[10] S. Suri, *Minimum link paths in polygons and related problems*, Ph.D. Thesis, The John Hopkins University, 1987.

[11] R.E. Tarjan and U. Vishkin, *An efficient parallel biconnectivity algorithm*, SIAM Journal on Computing, 14 (1985), pp. 862-874.

Optimal multi-packet routing on the torus

Michael Kaufmann[*]   Jop F. Sibeyn[†]

Abstract

We present new algorithms for k-k routing on mesh connected processor arrays with wrap-around connections. Using new techniques for the performance analysis we show that an original randomized four stage algorithm performs optimally: for all $k \geq 8$, k-permutations are routed on the torus with $k \cdot n/4 + \mathcal{O}((k \cdot n \cdot \log n)^{1/2})$ routing steps and queue size $k + o(k)$ with very high probability. We prove comparable results for routing in the cut-through mode: the same randomized algorithm routes packets consisting of k flits each (k arbitrary), with $k \cdot n/4 + 2 \cdot n + \mathcal{O}(k \cdot (n \cdot \log n)^{1/2})$ routing steps, with very high probability. The practical importance of this work is enhanced even more by the fact that the distribution of the packets only needs to be approximately a k-permutation.

1   Introduction

One of the main problems in the simulation of idealistic parallel computers by realistic ones is the problem of message routing through the sparse network of links which connect processing units, PUs, among each other. We consider the case that the PUs are connected by a two-dimensional mesh of communication links, of size $n \times n$ and additional wrap-around connections. This processor array can be seen as a torus. The additional wrap-around connections are a natural generalization of the standard grid and can be found in many real-world systems (e.g., ILLIAC IV, CAP-C5 (Fujitsu)). The links are bidirectional, one packet of information can go along a link in both directions in one routing step. The model of communication we use is the MIMD model. In this model, each PU can communicate with all its neighbors in a single step. The communication steps are synchronized. The PUs may store packets in a local queue.

We consider two different routing models: k-k routing and cut-through routing. The k-k routing problem, is the problem of transporting k packets from each PU in the mesh to k, not necessarily distinct, destination PUs, such that every PU finally receives k packets. In case $k = 1$, the problem is the permutation routing problem and has been attended most consideration in the past. In the k-k routing problem the packets coming from one PU have no relation to each other and can move independently of each other. In the cut-through model however, the k packets within a PU are considered as k flits arising from one big packet. The flits of a packet behave like a worm. All flits follow the first one, known as the head, to the destination. A worm may never be cut, i.e., at any given time, consecutive flits of a worm must be at the same or adjacent PUs. The goal of the research on routing problems is to construct

---

[*]Max-Planck-Institut für Informatik, 6600 Saarbrücken, Germany

[†]Department of Computer Science, University of Utrecht, P.O. Box 80.089, 3508 TB Utrecht, the Netherlands. Email: jopsi@cs.ruu.nl. The work of the author was financially supported by the Foundation for Computer Science (SION) of the Netherlands Organization for Scientific Research (NWO).

algorithms that bound the number of routing steps as well as the maximal used size of the queues. Furthermore, oblivious routing algorithms (for which the path of a packet through the mesh is independent of other packets) are preferred over adaptive algorithms. For an oblivious routing algorithm, it is possible to calculate the path of a packet during some preprocessing.

Many algorithms have been developed for the routing problems in both models for arrays without wrap-arounds. At the very beginning, 1-1 problems have been considered [6, 5, 11]. The case $k \geq 2$ gained more attention recently. Here the trivial lower bound is $k \cdot n/2$, which arises when half of the packets have to change sides. This bound is known as the *bisection* bound. Kunde and Tensi [7] achieved a bound of $5/4 \cdot k \cdot n + \mathcal{O}(k \cdot n/f(n))$ with queue size $\mathcal{O}(k \cdot f(n))$ using a deterministic algorithm. With randomization Rajasekaran and Raghavachari [13] could generalize [5] to get an algorithm with a $k \cdot n + o(k \cdot n)$ step bound and $\mathcal{O}(k)$ queue size. Recently Kunde [8] presented a deterministic algorithm which comes very close to the bisection bound: $\lceil k/2 \rceil \cdot n + \mathcal{O}(k \cdot n^{2/3})$. In this algorithm it is assumed that the k-k problem consists of k separated permutation problems that are known in advance. In [2] and [3] we present randomized algorithms which solve the general k-k routing problem in $k \cdot n/2 + \mathcal{O}((k \cdot n \cdot \log n)^{1/2})$ steps and $k + o(k)$ queue size (means optimally) with very high probability, improving Kunde's bounds slightly and not needing his assumptions. For the cut-through routing model, the literature is not so extensive, although this model and one of its variants, the wormhole model, is applied much more in practice. (PSC/2, NCUBE-2, Symult S2010, etc. [15]). Makedon and Simvonis [12] gave the first good worst case bounds for the cut-through model, which is also discussed by Leighton [10] for average-case behavior of the greedy algorithm. Rajasekaran and Raghavachari [13] show that their analysis for the general k-k routing problem also holds for the cut-through problem. The best bound is due to [2], where an almost optimal bound of $k \cdot n/2 + n/k + 3/2 \cdot n + \mathcal{O}((k \cdot n \cdot \log n)^{1/2})$ routing steps with very high probability is achieved.

An optimal algorithm for routing on a mesh with wrap-around connections immediately gives an optimal algorithm for routing on meshes without wrap-around connections. The opposite is far from true. E.g., for the 1-1 routing problem on the torus, the lower bound is $n - 1$. Using the basic algorithm of [5], permutations can be routed with $3/2 \cdot n + o(n)$ routing steps. The techniques of [5] and [11] to reduce the routing time to optimal on the mesh without wrap-around connections do not apply here. Routing on the torus has its own problems: On the torus all packets may have to go the maximal distance of n. This makes that there is much less slack than on the mesh without wrap-around connections. For the k-k routing problem on the torus we find something comparable: In this case, the bisection bound is $k \cdot n/4$. Directly applying the new techniques of [2] routes k-k permutations with $3/8 \cdot k \cdot n + \mathcal{O}((k \cdot n \cdot \log n)^{1/2})$ routing steps. This is not very satisfactory. We will show that an alternative 4-phase algorithm is optimal on both kinds of meshes. The number of routing steps is $k \cdot n/4 + \mathcal{O}((k \cdot n \cdot \log n)^{1/2})$ for k-k routing, and $k \cdot n/4 + 2 \cdot n + \mathcal{O}(k \cdot (n \cdot \log n)^{1/2})$ for cut-through routing. For the k-k routing k must be at least 8. So, for the values of k between 2 and 7 special algorithms might give better results as was done for the mesh without wrap-around connections in [4]. The results for the cut-through routing hold for all k. Thus, the cut-through routing problem is solved: the case $k = 1$ has no practical importance, and for $k \geq 2$ our algorithm is almost optimal. The algorithm is generalized for k-k routing and cut-through routing on meshes of arbitrary dimensions: On the d-dimensional mesh the algorithm consists of $2 \cdot d$ phases. If $k \geq 4 \cdot d$, then for the k-k routing every phase can be carried out with $k \cdot n/(8 \cdot d) + \mathcal{O}((k \cdot n \cdot \log n)^{1/2})$ routing steps, for a total of $k \cdot n/4 + \mathcal{O}(d \cdot (k \cdot n \cdot \log n)^{1/2})$. The results for the cut-through routing are similar. These results come close to the bisection bound. As already was shown in [8] this means that, although the diameter of the mesh increases with its dimension, the routing time does not increase as long as k is large enough. Another interesting feature of our algorithms is that the distribution of source destination pairs need not form exactly a k-k permutation. We only need that any subset of m PUs is sending and receiving at most $k \cdot m + \mathcal{O}((k \cdot n \cdot \log n)^{1/2})$ packets. This closely fits a practical communication desire.

Another practical situation is that the packets have to travel only a relatively small distance. This is the problem of 'routing with locality'. It is analysed in [14]. We will present a modified version of our algorithm that performs better and is more general. As long as the maximal distance packets have to go is globally known, the problem is rather easy. The most interesting algorithm is the one that performs optimally in a situation with local information only.

The remainder of this paper is organized as follows: Firstly we state the Chernoff bounds. Then, we shortly derive some new results on the routing in one-dimensional meshes. These results are used in the subsequent sections. In Section 4 we present an algorithm that routes k-k permutations optimally for $k \geq 8$. In Section 5 we show that this same algorithm is suited for routing in the cut-through mode. There is no restriction on the minimal number of flits per packet. After this, we generalize in Section 6 our techniques to k-k routing and cut-through in higher dimensions. Finally, we present in section 7 efficient algorithms for routing with locality.

2   Chernoff bounds

In the analysis we use the so-called Chernoff bounds. These bounds provide approximations to the probabilities in the tail ends of a binomial distribution.

Let $X = B(n, p)$ be the number of heads in n independent flips of a biased coin, the probability of a head in a single flip being p. Such an X has the binomial distribution. Using known results it is shown in [1] that

$$prob\{X \geq n \cdot p + h\} \leq e^{-h^2/(3 \cdot n \cdot p)}, \text{ for all } 0 < h < n \cdot p, \tag{1}$$

$$prob\{X \leq n \cdot p - h\} \leq e^{-h^2/(2 \cdot n \cdot p)}, \text{ for all } 0 < h < n \cdot p. \tag{2}$$

(1) and (2) give bounds on the probability of a large deviation of the number of heads from the expected number. The results of this paper hold "with very high probability":

Definition 1 *An event a happens with very high probability if $prob(a) \geq 1 - n^{-\alpha}$, for some constant $\alpha > 0$.*

Usually we will need that a polynomial number of independent binomial random variables, $X_1, \ldots, X_M$, $X_i = B(k \cdot n, p)$, $M \leq n^m$, for some constant m, are all bounded at the same time. (1) and (2) are so strong that this is no problem at all:

Lemma 1 *There is an $h = \mathcal{O}((p \cdot k \cdot n \cdot \log n)^{1/2})$, such that $p \cdot k \cdot n - h \leq X_i \leq p \cdot k \cdot n + h$, for all $1 \leq i \leq M$ at the same time, with very high probability.*

Proof: Let $h = (3 \cdot m \cdot p \cdot k \cdot n \cdot \log n)^{1/2}$. We show that the probability that one of the X_i exceeds the bound is very small: $P(\bigvee_{i=1}^{M}(X_i \leq p \cdot k \cdot n - h \text{ or } X_i \geq p \cdot k \cdot n + h)) \leq M \cdot (P(X_i \leq p \cdot k \cdot n - h) + P(X_i \geq p \cdot k \cdot n + h)) \leq e^{0.7 \cdot m \cdot \log n} \cdot 2 \cdot e^{-m \log n}$. □

3 One-dimensional routing

With aid of Lemma 1 we derive in this section some important results concerning the routing in a one-dimensional ring of PUs. These results serve us as a basis for the analysis of the two- and multidimensional case in subsequent sections. The PU with index k is denoted P_k.

3.1 Farthest-first priority strategy

Firstly, we consider routing packets in a one-dimensional ring of PUs of length n using the farthest-first conflict resolution strategy. We will use it later for the analysis of a k-k routing algorithm.

Define for a given distribution of packets over the PUs

$$h_r(i,j) = \#\{ \text{ packets passing from left to right through both } i \text{ and } j\} - 1.$$

and let T_r be the number of routing steps a distribution of packets requires for the routing to the right. T_l is defined analogously. This leaves open the possibility that $i > j$ and that the packets go around before they reach j.

Lemma 2 $T_r = \max\{ \max_{i<j}\{j - i + h_r(i,j)\}, \max_{j<i}\{n - j - i + h_r(i,j)\} \}$.

Proof: Let i_{max}, j_{max} be such that $h_r(i_{max}, j_{max}) + (j_{max} - i_{max}) \bmod n = \max_{i,j}\{h_r(i,j) + (j - i) \bmod n\}$. No packet starting in a PU with index i_{max} or smaller will reach $P_{j_{max}}$ before step $(j_{max} - i_{max}) \bmod n$. At most one packet reaches a PU from the right in any step. Thus T_r is at least as big as stated.

Now we prove the other side of the inequality. Consider the packets moving to or through some PU P_j from left to right. These packets are called j-packets. We will show that

Claim: The last j-packet reaches P_j after $\max_i\{(j - i) \bmod n + h_r(i,j)\}$ steps.

A j-packet can only be delayed by other j-packets. Thus, for the proof of the claim we can neglect all other packets. Let i_{max} be such that $h_r(i_{max}, j) + (j - i_{max}) \bmod n = \max_i\{h_r(i,j) + (j - i) \bmod n\}$. Define $\mathcal{P}_l = \{P_{j+1}, \ldots, P_{i_{max}}\}$, $\mathcal{P}_r = \{P_{i_{max}} + 1, \ldots, P_{j-1}\}$. One of these sets wraps around (unless $j = 1$ or n). Consider the following subdivision of the mesh (that was cut open to the right of P_j):

	A	B		P_j

i_{max}

A and B are sections of $\mathcal{P}_l$ and $\mathcal{P}_r$ respectively. The definition of i_{max} implies that for all such sections A, B,

$$\#B_j \ \leq \ \#B, \tag{3}$$

$$\#A_j \ \geq \ \#A. \tag{4}$$

Here is B_j is the set of j-packets in B. (3) holds because otherwise i_{max} would have been chosen larger; (4) holds because otherwise i_{max} would have been chosen smaller. By (3) the j-packets that start in $\mathcal{P}_r$ can move ahead of the j-packets coming from $\mathcal{P}_l$. This means that the j-packets from $\mathcal{P}_r$ reached P_j by step $(j - i_{max}) \bmod n - 1$, and that these packets will not delay the j-packets from $\mathcal{P}_l$. By (4) the j-packets from $\mathcal{P}_l$ flow into P_j continuously from step

$(j - i_{max})$ mod n on. Thus, the last of them reaches P_j by step $h_r(i_{max}, j) + (j - i_{max})$ mod n, as stated. That the packets from $\mathcal{P}_l$ really flow into P_j without interruption may require explanation: These packets will not reach $\mathcal{P}_r$ faster if some of them are moved to the left. Now, we can wipe the j-packets to the left in a special way: For all i from i_{max} down to the smallest index (possibly we have to go around to the large indices) of a PU that sends packets to P_j, all j-packets in P_i except for one, are moved to P_{i-1}. (4) assures that there is always a packet to leave behind. After the wiping all these packets lie as a long uninterrupted chain in $\mathcal{P}_l$, with possibly a pile at the left-most PU that sends a packet to P_j. The movement of the chain is not retarded and thus it moves every step one position to the right.

The claim gives us the latest time for packets to reach their destination PU. Since it holds for arbitrary j, maximizing over j gives the total time bound. This is exactly the bound given in the lemma. □

In [3] we also prove a version of this lemma for a mesh without wrap-around connection. Then

Lemma 3 $T_r = \max_{i<j}\{j - i + h_r(i, j)\}$.

Combining Lemma 1, Lemma 2 and its analogue for routing to the left, we get a result for randomized routing on a ring of PUs:

Lemma 4 *Routing k packets from every PU of a circular processor array with n PUs to a random destination requires at most $\max\{n/2, k \cdot n/8\} + \mathcal{O}((k \cdot n \cdot \log n)^{1/2})$ routing steps, with very high probability.*

Proof: Increasing the number of packets to route will not decrease the routing time. Thus, we may assume that $k \geq 4$. We are going to use Lemma 2, so we have to show that $h_r(i, j)$ is small enough for arbitrary i and j. It is easy to check that for given i and j the expected number of packets going from left to right through both of them is $k \cdot \sum_{l=1}^{n/2-(j-i)\bmod n} l/n \simeq k \cdot (n/2 - (j - i) \bmod n)^2/(2 \cdot n)$. The value is maximal when $j = i + 1$. For $k \geq 4$ this maximal value is not greater than $k \cdot n/8$. □

3.2 Arbitrary priority strategy

For the analysis of the cut-through routing algorithm of Section 5 we need strong results without assuming the farthest-first strategy. We will prove analogues of Lemma 2 and Lemma 4 for the case that upon a collision any packet may be selected for proceeding first.

Analogously to $h_r(i, j)$ we define $h_r(i) = \#\{$packets to be routed out of P_i to the left$\} - 1$. T_r is defined as before. Define i_{max} to be an index of a PU such that $h_r(i_{max}) = \max_i\{h_r(i)\}$.

Lemma 5 *After t steps, a packet is delayed in the PU where it resides at most $h_r(i_{max}) - t$ times.*

Proof: Consider a packet p in a PU P. There are two possibilities: (1) Until now P has been passing packets to the right continuously; (2) P has been idle during some earlier step. In the first case there are still at most $h_r(i_{max}) - t + 1$ packets that P has to pass to the right. In any of the following steps, P can pass p to the right or another packet. After at most $h_r(i_{max}) - t + 1$ times passing other packets, it must pass p. In the second case, there were no packets in P that had to go to the right during some earlier step. Hereafter, packets will not be delayed in P anymore: They are sent on by P, the step after they are received. □

Corollary 1 *A packet is delayed at most $h_r(i_{max})$ times.*

Summing the maximal path length and the maximal delay, we find the analogue of Lemma 2:

Lemma 6 $T_r \leq h_r(i_{max}) + n/2$.

Now, we can prove an analogue of Lemma 4.

Lemma 7 *Routing with the arbitrary conflict resolution strategy k packets from every PU of a circular processor array with n PUs to a random destination requires at most $k \cdot n/8 + n/2 + \mathcal{O}((k \cdot n \cdot \log n)^{1/2})$, routing steps, with very high probability.*

Proof: By Lemma 6 we only have to check that $\max_i\{h_r(i)\}$ is bounded. The expected value of $h_r(i)$ is $k \cdot \sum_{j=1}^{n/2} j/n \simeq k \cdot n/8$, for all i. Thus, by Lemma 1 $h_r(i) \leq k \cdot n/8 + \mathcal{O}((k \cdot n \cdot \log n)^{1/2})$, for all i at the same time, with very high probability. □

In comparison with the result of Lemma 4 we see an additional term of $n/2$ in the routing time of a randomization. This is inevitable.

Lemma 7 holds for the case that $k \cdot n$ packets are routed on a circular processor to a random destination independently of each other. In the case of cut-through routing we have to be careful: the k-flits that constitute a packet are routed to the same destination. This means that the bound on $h_r(i)$ given by Lemma 1 is slightly less sharp. We find

Lemma 8 *Routing in the cut-through mode one packet consisting of k flits each from every PU of a one-dimensional processor array with n PUs to a random destination requires at most $k \cdot n/8 + n/2 + \mathcal{O}(k \cdot (n \cdot \log n)^{1/2})$, routing steps, with very high probability.*

4   The algorithm

The algorithm is variant of the earlier algorithm for routing on the hypercube of Valiant & Brebner [16] for the 2-dimensional mesh, completed by an application of the overlapping technique due to [7]. In Valiant & Brebner's algorithm the packets are distributed randomly within the hypercube during Phase 1. In Phase 2 they are routed dimension by dimension to their destination. The overlapping technique consists in a partition of the problem in roughly two equal subproblems which can be solved simultaneously using the same algorithms but one is rotated by 90°. Our algorithm consists of four phases. The first two for randomizing the packets along the row and column, the last two for sending the packets to their destination along the row and column. The algorithm can also be seen as a variant of the algorithm of [5] with a much more intensive randomization of the packets. We will show that this algorithm performs optimally (apart from a small additional term). This is an amazing result: Consider a permutation under which all packets have to move over exactly n steps (e.g., the permutation under which the packets from $P_{i,j}$ have to go to $P_{(i+n/2)\bmod n,(j+n/2)\bmod n}$. In total the connections must be used $k \cdot n^3$ times. As every PU has 4 connections, such a permutation requires at least $n/4$ routing steps. Our algorithm solves the k-k routing problem with $k \cdot n/4 + \mathcal{O}((k \cdot n \cdot \log n)^{1/2})$. At first sight, that seems to be impossible because a lot of packets are sent in a wrong direction during the randomization, thereby wasting connection availability. However, packets are only randomized in a wrong direction when the permutation itself makes sub-maximal use of the connections. In this and subsequent sections, we assume that the mesh is indexed by column and row position, starting with $(0,0)$ at the lower-left position. The PU with index (i,j) will be denoted $P_{i,j}$.

Initially the packets are colored randomly green or blue with equal probability. Thereafter, the green packets are routed by

Proc GreenRoute;
1. randomize the packets along the row;
2. randomize the packets along the column;
3. send the packets along the row to the correct column;
4. send the packets along the column to the correct row.

The blue packets are routed 90° out of phase of the green packets by the analogous procedure BlueRoute. The algorithm for routing all packets at the same time will be denoted Green/BlueRoute.

The randomization really randomizes:

Lemma 9 *After phase 1 and 2 a packet has equal probability to be in any of the n^2 PUs.*

Proof: Consider a packet residing in $P_{i,j}$. This packet is after phase 1 and 2 in $P_{i',j'}$ iff it was randomized in phase 1 to $P_{i,j'}$ and in phase 2 from there to $P_{i',j'}$. The probability that this happens is $1/n^2$ independent of i', j'. □

Once we know that the packets are really randomly distributed, phase 3 is just an inverse of a randomization and can be performed with the same number of routing steps. Likewise for phase 4. Now, we can use Lemma 4 for all four phases. For $k \geq 8$ this gives

Theorem 1 *Routing k-k permutations with Green/BlueRoute on a mesh with wrap-around connections takes at most $k \cdot n/4 + \mathcal{O}((k \cdot n \cdot \log n)^{1/2})$ routing steps with very high probability.*

In the proof of Theorem 1 we neglected the fact that at the start of Phase 2, 3 and 4, the packets are not distributed in an entirely regular way over the mesh. This is not a serious problem: going back to the proof of Lemma 2 we see that we only need to be sure that any subset of PUs of size m sends and receives at most $k \cdot m + \mathcal{O}((k \cdot n \cdot \log n)^{1/2})$ packets.

4.1 The size of the queues

The expected number of packets that resides in a PU is bounded by $k + \mathcal{O}(1)$. For $k \geq n$, the Chernoff bounds give that then the maximal number of packets in any PU is bounded by $k + o(k)$. For small k we cannot bound the number of packets in a single PU. But the Chernoff bounds give that the number of packets in any m PUs is bounded by $k \cdot m + \mathcal{O}((k \cdot m \cdot \log n)^{1/2})$. Employing a technique of [14] the packets in m adjacent PUs can be smeared out. This takes $\mathcal{O}(m)$ steps.

So, with an additional $\mathcal{O}(m)$ steps the queue sizes can be bounded to $k + \mathcal{O}((k \cdot \log n/m)^{1/2})$. Taking $m = 1$ gives the above stated result. If $k \leq n^\alpha$, for some $\alpha < 1$, we can take $m = k \cdot \log n$. Resuming,

Theorem 2 *If $k \leq n^\alpha$, for some $\alpha < 1$, then the queue size can be bounded to $k + \mathcal{O}(1)$, with very high probability, at the expense of $\mathcal{O}(n^\alpha \cdot \log n)$ extra routing steps. For larger k the queue size is bounded by $k + o(k)$.*

Note that $\mathcal{O}(n^\alpha \cdot \log n) = o(n)$, under the given condition on k. This theorem is general and in the same way the queue size of all algorithms in this paper can be bounded.

5 Cut-through Routing

We will show that using Green/BlueRoute without any specific conflict resolution strategy routes k-k permutations with at most $k \cdot n/4 + 2 \cdot n + \mathcal{O}((k \cdot n \cdot \log n)^{1/2})$ routing steps with very high probability. We only assume that each PU always sends a packet if one is available. This general assumption immediately gives the same result (apart from a slightly higher additional term) for routing packets consisting of k flits each in the cut-through model.

Like in Section 4 all four phases of Green/BlueRoute take the time required for routing a randomization. Using Lemma 7, this gives

Theorem 3 *Green/BlueRoute with arbitrary queuing strategy takes at most $k \cdot n/4 + 2 \cdot n + \mathcal{O}((k \cdot n \cdot \log n)^{1/2})$ routing steps, with very high probability.*

This theorem holds for all values of k. For the cut-through routing we must use Lemma 8:

Theorem 4 *Using Green/BlueRoute for routing packets consisting of k flits each in the cut-through model takes at most $k \cdot n/4 + 2 \cdot n + \mathcal{O}(k \cdot (n \cdot \log n)^{1/2})$ routing steps, with very high probability.*

6 Higher dimensions

In this section we consider k-k and cut-through routing on meshes with wrap-around connections of dimension $d > 2$. As the diameter of an d-dimensional mesh with n^d PUs is $d \cdot (n - 1)$ one might expect that the routing time would grow with d. This is not the case when k is sufficiently large: we will be able to show that if $k \geq 4 \cdot d$, k-permutations can be routed by a $2 \cdot d$ phase generalization of Green/BlueRoute with $k \cdot n/4 + \mathcal{O}(d \cdot (k \cdot n \cdot \log n)^{1/2})$ routing steps. This almost matches the lower bound derived from the connection availability (Section 4) of $k \cdot n/4$. For all values of k the cut-through routing of packets consisting of k flits each can be performed with $k \cdot n/4 + d \cdot n + \mathcal{O}(d \cdot k \cdot (n \cdot \log n)^{1/2})$ routing steps.

The algorithm starts by randomly attributing to all packets a color l, $0 \leq l \leq d - 1$, with equal probability. It is not important whether all packets in a single PU get different colors or that the colors are attributed independently with equal probability. The packets of color l are randomized and routed by

       Proc WrapRoute(d, l);
    for every packet p of color l **do**

for $j := l$ to $l + d - 1$ do route p to a random destination along axis $j \bmod d$;
for $j := l$ to $l + d - 1$ do route p to the correct position on axis $j \bmod d$;

The packets of different colors move along different axis at all times. After the randomization every packet has equal probability to be in any PU. This is demonstrated analogously to Lemma 9. In total there are $2 \cdot d$ routing phases. Each of them takes as much time as a randomization on a k/d loaded circular one-dimensional processor array. Thus, by Lemma 4

Lemma 10 *On a mesh with wrap-around connections, all phases of WrapRoute(d) can be carried out with* $\max\{n/2, k \cdot n/(8 \cdot d)\} + \mathcal{O}((k \cdot n \cdot \log n)^{1/2})$ *routing steps, with very high probability.*

In total,

Theorem 5 *WrapRoute(d) routes k-permutations on the d-dimensional mesh with wrap-around connections with* $\max\{d \cdot n, k \cdot n/4\} + \mathcal{O}(d \cdot (k \cdot n \cdot \log n)^{1/2})$ *routing steps, with very high probability.*

It turns out, that apart from a lower order term, the routing time for k-permutations is independent of the dimension of the mesh as long as there are enough packets to assure a good loading. When there are less packets the diameter of the mesh, $d \cdot (n-1)$, will play a role. The result of Theorem 5 matches the bisection bound for $k \geq 4 \cdot d$.

For cut-through routing we get

Theorem 6 *WrapRoute performs cut-through routing on the d-dimensional mesh with wrap-around connections of packets consisting of k flits each, with at most* $k \cdot n/4 + d \cdot n + \mathcal{O}(d \cdot k \cdot (n \cdot \log n)^{1/2})$ *routing steps, with very high probability.*

7 Routing with respect to locality

In this section we describe how our algorithm works in an environment, where each packet only has to travel a short distance. We describe only the two-dimensional case, generalizations can be done in the same way as described above. We consider two cases: In the first case, which is easier to analyse but which is not directly applicable to practice, we assume that the maximum distance L any packet has to travel in one directions is known to all PUs. We say the problem has locality L. In the second case we assume that the PUs have only local information. The lower bound for routing with locality L is higher than might be expected: Consider a permutation under which all packets have to move exactly L steps to the right and L steps up. In total the $k \cdot n^2$ packets have to move $2 \cdot L \cdot k \cdot n^2$ steps. Every PU has four connections over which packets could be transmitted, but only two of them are useful for bringing packets closer to their destination. Thus, this routing will take at least $k \cdot L$ routing steps. Due to space limitations we left out some of the proofs and most explanation that can be found in [3].

7.1 Locality L

In this section, we analyse k-k routing when the maximal distance L a packet has to go in one direction is known to all PUs. Our three-phase algorithm resembling the basic algorithm of [5] achieves close to optimal. It proceeds as follows:

Proc LocalRoute(L);

1. **for every packet** p **in** $P_{i,j}$ **do**
   generate a random number $x_p \in \{L/2, \ldots, L/2\}$; route p to row $i + x_p$;
2. **for every packet** p **with destination** $P_{i,j}$ **do** route p to column j;
3. **for every packet** p **with destination** $P_{i,j}$ **do** route p to row i;

The analysis is based on Lemma 3. As we saw before, when the origins or the destinations of the packets are uniformly distributed, the routing time equals the maximum of the maximal distance over which a packet has to move during a phase and the maximal number of packets a connection has to transmit. In this way we find

Lemma 11 *Phase 1 takes at most* $\max\{L/2, k/8 \cdot L\} + \mathcal{O}((k \cdot L \cdot \log n)^{1/2})$ *routing steps.*

After phase 1 each packet resides in a random row within distance $L/2$ from the row where it started. The packets are still well-distributed: Any subset of m PUs contains at most $k \cdot m + \mathcal{O}((k \cdot m \cdot \log n)^{1/2})$ packets. Thus, during phase 2 no connection has to transmit more than $k \cdot L + \mathcal{O}((k \cdot L \cdot \log n)^{1/2})$ packets. More appealing randomization techniques, requiring less routing steps, do not have this essential property. Now we get (proof omitted)

Lemma 12 *Phase 2 takes at most* $\max\{L, k \cdot L\} + \mathcal{O}((k \cdot L \cdot \log n)^{1/2})$ *routing steps.*

The analysis of LocalRoute is completed with

Lemma 13 *Phase 3 takes at most* $\max\{3/2 \cdot L, k \cdot L\} + \mathcal{O}((k \cdot L \cdot \log n)^{1/2})$ *routing steps.*

Proof: In phase 3, a packet moves over distance $3/2 \cdot L$ at most (when it had to go L steps and was randomized in phase 1 to the extreme of the strip of size L around its origin). Consider the packets that have to pass through a PU $P_{i,j}$ from below. These packets must have their destination in one of the PUs $P_{i+i',j}$, for some $1 \leq i' \leq L$. There are $k \cdot L$ such packets. □

Introducing two colors and routing the green and blue packets 90° degrees out of phase:

Theorem 7 *k-k routing with locality L can be performed on a mesh with wrap-around connections with* $\max\{5/2 \cdot L, k \cdot L\} + L/2 + \mathcal{O}((k \cdot L \cdot \log n)^{1/2})$ *routing steps for $k < 8$ and for $k \geq 8$ with* $17/16 \cdot k \cdot L + \mathcal{O}((k \cdot L \cdot \log n)^{1/2})$ *routing steps, with very high probability.*

The queues are bounded by a smearing technique to $k + o(k)$ (see Section 4.1).

7.2 Local information only

Now, we consider the case that the locality value L is not known, but that each PU only knows the destination of its own packets.

For phase 2 and phase 3 it is not necessary to know L. An estimate of L is only necessary for knowing how large the strips must be in which the randomization of phase 1 is carried out. This randomization in strips of size ϵ is performed to prevent the queues getting long during phase 2. It is not necessary that we know the precise value of L. A good estimate of it can be just as useful. E.g., suppose that a PU can hold at most $2 \cdot k + \mathcal{O}(1)$ packets. Then, we can take any $\epsilon \geq L/2$: we only have to calculate the value of L within a factor two. This requires the transmission of much less packets than the calculation of the exact value of L. The connections that are not used for transmitting these information packets can be used for a preliminary randomization of the packets. In more detail, phase 1 of LocalRoute is replaced by

a. **for** every PU $P_{i,j}$ **do**

$L_{i,j} :=$ the largest move in one direction a packet residing in $P_{i,j}$ has to make;

route a packet with $L_{i,j}$ in all four directions;

if a packet with $L' \geq 2 \cdot L_{i,j}$ reaches $P_{i,j}$

then $L_{i,j} := L'$; route a packet with L' in all four directions

else eliminate the packet;

b. **for** every green packet p in $P_{i,j}$ **do**

generate a random number $x_p \in \{-8 \cdot n/k, \ldots, 8 \cdot n/k\}$;

route a packet (p, i, x_p) to row $i + x_p$;

c. **for** every PU $P_{i,j}$ **do** $L_{i,j} := 2^{\lfloor \log L_{i,j} \rfloor}$;

d. **for** every PU $P_{i,j}$ **do**

if $L_{i,j}/2 > 8 \cdot n/k$ **then** { the randomization is not yet completed }

for every green packet (p, i', x_p) residing in $P_{i,j}$ **do**

Route p to row $i' + x_p \cdot L_{i,j} \cdot k/(16 \cdot n)$;

The packets of step a are called information packets. No PU will transmit more than $\log n$ information packets over a single connection. Step a and b are coalesced with priority for the information packets. Step b randomizes the packets in strips of size $8 \cdot n/k$. Because every PU holds $k/2$ green packets on the average, this can be done with $n + \mathcal{O}((n \cdot \log n)^{1/2})$ routing steps. Thus, step a and step b can be routed together with $n + \mathcal{O}((n \cdot \log n)^{1/2})$ routing steps. Step c is executed after routing step n. After step c all PUs will have the same estimate L' satisfying $L/2 < L' \leq L$. If the randomization of step b was good enough, the packets can be routed to their destination immediately (phase 2 and phase 3), otherwise some additional randomization has to be performed first (step d).

Theorem 8 *A k-k routing problem with unknown locality L can be performed on a mesh with wrap-around connections with $\max\{n, k \cdot L/16\} + k \cdot L + \mathcal{O}(((n + k \cdot L) \cdot \log n)^{1/2})$ routing steps, with very high probability.*

For $k \geq 16 \cdot n/L$ this result is the same as the result of Theorem 7.

8 Conclusion

In this paper we presented a randomized algorithm for k-k and cut-through routing on meshes with wrap-around connections of arbitrary dimension. For $k \geq 4 \cdot$ *dimension of the mesh*, the algorithm is optimal for the k-k routing. For the cut-through routing, the algorithm is close to optimal for arbitrary k.

References

[1] Hagerup, T., and Rüb, C. 'An efficient guided tour of Chernoff bounds,' Inf. Proc. Lett., 33 (1990), 305-308.

[2] Kaufmann, M., Rajasekaran, S., and Sibeyn, J., 'Matching the Bisection Bound for Routing and Sorting on the Mesh,' preprint, submitted to SPAA '92.

[3] Kaufmann, M., and Sibeyn, J., 'Randomized Multi-Packet Routing on Meshes,' Techn. Rep. RUU-CS-91-48, Dep. of CS, Utrecht University, Utrecht, 1991.

[4] Kaufmann, M., and Sibeyn, J., 'Multi Packet Routing in Lightly Loaded Arrays, draft (1991).

[5] Krizanc, D., Rajasekaran, S., and Tsantilas, T. 'Optimal Routing Algorithms for Mesh-Connected Processor Arrays,' Proc. VLSI Algorithms and Architectures: AWOC, 1988. Springer-Verlag LNCS #319, pp. 411-22.

[6] Kunde, M. 'Routing and Sorting on Mesh-Connected Processor Arrays,' Proc. VLSI Algorithms and Architectures: AWOC 1988. Springer-Verlag LNCS #319, pp. 423-33.

[7] Kunde, M., and Tensi, T. 'Multi-Packet Routing on Mesh Connected Arrays,' Proc. ACM Symposium on Parallel Algorithms and Architectures, SPAA 89, ACM Press, pp. 336-343.

[8] Kunde, M., 'Concentrated Regular Data Streams on Grids: Sorting and Routing Near to the Bisection Bound,' Proc. IEEE Symposium on Foundations of Computer Science, 1991, pp. 141-150.

[9] Kunde, M. 'Balanced Routing: Towards the Distance Bound on Grids,' Proc. 1991 ACM Symposium on Parallel Algorithms and Architectures, SPAA 91, ACM Press, pp. 260-271.

[10] Leighton, T., 'Average Case Analysis of Greedy Routing Algorithms on Arrays,' Proc. ACM Symposium on Parallel Algorithms and Architectures, SPAA 90, ACM Press, pp. 2-10.

[11] Leighton, T., Makedon, F., and Tollis, I.G. 'A $2n - 2$ Step Algorithm for Routing in an $n \times n$ Array With Constant Size Queues,' Proc. ACM Symposium on Parallel Algorithms and Architectures, SPAA 89, ACM Press, pp. 328-35.

[12] Makedon, F., Simvonis, A. 'On bit Serial packet routing for the mesh and the torus.' Proc. 3rd Symposium on Frontiers of Massively Parallel Computation, 1990, pp. 294-302.

[13] Rajasekaran, S., and Raghavachari, M. 'Optimal Randomized Algorithms for Multipacket and Cut Through Routing on the Mesh,' presented at the 3rd Symposium on Parallel and Distributed Computing 1991.

[14] Rajasekaran, S., and Tsantilas, T. 'An Optimal Randomized Routing Algorithm for the Mesh and A Class of Efficient Mesh like Routing Networks,' Proc. 7th Conference on Foundations of Software Technology and Theoretical Computer Science, 1987. Springer-Verlag Lecture Notes in Computer Science #287, pp. 226-241.

[15] Trew, A., and Wilson ,G. 'Past, Present, Parallel - A survey of Available Parallel Computing Systems,' Springer-Verlag 1988.

[16] Valiant, L., and Brebner, G.J., 'Universal schemes for parallel communication,' Proc. 13th ACM Symp. on Theory of Comp., 1981, pp 263-277.

Parallel Algorithms for Priority Queue Operations[*]

Maria Cristina Pinotti[1] and Geppino Pucci[2]

[1] Istituto di Elaborazione della Informazione, CNR, Pisa, Italy
[2] Dipartimento di Informatica, Università di Pisa, Pisa, Italy

Abstract. This paper presents parallel algorithms for priority queue operations on a p-processor EREW-PRAM. The algorithms are based on a new data structure, the Min-path Heap (MH), which is obtained as an extension of the traditional binary-heap organization. Using an MH, it is shown that insertion of a new item or deletion of the smallest item from a priority queue of n elements can be performed in $O(\frac{\log n}{p} + \log \log n)$ parallel time, while construction of an MH from a set of n items takes $O(\frac{n}{p} + \log n)$ time. The given algorithms for insertion and deletion achieve the best possible running time for any number of processors p, with $p \in O(\frac{\log n}{\log \log n})$, while the MH construction algorithm employs up to $\Theta(\frac{n}{\log n})$ processors optimally.

Keywords. Analysis of Algorithms, Data Structures, Heaps, Parallel Algorithms.

1 Introduction

A *Priority Queue* (PQ) is an abstract data type storing a set of integer-valued items and providing operations such as insertion of a new item and deletion of the smallest stored item. In this note we introduce the *Min-path Heap* (MH) data structure. We employ this new structure to develop efficient parallel algorithms for the basic PQ operations of insertion, deletion and construction on the EREW-PRAM [5] model of computation.

Several parallel implementations of PQ's can be found in the literature. The first approach to date is due to Biswas and Browne [2], subsequently improved by Rao and Kumar in [7]. In their schemes, $p \in O(\log n)$ processors concurrently access a binary heap of n elements by acquiring locks on the nodes of the heap. Insertions and deletions are executed in a pipelined fashion, thus increasing the throughput of the structure from one to p simultaneous operations. However, the time requirement for a single insertion or deletion remains $O(\log n)$. More recent papers deal with the problem of speeding up a *single* heap operation. In this direction, optimal parallel algorithms for heap construction have been devised for the PRAM model in [6] and [8]. As to insertion and deletion, we are only aware of the implementation devised by Zhang in [9] which requires $O(\log n \log \log n)$ work and $O(\log \log n)$ time with

---

[*] This work has been supported by the C.N.R. project "Sistemi Informatici e Calcolo Parallelo". Part of this research was done while G. Pucci was visiting the International Computer Science Institute, Berkeley, California.

$p = \log n$ PRAM processors. Note that the scheme fails to attain linear speedup by a factor of $O(\log \log n)$.

In the following sections we provide optimal parallel algorithms for PQ operations based on the MH data structure. Our results are the following. Let M be an MH of n elements stored in the shared memory of a p processor EREW-PRAM. We show how to insert a new item or delete the smallest item from M in parallel time $O\left(\frac{\log n}{p} + \log \log n\right)$. Moreover, we adapt the above referenced algorithms for parallel heap construction so that M can be built from a set of n elements in time $O\left(\frac{n}{p} + \log n\right)$. Our insertion and deletion algorithms achieve the best possible running time for any number of processors p, with $p \in O(\frac{\log n}{\log \log n})$, while the MH construction algorithm employs up to $\Theta(\frac{n}{\log n})$ processors optimally.

2 Parallel Algorithms for MH Operations

Our MH data structure is obtained as an extension of the traditional *binary heap* organization. Recall that a binary heap H is a complete binary tree (stored in vectorial form) where each node i contains an item, $H[i]$, whose value is less than the values $H[2i]$ and $H[2i+1]$ stored at its children $2i$ and $2i+1$[1]. For any node i of H, its *min-path* μ_i is the set of nodes defined by the following recurrence:

1. i belongs to μ_i.
2. Let a non leaf node j belong to μ_i. If j has only one child u, then u belongs to μ_i. Otherwise, if u and v are the children of j and $H[u] < H[v]$, then u belongs to μ_i.

Informally, μ_i is the unique path in H from i to a *target leaf* of address L_i, such that each internal node on the path stores an item whose value is less than the one stored at its sibling. Note that since H is stored as a vector, we can easily determine the addresses of all the nodes on μ_i from L_i. More precisely, if $h \geq 1$ is the height of H and i is at level k, $1 \leq k \leq h$, the nodes on μ_i have addresses L_i **div** 2^j, with $0 \leq j \leq h - k$. The importance of min-paths for the realization of fast parallel algorithms for PQ operations will be made clear in subsection 2.2.

We have just shown that in order to have fast access to min-path information it suffices to maintain, for each node i of H, the address of its target leaf L_i. Therefore, we define a *Min-path Heap* M to be a data structure whose representation consists of two vectors:

- M_H, where the items are stored in a binary heap fashion.
- M_L, with $M_L[i]$ storing L_i, that is, the address of the target leaf of node i.

In addition to restoring the heap order on M_H, insertion and deletion algorithms for an MH M have to update the min-path information stored in M_L. We also associate M with an integer variable, N_M, denoting the number of nodes currently

---

[1] For the sake of simplicity, we shall assume that all the items stored or to be inserted in an MH are distinct. The handling of duplicates requires some trivial modifications to the algorithms, whose complexities remain unaltered.

stored in M. Note that an MH induces only a constant factor increase in space over the traditional binary heap organization. Moreover, its representation is simple and compact (compare it with the structure in [9], where some nodes have to store a table of $O(\log \log n)$ entries).

In the following sections we will sometimes refer to M, M_H and M_L using the classical binary tree terminology. For instance, we will say that $h = \lfloor \log N_M \rfloor + 1$ is the *height* of M, M_H or M_L and will use terms like *leaf* or *sibling* to denote particular locations in the vectors.

In order to describe the EREW-PRAM algorithms for the basic operations on an MH, we introduce the following conventions. Let S be a set of processor indices. The statement

for $i \in S$ do in parallel *statement list* **endfor**

denotes $|S|$ parallelizable executions of the statement list, one execution for each index $i \in S$. If a statement is not within the scope of a **for** ... **endfor** construct, it is executed in parallel by all the active processors. Within a **for** ... **endfor** construct, the statement

$$P_i : \text{BROADCAST}(X = x)$$

denotes the computation needed to broadcast the value x from the processor P_i in charge of the ith instance of the statement list to all the active processors. These processors will store the received value in their local variable X. This operation can be realized in time $O(\log p)$ on an EREW-PRAM, where p is the number of active processors. Finally, if M is an MH of height h and L is the address of one of its leaves, the function

$$\text{SIBLING}(M, L, i) = \textbf{ let } K = L \textbf{ div } 2^{h-i} \textbf{ in } K + (-1)^K \textbf{ mod } 2$$

returns the address of the sibling of the ith node on the path in M from the root to L.

2.1 Insertion

The algorithm for inserting a new item I in an MH M of height h proceeds in two phases:

1. The processors determine the position that item I has to occupy in the *insertion path* μ_I from the root to the first vacant leaf of M_H (which has address $N'_M = N_M + 1$) so that the heap order in M_H is not violated. Note that the ith node on the insertion path, starting from the root, has address $n_i = N'_M \textbf{ div } 2^{h-i}$ for $1 \le i \le h$. Once such position n_k has been determined, all the items stored in $M_H[n_j]$, $k \le j \le h-1$ are shifted to position $M_H[n_{j+1}]$ and I is stored in $M_H[n_k]$.

2. The processors recompute the new target leaves for the nodes of μ_I. For nodes n_j, $k \le j \le h$, it must be $M_L[n_j] = N'_M$, as in the updated M_H we have $M_H[\text{SIBLING}(M, N'_M, j)] > M_H[n_j]$, for $k + 1 \le j \le h$, because of the heap property and the shift performed on μ_I. For nodes n_j, $1 \le j \le k-1$, the target leafs may be different only if their min-path μ_{n_j}, prior to the insertion, went

through n_k or SIBLING(M, N'_M, k) (this can be easily checked by comparing $M_L[n_j]$ with $M_L[n_k]$ and $M_L[\text{SIBLING}(M, N'_M, k)]$). The new target leaf for such nodes is N'_M if $I < M_H[\text{SIBLING}(M, N'_M, k)]$ and $M_L[\text{SIBLING}(M, N'_M, k)]$ otherwise.

Note that the above strategy still yields a valid MH when we insert an item starting from any empty leaf of address different from N'_M. In particular, the deletion algorithm described in the following subsection creates a "hole" in the structure at the target leaf of the root, $L = M_L[1]$. The hole is then refilled by performing an insertion starting from L.

The following procedure INSERT implements the above ideas. Parameter L is the address of a leaf, while parameter I is the item to be inserted. INSERT uses the auxiliary vectors V_H and V_L to perform operations on the elements stored along the insertion path μ_I and their target leaves.

procedure INSERT(L, I):

    $h := \lfloor \log L \rfloor + 1$; { height of M_H }

    for $i \in \{1, \ldots, h-1\}$ **do in parallel**

        $V_H[i] := M_H[L \text{ div } 2^{h-i}]$;

        $V_L[i] := M_L[L \text{ div } 2^{h-i}]$;

        {copy the elements and target leaves stored along μ_I}

    endfor;

    for $i \in \{1\}$ **do in parallel**

        $V_H[0] := -\infty$

        $V_H[h] := +\infty$

        {these two dummy elements are needed for the next parallel steps}

    endfor;

    for $i \in \{1, \ldots, h\}$ **do in parallel**

        if $V_H[i] > I$ and $V_H[i-1] < I$

            then P_i : BROADCAST(Pos $= i$)

            {i is the position where I must be inserted}

        fi

    endfor;

    for $i \in \{\text{Pos}, \ldots, h-1\}$ **do in parallel**

        TEMP $:= V_H[i]$; $V_H[i+1] := TEMP$;

        { shift the items greater than $I \ldots$ }

        $V_L[i+1] := L$;

        { $\ldots$ and set their target leaves to L}

    endfor;

    for $i \in \{\text{Pos}\}$ **do in parallel**

        if $i > 1$ **then**

            if $V_H[i] \geq M_H[\text{SIBLING}(M, L, i)]$

                then P_i : BROADCAST($L_1 = M_L[\text{SIBLING}(M, L, i)]$)

                else P_i : BROADCAST($L_1 = V_L[i]$)

            fi;

            { if $V_L[j] = L_1$ for $j <$ Pos, then $V_L[j]$ must be set$\ldots$}

            if $I < M_H[\text{SIBLING}(M, L, i)]$

                then P_i : BROADCAST($L_2 = L$)

```
      else P_i : BROADCAST(L_2 = M_L[SIBLING(M, L, i)])
   fi
   { ... to L_2}
   fi;
   V_H[i] := I; V_L[i] := L;
endfor;
for i ∈ {1, ..., Pos − 1} do in parallel
   if V_L[i] = L_1
      then V_L[i] := L_2
   fi
endfor;
for i ∈ {1, ..., h} do in parallel
   M_H[L div 2^{h−i}] := V_H[i];
   M_L[L div 2^{h−i}] := V_L[i];
   { copy the updated path back}
endfor
end INSERT.
```

To insert a new item I, we first increment N_M and then call INSERT(N_M,I). The time complexity of INSERT on an MH of n elements and with $p \leq \log n$ processors is determined by the BROADCAST operations (time $O(\log \log n)$) and by the parallel execution of constant-time operations on at most $O(\log n)$ nodes (time $O\left(\frac{\log n}{p}\right)$) for a total time

$$C_I \in O\left(\frac{\log n}{p} + \log \log n\right) .$$

It should be noted that the above procedure INSERT can be employed to provide an implementation of the useful *decrease-key* operation [4]. In MH terms, *decrease-key* is given a pointer to a node k of M_H and a value v smaller than $M_H[k]$. The algorithm sets $M_H[k]$ to v and then re-establishes the heap property on M_H. Such readjustment can be obtained by simply considering k as a leaf node and running a slight variant of INSERT with parameters k and v. The details of the algorithm are omitted for the sake of brevity.

2.2   Deletion

The algorithm for deleting the root of an MH M proceeds in three phases:

1. Let $\mu_1 = \{n_1 = 1, ..., n_h = L = M_L[1]\}$ be the min-path of the root of M. The root item $M_H[1]$ is returned and the target leaf of the root $L = M_L[1]$ is broadcast to all the processors.
2. Nodes $n_2, ..., n_h$ are shifted one position above, that is, $M_H[n_i] := M_H[n_{i+1}]$ (for technical reasons, we set $M[n_h] = +\infty$). Note that this operation restores the heap order in M_H, but disrupts the target leaf information in M_L for the nodes in μ_1.
3. The target leaves for the nodes on μ_1 are recomputed and the "hole" in position L is filled by invoking INSERT(L,$M_H[N_M]$). Finally, N_M is decremented.

It remains to explain how to recompute $M_L[n_i]$, $1 \leq i \leq h$, once μ_1 is shifted upwards. Consider the new min-paths for nodes n_i in the updated structure. Starting from the root and proceeding along the nodes of μ_1, the min-path will follow the same route as before if the new values stored at nodes n_i are still smaller than the ones stored at their siblings. However, whenever we reach a node n_k whose sibling SIBLING(M, L, k) contains now a smaller value, the min-path "deviates" and reaches the target leaf $M_L[\text{SIBLING}(M, L, k)]$. This observation suggests the following strategy to rebuild M_L efficiently in parallel. In the following, RANK1, RANK2, RANK3 and RANK4 are auxiliary vectors of h positions.

1. Each value $M_H[n_i]$, $2 \leq i \leq h$ is compared with $M_H[\text{SIBLING}(M, L, i)]$. If $M_H[n_i]$ is smaller, then RANK1$[i]$ is set to 0, otherwise RANK1$[i]$ is set to 1 (the values 1 indicate a "deviation" of the min-path). Note that at least RANK1$[h]$ will be initialized to 1, as we set $M_H[L]$ to $+\infty$.
2. Prefix sums are computed on input RANK1. The results are stored in RANK2.
3. (*compaction*) For each position i, if RANK1$[i] = 1$ (i.e., the min-path deviates at n_i) and RANK2$[i] = j$ then RANK3$[j]$ is set to i (RANK3$[j]$ is the address of the jth deviation).
4. (*target leaf assignment*) For $1 \leq i \leq h-1$ let $j_i =$ RANK2$[i]+1$ and $k_i =$ RANK3$[j_i]$. $M_L[n_i]$ is set to $M_L[\text{SIBLING}(M, L, k_i)]$, which is the target leaf of the sibling of the first node n_{k_i}, following n_i, where the the min-path deviates.

Note that in Step 4 the same cell of vector RANK3 could be accessed concurrently by two processors associated to nodes n_i and n_j with RANK2$[i] = $ RANK2$[j]$. This problem is easily overcome by computing the vector RANK4$[i] =$ RANK3[RANK2$[i]+1]$ by means of a simple prefix operation, whose description is omitted for the sake of brevity.

The following procedure DELETEMIN implements the above strategy. In the procedure, we use the statements

PREFIX-SUMS(RANK1,RANK2) and COMPUTE(RANK2,RANK3,RANK4)

respectively to denote the prefix-sums computation with input RANK1 and output RANK2 and the creation of vector RANK4$[i] =$ RANK3[RANK2$[i] + 1]$. These operations can be realized on an EREW-PRAM in $O\left(\frac{\log n}{p} + \log\log n\right)$ time [5].

procedure DELETEMIN:
    for $i \in \{1\}$ **do in parallel**
        return $M_H[1]$;
        $P_i : \text{BROADCAST}(L = M_L[1])$
    endfor;
    { return the min value and distribute the address of the target leaf of the root}
    $h := \lfloor \log L \rfloor + 1$; { length of μ_1}
    for $i \in \{1, \ldots, h-1\}$ **do in parallel**
        $V_H[i] := M_H[L \text{ div } 2^{h-i-1}]$;
        {copy and shift the elements stored along μ_1}
    endfor;
    for $i \in \{h\}$ **do in parallel**

```
    V_H[h] := +∞; RANK1[1]:= 0;
  endfor;
  for i ∈ {2,...,h} do in parallel
    if V_H[i] < M_H[SIBLING(M, L, i)]
      then RANK1[i] := 0
      else RANK1[i] := 1
    fi;
    PREFIX-SUMS(RANK1,RANK2);
    if RANK1[i] = 1
      then RANK3[RANK2[i]] := i
    fi;
    {compact the indices of the nodes where the min-path deviates}
  endfor;
  for i ∈ {1...,h-1} do in parallel
    COMPUTE(RANK2,RANK3,RANK4);
    {RANK4[i] =RANK3[RANK2[i] + 1]}
    V_L[i] := M_L[SIBLING(M, L,RANK4[i])]
    {these are the new addresses of the target leaves of nodes in μ₁}
    M_H[L div 2^{h-i}] := V_H[i];
    M_L[L div 2^{h-i}] := V_L[i];
    { copy the updated path back}
  endfor;
  for i ∈ {1} do in parallel
    I := M_H[N_M];
    M_H[N_M] := M_L[N_M] := +∞
  endfor;
  INSERT(L, I);
  {fill the hole in M_H[L]}
  N_M := N_M - 1
end DELETEMIN.
```

The time complexity of DELETEMIN on an MH of n elements is determined by the broadcast and prefix steps, and by the parallel execution of constant-time operations on $O(\log n)$ nodes, for a total time complexity

$$C_D \in O\left(\frac{\log n}{p} + \log\log n\right) \ .$$

2.3 Construction

We are finally left with implementing MH construction. Let $S[1,\ldots,n]$ be the set of n elements to be stored in an MH M. We first build the vector M_H by applying one of the optimal parallel algorithms for heap construction proposed in the literature (see [6] and [8]). The vector M_L is subsequently created by first initializing $M_L[i] = i$ for each leaf i and then computing $M_L[j]$ for any internal node j. More precisely, if $M_L[2j]$ and $M_L[2j + 1]$ have been computed, $M_L[j]$ is set to $M_L[k]$, where $k \in \{2j, 2j+1\}$ is such that $M_H[k] = \min\{M_H[2j], M_H[2j + 1]\}$.

Construction is performed by the following procedure CONSTRUCT_MH(S, M). In the procedure, the statement

$$\text{BUILD}(S, M_H)$$

denotes the invocation of an optimal parallel heap construction scheme which builds M_H out of S.

procedure CONSTRUCT_MH(S, M):
  $N_M := |S|;\ h := \lfloor \log N_M \rfloor + 1;$

  $\{M_H$ construction:$\}$
  BUILD(S, M_H);

  $\{M_L$ construction:$\}$
  count:=2;
  while *count*> 0 **do**
    $\{$determine the target leaves of the nodes in the last two levels$\}$
    if $h > 0$
      then for $i \in \{2^{h-1}, \ldots, 2^h - 1\}$ **do in parallel**
        if $2i \leq N_M$
          then if $2i + 1 \leq N_M$
            then if $M_H[2i] < M_H[2i + 1]$
              then $M_L[i] := M_L[2i]$
              else $M_L[i] := M_L[2i + 1]$
            fi
            else $M_L[i] := 2i$
          fi
          else $M_L[i] := i$
        fi
      endfor
    fi;
    $h := h - 1;$ *count*:=*count*-1
  endwhile;
  while $h > 0$ **do**
    for $i \in \{2^{h-1}, \ldots, 2^h - 1\}$ **do in parallel**
      if $M_H[2i] < M_H[2i + 1]$
        then $M_L[i] := M_L[2i]$
        else $M_L[i] := M_L[2i + 1]$
      fi
    endfor;
    $h := h - 1$
  endwhile
end CONSTRUCT_MH.

Let us analyze the running time of the above procedure. By using the previously referenced schemes in [6] and [8], the M_H construction can be executed in

$O\left(\frac{n}{p} + \log n\right)$ time. As to the M_L construction, the algorithm is essentially a min-computation performed along a complete binary tree of n nodes, thus requiring $O\left(\frac{n}{p} + \log n\right)$ time. Therefore, the overall time complexity of the procedure is

$$C_C \in O\left(\frac{n}{p} + \log n\right) .$$

3   Conclusions

The *Min-path Heap* (MH) data structure introduced in the previous sections provides an optimal implementation of priority queues on a p-processor EREW-PRAM. We have devised insertion and deletion algorithms for an MH M of n elements which require $O\left(\frac{\log n}{p} + \log \log n\right)$ time and have adapted known parallel heap-construction schemes to build M in $O\left(\frac{n}{p} + \log n\right)$ time. All the algorithms are extremely simple and the orders of magnitude do not hide "big" constants. Moreover, the space requirement of M is only $2n$ memory cells, arranged in two vectors of n locations each.

It has to be noted that the number of processors that can be profitably exploited by our algorithms is (necessarily) small ($p \in O(\log n)$). However, the current (or even foreseeable) technology for the construction of parallel machines with shared memory is applicable only to systems with "few" processors [5]. Our simple and efficient algorithms are suitable for an optimal exploitation of such "coarse grain" parallelism.

For the above systems, MH structures can be employed to optimally speed up those sequential applications which make use of binary heaps and whose time complexity is determined by the cost of the heap operations. Consider, for instance, *Heap-Sort* or the implementation of the *LPT* heuristic for scheduling [3]. The use of MH's yields optimal $O\left(\frac{n \log n}{p}\right)$ time parallel algorithms for the above problems, for any number $p \in O\left(\frac{\log n}{\log \log n}\right)$ of processors. As a final example, consider the straightforward parallelization of Dijkstra's algorithm for computing a rooted *Shortest Path Tree* (SPT) of a weighted directed graph. The complexity of the algorithm is dominated by the time needed for n deletions and $O(m)$ *decrease-key* operations on a priority queue of $O(n)$ elements [4]. The use of an MH yields a parallel SPT algorithm with running time $O(\frac{m \log n}{p})$ with $p \in O(\frac{\log n}{\log \log n})$. For this range of processors and $m \in O(\frac{n \log n}{\log \log n})$ this simple algorithm achieves a better processor-time product than the other parallel SPT algorithms in the literature [1, 4].

References

1. B.Auerbuch and Y.Shiloach, New Connectivity and MSF Algorithms for Ultracomputer and PRAM, in: *Proc. of the 1983 Int. Conf. on Parallel Processing* (1983) 298-319.
2. J.Biswas and J.C.Browne, Simultaneous Update of Priority Structures, in: *Proc. of the 1987 Int. Conf. on Parallel Processing* (1987) 124-131.

3. T.H.Cormen, C.E.Leiserson and R.L.Rivest, *Introduction to Algorithms* (MIT Press, Cambridge Mass., 1990).

4. J.M.Driscoll, H.V.Gabow, R.Shrairman and R.E.Tarjan, Relaxed Heaps: An Alternative to Fibonacci Heaps with Applications to Parallel Computation, *Communications of the ACM* **31**(11) (1988) 1343-1354.

5. R.M.Karp and V.Ramachandran, Parallel Algorithms for Shared-Memory Machines, in: J.van Leeuween, ed., *Handbook of Theoretical Computer Science, Volume A: Algorithms and Complexity* (Elsevier, Amsterdam, 1990) 870-941.

6. S.Olariu and Z.Wen, An Optimal Parallel Construction Scheme for Heap-like Structures, in: *Proc. Twenty-eight Allerton Conf. on Communication, Control, and Computing* (1990) 936-937.

7. V.N.Rao and V.Kumar, Concurrent Access of Priority Queues, *IEEE Trans. on Computers* **C-37**(12) (1988) 1657-1665.

8. V.N.Rao and W.Zhang, Building Heaps in Parallel, *Information Processing Letters* **37** (1991) 355-358.

9. W.Zhang and R.Korf, Parallel Heap Operations on EREW PRAM, in: *Proc. Sixth Int. Parallel Processing Symp.* (1992) 315-318.

This article was processed using the LaTeX macro package with LLNCS style

Heap Construction in the Parallel Comparison Tree Model

Paul F. Dietz[1]

Department of Computer Science

University of Rochester

Rochester, NY 14627

dietz@cs.rochester.edu

Abstract I show how to put n values into heap order in $O(\log \log n)$ time using $n/\log \log n$ processors in the parallel comparison tree model of computation, and in $\tilde{O}(\alpha(n))$ time on $n/\alpha(n)$ processors, in the randomized parallel comparison tree model, where $\alpha(n)$ is an inverse of Ackerman's function. I prove similar bounds for the related problem of putting n values into a min–max heap.

1 Introduction

I consider the following problem, which I will call the *heap problem*. Given n values $x_1, \ldots, x_n$ drawn from some totally ordered set, place the x_i into a *heap*. That is, find some permutation π on $\{1, \ldots, n\}$ such that $x_{\pi(i)} \geq x_{\lfloor \pi(i)/2 \rfloor}$ (for $\pi(i) \geq 2$). There is a straightforward linear time sequential algorithm for this problem [1]. A heap can be considered to be a complete binary tree; the height of a heap of n elements is $\lceil \log_2(n + 1) - 1 \rceil$.

More generally, a *min–heap (max–heap) of degree d* is a permutation π on $\{1, \ldots, n\}$ such that $x_{\pi(i)}$ is greater (less) than or equal to $x_{\lfloor \pi(i)/d \rfloor}$ for all $\pi(i) \geq d$. A heap of degree d can be thought of as a complete d-ary tree. For brevity, define

$$S(d, h) = \frac{d^h - 1}{d - 1} \tag{1}$$

and

$$H(d, n) = \lfloor \log_d((d - 1)n) \rfloor \tag{2}$$

to be the number of nodes in a maximal d-ary tree of height $h - 1$ and the height of a complete d-ary tree containing n nodes, respectively.

---

[1] This work supported by NSF grant CCR–8909667.

A related problem is to put the n values into a *min–max heap*. In this "min–max heap problem", an element stored at odd depth in the heap must be greater than or equal to all its descendants; an element at even depth must be less than or equal to all descendants. Min–max heaps are used to implement double ended priority queues [3].

The algorithms for these problem will be expressed in the (randomized) parallel comparison tree (PCT) model. In this model, there is a fixed tree for each value of n. Each node of the tree performs $p(n)$ comparisons between the x_i and performs a $2^{p(n)}$-way branch. Leaf nodes contain the desired permutation π. The time of a program is the depth of the deepest leaf. In the randomized PCT model, there are, additionally, nodes at which $2^{p(n)}$-way randomized choices can be made.

The notation $\tilde{O}(g(n))$ means the class of random variables that are bounded by a constant time $g(n)$ "with high probability"; more precisely, $f(n) = \tilde{O}(g(n))$ if for all $d > 0$ there is some $c > 0$ and some n_0 such that for all integer $n > n_0$,

$$\Pr[f(n) < cg(n)] > 1 - n^{-d}. \tag{3}$$

The statement that the running time of programs in the randomized PCT model is $\tilde{O}(T(n))$ means that there is some random variable $f(n)$ in this class such that the running time on any particular input (also a random variable) is dominated by $f(n)$ (in the sense that the probability the running time is less than t is at least as great as the probability that $f(n)$ is less than t, for all t).

Valiant [10] has shown that computing the minimum of a set requires $\Omega(\log \log n)$ time on n processors in the PCT model. Since the root of a heap is the minimum element, the same lower bound applies to the heap problem. I show that this bound is tight.[2]

Without loss of generality assume the x_i's are distinct. Let x be some element in $X = \{x_1, \ldots, x_n\}$. The *rank* of x is

$$rank(x) = |\{x_i | x_i \leq x\}|. \tag{4}$$

The *selection problem* is to find the element of X of rank i. Reischuk [9] showed that this could be done in $\tilde{O}(1)$ time in the randomized PCT model. Ajtai *et al.* [2] showed that there is an $O(\log \log n)$ time algorithm in the deterministic parallel comparison tree mode; Azar and Pippenger improved this to use $O(n/\log \log n)$ processors [4].

Rao and Zhang [8] have found an $O(\log n)$ time algorithm for the heap problem of degree 2 on $n/\log n$ processors in the EREW PRAM model. Their algorithm naturally extends to the heap problem of degree $d > 2$. Carlsson and Zhang [6] have an upper bound for the heap problem and the min–max heap problem in the PCT model of

---

[2] In this paper, all logarithms are assumed to be base 2 unless otherwise indicated.

$O((\log\log n)^2)$. These bounds are obtained by divide-and-conquer algorithms in which the division step uses AKSS selection to find the element of rank $n^{1/2}$. Their algorithms achieve $\tilde{O}(\log\log n)$ time in the randomized PCT model if Reischuk's selection algorithm is used.

Berkman, Matias and Vishkin [7] have recently shown that a sequence $x_1, \ldots, x_n$ can be preprocessed in $\tilde{O}(\alpha(n))$ time so that queries of the form $\max\{x_i, \ldots, x_j\}$ can be answered in $O(\alpha(n))$ time (all on $n/\alpha(n)$ processors). Vishkin [11] describes a class of algorithms called "structurally parallel algorithms" that admit $\tilde{O}(\alpha(n))$ time algorithms with linear work. The randomized algorithm presented in this paper is another example of an algorithm of this type.

2 Deterministic Algorithms for the Heap Problem

This section gives an $O(\log\log n)$ time algorithm for the heap problem of degree $d \geq 2$ on $O(n/\log\log n)$ processors.

The deterministic algorithm uses a basic step, called partition by rank, that works as follows. Given a set S of size n, "partition by rank about m" finds the element x in S of rank m, then compares all elements of S against x and partitions S into the elements of rank at most k, S_1, and the elements of rank more than k, S_2. This step can be done in $O(\log\log n)$ time on $n/\log\log n$ processors using the selection algorithm of Azar and Pippenger.

Lemma 1 The heap problem can be solved on $n/\log\log n$ processors in $O(\log\log n)$ time iff it can be solved on n processors in this time bound for $n = S(d,r)$ for some integer $r > 0$.

Proof: I reduce the $n/\log\log n$ processors problem to the problem on n processors. Let S be the set of inputs.

1. If there is no positive integer r such that $n = S(d,r)$ then let $r = H(d,n)$.

   (a) Partition S by rank about $S(d,r)$ into sets S_1 and S_2.

   (b) Construct, recursively, a heap of degree d using S_1. The elements of S_1 are made leaves of this heap. Stop.

2. Otherwise, let r be as defined above. Let $n' = n - d^{r-1} = S(d, r-1)$. Note that $n' < n/d$. Partition by rank about n' into sets S_3 and S_4. The elements of of S_4 are precisely those elements in the last row of the heap. Put them there in arbitrary order.

3. We now must put the elements of S_3 into heap order. If $d \geq \log \log n$, then by assumption S_3 can be put into heap order in $O(\log \log n)$ time, and stop. Otherwise, let $s = H(d, \lceil (\log \log n)/d \rceil)$. If $s > r - 1$ then put S_3 into heap order by a sequential algorithm; stop. Otherwise, let $m = S(d, r - s - 1)$. Note that $S(d, s)$ is $\Omega((\log \log n)/d)$ and $O(\log \log n)$, and that m is $O(n/\log \log n)$.

4. Partition S_3 by rank about m into S_5 and S_6.

5. The set S_6 is broken arbitrarily into subsets of size $S(d, s) = O(\log \log n)$. Each subset is put into heap order in $O(\log \log n)$ time by a single processor using the well-known sequential algorithm.

6. In $O(\log \log n)$ time put S_5 into heap order using m processors.

7. The heaps found in step 5 are made children of the leaves of the heap found in step 6.

This algorithm runs in $O(\log \log n)$ time. It is clearly correct, since all the elements in the bottom row are greater than all the higher elements, and all the element of the small heaps are greater than all the elements of the top heap built in step 5. ∎

Lemma 2 The heap problem can be solved in $O(\log \log n)$ time on $n \log_d n$ processors.

Proof: The algorithm does the following. Without loss of generality, assume $n = S(d, r)$ for some positive integer r (if not, partition by rank to separate off the last row as in the previous lemma).

1. Find the elements of rank $S(d, i)$, $i = 1, \ldots, r$, using the algorithm of Azar and Pippenger. Call these elements $y_1, \ldots, y_r$.

2. For each x_i, compare against each y_j. This requires $nr = \Theta(n \log_d n)$ comparisons. For each x_i, let $d_i = \min\{j \mid x_i \leq y_j\}$.

3. x_i is placed into the heap at depth $d_i - 1$.

This algorithm is correct, since all elements of depth $i - 1$ are at most y_i, while all elements in depth i are greater than y_i. Also, there are exactly d^i elements of depth i, so the heap is correctly filled. The algorithm takes $O(\log \log n)$ time (step 1). Step 2 is performed in constant time on $O(n \log_d n)$ processors, since there are $O(\log_d n)$ rows in the heap, while step 3 requires no comparisons at all. ∎

Lemma 3 The heap problem of degree d be solved in $O(\log \log n)$ time on n processors.

Proof: The following algorithm uses n processors. As before, without loss of generality, assume $n = S(d, r)$ for some positive integer r.

1. If $d > \log n$,

   (a) Partition by rank about $S(d, r - 1)$ into S_1 and S_2.

   (b) Noting that $|S_1| < n/\log n$, place S_1 into heap order using the algorithm of lemma 2.

   (c) Place the element of S_2 into the rth row of the heap in arbitrary order.

2. Otherwise, let $s = H(d, \lceil \log n \rceil)$.

   (a) Partition by rank about $S(d, r - s)$ into S_3 and S_4.

   (b) Noting that $|S_3| < n/\log n$, place S_3 into heap order using the algorithm of lemma 2.

   (c) Arbitrarily partition S_4 into disjoint sets of size $S(d, s) = O(\log^2 n)$. Apply the algorithm of Rao and Zhang [8] to each subset.

   (d) Make each of these small heaps a subheap of the heap constructed in step 2.b.

This algorithm correctly constructs a heap, since all the elements in S_2 (S_4) are greater than all the elements in S_1 (S_3). The running time of this algorithm is $O(\log \log n)$, since steps 1.a, 1.b, 2.a, 2.b and 2.c can be performed in $O(\log \log n)$ time (lemma 2) and steps 1.c and 2.d can be performed in constant time. ∎

Theorem 4 The heap problem can be solved on $n/\log \log n$ processors in $O(\log \log n)$ time in the parallel comparison tree model.

Proof: Immediate from lemmas 1 and 3. ∎

3 Randomized Algorithms for the Heap Problem

The algorithms given in the previous section can be significantly improved if we are allowed to use randomization.

The following definitions are taken from [5]. Let $f(n)$ be some function on integers such that $f(n) < n$ for all $n > 0$. Define $f^{(0)}(n) = n$ and $f^{(i+1)}(n) = f(f^{(i)}(n))$ for all integer $i > 0$. Define

$$* f(n) = \min\{k | f^{(k)}(n) \le 1\}. \tag{5}$$

Let $d \geq 2$ be some integer. Define $I_0(n) = n - d$, and define $I_{k+1}(n) = *I_k(n)$. Therefore, $I_1(n) = \lfloor n/d \rfloor$, $I_2(n) = \lfloor \log_d n \rfloor$, $I_3(n) = \lfloor \log_d^* n \rfloor$, and so on. Define $\alpha(n) = \min\{k | I_k(n) \leq k\}$.[3]

Berkman and Vishkin [5] gave several examples of parallel algorithms that ran in $\alpha(n)$ time on CRCW PRAMs. Their algorithms employed a decomposition called a *recursive *-tree*. For the problems they considered, they showed that, for each m, $2 \leq m \leq \alpha(r)$, there was an algorithm using $O(I_m(n))$ time on mn processors, (a simple trick reduced this to $n/I_m(n)$ processors). See also [11] for a review of fast parallel algorithms.

I define a sequence of algorithms for $m = 2, 3, \ldots, \alpha(r)$. The algorithms operate by decomposing the heap into *layers*. As before, without loss of generality let $n = S(d, r)$, $r > 1$ an integer. The heap is decomposed into $I_m(r)+1$ layers, numbered $1, \ldots, I_m(r)+1$. Layer i contains $I_{m-1}^{(i-1)}(r) - I_{m-1}^{(i)}(r)$ successive rows of the heap, with the rows in layer i directly above those in layer $i + 1$.

For sake of generality, I consider a slightly more general problem: given a positive integer r and $n = kS(d, r)$ elements, k a positive integer, put the elements into k heaps of height r. The mth algorithm for this problem will use $nI_m(r)$ processors, and will take $\tilde{O}(m)$ time.

1. If $m = 1$ or $r \leq 2$, then use the deterministic algorithm in the previous section, on $O(nr)$ processors, but use the Reischuk algorithm for selection. This takes constant time with high probability.

2. Otherwise, let $s = I_m(r) + 1$, $k = n/S(d, r)$ and let $j_i = kS(d, r - I_{m-1}^{(i)}(r))$, $i = 0, \ldots, s$. Find the elements $x_1, \ldots, x_{s-1}$, where $rank(x_i) = j_i$, using Reischuk's algorithm.

3. Partition the elements of the set into those of rank more than j_{i-1} and at most j_i, $i = 1, \ldots, s$.

4. The elements in the last group fall into the last row of the heaps. Place them there in arbitrary order.

5. Recursively apply the algorithm of level $m - 1$ to each of the other layers.

It is easily seen that Reischuk's algorithm is used to find only the elements of rank $kS(d, j)$, $j = 1, \ldots, r - 1$, so it is invoked $r - 1$ times.

Lemma 5 If $m \leq \alpha(n)$, then any subproblem solved by this algorithm has $\Omega(n^{1/2})$ elements.

---

[3] In [5], these functions are defined with $d = 2$. For larger d, the functions grow more slowly still.

Proof: The smallest subproblem produced by a level m algorithm is the first, which has $\Omega(n/I_{m-1}(r))$ elements. Therefore, m levels of recursion yields a subproblem with at least $\Omega(n/(\log_d n)^m)$ elements, which is $\Omega(n^{1/2})$. ∎

Lemma 6 The level m algorithm $(m \le \alpha(r))$ with $nI_m(r)$ processors uses $\tilde{O}(m)$ time.

Proof: For any k, there is some c such that Reischuk's algorithms uses at most c time with probability at least $1 - n^{-2k}$ time for almost all n. By lemma 5, all subproblems invoked by a level m algorithm are of size at least $n^{-1/2}$; therefore, each selection runs in time at most c with probability at least $1-n^{-k}$. Since there are $r-1$ selections performed, the probability that they each run in this time bound is at least $1 - (r-1)n^{-k} > 1 - n^{-(k-1)}$ for almost all n. Therefore, for any l we can find a c such that the selections all use at most c time with probability at least $1 - n^{-l}$.

Since the selections are the only parts of the algorithm employing randomization, it suffices to prove that (1) enough processors are allocated to do the selections in constant time, and (2) the rest of the algorithm uses $O(m)$ time.

We show, inductively, that the lemma is true for $m = 1, \ldots, \alpha(r)$. For $m = 1$, the algorithm of the previous section performs r selections, each on $nI_1(r)/r = \Theta(n)$ processors. Partitioning the inputs among the rows takes constant time. Therefore, the algorithm itself takes constant time.

For the induction step, assume the level $m-1$ algorithm takes at most $c(m-1)$ time on $nI_{m-1}(r)$ processors.

1. Step 2 in the level m algorithm performs $s - 1$ selections on n elements using $n(s-1)$ processors, and takes constant time.

2. Step 3 performs $n(s-1)$ comparisons, which again take constant time.

3. Step 4 performs no comparisons, and requires no time in the randomized PCT model.

4. Step 5 recursively assigns n processors to each layer. We must show that each layer (save the last) receives at least $I_{m-1}(r_i)$ processors per element, where r_i is the number of rows in the ith layer, $i = 1, \ldots, s - 1$. The ith layer has fewer than $kS(d, r - I_{m-1}^{(i)}(r))$ elements. It is assigned $n = kS(d, r)$ processors. Now, $r_i < I_{m-1}^{(i-1)}(r)$, so $I_{m-1}(r_i) \le I_{m-1}^{(i)}(r)$. Let $p_i = I_{m-1}^{(i)}(r)$. It suffices to show that

$$S(d, r) \ge d^{r-p_i} p_i. \tag{6}$$

For $r \ge 1$, $d \ge 2$ and p_i a nonnegative integer, equation 6 always holds. Therefore, by the induction hypothesis, step 5 requires $c(m-1)$ time.

With a proper choice for the constant c, we conclude that the level m algorithm runs in $\tilde{O}(m)$ time. ∎

Corollary 7 The heap problem can be solved in $\tilde{O}(\alpha(n))$ time using $n\alpha(n)$ processors.

Theorem 8 The heap problem can be solved in $\tilde{O}(\alpha(n))$ time on $n/\alpha(n)$ processors in the randomized PCT model.

Proof: Reduce the problem to that with $n\alpha(n)$ processors by selecting an element x of rank $S(d, j) = \Theta(n/(\alpha(n))^2)$. Partition the elements about x. For those of lower rank, apply the previous algorithm. This uses $O(n)$ processors and $\tilde{O}(\alpha(n))$ time. For the rest, partition into subproblems of size $\Theta(\alpha(n)^2)$ and solve using the algorithm of Rao and Zhang. This takes $O(\log \alpha(n))$ time. ∎

We can also show that, for constant k, heaps of degree $d \geq I_k(n)$ can be constructed in constant time with high probability on n processors in the randomized PCT model. In particular, heaps of degree $O(\log n)$ or $O(\log^* n)$ can be constructed in constant time.

4   Min-Max Heaps

The algorithms for heaps can be modified to apply to min–max heaps. I outline here the necessary modifications. In this section, n may be any positive integer, and $x_1, \ldots, x_n$ are the (distinct) input values.

In a min–max heap, call the values stored at even depth the *min-values*; the values at odd depth, the *max-values*.

Lemma 9 Let $i = H(d, n+1)$. A min–max heap of size n has $L(n)$ min-values, where

$$L(n) = \begin{cases} S(d^2, (i+1)/2) & \text{if } i \text{ is odd} \\ n - dS(d^2, i/2) & \text{if } i \text{ is even.} \end{cases} \tag{7}$$

Proof: If i is odd, the last complete row of the heap contains min-values; if i is even, the last complete row contains max-values. The lemma follows immediately from the definition of $S(d, h)$. ∎

I now describe an algorithm for building a min–max heap of degree d in parallel.

1. Let $i = \lfloor \log n \rfloor$. Partition inputs by rank about $L(n)$ into S_1 and S_2.

2. Build a min–heap of degree d^2 using the elements of S_1 and $L(n)$ processors. The elements of this heap become the elements of even depth in the min–max heap.

3. Partition S_2 into d sets of size differing by at most 1 (with the larger heaps first). Convert each into a max–heap using a linear number of processors. The elements of the ith heap becomes the descendants of odd depth of the ith child of the root in the min–max heap.

Lemma 10 This algorithm correctly produces a min–max heap.

Proof: Observe that the resulting tree is a complete d-ary tree. All descendants of a min-value are less than that min-value, since either they are max-values and therefore of higher rank, or are min-values and have been placed into heap order with all descendants. A similar argument shows that max-values are maximum in their subtrees. ∎

Lemma 11 Min–max heaps of degree d can be computed in $O(\log\log n)$ time on $n/\log\log n$ processors in the PCT model.

Proof: The first step in the algorithm can be performed in $O(\log\log n)$ time on $n/\log\log n$ processors by the algorithm of Azar and Pippenger. The remaining two steps can be performed in $O(\log\log m)$ time on $n/\log\log m$ processors, where m is the maximum size of a subheap. By Brent's theorem, these steps can also be performed in $O(\log\log n)$ time on $n/\log\log n$ processors. ∎

Lemma 12 Min–max heaps of degree d can be constructed in $\tilde{O}(\alpha(n))$ time on $n/\alpha(n)$ processors in the randomized PCT model.

Proof: The first step is performed in $\tilde{O}(\alpha(n))$ time on $n/\alpha(n)$ processors by Reischuk's selection algorithm.

If $d > n/2$, the min–heap constructed in the second step has height at most 1 and can be constructed simply by finding the minimum element of S_1. This can be done in $\tilde{O}(\alpha(n))$ time on $n/\alpha(n)$ processors. The max–heap in step (3) have height zero, and so can be trivially constructed with no comparisions.

Otherwise, if $d \leq n/2$, the sizes of both S_1 and S_2 are at least $n/2$, and the generalized randomized algorithm of section 3 can find the min–heaps and max–heaps in $O(\alpha(n))$ time on $n/\alpha(n)$ processors (theorem 8). ∎

As for ordinary heaps, for constant k, min–max heaps of degree $d \geq I_k(n)$ can be constructed in constant time with high probability in the randomized PCT model.

5 Summary

This paper has described algorithms for putting data into heap order in the (randomized) parallel comparison tree model. The deterministic algorithm is optimal in time and work, achieving $O(\log\log n)$ time on $n/\log\log n$ processors. The randomized algorithm achieves $\tilde{O}(\alpha(n))$ on $n/\alpha(n)$ processors. Algorithms were also presented for putting elements into min–max heap order in the same time and work bounds.

It would be interesting to determine if the second algorithm can be implemented on the CRCW PRAM model in the same time bound. It would also be interesting to either improve the ¡randomized algorithm presented here or to prove it to be time–optimal for n processors.

6 Acknowledgements

I would like to thank Joel Seiferas, Danny Krizanc and Lata Naranyan for their interest and helpful comments.

References

[1] Alfred V. Aho, John E. Hopcroft, and Jeffrey D. Ullman. *The Design and Analysis of Computer Algorithms*. Addison-Wesley, New York, 1974.

[2] M. Ajtai, J. Komlós, W. L. Steiger, and E. Szemerédi. Optimal parallel selection has complexity $O(\log\log n)$. *Journal of Computer and System Sciences*, 38:125–133, 1989.

[3] M. Atkinson, J. Sack, N. Santoro, and T. Strothotte. Min–max heaps and generalized priority queues. *Communications of the ACM*, 29(10):996–1000, October 1986.

[4] Y. Azar and N. Pippenger. Parallel selection. *Discrete Appl. Math.*, 27:49–58, 1990.

[5] Omer Berkman and Uzi Vishkin. Recursive *-tree parallel data-structure. In *Proc. 30th Ann. IEEE Symp. on Foundations of Computer Science*, pages 196–202, October 1989.

[6] Svante Carlsson and Jingsen Zhang. Parallel complexity of heaps and min-max heaps. Technical Report LU–CS–TR:91-77, Department of Computer Science, Lund University, Lund, Sweden, August 1991.

[7] Y. Matias O. Berkman and U. Vishkin, July 1991. Unpublished manuscript.

[8] N. S. V. Rao and W. Zhang. Building heaps in parallel. *Info. Proc. Lett.*, 37:355–358, 1991.

[9] Rüdiger Reischuk. Probabilistic parallel algorithms for sorting and selection. *SIAM J. On Computing*, 14(2):396–409, May 1985.

[10] Leslie G. Valiant. Parallelism in comparison problems. *SIAM J. On Computing*, 4(3):348–355, September 1975.

[11] Uzi Vishkin. Structural parallel algorithmics. In *ICALP 91*, pages 363–380, July 1991.

Efficient Rebalancing of Chromatic Search Trees

Joan Boyar*[1] and Kim S. Larsen[2]

[1] Odense University, Loyola University Chicago
[2] Aarhus University

Abstract. In PODS'91, Nurmi and Soisalon-Soininen presented a new type of binary search tree for databases, which they call a *chromatic* tree. The aim is to improve runtime performance by allowing a greater degree of concurrency, which, in turn, is obtained by uncoupling updating from rebalancing. This also allows rebalancing to be postponed completely or partially until after peak working hours.

The advantages of the proposal of Nurmi and Soisalon-Soininen are quite significant, but there are definite problems with it. First, they give no explicit upper bound on the complexity of their algorithm. Second, some of their rebalancing operations can be applied many more times than necessary. Third, some of their operations, when removing one problem, create another.

We define a new set of rebalancing operations which we prove give rise to at most $\lfloor \log_2(N+1) \rfloor - 1$ rebalancing operations per insertion and at most $\lfloor \log_2(N+1) \rfloor - 2$ rebalancing operations per deletion, where N is the maximum size the tree could ever have, given its initial size and the number of insertions performed. Most of these rebalancing operations, in fact, do no restructuring; they simply move weights around. The number of operations which actually change the structure of the tree is at most one per update.

1 Introduction

In [6], Nurmi and Soisalon-Soininen considered the problem of fast execution of updates in relations which are laid out as dictionaries in a concurrent environment. A *dictionary* is a data structure which supports the operations *search*, *insert*, and *delete*. Since both insertion and deletion modify the data structure, they are called the *updating operations*. For an implementation of a dictionary, Nurmi and Soisalon-Soininen propose a new type of binary search tree, which they call a *chromatic* tree.

One standard implementation of a dictionary is as a *red-black* tree [3], which is a type of balanced binary search tree. However, often when a data structure is accessed and updated by different processes in a concurrent environment, parts of the structure have to be *locked* while data items are changed or deleted. In the case of red-black trees of size n, an update requires locking $O(\log_2(n))$ nodes, though not necessarily simultaneously [3], in order to rebalance the tree. No other users can access the subtree below a node which is locked. Since the root is often one of the nodes locked, this greatly limits the amount of concurrency possible.

This leads Nurmi and Soisalon-Soininen to consider a very interesting idea for making the concurrent use of binary search trees more efficient: uncouple the updating (insertion and deletion) from the rebalancing operations, so that updating

---

* Some of this work was done while visiting Aarhus University.

becomes much faster. The rebalancing can then be done by a background process, or it can be delayed until after peak working hours. Nurmi and Soisalon-Soininen call this new data structure a *chromatic tree*. Another important property of this data structure is that each of the updating and rebalancing operations can be performed by locking only a small, constant number of nodes, so considerable parallelism is possible. In [6], one locking scheme is described in great detail. Other possibilities that help avoid some of the locking are described in [5].

The idea of uncoupling the updating from the rebalancing operations was first proposed in [3], and has been studied in connection with AVL trees [1] in [4, 7]. This idea has also been studied, to some extend, in connection with B-trees [2]. A summary of this, along with references, can be found in [5].

In this paper, we consider chromatic trees and propose different rebalancing operations which lead to significantly more efficient rebalancing. Suppose the original search tree, T, has $|T|$ nodes before k insertions and s deletions are performed. Then $N = |T| + 2k$ is the best bound one can give on the maximum number of nodes the tree ever has, since each insertion creates two new nodes (see below). In [6], it is shown that if their rebalancing operations are applied, the tree will eventually become balanced. In contrast, with these new operations, the tree will become rebalanced after at most $k(\lfloor \log_2(N + 1) \rfloor - 1) + s(\lfloor \log_2(N + 1) \rfloor - 2) = (k + s)(\lfloor \log_2(N + 1) \rfloor - 1) - s$, rebalancing operations which is $O(\log_2(N))$ for each update. In any balanced binary search tree with N nodes, a single insertion or deletion would require $\Theta(\log_2(N))$ steps in the worst case simply to access the item, so the rebalancing is also efficient. Most of these rebalancing operations, however, do no restructuring; they simply move weights around. The total number of operations which actually change the structure of the tree is at most equal to the number of updates. Since it is only when the actual structure of the tree is being changed that a user who is searching in the tree should be prevented from accessing certain nodes, this should allow a considerable degree of concurrency.

2 Chromatic Trees

The definition used in [6] for a chromatic tree is a modification of the definition in [3] for red-black trees. In this section, we give both of those definitions. The binary search trees considered are *leaf-oriented* binary search trees, so the keys are stored in the leaves and the internal nodes only contain *routers* which guide the search through the tree. The router stored in a node v is greater than or equal to any key in the left subtree and less than any key in the right subtree. The routers are not necessarily keys which are present in the tree, since we do not want to update routers when a deletion occurs. The tree is a *full* binary tree, so each node has either zero or two children.

Each edge e in the tree has an associated nonnegative integer weight $w(e)$. If $w(e) = 0$, we call the edge *red*; if $w(e) = 1$, we say the edge is *black*; and if $w(e) > 1$, we say the edge is *overweighted*. The *weight* of a path is the sum of the weights on its edges, and the *weighted level* of a node is the weight of the path from the root to that node. The weighted level of the root is zero.

Definition 1. A full binary search tree T with the following balance conditions is a *red-black* tree:

  B1: The parent edges of T's leaves are black.
  B2: All leaves of T have the same weighted level.
  B3: No path from T's root to a leaf contains two consecutive red edges.
  B4: T has only red and black edges.

The definition of a chromatic tree is merely a relaxation of the balance conditions.

Definition 2. A full binary search tree T with the following conditions is a *chromatic tree.*

  C1: The parent edges of T's leaves are not red.
  C2: All leaves of T have the same weighted level.

Insertion and deletion are the updates that are allowed in chromatic trees (and dictionaries in general). As for search trees in general, the operations are carried out by first searching for the element to be deleted or for the right place to insert a new element, after which the actual operation is performed. How this is done can be seen in the appendix. The lower case letters are names for the edges, and the upper case letters are names for the nodes. The other labels are weights. We do not list symmetric cases.

In ordinary balanced search trees, rebalancing is performed immediately after the update, moving from the leaf in question towards the root or in the opposite direction. In a chromatic tree, the data structure is left as it is after an update and rebalancing is taken care of later by other processes. The advantages of this are faster updates and more parallelism in the rebalancing process.

The maximum depth of any node in a red-black tree is $O(\log_2(n))$, but a chromatic tree could be very unbalanced. We follow [6] in assuming that initially the search tree is a red-black tree, and then a series of search, insert, and delete operations occur. These operations may be interspersed with rebalancing operations. The rebalancing operations may also occur after all of the search and update operations have been completed; our results are independent of the order in which the operations occur. In any case, the search tree is always a chromatic tree, and after enough rebalancing operations, it should again be a red-black tree.

A chromatic tree can have two types of problems which prevent it from being a red-black tree. First, it could have two consecutive red edges on some root-to-leaf path; we call this a *red-red conflict*. Second, there could be some overweighted edges; we call the sum $\sum_e \max(0, w(e) - 1)$ the amount of overweight in the tree. The rebalancing operations must be able to handle both of these problems.

The proposal in [6] is quite successful in uncoupling the updating from the rebalancing operations and in making the updates themselves fast. The problem is that if the tree is large and is updated extensively, the number of rebalancing operations that might be applied before the tree is red-black again could be very large. In addition, their operations, when removing overweight, can create new red-red conflicts. Although they give a proof of termination, they give no specific bound on the number of times their operations can be applied.

3   New Rebalancing Operations

The rebalancing operations are shown in the appendix. The seven weight decreasing operations are referred to as (w1) through (w7). The order in which operations are applied is unrestricted, except that an operation cannot be applied if the conditions shown are not met. When an operation which alters the actual structure of the tree occurs, all of the nodes being changed must be locked. With the other operations, the weights being changed must be locked, but users *searching* in the tree could still access (pass through) those nodes.

The rebalancing operations alter chromatic search trees in a well-defined way: it is clear how the subtrees not shown should be attached. For example, consider the sixth weight decreasing operation. The subtree which was below edge b should remain below edge b, and the subtree below edge e should remain below edge e. In addition, the left subtree below edge d should become the right subtree of the new edge c, and the right subtree below edge d should become the left subtree below the new edge d. Note that, even though this is not shown in the appendix, all operations, except the first four weight decreasing operations, are applicable when the edge a is not present because the operation is occurring at the root. In this case, there is obviously no need to adjust weight w_1. In practice, operations (w1) and (w7) would obviously be altered to allow the shifting of more than one unit of overweight at a time. However, this would not improve the worst case analysis.

In order to discuss these operations, we need the following definitions. Suppose that e is an edge from a parent u to a child v and that e' is an edge from the same parent u to another child v'. Then we call e' the *sibling edge* of e. We use the terms *parent edge* and *parent node* to refer to the edge or node immediately above another edge or node.

We will now briefly describe a situation in which the operations from [6] can be applied many more times than is necessary if the order chosen turns out to be unlucky. Consider the red-balancing operations. Nurmi and Soisalon-Soininen have similar operations, but they do not require that w_1 be at least one. One can show that, with their original operations, $\Omega(k^2)$ red-balancing operations can occur, regardless of the original size of the tree. To see this, consider k insertions, each one inserting a new smallest element into the search tree. This will create a sequence of k red edges and $k-1$ red-red conflicts. Now start applying the first red-balancing operation to the left-most red-red conflict. The same bottom edge will take part in $k-1$ operations. Then, below the final sibling to that edge, there will be a sequence of $k-2$ red edges in a string going to the left. The bottom red edge in the left-most red-red conflict will now take part in $k-3$ red-balancing operations. Continuing like this, a total of $\Omega(k^2)$ operations will occur. This is fairly serious since the red-balancing operations are among those which change the actual structure of the tree and thus necessitate locks which prevent users who are simply searching in the tree from accessing certain nodes. In contrast, with our new operations, the number of these restructuring operations is never more than the number of updates.

With the modifications we have made, applying one of the red-balancing operations decreases the number of red-red conflicts in the tree. This greatly limits the number of times they can be applied. Furthermore, as opposed to the operations proposed in [6], none of our overweight handling operations can increase the number of red-red conflicts. We avoid this by increasing the number of distinct rebalancing operations allowed. In some cases, this implies that we lock two more nodes than they would have. However, these modified operations significantly improve the worst-case number of rebalancing operations.

The following lemma shows that these operations are sufficient for rebalancing any chromatic tree, given that the process eventually terminates.

Lemma 3. *Suppose the tree T is a chromatic tree, but is not a red-black tree. Then at least one of the operations listed in the appendix can be applied.*

Proof. Suppose T contains an overweight edge e. If e's sibling edge, f, is overweighted, then (w7) can be applied. If f has weight one, then (w5), (w6), or the push operation can be applied. So, assume that f has weight zero. If none of the operations (w1), (w2), (w3), or (w4) can be applied, then at least one of f's children must also have weight zero. Hence, if neither a push nor a weight decreasing operation can be applied, there must be a red-red conflict.

Suppose T contains a red-red conflict. Consider a red-red conflict which is closest to the root. Let e_1 be the bottom edge and e_2 the top edge. The parent of e_2 cannot be red since otherwise there would be another red-red conflict closer to the root. If neither of the red-balancing operations can be applied, then the sibling of e_2 must be red. Hence the blacking operation can be applied.

Therefore, if a chromatic tree is not a red-black tree, there is always some operation which can be applied.

In the remainder of this paper, we prove bounds on the number of times these operations can be applied. Thus, after a finite number of operations applied in any order, a chromatic tree T will become a red-black tree.

4 Complexity

If some of the operations are done in parallel, they must involve edges and nodes which are completely disjoint from each other. The effect will be exactly the same as if they were done sequentially, in any order. Thus throughout the proofs, we will assume that the operations are done sequentially. At time 0, there is a red-black tree, at time 1 the first operation has just occurred, at time 2 the second operation has just occurred, etc.

In order to bound the number of operations which can occur, it is useful to follow red-red conflicts and units of overweight as they move around in the tree due to various operations. In order to do so, we notice that each of the rebalancing operations preserves the number of edges, and one can give a one-to-one mapping from the edges before an operation to those after. Thus one can talk about an edge e over time, even though its end points may change during that time. In the appendix, the one-to-one correspondence has been illustrated by the naming of the edges.

We define a fall from an edge e to be a path from e down to a leaf.

Definition 4. A *fall* from an edge e at time t is a sequence of edges $e_1, \ldots, e_k$, $k \geq 1$, in the tree at time t, such that $e_1 = e$, e_k is a leaf edge, and for each $2 \leq i \leq k$, e_i is a child of e_{i-1}. The *weight* of this fall is the sum $\Sigma_{1 \leq i \leq k} w(e_i)$.

Because of balance condition C2 of Definition 2, all of the falls, from any given edge e in a chromatic search tree, have the same weight. Clearly, if the tree is red-black and an edge e has heavy falls, then there is a large subtree below e. In a chromatic tree, however, an edge could have heavy falls because many edges below it have been deleted and have caused edges to become overweighted. In this case, e may not have a large subtree remaining. It will be useful, though, to somehow count those edges below e which have been deleted. Edges are inserted and deleted and we want to associate every edge which has ever existed with edges currently in the tree.

Definition 5. Any edge in the tree at time t is *associated* with itself. When a node is deleted, the two edges which disappear, and all of their associated edges, will be associated with the edge which was the parent edge immediately before the deletion.

Thus, every edge that was ever in the tree is associated with exactly one edge which is currently in the tree.

Definition 6. Define an *A-subtree* (associated subtree) of an edge e at a particular time t to be the set of all the edges associated with any of the edges currently in the subtree below e together with the edges currently associated with e.

Lemma 7. *If an edge e has falls of weight W at time t, then there are at least $2^W - 1$ edges in the A-subtree of e at time t.*

Proof. The proof will be by induction on the time.

Base case:

We look at the situation at time 0. The A-subtree of e is simply e together with its subtree.

We will show that the subtree is large enough using induction on W. If $W = 1$, everything is fine because there is at least one edge and $2^W - 1 = 1$.

Suppose W is greater than one. Consider any fall from e. As the tree is red-black at time 0, all edges have weight zero or one. Among the edges on this fall which have weight one, let g be the one which is closest to the root. The edge g must have two children, f_1 and f_2. Then f_1 and f_2 both have falls of weight $W - 1$. Let S_1 be the subtree of f_1 and S_2 be the subtree of f_2. By the induction hypothesis, S_1 and S_2 each have at least $2^{W-1} - 1$ edges. The subtree of e contains the disjoint subtrees S_1 and S_2, along with the edge e, so it contains at least $(2^{W-1} - 1) + (2^{W-1} - 1) + 1 = 2^W - 1$ edges.

Induction step:

Assume that $t \geq 1$ and consider the possible operations at time t individually.

Insertion: If e is one of the edges just added, then $W = 1$, and $2^W - 1 = 1$. Since the A-subtree of e contains the edge e, it has enough edges. No other falls change weight, and no A-subtrees decrease in size.

Deletion: Suppose that e is the parent edge from the deletion. At time $t - 1$, a fall of weight W also existed, so a sufficiently large A-subtree existed then. Everything that was in the A-subtree of e at time $t - 1$ is still there, so the A-subtree is large enough. The argument is similar for edges above e. For the remaining edges, there is nothing to show.

Other Operations: These operations preserve the properties of chromatic trees, so any two falls from a particular edge have the same weight. They can change the A-subtrees of some edges, but no A-subtrees of edges with weight greater than one are altered. In addition, no edge with weight greater than one has heavier falls after one of these operations than it did before. Operations having these properties cannot make the lemma fail. To see this, assume to the contrary that it fails at time t. Among those edges which no longer have large enough A-subtrees, choose an edge e which is furthest from the root of the tree. As argued above, e can only have weight zero or one. Consider one of the falls from e. This fall must have weight greater than one, or there is no problem, so the edge must have two children, say f_1 and f_2. Both of these children have falls of weight $W - 1$ or W, so, as they are further from the root than e, they have A-subtrees of size at least $2^{W-1} - 1$. Since the A-subtree of e contains both of these A-subtrees and the edge e, it contains at least $2^W - 1$ edges, contradicting the assumption. Therefore, all the A-subtrees are still large enough.

Recall that $N = |T| + 2k$ is the bound on the maximum number of nodes the tree ever has. Let $M = \lfloor \log_2(N + 1) \rfloor - 1$. In the following theorem, we prove that no edge can have falls of weight greater than M. In proofs to come, this number will turn up frequently.

Theorem 8. *If the falls from edge e have weight W at any time $t \geq 0$, then $W \leq M$.*

Proof. Lemma 7 says that if e has falls of weight W, then the A-subtree of e contains at least $2^W - 1$ edges. As T is chromatic, e's sibling edge also has falls of weight W, implying that the A-subtree of e's sibling also contains at least $2^W - 1$ edges. Thus, there must have been at least $2^{W+1} - 2$ distinct edges in the tree, though not necessarily all at the same time. The total number of edges in T through time t is bounded by $N - 1$, so $2^{W+1} \leq N + 1$, from which the theorem follows.

In the following sections, we look at the types of operations individually and bound the number of times they can be applied. Theorem 8 is used to bound the number of times the blacking operation and the push operation can be applied. The other operations are much easier to handle; the theorem is unnecessary for bounding the number of times they are applied.

5 Red-Red Conflicts

The only operation which increases the number of red-red conflicts is the insertion; each insertion increases this total by at most one. The edge above the top edge in the red-red conflict will be called the parent edge.

The blacking operation is only applied when at least one of the child edges is involved in a red-red conflict. That red-red conflict is eliminated, but the operation could create another red-red conflict involving the parent edge. If only one of the child edges was involved in a red-red conflict before the operation, but a new red-red conflict was created, one can identify the new red-red conflict with the old one. If both child edges were involved in red-red conflicts, if a new red-red conflict was created, one can identify the new red-red conflict with the old one the left child edge was involved in. With each of the other operations which move the location of a red-red conflict in the process of rebalancing, one can always identify the new red-red conflict with a previous one; the lower edge in each new red-red conflict was also involved in a previous red-red conflict. Thus, one can follow the progress of a red-red conflict.

Lemma 9. *No red-red conflict can be involved in more than $M - 1$ blacking operations.*

Proof. Suppose a particular red-red conflict is involved in blacking operations at times $t_1, t_2, \ldots, t_r$, where $t_i < t_{i+1}$ for $1 \leq i \leq r - 1$. Let e_i be the lower edge of the red-red conflict at time $t_i - 1$, let f_i be the higher edge, and let g_i be the parent edge for the blacking operation. At time t_i, the edge f_i has just been made black and the edge g_i has just been made red (though, if $i = r$, then g_i may not have been made red). If $i \neq r$, then g_i becomes the lower edge of the red-red conflict. Clearly, falls from g_1 will have weight W for some $W \geq 2$, as the weight of g_1 is at least one at time $t_1 - 1$ and as there will always be a leaf edge of weight at least one somewhere under e_1.

It follows from the above that g_i is e_{i+1}, since it becomes the lower edge of the red-red conflict. We show that the weights of falls from e_{i+1} do not change from time t_i through time $t_{i+1} - 1$. Notice that all of the operations preserve the weights of falls beginning above or below the location where the operation takes place; this is necessary, of course, in order to maintain condition C2 of Definition 2. This means that for the weight of the falls from e_{i+1} to change, e_{i+1} has to be involved in the operation which causes the weight change. We now discuss the different operations.

The operation in question cannot be the blacking operation as, by assumption, this red-red conflict is not involved in a blacking operation (again) until time t_{i+1}. An insertion cannot involve a red-red conflict. Any red-red conflict involved in a deletion operation, (w1), (w2), (w7), or a push would disappear, which, by assumption, it does not. For the operations (w5) and (w6), e_{i+1} could only be the top edge, a, which clearly does not have the weight of its falls changed. For the operations (w3) and (w4), e_{i+1} could only be the edge b, but then the red-red conflict would disappear. Finally, for the red-balancing operations, e_{i+1} can only be the edge b, but this is impossible, since the red-red conflict would disappear.

We have proved that the weight of falls from e_{i+1} remains the same from time t_i through time $t_{i+1} - 1$. At time t_{i+1}, the blacking operation has occurred, so the weight of f_{i+1} has changed from zero to one. Thus, the weight of falls from g_{i+1}, which is also e_{i+2}, is exactly one more than the weight of falls from e_{i+1}.

By induction, it follows that at time t_r, the weight of falls from f_r is at least $W + r - 1$ which is at least $r + 1$, as $W \geq 2$. By Theorem 8, we find that $r \leq M - 1$.

Because the blacking operation can only be used when there is a red-red conflict, we obtain the following:

Corollary 10. *At most $k(M - 1)$ blacking operations can occur.*

It is easy to bound the number of times the red-handling operations can be applied:

Lemma 11. *At most k red-balancing operations can occur.*

Proof. Each red-balancing operation removes a red-red conflict. As only the insertion operation increases the number of red-red conflicts, and it can increase this number by at most one, it follows that the number of red-balancing operations is bounded by the number of insertions.

6 Overweight

The only operation which can increase the total amount of overweight in the tree is the deletion, and each deletion increases this overweight by at most one. Each of the weight decreasing operations decreases the total amount of overweight in the tree. The push operation decreases the overweight of some edge, but not necessarily the total amount of overweight (when $w_1 = 0$, the total amount of overweight in the tree is decreased). In the following, we only refer to the push operation as a push if it fails to decrease the total amount of overweight in the tree. Otherwise, we simply consider it to be one of the overweight decreasing operations.

For the sake of the following proof, we equip every edge with a *queue* for its overweight. Whenever new overweight is created by a deletion, we say that a new *unit* of overweight has been created. We assume that units of overweight are *marked* such that they can be distinguished. Whenever a unit of overweight is removed from an edge, we assume that it is the first unit of overweight in the queue which is removed. Likewise, when a unit of overweight is moved onto an edge, we assume that it enters the queue as the last element. When a deletion occurs, several units of overweight can be moved to a new edge. When this occurs, the order they had in their previous queue should be preserved. If a new overweight unit is created, it enters the queue as the last element.

Lemma 12. *At most $s(M - 2)$ push operations can occur.*

Proof. Suppose a particular unit of overweight u is moved up by a push operation at times $t_1, t_2, \ldots, t_r$, where $t_i < t_{i+1}$ for $1 \le i \le r - 1$. Just before the first push operation at time $t_1 - 1$, u sits on an overweighted edge (the edge b), so this edge has falls of weight at least two.

Assume that at time $t_i - 1$, u sits on an edge with falls of weight W. At time t_i, it has been pushed up (onto the edge a). We will argue that until time $t_{i+1} - 1$, this edge will have falls of weight at least $W + 1$. At time $t_i - 1$, the edge a must have falls of weight $W + w_1$. As it has weight w_1, it has exactly $w_1 - 1$ units of overweight in its queue. So, without removing u from the queue, we can only remove $w_1 - 1$ units of overweight. This can decrease the weight of the edge, and thus the weight of falls from this edge, by $w_1 - 1$, down to $W + 1$.

Units of overweight can also be moved onto another edge when a deletion occurs. These units of overweight will enter the queue of edge a and there will be $w_1 - 1$ (if $w_1 \ge 1$) units of overweight ahead of them in the queue. But the weight of the falls from this edge is also w_1 larger than the weight of falls from edge b, so removing the units of overweight ahead of these new units can never result in a lighter fall than there was from edge b. The case $w_1 = 0$ is trivial.

Now, we need only observe that the weight of falls from an overweighted edge is never decreased by any operation unless a unit of overweight is removed from the queue at the same time; and then it is decreased by at most one.

By induction, we have obtained that at time t_r, the weight of falls from edge a in the last push operation is at least $r + 2$. By Theorem 8, we find that $r \leq M - 2$, from which the result follows.

It is easy to bound the number of times weight decreasing operations can be applied:

Lemma 13. *At most s weight decreasing operations can occur.*

Proof. Only deletions introduce overweight and at most one unit is added each time. As weight decreasing operations remove one unit of overweight when they are applied, at most s such operations can occur.

7 Conclusion

Recall that $M = \lfloor \log_2(N + 1) \rfloor - 1$.

Theorem 14. *If a red-black tree initially has $|T|$ nodes and it later has k insertions, s deletions, and a number of rebalancing operations, then this number of rebalancing operations is no more than $(k + s)(\lfloor \log_2(N + 1) \rfloor - 1) - s$, where $N = |T| + 2k$ is the obvious bound on the number of nodes in the tree. Furthermore, the number of rebalancing operations which change the structure of the tree is at most $k + s$.*

Proof. Let us summarize how many times each type of operation can be used:

Operation	Bound	From
blacking	$k(\lfloor \log_2(N + 1) \rfloor - 2)$	Corollary 10
red-balancing	k	Lemma 11
push	$s(\lfloor \log_2(N + 1) \rfloor - 3)$	Lemma 12
weight decreasing	s	Lemma 13

The results follow by adding up the bounds.

References

1. Adel'son-Vel'skiĭ, G. M., Landis, E. M.: An algorithm for the organisation of information. Dokl. Akad. Nauk SSSR **146** (1962) 263–266 (in Russian; English translation in Soviet Math. Dokl. **3** (1962) 1259–1263).
2. Bayer, R., McCreight, E.: Organization and maintenance of large ordered indexes. Acta Inf. **1** (1972) 97–137.
3. Guibas, L. J., Sedgewick, R.: A dichromatic framework for balanced trees. IEEE FOCS (1978) 8–21.
4. Kessels, J. L. W.: On-the-fly optimization of data structures. Comm. ACM **26** (1983) 895–901.
5. Kung, H. T., Lehman, Philip L.: A concurrent database manipulation problem: binary search trees. Very Large Data Bases (1978) 498 (abstract; full version in Concurrent manipulation of binary search trees. ACM TODS **5** (1980) 354–382).
6. Nurmi, O., Soisalon-Soininen, E.: Uncoupling updating and rebalancing in chromatic binary search trees. ACM PODS (1991) 192–198.
7. Nurmi, O., Soisalon-Soininen, E., Wood D.: Concurrency control in database structures with relaxed balance. ACM PODS (1987) 170–176.

Appendix

Insertion.

Comment: the leaf B is inserted to the right of the leaf A.

Deletion.

Comment: the leaf A is deleted.

Blacking.

Restriction: at least one edge must be in a red-red conflict.

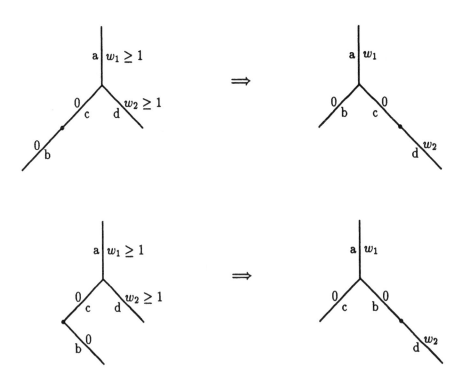

Red-balancing.

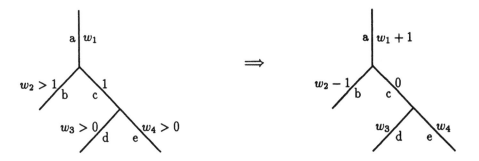

Push.

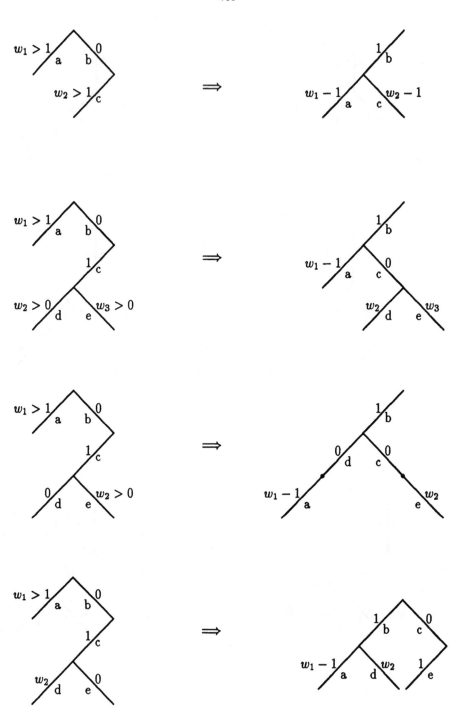

Weight decreasing (1-4).

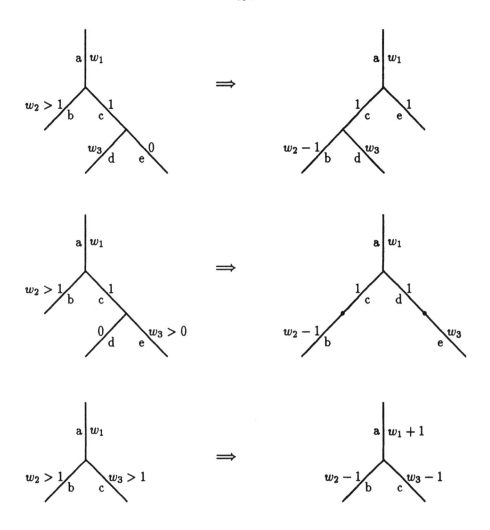

Weight decreasing (5-7).

This article was processed using the LaTeX macro package with LLNCS style

The Complexity of Scheduling Problems with Communication Delays for Trees

Andreas Jakoby

Rüdiger Reischuk

Technische Hochschule Darmstadt [1]

Abstract

Scheduling problems with interprocessor communication delays have been shown to be $\mathcal{NP}$-complete ([PY88],[C90],[CP91],[P91]). This paper considers the scheduling problem with the restriction that the underlying DAGs are trees and each task has unit execution time. It is shown that the problem remains $\mathcal{NP}$-complete for binary trees and uniform communication delays. The same holds for complete binary trees, but varying communication delays. On the other hand, by a nontrivial analysis a polynomial time algorithm is obtained that solves the problem for complete k-ary trees and uniform communication delays.

1.   Introduction

A directed acyclic graph can be used to represent the dependencies among elementary computation steps in a given algorithm. We will call such a graph a **process graph**. Let us assume that each node of the process graph has the same execution time which will be taken as the unit. If one wants to implement an algorithm as fast as possible on a synchronous multiprocessor architecture, one simply has to schedule the nodes of its process graph on the available set of processors.

In general, there will be a delay due to the necessary communication between processors. Typically, a single communication step exceeds the time of an elementary computation step by far. If processor P_a has computed an intermediate result z in the t-th step and this is needed by some other processor P_b as input for its computation one cannot assume that z is available for P_b already at step $t+1$. This leads to the following scheduling problem.

Definition 1 A **process graph** (G, δ) is a DAG $G = (V, A)$ with a delay function $\delta : A \rightarrow \mathbb{N}$. Let (G, δ) be a process graph and $P = \{P_1, P_2, \ldots\}$ a set of processors. A **schedule** S for (G, δ) and P is a set of triples $S \subset V \times P \times \mathbb{N}$ such that the following conditions are fulfilled:

---

[1] Institut für Theoretische Informatik, Alexanderstraße 10, 6100 Darmstadt, Germany
email: reischuk@iti.informatik.th-darmstadt.de

1. For each task $a \in V$ there is at least one processor P_i and a time t such that P_i processes x at t.

2. Each P_i does not process different nodes x, y at the same time t.

3. If x is executed at time t by processor P_i and y is a direct predecessor of x then either

   (a) y is processed by P_i by time $t - 1$ or

   (b) y is processed by some other processor P_j by time $t - 1 - \delta(y, x)$.

Let S_i be the **schedule** S restricted to processor P_i: $S_i := \{(x, t)|(x, P_i, t) \in S\}$ and $T(S_i)$ the finishing time of P_i: $T(S_i) := 1 + \max\{t| \exists x \in V : (x, t) \in S_i\}$. $S_i(t)$ denotes the node of G that is computed by P_i at step t. The **time span** or **length** of S is given by $T(S) := \max_{1 \leq i \leq p} T(S_i)$. A schedule S is called **optimal** for (G, δ) and P, if $T(S)$ is minimal for all schedules S' for (G, δ) and P. Let T_{opt} be the optimal schedule length:

$$T_{opt}((G, \delta), P) := \min\{T(S)|S \text{ is a schedule for } (G, \delta) \text{ and } P\} .$$

In the following the most interesting case of maximal parallelism is considered, that means the set of processors P can be chosen arbitrarily large. We therefore simply denote the minimal scheduling time by $T_{opt}(G, \delta)$. Depending on the type of communication delays the following variants of this scheduling problem can be distinguished.

DELAY SCHEDULING (DS):
Given a process graph (G, δ), and a deadline $T^* \in \mathbb{N}$, decide:
Is there a schedule S with $T(S) \leq T^*$, that means $T_{opt}(G, \delta) \leq T^*$?

UNIFORM DELAY SCHEDULING (UDS):
DS restricted to inputs (G, δ) with $\delta \equiv \tau$ for some $\tau \in \mathbb{N}$ that may depend on G and increase with the number of nodes of the process graphs.

CONSTANT DELAY SCHEDULING (CDS):
DS restricted to inputs (G, δ) with $\delta \equiv c$ for a universal constant c independent of the graph G. $\qquad\qquad\Box$

The UDS problem is defined by Papadimitriou and Yannakakis in [PY88] and it is shown

Theorem: [PY88] UDS is $\mathcal{NP}$-complete.

In that paper also a simple algorithm that approximates the optimal schedule length within a factor of 2 is described. The accuracy of the approximation follows from a simple lower bound that can be derived for the minimal schedule length. An extension of the algorithm can be given that also solves the more general DS-problem within the same factor.

In [JLS89] Jung, Kirousis and Spirakis describe a strategy based on dynamic programming for computing a schedule of minimal length. The runtime grows polynomial in the size of the DAGs, but exponential in the communication delay. The algorithm as stateds seems to have a minor mistake. A corrected version yields a time bound $O(n^{\tau+2})$, where n is the number of nodes of the process graph (G, δ) and $\delta \equiv \tau$ the uniform communication delay. This result implies that the CDS problem can be solved in deterministic polynomial time. Furthermore, the communication delay necessarily has to grow substantially with the number of nodes for any $\mathcal{NP}$-reduction to UDS (unless $\mathcal{P} = \mathcal{NP}$). In a subsequent paper [JR92] we will give improved results for deriving and estimating strategies to compute optimal schedules.

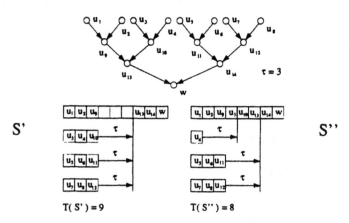

Figure 1: *A simple example of scheduling a complete binary tree with delay $\tau = 3$. The first schedule S' of length 9 is constructed by the algorithm described in [PY88]. S'' of optimal length 8 is obtaind by our new algorithm.*

Jung et al. also improve the lower bound of [PY88] for the minimal schedule length for graph that are complete binary trees, namely from about $\log n \cdot \tau / \log \tau$ to about $2 \log n \cdot \tau / \log \tau$. They conclude that this matches the upper bound given in [PY88] at the constant level. Does this mean that the algorithm of [PY88] constructs optimal schedules when restricted to complete binary trees? The answer must be no. In general, an optimal schedule is shorter and does not have the simple symmetric structure common to all schedules constructed by this algorithm. Since the lower and the upper bound as stated are a little floppy – they treat τ and $\tau+1$ as equal and do not differentiate between rounding up or rounding down $\log \tau$ – the bound $2 \log n \cdot \tau / \log \tau$ does not even give the correct constant factor, by which $\log n$ has to be multiplied. For example, for $\tau = 11$ the bound claims a factor $6.359\ldots$, but the correct factor in the asymptotic limit is 5, which will follow from our results below.

Let us finally mention that the DS problem can also be solved in polynomial time for any delay function δ that takes arbitrary values in the real interval $[0, 1]$. This is shown in the paper [CP91] by Chrétienne and Picouleau. Here, it is essential that all occuring delays are smaller than an elementary processing step.

2. New Results

The delay scheduling problem is of great practical importance. Therefore it seems useful to fully understand the influence of the different parameters like type of process graphs and type of delays. In particular, which properties give rise to $\mathcal{NP}$-completeness? For example as already mentioned, an efficient approximation of an optimal schedule within a factor 2 is possible, but we do not know of any polynomial time scheme that for arbitrary graphs guarantees a factor $\alpha < 2$.

We continue this line of reseach for the delay scheduling problem and concentrate on the underlying process graphs. It has been shown for many $\mathcal{NP}$-complete optimization on graphs that very fast solutions are possible if one restricts to special types of graphs like series-parallel or bounded tree-width graphs (for a recent overview see [R91]). In particular, almost every problem becomes simple on trees. The well studied unit-execution time multiprocessor scheduling problems without delay, that means $\delta \equiv 0$, is known to be $\mathcal{NP}$-complete for arbitrary graphs, but can be solved in polynomial time on forests [GJ79]. Note that in this case the number of processors should be bounded. Otherwise the problem becomes trivial with minimal schedule length equal to the depth of the DAG.

If one extends this scheduling problem such that each task has its individual deadline the situation is as follows. $\mathcal{NP}$-complete even for DAGs that are trees, but in $\mathcal{P}$ for reversed trees ([GJ79]). In the following tree alway means an intree, that is all edges run in direction from the leaves to the root of the tree. In a reversed or outtree the direction is switched. In this paper we want to estimate the complexity of the delay scheduling problem when restricted to trees of bounded degree. The $\mathcal{NP}$-completeness reduction of UDS given in [PY88] generates DAGs of unbounded indegree. The degree may become close to the size of the graph. Our results improve this situation considerably. (1,2)-trees and complete binary trees are the simplest types of such trees. The main result of this paper is that the delay scheduling problem even restricted to such simple graphs does not become easier.

Theorem 1 *Restricted to (1,2)-trees UDS remains $\mathcal{NP}$-complete.*

To prove the $\mathcal{NP}$-hardness we will construct a reduction of the X3C-problem. Nodes of degree 1 that occur in the (1,2)-trees obtained by this redurction can even be replaced by nodes of degree 2. Therefore the same result holds for binary trees, but typically these trees are not completely balanced. The reduction can be modified to yield the same result for complete binary trees, but in this case we have to use more than one value for the communication delays between nodes.

Theorem 2 *Restricted to complete binary trees DS remains $\mathcal{NP}$-complete.*

As a positive result, a polynomial time algorithm will be given that finds an optimal schedule for a given complete k-ary tree.

At first glance it might seem easy to solve UDS on complete trees because of their symmetry and regularity. But this is not the case. We do not even know how to give a direct recurrence relation for the minimal schedule of a complete tree with n nodes that can be evaluated in time polynomially bounded in n. An examination of the (1,2)- trees constructed in the proof of Theorem 1 reveals some of the difficulties in finding an optimal schedule for complete trees.

By a detailed case analysis we show that any optimal schedule for a complete tree can be transformed into one of a special *normal form* without increasing its length. We will describe a polynomial time algorithm that construct schedules of this special form. For such schedules one can even write down a recurrence relation for their length of moderate complexity.

Theorem 3 *Restricted to complete trees UDS can be solved in $O(n^2 \log n)$ where n is the number of nodes of the process graph (G, δ).*

This paper does not contain detailed proofs. In particular, the verification of the algorithm for complete trees is only sketched (for a complete analysis of it see [J91]).

3. The $\mathcal{NP}$-Completeness Proofs

Let us first observe that the DS problem for a graph $G = (V, A)$ with delay δ and deadline T^* can be solved by a polynomial time bounded NTM as follows:

1. Guess a set $S \subseteq V \times P \times \mathbb{N}$ of triples of size at most $|V|^2$ where P is a set of $|V|$ processors.

2. Verify that S is a schedule for G and δ.

3. Verify that $T(S)$ is not larger then T^*.

The correctness follows from the fact that for a process graph of size n there are always optimal schedules that use at most n processors with the total number of processor steps being bounded by n^2.

Lemma 1 *The DS problem can be solved by a NTM in polynomial time.* ∎

The $\mathcal{NP}$-hardness of the delay scheduling problem will be proved by a reduction of the X3C problem (exact cover by 3-sets, see [GJ79]).

Definition 2 (X3C problem) *Given a set* $M = \{1, \ldots, 3m\}$ *and a collection* $K = K_1, \ldots, K_k$ *of 3-element subsets of* M, *answer the question: Does* K *contain an exact cover for* M, *i.e., a subcollection* $K' \subseteq K$ *such that every element of* M *occurs in exactly one member of* K' ?

Proof of Theorem 1: We make the following reduction of X3C.
A set $M = \{1, \ldots, n\}$ with $|M| = 3m = n$ and a collection $K = K_1, \ldots, K_k$ of 3-element subsets of M are the inputs of the X3C-problem. The idea of the reduction to an instance of UDS is as follows. We build a tree B with a set of subtrees D_i that correspond bijectively to the elements of K. B has the property that a schedule S can only finish by a deadline T^* if and only if the processor that computes the root w of B also computes m subtrees D_i that corresponds to pairwise disjoint elements of K. This is achieved by the following properties of D_i :

> For each $j \in K_i$ D_i contains a subtree U_j with root u_1 and t_j special nodes such that if a processor computes u_1, it also has to compute these special nodes. The numbers t_j are chosen such that $t_j < t_{j+1}$ for all j (in this construction $t_j = j$ will suffice). The processing of these t_j nodes can be started at time $\alpha_j := 2\tau + \sum_{l=j+1}^{3m} t_l$. This property will be achieved by **delay trees** Δ_t as defined in figure 2.

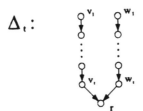

Figure 2: A delay tree Δ_t delays the node r for $t + \tau$ steps if $t \geq \tau$.

The details of this construction are described in the following figures. We need some additional parameters, which will be chosen as:

$$d := 8k, \qquad \gamma := \alpha_0 + m(d+3), \qquad \beta := \gamma + 2k.$$

1. The tree B in figure 3 represents the X3C-problem with $k = |K|$. D_i represents one element K_i of K. F is a binary tree with k leaves.

2. The subtree D_i in figure 4a represents the subset $K_i = \{j_1, j_2, j_3\}$. The 3 elements of K_j are represented by U_{j_1}, U_{j_2} and U_{j_3}.

3. The subtree U_j in figure 4b represents one element j of K_i.

4. The communication delay is chosen as $\tau := (m-1)(7+d) + k + m$ and the deadline as $T^* := \beta + 3k - 1$.

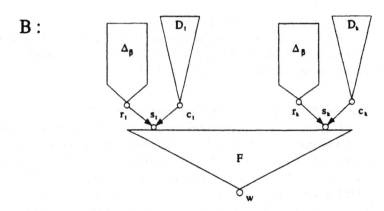

B :

Figure 3: *A binary tree representing an X3C problem with k subsets.*

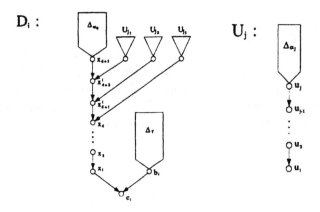

D_i :

U_j :

Figure 4: a) *A subgraph D_i with subtrees U_{j_1}, U_{j_2} and U_{j_3} .*
b) *The graph U_j that represents the element j .*

The following lemmata describe some important properties of the delay trees Δ_t and the final tree B .

Lemma 2 $T_{opt}(\Delta_t, \tau) = t + \tau + 1$ *for* $t \geq \tau$.

Proof: The lower bound follows easily by differenciating between schedules, in which a single processor computes the whole delay tree or not, where in the later case at least a delay τ occurs. Observe that the time bound $t + \tau + 1$ can be achieved by a strategy that uses 3 processors P, P', P''. P', resp. P'' processes the 2 chains in parallel in t steps. After $t + \tau$ steps P computes the root r . ∎

Lemma 3 *Let (B, τ) be a tree representing an instance $(\mathcal{M}, \mathcal{K})$ of the X3C-problem, and let S be an optimal schedule for (B, τ) . Then it must hold: If processor P_w computes the root w , it also computes all b_i , c_i and r_i .* ∎

Lemma 4 *Let* (B, τ) *be a process tree representing an instance* $(\mathcal{M}, \mathcal{K})$ *of the X3C-problem and the deadline* T^*, *then* $T_{opt}(B, \tau) \geq T^*$. ∎

To finish the proof of theorem 1 it remains to show the following lemma.

Lemma 5 *Let* (B, τ) *and* T^* *represent the X3C-problem for a set* $\mathcal{M}$ *and subsets* $\mathcal{K}$. *Then* (B, τ) *can be scheduled within time* T^* *if and only if* $\mathcal{K}$ *has a subset* $\mathcal{K}'$ *that is an exact cover of* $\mathcal{M}$.

Proof: Figure 5 and 6 illustrate properties of B and τ that imply the claim. ∎

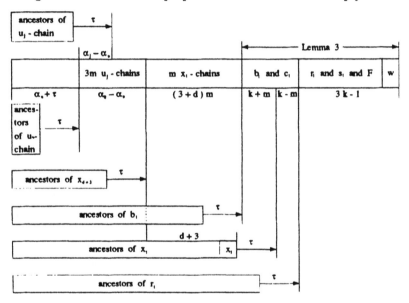

Figure 5: *This diagram describes the schedule* S *restricted to the processor* P_w *that computes the root of* B.

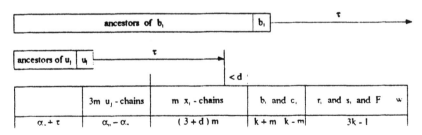

Figure 6: *A node* u_1 *or* b_i *canot be managed by a processor different from* P_w, *if the schedule obeys the deadline* T^*.

These lemmata prove Theorem 1. ∎

The proof of Theorem 2 goes along the same lines. Technically, it is more involved in order to achieve depth $O(\log n)$ for the complete binary tree constructed, where $n = |M|$ and M and K are the inputs of the X3C problem.

Proof of Theorem 2: The construction is illustrated by figure 7. Each subset $K_i = \{j_1, j_2, j_3\} \in K$ is represented by a subtree D_i. The subtree with root j_l consisting of the pieces E_{j_1} and C_{j_1} represents the element j_l.

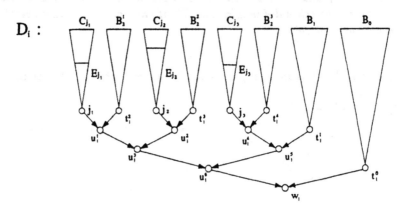

Figure 7: *The modified subgraphs D_i for the proof of Theorem 2.*

A B-tree (resp. a C-forest) delays the computation of the direct descendent of its root for a suitable number of steps like the Δ-trees constructed in the proof of theorem 1. The roots of the D_i-trees are the sources of a complete binary tree which represents the X3C problem like the B-tree in the proof of theorem 1. In the following a 0-tree of depth d is a complete binary tree with no communication delay, that means $\delta = 0$. D_i can be constructed by the following strategy:

1. Build a tree D_i' with subgraphs $E_{j_1}', C_{j_1}', E_{j_2}', C_{j_2}', E_{j_3}', C_{j_3}', B_2^{1'}, B_2^{2'}, B_2^{3'}, B_1'$ and B_0' where all leavs of D_i' have the same depth. The subtrees of D_i' have the following number of nodes

$$N(E_{j_i}) := j_i \qquad\qquad N(B_2^1) := \sum_{l=1}^{n+1} l$$
$$N(B_2^2) := m + \sum_{l=1}^{n+1} l \qquad\qquad N(B_2^3) := 3m + \sum_{l=1}^{n+1} l$$
$$N(B_1) := 4m + \sum_{l=1}^{n+1} l \qquad\qquad N(B_0) := 6m + \sum_{l=1}^{n+1} l$$

and each tree C'' of a C'-forest

$$N(C_{j_i}'') := \sum_{l:=j_i+1}^{n+1} l.$$

2. Between the root of a B'-tree (resp. a tree of a C'-forest) and its successors there is no communication delay, that means $\delta = 0$. The communication delay between u_i^6 and w_i is equal to $3m - 2$. All other edges have a communication delay that is equal to the number of nodes of the total graph.

3. Generate the complete binary tree D_i by using D'_i and 0-trees.

Oberserve that for the communication delay only 3 different values are needed. ∎

4. A Polynomial Time Algorithm for Complete Trees

In this section an algorithm will be described, that generates an optimal schedule if the communication delay is fixed and the DAG is a complete k-ary tree. The correctness proof is rather long. We will only describe the essential steps in a sequence of lemmata without proofs.

Proof of Theorem 3: Let (G, τ) with $G = (V, A)$ and $\tau \in \mathbb{N}$ be the inputs of UDS where G is a complete k-ary tree. For $x \in V$ let $G(x)$ be the maximal subtree of G with root x. We define the set $\mathcal{M}$ of tuples

$$(x, \text{depth}(x), S[x], T(S[x])) ,$$

where x is a node of G and $S[x]$ is an optimal schedule for $(G(x), \tau)$ restricted to the processor that computes x. A subset M of $\mathcal{M}$ is constructed that contains for each $x \in V$ at least one tupel $(x, \text{depth}(x), S[x], T(S[x]))$. This is done in a sequence of stages $M_0, M_1, \ldots, M_d = M$, where $d = \text{depth}(G)$. Define

$$M_0 := \{(x, 0, (x, 0), 1) | x \in V \wedge \text{depth}(x) = 0\} .$$

Given $M = M_{i-1}$ the following procedure **Schedule**(x, CP, t) finds for x with $\text{depth}(x) = i$ an element $(x, \text{depth}(x), S[x], T(S[x]))$ of M_i. **Schedule** uses 3 parameters: the node x, a set CP, which is a prefix of the schedule for the processor P_x that is chosen to compute x, and a step number t, which denotes the time when the last element of the current CP is finished. These variables are initialized to $CP := S[z_0]$ and $t := T(S[z_0])$, where z_0 is an arbitrary direct predecessor of x.

Schedule(x, CP, t) returns a tupel (CP', t') such that CP' is the complete schedule for P_x and t' is the time, when x will be finished. If **Schedule**(x, CP, t) finishes with (CP', t'), the tupel $(x, \text{depth}(x), CP', t')$ is added to M.

```
Schedule( x , CP , t )
        Γ := {z | z is a direct predecessor of x, z ≠ z₀, and for all tupels
                 (z, depth (z), S[z], T(S[z])) ∈ M holds: T(S[z]) + τ > t}
        WHILE Γ ≠ ∅  DO
             BEGIN
             choose y ∈ Γ arbitrarily
             (CP,t) := Schedule(y, CP, t)
             Γ := Γ \ {y}
             Γ := Γ \ {z | (z, depth(z), S[z], T(S[z])) ∈ M  and  T(S[z]) + τ < t}
             END
        RETURN (CP ∪ {(x,t)}, t + 1)
END
```

From the set M_d a schedule for G can be derived easily. For each node x of G choose a different processor P_x and let $S[x]$ be the schedule for P_x.

Definition 3 *A schedule S for a tree (G, δ) is in* **normal form** *if S fulfills the following three conditions:*

A1: *For any subtree B of G: If processor P_r at time j computes the root w of B, that means $w = S_r(j)$, and $S_r(i)$ is also a node of B all nodes computed by P_r between time i and j also belong to B.*

A2: *If processor P_r computes a node x at step j, no other processor has computed x by time $j - \tau - 1$.*

A3: *No processor computes a node several times.* □

Lemma 6 *The algorithm described above always generates a schedule S in normal form.* ∎

We can show that this normal form condition is no restriction in case of complete trees. This property is not trivial since it does not hold for arbitrary trees, even for binary trees.

Lemma 7 *For every complete k-ary tree (G, τ) there exists an optimal schedule S that is in normal form.*

This claim can be shown by a rather long and detailed case analysis. It seems difficult to give a simple argument for this property. ∎

The next lemma shows that all schedules in normal form have the same length.

Lemma 8 *Let $\mathcal{F}$ be the set of schedules in normal form for a complete k-ary tree (G, τ). Then for all $S, S' \in \mathcal{F}$ holds: $T(S) = T(S')$.* ∎

Lemma 7 and Lemma 8 imply that all schedules S in $\mathcal{F}$ are optimal. The algorithm described above generates a schedule S in normal form (lemma 6). Therefore S is an optimal schedule. The time complexity can be estimated as follows.

Lemma 9 *The algorithm constructs an optimal schedule S for a complete k-ary tree (G, τ) within $O(|V|^2 \log(|V|))$ steps.* ∎

This analysis of the structure of an optimal schedule in normal form for complete trees implies that the length of such an schedule can be estimated by the following recurrence equations. Without this result we could only deduce a much more involved recurrence.

Corollary 1 *Let* $f_{\tau,k}(h) = T_{opt}(G,\tau)$ *be the minimal time to schedule a complete k-ary tree of height h and delay τ on an unbounded set of processors. Then*

$$f_{\tau,k}(h) = \begin{cases} 1 & \text{for } h = 0 \\ f_{\tau,k}(h-1) + g_{\tau,k}(h) & \text{for } h > 0, \end{cases}$$

where

$$g_{\tau,k}(h) = u_{\tau,k}(h-1, k-1, \tau)$$

and $u_{\tau,k}(h,s,t) = 1$ *if any of the arguments is 0, else*

$$u_{\tau,k}(h,s,t) = u_{\tau,k}\left(h-1,\ k,\ |t - g_{\tau,k}(h-1)|^+\right)$$
$$+ u_{\tau,k}\left(h,\ s-1,\ \left|t - u_{\tau,k}(h-1,k,|t-g_{\tau,k}(h-1)|^+)\right|^+\right),$$

where $|x|^+ := \max\{0,x\}$. ∎

Observe that $g_{\tau,k}(h)$ and hence also $f_{\tau,k}(h)$ can be computed in time $\exp O(h)$, that means polynomial in the size of the graph for fixed degree k. This also implies that the decision problem for complete trees belongs to $\mathcal{P}$.

5. Conclusion

We have shown that the delay scheduling problem remains difficult, even for trees with a very simple structure. Only for completely symmetric trees a nontrivial strategy could be exhibited to compute an optimal schedule efficiently. We have also made a detailed analysis of the complexity of this problem with respect to the size of the delays. These results will be reported in [JR92]. The following table gives an overview on the complexity in the different situations.

	DAGs	binary trees	complete trees
CDS	$\mathcal{P}$-complete [JKS89,JR92]	$\mathcal{P}$	$\mathcal{P}$
UDS	$\mathcal{NP}$-complete [PY88]	$\mathcal{NP}$-complete [Th. 1]	$\mathcal{P}$ [Th. 3]
DS	‖	‖	$\mathcal{NP}$-complete [Th. 2]

6. References

[C90] P. Chrétienne, Complexity of Tree-Scheduling with Interprocessor Communication Delays, MASI Report 90.5, 1990

[CP91] P. Chrétienne and C. Picouleau, The Basic Scheduling With Interprocessor Communication Delays, MASI Report 91.6, 1991

[GJ79] M. R. Garey and D. S. Johnson, Computers and Intractability, A Guide To the Theory of $\mathcal{NP}$-Completeness, Freeman 1979.

[J91] A. Jakoby, Prozess-Prozessor-Allokationsprobleme für Baumförmige Datenflussgraphen, Diplomarbeit, Technische Hochschule Darmstadt, 1991

[JKS89] H. Jung, L. Kirousis, P. Spirakis, Lower Bounds and Efficient Algorithms for Multiprocessor Scheduling of DAGs With Communication Delays, Proc. 1. SPAA, 1989, 254 - 264

[JR92] A. Jakoby, R. Reischuk, Efficient Strategies for Bounded Delay Scheduling, Technical Report, Technische Hochschule Darmstadt, 1992

[P91] C. Picouleau, Two New $\mathcal{NP}$-complete Scheduling Problems With Communication Delays and Unlimited Number of Processors, MASI Report 91.24, 1991

[PY88] C. Papadimitriou and M. Yannakakis, Towards an Architecture-Independent Analysis of Parallel Algorithms, Proc. 20. STOC, 1988, 510-513

[R91] R. Reischuk, Graph Theoretical Methods for the Design of Parallel Algorithms, Proc. 8. FCT, 1991, 61 - 67

The List Update Problem and the Retrieval of Sets*

Fabrizio d'Amore and Vincenzo Liberatore

Università di Roma "La Sapienza",
Dipartimento di Informatica e Sistemistica,
via Salaria 113, I-00198 Roma, Italy

Abstract. We consider the list update problem under a sequence of requests of *sets* of items, and for this problem we investigate the competitiveness features of two algorithms. We prove that algorithm Move-Set-to-Front (MSF) is $(1 + \beta)$-competitive, where β is the size of the largest requested set, and that a lower bound is roughly 2. We provide an upper bound to the MSF competitive ratio by relating it to the size n of the list, obtaining that the algorithm is $(1 + n/4)$-competitive in general, and $O(\sqrt{n})$-competitive if we add a not too restrictive constraint to the sizes of the requested sets.

We are in touch with two more problems. The first one generalizes the list update problem under a sequence of requests of sets by considering weighted lists, where a weight representing a visiting cost is associated with each item. For this case we give a competitiveness result as well.

The second one is a variant, where the list is searched to retrieve whichever element of the currently requested set (the first that can be found in the list). For this problem we provide negative results.

1 Introduction

In the last two decades the list update problem has been studied under several formulations and different aspects: first, the problem of studying the performance of linear search under self-adjusting algorithms has been faced (for a complete bibliography see [5]); then, in [13] these issues have been formalized into the list update problem by providing a suitable cost model. In that article the authors studied the problem by means of amortized analysis, furthermore they made use of "comparative" techniques. This is actually one of the first competitiveness analyses, even if at that time the term "competitive" [8] had not been introduced yet. Moreover, the authors introduced a method of analysis of practical use in many cases, and which has been very often exploited in subsequent papers dealing with on-line algorithms.

The traditional list update problem consists in rearranging the items of a list during the processing of a sequence of operations on the list in order to minimize the processing costs of subsequent operations. So the very question is finding on-line rearrangement rules, also called permutation algorithms [5], with high performance.

---

* Work partially supported by the ESPRIT II Basic Research Actions Program Project no. 3075 "ALCOM", and by the Italian MURST National Project "Algoritmi, Modelli di Calcolo e Strutture Informative".

For the traditional list update problem, it has been shown [13] that algorithm Move-to-Front (after accessing an item, move it to the front of the list without changing the relative ordering of the other items) is 2-competitive against the optimum off-line algorithm, and that no deterministic on-line algorithm can achieve a better competitive ratio [9, 6]. For the same problem randomized algorithms have been presented too [7]: for example BIT (a randomly initialized bit is associated with any item: complement this bit whenever the item is accessed, and if an access causes the bit to be changed to 1, then move the item to the front of the list) has been shown to be 1.75-competitive.

Later on, some generalizations of the list update problem have been considered, such as the P^d model in [11, 7] and the weighted list model [1, 2]. For these generalization several algorithms, with different competitiveness features, have been provided.

In this paper we are concerned with the generalization of the list update problem to the case in which the list is searched in order to retrieve a *set* of elements rather than just one item at a time. We study this problem under two different (but related) cost models: the standard one [13], and the wasted work one [3, 7]. To the best of our knowledge, this generalization has not been investigated till now.

Also in this case we are mainly interested in designing efficient permutation algorithms, though now the rules should rearrange the list with the goal of minimizing the overall cost of processing a sequence of requests of sets rather than of single items.

Broadly speaking, list update problems, generalized in suitable ways, are tools one can use in order to design algorithms for several on-line problems, such as the scheduling of the operations of a sequential machine. For example, weighted lists can model situations where different operations have different costs, such as in the case of visiting a tree by depth first searches [1, 2]. Also, in the case of searches for sets, we can model situations where in order to serve a request, one has to perform several operations that may be executed in an unordered fashion, i.e. without constraints that force a certain operation to precede some other operations. As an example, we conjecture this approach is useful while modeling techniques for visiting self-adjusting AND-OR trees, in order to obtain competitiveness results.

The main result of this paper is that the deterministic algorithm Move-Set-to-Front is $(1 + \beta)$-competitive, where β is the maximum size among all the requested sets. The proving technique provides innovative tools that can be useful in a wide range of applications. We also provide other upper and lower bounds to the competitive ratio of any deterministic on-line algorithm for the list update problem with retrieval of sequences of sets.

Moreover, we show that algorithm BIT-for-Sets is $(1 + (3/4)\beta)$-competitive, and that, while searching a weighted list for a sequence of sets, algorithm Move-Set-to-Front is still competitive.

The rest of the paper is organized as follows. In the next section some basic concepts and notations are introduced.

In section 3 we provide upper and lower bounds to the competitive ratios of wide classes of on-line algorithms for the list update problem with retrieval of sequences of sets.

In section 4 we study algorithm Move-Set-to-Front and prove its competitiveness.

Furthermore, in this section we correlate the competitive ratio to the size of the list.

In section 5, we consider two variants of the studied problem: the first one consists in the case of weighted lists, the second one is a variant where the list is searched to retrieve whichever element of the currently requested set (the first that is found). For this problem we provide negative results.

Finally, in section 6 we draw some conclusions and address future work and open questions.

2 Preliminaries

The *list update problem with retrieval of sets* consists in storing the items of a set S as an unsorted list $\mathcal{L}$ which only supports sequential access, and in on-line updating the list while serving a sequence $\underline{r} = (r_1, \ldots, r_m)$ of requests, each of which can consist in finding, inserting or deleting a *set* of items, i.e. a subset of S.

We are interested in developing on-line algorithms for updating the list in order to efficiently process $\underline{r}$. Updates operations consist in inserting a new item, deleting an item, or rearranging the list after any access, by means of some permutation algorithm [5]. We shall denote by $\mathcal{L}_j$ the list after j-th request has been served.

If x and y are items of $\mathcal{L}_j$ we shall write that $x \prec_j y$ (x *precedes* y) to denote that x is stored in $\mathcal{L}_j$ before y (however we will omit the subscript j when ambiguities do not arise).

Since both insertions and deletions can be handled as special cases of access operations [11] we shall consider sequences $\underline{r}$ that consist only of accesses to a list of a fixed-size n.

Consequently, the generic request r_j can be expressed as a subset of S. In order to serve request r_j, list $\mathcal{L}_{j-1}$ is sequentially searched for finding *all* the items belonging to r_j.

Henceforward, we shall denote by β_j the size of r_j, and by β the maximum among all β_j's. Another quantity we will consider is the *amortized* (averaged on the sequence) value of the sizes, defined as $\beta_{am} = \sum_{j=1}^{m} \beta_j / m$.

For the list update problem with retrieval of sets we consider two cost models. First, with regard to the special case in which r_j is simply a singleton, say the i-th item of the list, the cost of finding r_j is i in the *standard model* and $i - 1$ in the *wasted work model*. So, in the first model we consider all the cost paid in order to answer a request, while in the latter we do not take account of useful costs.

If r_j is not a singleton, then the cost of retrieving all its elements is *not* the sum of all the costs which we would pay by separately searching all these elements, as singletons. In fact, we assume that a special function is available:

$$\theta_{\mathcal{L}} : \mathcal{P}(S) \times S \to \{\text{no}, \text{yes\_continue}, \text{yes\_stop}\},$$

where $\mathcal{P}(S)$ is the power set of S, and use $\theta_{\mathcal{L}}$ in the following fashion. While sequentially searching $\mathcal{L}$, if x is currently the item under inspection,

- $\theta_{\mathcal{L}}(r_j, x) = \text{yes\_continue}$ if $x \in r_j$ and there is some other item in r_j to be found in $\mathcal{L}$;
- $\theta_{\mathcal{L}}(r_j, x) = \text{yes\_stop}$ if $x \in r_j$ and all the elements in r_j have been already found; and, finally,

– $\theta_{\mathcal{L}}(r_j, x) = $ no if $x \notin r_j$ (obviously, in this case the search continues).

So, $\theta_{\mathcal{L}}$ is the function the algorithms uses to test whether the current item under inspection x belongs to the answer. In the sequel, we shall assume that the cost of computing function $\theta_{\mathcal{L}}$ is 1, but later on we shall renounce this hypothesis and examine some more general cases.

Hence, in the standard model, the cost C_j of finding r_j is the cost l_j of finding the element of r_j that in $\mathcal{L}$ is preceded by all the other elements of r_j. In the wasted work model, C_j is $l_j - \beta_j$, according to the spirit of not considering useful costs.

Permutation algorithms are allowed to exchange adjacent items. There are two kinds of exchanges, *free* and *paid* ones. The elements recognized as belonging to r_j can be moved by means of free exchanges closer to the front of the list (at no cost) without changing their relative ordering. All other exchanges are said paid and have unitary cost.

The generalized list update problem reduces to the well-known list update problem [13] when all requests are singletons. In this case, the cost models match those presented in [13] (standard) and [3, 7] (wasted work).

In the case of *weighted* lists [1, 2] there is a total function $w : \mathcal{S} \to \mathbb{R}^+$ which expresses the cost $w(x)$ that $\theta_{\mathcal{L}}$ is charged for the inspection of item x. However, in what follows, when a precise specification is missing, we refer to unweighted lists.

Typically, the performance of an on-line list update rule is analyzed in comparison to that of an off-line algorithm, i.e. an algorithm that is able to perform some optimizations by virtue of its knowledge about future requests. In practice, in order to make the comparison, it is convenient to introduce the concept of *adversary*, that has the power of *creating* the sequence $\underline{r}$ to be served by the on-line algorithm and of making use of an off-line algorithm H which can take advantage on the grounds of the knowledge on $\underline{r}$. Of course, among these off-line algorithms there is also the optimal one, which minimizes the overall cost of processing $\underline{r}$.

Let α be a function $\alpha : \mathbb{R} \to \mathbb{R}$ of the type $\alpha(x) = c \cdot x + e$. A deterministic on-line algorithm G is said to be α-competitive against the adversary that uses algorithm H if for any sequence $\underline{r}$ $G(\underline{r}) \le \alpha(H(\underline{r}))$, where $G(\underline{r})$ and $H(\underline{r})$ are the costs on $\underline{r}$ respectively paid by G and H [8, 4]. The constant e does not depend on the sequence $\underline{r}$ but only on the handled list [6].

If H is the static algorithm, namely the one that does not actually rearrange the list (so it never changes the initial arrangement), then the adversary that uses H is said *static*. Among all the initial orderings for the items of the list there is the *optimal static* one, i.e. the one that minimizes the overall cost of processing a sequence $\underline{r}$ while using the static algorithm.

If G is a randomized algorithm, G is said to be α-competitive if $E\{G(\underline{r})\} \le \alpha(H(\underline{r}))$, where $E\{G(\underline{r})\}$ is the cost of G averaged over all the random choices that G makes while processing $\underline{r}$ [10, 4].

In practice, both deterministic and randomized competitive algorithms are said to have competitive ratio c or to be c-competitive [10].

If c is a competitive ratio for G, then $c + \varepsilon$, for any positive real ε, is also an admissible competitive ratio, because $G(\underline{r}) < (c + \varepsilon) \cdot H(\underline{r}) + e$.

It is worth observing that while analyzing the performance of a randomized algorithm G, it makes sense to refer to different kinds of adversaries, which differ on the basis of their actual power [4]. The *oblivious* adversary is required to generate $\underline{r}$ before G starts working: this is the *weak* adversary since it cannot see the random choices made by G; so, intuitively, randomization can help against the oblivious adversary.

Two more kinds of adversaries are said *adaptive* and are more powerful since they are allowed to generate next request after G has served the current one: in this way, they can force G to pay an higher cost. If an adaptive adversary uses an on-line algorithm it is said *medium*; if it is allowed to use the optimum off-line algorithm it is said *strong*. Randomization cannot help against strong adversaries [4], and, for the list update problem, not even against medium adversaries [7].

In order to compare the performance of two given algorithms a common technique consists in using a *potential function* [12] whose value at any time measures a suitable quantity which reflects the difference between the states of the lists handled by the two algorithms [7, 13].

Finally, concerning with the cost models we adopted, it is worth observing that if we supposed that the free exchanges could be used to modify the relative ordering of the items belonging to r_j, the adversary would unfairly profit by this. In fact, while in the model we adopted the adversary is charged $O(n^2)$ for any total reorder, and on-line algorithms have no cost, in the model with free rearrangements, the adversary at any time could request $r_j = S$ in order to completely reorganize the list at no cost, and for this request both the adversary and the on-line algorithm would be charged the same cost. In addition, we would assume that the rearrangement cost paid by the adversary is null even for very large requested sets.

However, we shall see in section 4 (Corollary 5) that there exists an on-line algorithm which, in the standard model, is so powerful that one can allow the adversary to reorganize the list by free exchanges without worsening the competitive ratio.

3 Lower and Upper Bounds

In this section we provide a lower and an upper bound to the competitive ratio, which hold for any on-line deterministic algorithm under no restrictive hypotheses, in the standard cost model.

With regard to the lower bound, it is obtained against a static off-line adversary. It is well-known that for the traditional list update problem the optimal static ordering is any ordering according to non-increasing frequencies (NIF) of access [3, 5]. However, this result does not hold in the case we are studying, as the following counterexample shows.

Let S be $\{a, b, c, d\}$, and $\underline{r} = (\{a\} \{a, c\} \{a, c\} \{b\} \{b, d\} \{b, d\})$. According to a NIF ordering, such as for example $(a\ b\ c\ d)$, a and b precede c and d and then a total cost of 17 is charged (in the standard model) for processing $\underline{r}$. On the other side, the (non-NIF) static ordering $(a\ c\ b\ d)$ has a total processing cost equal to 16.

The lower bound against an adversary that uses a static NIF ordering algorithm is roughly 2, as the following theorem states.

Theorem 1. *In the standard model, the competitive ratio of any on-line deterministic algorithm G for the list update problem with searches for sets cannot be less than $2n/(n + \beta_{\mathrm{am}})$.*

Proof. The proof is only sketched. The adversary H makes the j-th request by asking the last element x of the list maintained by G. The other $\beta_j - 1$ items of the requested set are the first ones of list handled by H if x is not among these ones. Otherwise H requests the first β_j elements in its own list.

The thesis follows keeping into account the scheme by Karp and Raghavan [9] to prove lower bounds on competitive ratios, and after some algebra.

The lower bound in Theorem 1 specializes in the traditional one [9] if $\beta_{am} = 1$.

Concerning to the upper bound, the following theorem gives a result for a wide class of algorithms.

Theorem 2. *Let G be a deterministic on-line algorithm that never makes paid exchanges. Then in the standard model there exists at least one admissible value for the competitive ratio of G that is less than or equal to n/β_{am}.*

Proof. In the worst case, the adversary H requests at each step a set in which one element is in the last position in the list G manages, while the whole set is located in the first β_j positions in H's list.

Since G does not make paid exchanges, its global cost is $n \cdot m$; on the other hand the cost of H is $\sum_{j=1}^{m} \beta_j$.

The curves which lower and upper bound the competitive ratio are drawn in Fig. 1.

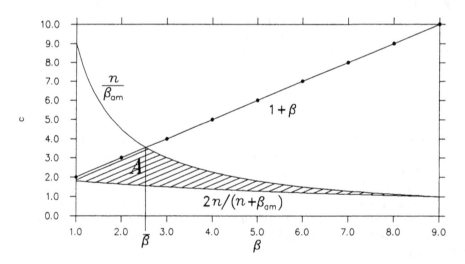

Fig. 1. Curves illustrating the upper and lower bounds. They are drawn for $n = 9$ and $\beta_j = \beta_{am}$, for any j

4 Algorithms

In this section, we study two algorithms for updating lists that, even if they will be shown to have competitive ratios depending on β, do not depend on β, i.e. they manage the list in a uniform way regardless of the value of β.

The first algorithm generalizes the well-known Move-to-Front, and we name it Move-Set-to-Front (MSF). It consists in moving to the front of the list any accessed set of items, without changing neither their relative ordering nor that of the other items.

We shall prove that MSF is $(1 + \beta)$-competitive against the optimum off-line algorithm, and shall make the comparison on the basis of a potential function that depends on the number of inversions between the lists handled by the two algorithms.

Let τ be a permutation of S. In particular, let ι be the identity permutation, i.e. the one such that $\iota(x) = x$, for any $x \in S$.

Now we give a the definition of *inversion* which generalizes that provided by Sleator and Tarjan [13]. Given a permutation τ and two lists $\mathcal{L}_H$ and $\mathcal{L}_G$, an inversion is an ordered pair $(y, x) \in S^2$ such that $\tau(x) \prec \tau(y)$ in $\mathcal{L}_H$ and $y \prec x$ in $\mathcal{L}_G$. When $\tau = \iota$ the usual definition of inversion is obtained.

Note that the number of inversions depends on the order the two lists are considered and on the used permutation, and two different permutations τ_1 and τ_2 generally yield different results.

Throughout our analysis we shall make use of the concept of *current* permutation, i.e. the permutation which we shall be referring inversions to. While changing the current permutation from τ_1 to τ_2, a variation in the number of inversions occurs. Let $h(x)$ be the number of items that precede x in $\mathcal{L}_G$ and are located between $\tau_1(x)$ and $\tau_2(x)$ in $\mathcal{L}_H$. In a similar way, let $t(x)$ the number of items that follow x in $\mathcal{L}_G$ and are located between $\tau_1(x)$ and $\tau_2(x)$ in $\mathcal{L}_H$. When the current permutation passes from τ_1 to τ_2, x is said to move *forward* if $\tau_2(x) \prec \tau_1(x)$, or *backward* if $\tau_1(x) \prec \tau_2(x)$.

It is easily seen that, for any item x, if we only consider its movement while the rest of the list remains fixed, the following holds:

Lemma 3. *For any fixed x, the increase in the number of inversions of type (x, y) plus those of type (z, x), for any y and z belonging to S, is given by*

$$\begin{cases} h(x) - t(x) \text{ if } x \text{ moves forward} \\ t(x) - h(x) \text{ if } x \text{ moves backward} \end{cases}$$

Now, regarding to the problem of updating a list under sequences of requests of sets, we proof that MSF is competitive with respect to any algorithm H.

Theorem 4. *MSF is $(1+\beta)$-competitive both in the standard and in the wasted work model.*

Proof. First of all, we introduce some symbols. Let $\mathcal{L}_H$ and $\mathcal{L}_{\text{MSF}}$ be the lists updated by H and MSF, respectively. Let us assume that the j-th request consists in set B, and let us denote its elements by $a_1, \ldots, a_{\beta_j}$, ordered according to the ordering of $\mathcal{L}_H$. Furthermore, for our convenience, we shall denote by $b_1, \ldots, b_{\beta_j}$, the same elements ordered according to the ordering of $\mathcal{L}_{\text{MSF}}$.

Let τ_j be the following permutation:

$$\begin{cases} \tau_j(x) = a_l \text{ if } x \in B \text{ and } x = b_l \\ \tau_j(x) = x \text{ if } x \notin B \end{cases}$$

The permutation τ_j leaves in their place items not belonging to B and rearranges those belonging to B in such a way that inversions with respect to ι between two items both belonging to B are not inversions with respect to τ_j. In other words, the ordering of B obtained by applying τ_j to $\mathcal{L}_{\mathrm{MSF}}$ is the same as that in $\mathcal{L}_H$.

At any time the current permutation is an element in $\{\tau_j | 1 \le j \le m\} \cup \{\iota\}$. In what follows an inversion will be always meant with respect to the current permutation, $\mathcal{L}_{\mathrm{MSF}}$ and $\mathcal{L}_H$. At the very beginning the current permutation is ι.

The proof makes use of a potential function, which is defined as the number of inversions between $\mathcal{L}_{\mathrm{MSF}}$ and $\mathcal{L}_H$. The proving technique basically consists in considering a current permutation, which we modify in order to more easily evaluate the amortized costs associated with the various operations made by the two algorithms.

So, we shall calculate the variations in the number of inversions due to changes in $\mathcal{L}_{\mathrm{MSF}}$, in $\mathcal{L}_H$ (as usual, see [13, 7]), and in the current permutation.

In order to analyze and compare the behavior of the two algorithms, we consider the sequence of operations, made by two algorithms, to serve the j-th request. We may assume they are made in the following order:

1. H makes its paid exchanges;
2. MSF and H access to B and MSF makes its free exchanges;
3. H makes its free exchanges.

In order to compute the amortized cost of the steps above we will consider two additional steps, the former between step 1 and step 2, the latter between step 2 and step 3. In the first one step we modify the current permutation and in the second one resume the earlier one. Thus, in our analysis, we have actually five phases and we will refer to them as a, b, c, d, and e, where phases a, c, and e respectively correspond to steps 1, 2, and 3 above, and phases b and d are the following:

b. the current permutation changes from ι to τ_j;
d. the current permutation changes from τ_j to ι.

An amortized cost (possibly due only to a change in the potential) is associated with each of the five phases. The total amortized cost to serve the j-th request is the sum of the amortized costs of each phase:

$$a_j = t_j + \Phi_j - \Phi_{j-1} =$$

$$= t_a + t_b + t_c + t_d + t_e + (\Phi_j - \Phi_d) + (\Phi_d - \Phi_c) + (\Phi_c - \Phi_b) + (\Phi_b - \Phi_a) + (\Phi_a - \Phi_{j-1})$$

where t_j is the actual total time for MSF to process the j-th request, t_a, t_b, t_c, t_d and t_e are the actual costs associated with the five phases, Φ_{j-1} and Φ_j are the potential respectively at the beginning and at the end of the processing of the j-th request, Φ_a, Φ_b, Φ_c and Φ_d are the potentials at the end of the corresponding phases. It is easily seen that $t_a = t_b = t_d = t_e = 0$.

In phase a, a paid exchange can increase the number of inversions at most by 1, thus $\Phi_a - \Phi_{j-1}$, which is the amortized cost of MSF in this phase, is less than or equal to the cost paid by H for making its exchanges. Disregarding in our analysis the costs paid by both H and MSF we can obtain an upper bound to the cost ratio.

Now we show that the new inversions created in phase b by the modification of the current permutation do not give rise to a growth of the potential. In fact, inversions between elements in B are gotten rid of, those between elements of $S - B$ are not modified.

It remains to show that neither the inversions in which one item belongs to $S - B$ and the other to B give rise to an increasing in the potential.

Any portion of the sublist of $\mathcal{L}_H$ between a_1 and a_{β_j} (which is the largest list portion involved in elements movements) is spanned the same number of times by forward and backward movements. The sublist may be divided up into *stretches*, where we define stretch any maximal portion of the sublist that completely lies within the range of action of one backward movement.

Let σ be a stretch, x the item whose backward movement spans σ and $y_1, \ldots, y_p$ the items that balance on the same stretch σ. Fig. 2 illustrates an example.

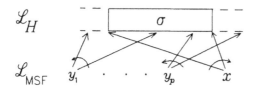

Fig. 2. Spanning of a stretch

Some of these stretches may overlap. This implies that for the items in the overlapping substretches, the following part of the proof must be repeated once per each stretch which the items belong to.

The number of inversions in which one item belongs to $S - B$ and the other belongs to $\{x, y_1, \ldots, y_p\}$ is (from Lemma 3):

$$t_\sigma(x) - t_\sigma(y_1) - \ldots - t_\sigma(y_p) + h_\sigma(y_1) + \ldots + h_\sigma(y_p) - h_\sigma(x) ,$$

where t_σ and h_σ are defined in the same way as t and h except for the fact that they now are referred to the two lists $\mathcal{L}_{\text{MSF}}$ and $\mathcal{L}_{H,\sigma,B}$, which is obtained from $\mathcal{L}_H$ by only considering elements outside B belonging to σ.

Now, $y_1, \ldots, y_p$ precede x in $\mathcal{L}_{\text{MSF}}$, otherwise τ_j, by its definition, would have assigned to each of them an item that follows x rather than one that precedes x. Since the spanned stretch is the same (σ), it holds:

$$t_\sigma(x) \leq \sum_{q=1}^{p} t_\sigma(y_q)$$

$$h_\sigma(x) \geq \sum_{q=1}^{p} h_\sigma(y_q) .$$

Summarizing, in phase b, inversions between items in B are deleted because τ_j makes $\mathcal{L}_{\text{MSF}}$ and $\mathcal{L}_H$ more similar. Other inversions do not give rise to any trouble since they are at least counterbalanced.

A similar reasoning can be made in order to analyze phase d, in which the current permutation is reset to the identity. In fact, in this phase, exactly the inversions (between elements of B) that were destroyed in phase b are re-introduced, since MSF does not change the relative ordering among the elements of B. The inversions between elements of $S - B$ remain the same. The inversions in which one item belongs to $S - B$ and the other to B globally give a null contribution to the potential since backward and forward movements exactly balance, and, in the part of the list each movement spans, the movement itself involves each item in $S - B$ (note that $h_\sigma(\cdot)$ is null). In other words, after a permutation change, a modification in the ordering of the first β_j items of the list (where B's elements now lie) does not increase the number of inversions between that portion and the remaining part of the list. So, $(\Phi_d - \Phi_c) + (\Phi_b - \Phi_a) \leq 0$.

During phase c, let i be the position of $a_{\beta_j} = \tau_j(b_{\beta_j})$ in $\mathcal{L}_H$, k the position of b_{β_j} in $\mathcal{L}_{\text{MSF}}$, x_i the number of list elements that precede b_{β_j} in $\mathcal{L}_{\text{MSF}}$ and follow a_{β_j} in $\mathcal{L}_H$. As seen in [13], we have that $t_c = k$, the change in the potential due to free exchanges on b_{β_j} only is $\Delta\Phi|_{\beta_j} \leq (k - \beta_j - x_i) - x_i$, and $k - x_i \leq i$. It follows that $t_c + \Delta\Phi|_{\beta_j} \leq 2i - \beta_j$. The moving of the other $\beta_j - 1$ items to the front of the list gives rise to at most $(\beta_j - 1)(i - \beta_j)$ inversions.

During phase e, the free exchanges made by H give rise to a unit decreasing of the potential.

On the whole, disregarding changes in the potential due to the paid exchanges made by H, we obtain

$$a_j \leq (1 + \beta_j)(i - \beta_j) + \beta_j . \tag{1}$$

Since H is charged i in the standard model and $i - \beta_j$ in the wasted work one, the thesis follows.

Corollary 5. *MSF is $(1+\beta)$-competitive in the standard model even if the adversary is allowed to change the relative ordering of the requested set items by means of free exchanges.*

Proof. The proof is the same of that of the previous theorem except for phase e. H creates at most

$$(\beta_j - 1) + (\beta_j - 2) + \ldots + 1 = \frac{\beta_j^2 - \beta_j}{2}$$

inversions between elements of B. Thus

$$a_j \leq (1 + \beta_j)i - \frac{1}{2}(\beta_j^2 - \beta_j)$$

As a remark to the given proofs, we observe that in the case of MSF the pairwise independence property (pointed out in [3]) still holds, but it cannot be used in the same way as in [6, 7] to analyze the competitiveness properties because the total cost depends not only on the relative position of each pair of items, but also on the current position of any other item of the searched set. The straightforward use of the pairwise independence property would make us count many times the same interword (unsuccessful) comparison, once per each element of B that follows the non-searched item.

The results of Theorems 1, 2 and 4 are compared in Fig. 1. If $\beta = \beta_{\mathrm{am}}$, i.e. the cardinality of the requested sets is always the same, MSF's infimum competitive ratio, in the standard model, is located inside area A.

Now we relate the competitive ratio of MSF to n, the size of the list.

First, we show by an example that the ratio depends on n, even against a static algorithm.

The adversary H serves the sequence without changing the ordering of the initial list $\mathcal{L}_0$. H builds $\underline{r}$ by requesting at first the last $n - \sqrt{2n}$ (assume $\sqrt{2n}$ is an integer) items of $\mathcal{L}_0$, then one at a time the first $\sqrt{2n}$ items of $\mathcal{L}_0$, as singletons. It is easy to verify that this is possible if and only if $\sqrt{2n} \leq n$, namely, $n \geq 2$, that is to say, in every meaningful case.

If we denote $\sqrt{\frac{n}{2}}$ by x, the costs of the two algorithms over sequence $\underline{r}$ are:

$$H(\underline{r}) = n + \frac{n\frac{n}{x}+1}{x}\frac{}{2} = \frac{n(n+x+2x^2)}{2x^2},$$

$$\mathrm{MSF}(\underline{r}) = n + \frac{n}{x}n = \frac{n(n+x)}{x}.$$

Hence, in the standard model, the cost ratio of MSF is

$$\frac{2x(n+x)}{n+x+2x^2} > \sqrt{\frac{n}{2}}.$$

The sequence $\underline{r}$ yields an example in which not any NIF ordering is an optimal static one.

Now, with reference to Fig. 1, denoting by $\overline{\beta}$ the abscissa (which is roughly $\sqrt{n}$) of the intersection between the curves which represent the two upper bounds to the competitive ratio, the following theorem states that if β and β_{am} lie on the same part with respect to $\overline{\beta}$ then MSF is $O(\sqrt{n})$-competitive.

Theorem 6. *Let* $\overline{\beta} = 2n/(1 + \sqrt{4n+1})$ *and assume the standard model. If* $(\beta - \overline{\beta})(\beta_{\mathrm{am}} - \overline{\beta}) \geq 0$ *then the competitive ratio of* MSF *is*

$$1 + 2n/(1 + \sqrt{4n+1}) = O(\sqrt{n}),$$

otherwise the ratio is $(1 + n/4)$.

Proof. If $(\beta - \overline{\beta})(\beta_{\mathrm{am}} - \overline{\beta}) \geq 0$ then $\min\{1 + \beta, n/\beta_{\mathrm{am}}\}$ is an admissible competitive ratio for MSF. The worst case occurs when $\beta_{\mathrm{am}} = \beta$ and

$$1 + \beta = n/\beta \tag{2}$$

(see Fig. 1), that is when $\beta = \overline{\beta}$, and this implies $c = 1 + \overline{\beta}$. If $(\beta - \overline{\beta})(\beta_{\mathrm{am}} - \overline{\beta}) < 0$, the cost ratio is (from (1))

$$1 + \beta_j - \frac{\beta_j^2}{i}.$$

The maximum is in correspondence of $\beta_j = i/2$. Thus, the competitive ratio is at most $1 + i/4 \leq 1 + n/4$.

So far, we have supposed that the examination of one item had unit cost. Suppose now that the requested set r_j is represented by means of a sequential list or a heap or another data structure. In this case, the cost charged to one computation of $\theta_{\mathcal{L}}$ depends on the chosen representation of r_j and on the list state. Let γ be an upper bound to such a cost. For example, for a linear list $\gamma = \beta$. Now suppose that the adversary pays 1 per item test. On any sequence of requests, MSF and the adversary H carry out the same operations they do when unit cost is charged. Now MSF pays at most γ times the cost it should pay in case of unit cost, while H pays the same cost. Hence, MSF is $((1 + \beta)\gamma)$-competitive. Furthermore, the $O(\sqrt{n})$ upper bound given in Theorem 6 still holds under the same hypotheses, since now the worst case equation (2) becomes $(1 + \beta)\gamma = (n/\beta)\gamma$, that is in practice the same as that given in the proof of Theorem 6.

Now we consider a different algorithm, which is randomized and is devised from BIT [7]: we call it "BITS" (BIT-for-Sets). It associates a bit with each element in the list, and the n bits are initialized at random. Whenever one accesses set B, the bit of the last element of B in BITS' list is complemented, and if it changes to 1, the accessed set is moved to the front of the list, otherwise it remains unchanged.

Theorem 7. *BITS is $(1+(3/4)\beta)$-competitive both in the standard and in the wasted work model.*

Proof. We omit the proof here.

5 Extensions

Now we consider the case in which the list is weighted. Assume that a paid exchange between x and y, with $y \prec x$ just before the exchange, costs $w(y)$. This is a "minimum cost" assumption, similar to those in [13] for non-decreasing cost functions and in [11] for the purpose of showing that free exchanges are not necessary in the standard model for the traditional list update problem.

Let be:
$$W = \max_{1 \leq j \leq m}\{\textstyle\sum_{x \in r_j} w(x)\} ,$$
$$w_{\max} = \max_{x \in S}\{w(x)\}$$
$$w_{\min} = \min_{x \in S}\{w(x)\} .$$

We give the following:

Theorem 8. *MSF is $\max\{1+W/w_{\min}, w_{\max}/w_{\min}\}$-competitive both in the standard and in the wasted work model.*

Proof. We omit the proof which, however, is similar to that of Theorem 4, but requires a cunning choice for the potential function.

Another interesting problem is the retrieval of one element (whichever) of the currently requested set r_j. In this case, the list maintenance algorithm can answer to the request returning the first element of r_j that it finds in the list.

The cost model is the same as that introduced in section 2, except for the fact that only the (first) found item may be moved at any distance forward in the list.

Any other exchange costs 1. For this problem, the proving scheme in [9] yields a lower bound of 2. Furthermore, it can be easily show that, for any fixed size of the list n, the actual lower bound is $2 - 2/n$.

On the other hand, for this problem, not only no competitive on-line algorithm is known, but in addition we claim that a very general class of algorithms are not c-competitive for any fixed c, even if $\beta_j = 2$ for any j.

Theorem 9. *For any on-line deterministic algorithm G which, after the first request has been processed, never makes a paid exchange on the item in a given position ν, and for any fixed constant c, there exist a sequence of requests $\underline{r}$ such that $\beta_j = 2$ for any j and an algorithm H such that in the wasted work model $G(\underline{r}) > c \cdot H(\underline{r})$.*

Proof. We do not prove the theorem here.

6 Final Remarks

In this paper we have considered a few generalizations and extensions to the traditional list update problem.

The first consists in updating an unweighted list of items while searching for *sets* of items. For this problem we have provided a lower bound to the competitive ratio of any on-line deterministic algorithm, and two algorithms, the first one deterministic, the second one randomized, for which the competitive ratio is a linear function of β, the size of the largest requested set. Moreover, we have upper bounded the competitive ratio of MSF algorithm by relating β to the size of the list.

These results well extend to the case of *weighted* lists, where the cost model generalizes the traditional one [1, 2] because we also consider paid exchanges.

Furthermore, we have been in touch with the particular case where we are only interested in finding whichever element belonging to the set currently specified by the sequence of requests. However, for this case, we have been only able to provide negative results.

Currently, we are working to define on-line maintenance algorithms for AND-OR trees and DAGs under several visiting algorithms, by exploiting the results obtained in this paper. This work should lead to the design of competitive algorithms.

There are some open questions concerning the list update problem under searches for sequence of sets. Of course, the first one is either to devise an algorithm whose upper bound matches the lower bound stated in Theorem 1 or to prove that MSF achieves the best possible competitive ratio (i.e. MSF is strongly competitive [8]). It is known that one can use dynamic programming to develop an optimum algorithm [10]. On the other hand, some properties of the optimum algorithm for the traditional list update problem (see [11]), no longer hold.

The properties of the optimal static ordering deserve study as well.

References

1. d'Amore, F., Marchetti-Spaccamela, A., Nanni, U.: Competitive Algorithms for the Weighted List Update Problem. *Proc. 2nd Workshop on Algorithms and Data Structures*, Ottawa, Canada, August 1991, (*Lecture Notes in Computer Science*, **519**, 1991, 240–248).

2. d'Amore, F., Marchetti-Spaccamela, A., Nanni, U.: The Weighted List Update Problem and the Lazy Adversary. *Theoretical Computer Science*, (to appear).

3. Bentley, J.L., McGeoch, C.C.: Amortized Analyses of Self-Organizing Sequential Search Heuristics. *Communications of the ACM*, **28**, 4, April 1985, 404–411.

4. Ben-David, S., Borodin, A., Karp, R., Tardos, G., Wigderson, A.: On the Power of Randomization in Online Algorithms. *Proc. 22nd Annual ACM Symposium on Theory of Computing*, Baltimore, MD, May 1990, 379–386.

5. Hester, J.H., Hirschberg, D.S.: Self-Organizing Linear Search. *ACM Computing Surveys*, **17**, 3, September 1985, 295–311.

6. Irani, S.: Two Results on the List Update Problem. Tech. Rep. TR-90-037, Computer Science Division, U.C. Berkeley, CA, August 1990.

7. Irani, S., Reingold, N., Westbrook, J., Sleator, D.D.: Randomized Competitive Algorithms for the List Update Problem. *Proc. 2nd ACM-SIAM annual Symp. on Disc. Algorithms*, San Francisco, CA, January 1991, 251–260.

8. Karlin, A.R., Manasse, M.S., Rudolph, L., Sleator, D.D.: Competitive Snoopy Caching. *Algorithmica*, **3**, 1988, 79–119.

9. Karp, R., Raghavan, P.: Private communication reported in [6].

10. Manasse, M.S., McGeoch, L.A., Sleator, D.D.: Competitive Algorithms for On-line Problems. *Proc. 20th Annual ACM Symposium on Theory of Computing*, Chicago, IL, May 1988, 322–333.

11. Reingold, N., Westbrook, J.: Optimum Off-line Algorithms for The List Update Problem. Tech. Rep. YALEU/DCS/TR-805, Department of Computer Science, Yale U., August 1990.

12. Sleator, D.D.: Private communication reported in:
Tarjan, R.E.: Amortized Computational Complexity. *SIAM J. Alg. Disc. Meth.*, **6**, 2, April 1985, 306–318.

13. Sleator, D.D., Tarjan, R.E.: Amortized Efficiency of List Update and Paging Rules. *Communications of the ACM*, **28**, 2, February 1985, 202–208.

This article was processed using the LaTeX macro package with LLNCS style

GKD-Trees: Binary Trees that Combine Multi-dimensional Data Handling, Node Size and Fringe Reorganization

Walter Cunto[1] and Vicente Yriarte[2]

[1] Centro Científico IBM de Venezuela A.P. 64778, Caracas, Venezuela
[2] Departamento de Computación, Universidad Simón Bolívar A.P. 89000
Caracas, Venezuela

Abstract. We introduce a new multi-dimensional data structure based on binary trees called gkd-tree that generalizes several previously defined data structures, including the widely accepted kd-trees. Up to the present, gkd-trees are the fastest on-line improvement over kd-trees due to the combined effects of storing more than one element per node and making local reorganizations during insertions. The improvement is such that gkd-tree's performance approaches very rapidly that of balanced trees. Space performance is guaranteed to be at least 83% for any node size and local reorganization effort. Because of its competitive time and space performances, the proposed data structure is also practical for secondary storage devices.

1 Introduction

The binary search tree needs no presentation. It is a fundamental data structure that can handle uni- and multi-dimensional data. Bentley's kd-trees [1] is one of the preferred binary search tree variants for multi-dimensional data manipulation.

This paper introduces a new data structure called *gkd-tree*—which stands for generalized kd-trees—that enhances the performance of kd-trees by combining three different design criteria:

(*i*) Multi-dimensional data handling
(*ii*) Node size
(*iii*) On-line reorganization of fringe subtrees.

More precisely, data with $k \geq 1$ dimensions are handled in a kd-tree fashion [1]; that is, the $(j \oplus k)$th coordinate drives branching in a node of height $j > 0$, where $j \oplus k$ is k if k divides j and, $j \bmod k$ otherwise. Nodes may hold up to $2\ell+1$, $\ell \geq 0$, elements. Furthermore, this structure reorganizes fringe subtrees to gain balance and compactness when they reach size $(2t + 1)(2\ell + 1)$, with $t \geq 0$ denoting the reorganization parameter. Figure 1 depicts a gkd-tree with data dimension $k = 2$, node size 3 ($\ell = 1$) and balanced fringe subtrees of size 3 ($t = 1$). Figure 2 illustrates the partition over the domain induced by the structure given in Fig. 1.

Some data structures that appear in the literature are special cases of gkd-trees:

(*i*) Classical binary search trees [2] are gkd-trees with $k = 1$, $\ell = 0$ and $t = 0$

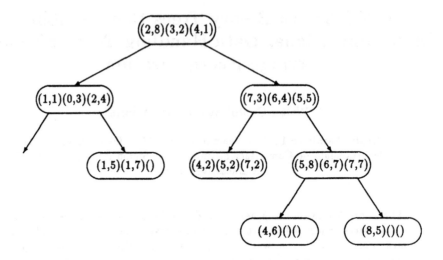

Fig. 1. Example of a gkd-tree with $k = 2$, $\ell = 1$ and $t = 1$

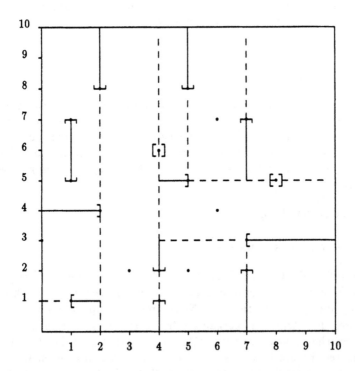

Fig. 2. Space partition according to the gkd-tree given in Fig. 1

(ii) When $k = 1$, $\ell = 1$ and $t \geq 1$, gkd-trees are the uni-dimensional binary search trees defined by Poblete and Munro [3] that keep fringe subtrees of size $2t + 1$, $t \geq 0$, balanced during the insertion of a new element in the structure

(iii) By setting $k = 1$, $\ell \geq 1$ and $t \geq 1$, we obtain the generalized binary search trees (gbs-trees) of Cunto and Gascón [4], which improve the time performance of the previous trees while showing a good space performance

(iv) Bentley's kd-trees [1] are derived from gkd-trees when $k \geq 1$, $\ell = 0$ and $t = 0$

(v) When $k \geq 1$, $\ell = 0$ and $t \geq 0$, the new structure becomes a kdt-tree: a kd-tree improved by fringe reorganizations [5]. This structure introduces the first practical dynamic reorganization scheme for kd-trees.

Figure 3 illustrates how all these data structures described above are embedded into the gkd-tree concept.

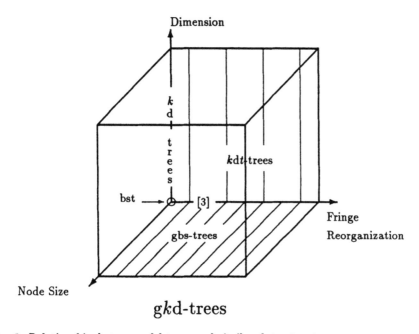

Fig. 3. Relationship between gkd-trees and similar data structures

The new data structure provides the best speed-up, at the present time, applicable to kd-trees while adding only a small overhead during insertions. Furthermore, we prove that the average and the variance of the number of element comparisons performed by partial queries with $s < k$ specified coordinates over a randomly generated gkd-tree with N elements are $C_1 N^{1-s/k+\theta(s,k,\ell,t)}(1 + o(1))$ and $C_2 N^{2(1-s/k+\theta(s,k,\ell,t))}(1 + o(1))$ respectively—the constants C_1 and C_2 are positive and dependent on k, ℓ and t. Also, the value of $\theta(s,k,\ell,t)$ and C_2 approach zero rapidly as ℓ or t grows; thus proving that the large variance observed in standard kd-trees is reduced substantially. The performance of total match queries (searches with all the coordinates specified) and insertions also improves substantially. Moreover, the space performance of gkd-trees is guaranteed to be at least 69% for any possible node size l even though no reorganization is applied ($t = 0$). It increases above 83% when fringe subtrees of three nodes are reorganized ($t = 1$). Average space performance approaches 100% if the value of t increases.

Gkd-trees can be adapted for secondary storage devices. Clearly, the criterion of node size matches the kind of data-block transfer of such devices. The proposed data structure is highly compact since it avoids indexing. Compared with indexed file schemes [6], gkd-tree partial match queries are competitive when the number of specified coordinates, s, is greater than $k/2$.

2   The Data Structure and Algorithms

Consider a k-dimensional domain $D = D^1 \times D^2 \times \ldots \times D^k$. Let each coordinate have an associated total order. Thus, the ith-order, $1 \leq i \leq k$, over D is the lexicographic order obtained with elements of D cyclically shifted $i - 1$ positions to the left. For each $e \in D$, e^j denotes the $(j \oplus k)$th coordinate of e.

A gkd-tree with parameters $k > 1$, $t \geq 0$ and $\ell \geq 0$, and built from elements taken from D, is a binary tree such that nodes at level $j > 0$ hold $a_1, a_2, \ldots, a_m$, $m \leq 2\ell + 1$, elements and

(i) Internal nodes—those that point to at least one non-empty node—hold exactly $2\ell + 1$ elements

(ii) Elements in a node are sorted according to the $(j \oplus k)$th-order, and elements in the left and right subtrees are smaller and larger respectively than elements in the node

(iii) If the node points to a non-empty right subtree, the median element in the node and its right adjacent have different $(j \oplus k)$th coordinate values; that is, $a_{\ell+1}^j < a_{\ell+2}^j$. This rule handles repetitions of coordinate values without the need for a third pointer in the structure.

When a subtree reaches a size of $(2t + 1)(2\ell + 1)$ elements, it is reorganized so that it is as balanced and compact as possible according to the previous rules.

2.1   Searching in a Gkd-tree

Given a search pattern u, the algorithm starts at the root and traverses the tree recursively to the bottom. If a node at level $j > 0$ is reached, the search distinguishes two cases:

(i) If the coordinate value u^j is not specified, the algorithm searches for u among the elements in the node and recursively in both subtrees

(ii) Otherwise, the algorithm searches for u^j among the a_s^j, $1 \leq s \leq m$. This search can be unsuccessful or successful. In the first case:

    1. If $a_s^j < u^j < a_{s+1}^j$ with $1 \leq s < m$, the search stops

    2. If $u^j < a_1^j$, the search continues in the left subtree

    3. If $u^j > a_{2\ell+1}^j$, the search continues in the right subtree.

In the second case, the algorithm checks all the elements in the node with the $(j \oplus k)$th coordinate equal to u^j and continues the search in the left subtree if $a_1^j = u^j$.

Note that the algorithm handles total and partial match queries in the same way. The main difference between these search operations is that the search path in total match queries is linear from the root to a node.

2.2 Insertion in a Gkd-tree

To insert a new element e in the structure, the algorithm performs a total match query. If the search stops because a nil pointer is reached, the node is created and e is inserted. If the search reaches at a node with less than $2\ell + 1$ elements, e is inserted in the proper place. During the search, if an internal node nr is traversed, the algorithm performs the following actions:

(i) The subtree with root nr is reorganized if it contains $(2t + 1)(2\ell + 1) - 1$ elements

(ii) If $e^j < a_1^j$, the insertion continues in the left subtree

(iii) Similarly, if $e^j > a_{2\ell+1}^j$, the insertion continues in the right subtree

(iv) In the case of e^j falling in an intermediate position, e is inserted in the node and a_1 or $a_{2\ell+1}$ is displaced and inserted in the left or right tree respectively. Displacement of either extremes favors tree balance and obeys the restriction on the middle node element $(a_{\ell+1}^j < a_{\ell+2}^j)$.

The reorganization selects the median of the sequence $f_1, f_2, \ldots, f_{(2t+1)(2\ell+1)}$ taken from the fringe subtree and sorted according to the $(j \oplus k)$th order. This median and its closest ℓ smaller and ℓ greater elements conform the root node of the reorganized subtree. The process is recursively applied in the lower levels of the subtree.

Some care must be taken when $f_{2t\ell+t+\ell+1}^j = f_{2t\ell+t+\ell+2}^j$ since an even balance is not possible. In such cases, let z be the largest position in the sequence, sorted according the $(j \oplus k)$th-order, such that $f_{2t\ell+t+\ell+1}^j = f_z^j$. If $(2t + 1)(2\ell + 1) - z \geq \ell$, f_z is placed in position $\ell + 1$ of the root; otherwise, $f_{(2t+1)(2\ell+1)} - z$ is placed. This undesired event may occur with a probability smaller than $1/|D^j|^{z - (2t\ell+t+\ell+1)}$ if the values of the domain D^j are equally likely. Also, it will have only a negligible effect on the time performance since the criterion of nodes with size greater than 1 improves performance independently of the fringe reorganization criterion.

Figure 4 illustrates a sequence of insertions in a gkd-tree.

2.3 Deletion in a Gkd-tree

If the element to be deleted, e, belongs to an external node, it is removed. The node is physically removed if e is the only element in the node. When e belongs to an internal node nr at level $j > 0$, let T_L and T_R be the left and right subtree of nr. In this case, the algorithm considers three cases:

(i) If T_R is empty or, T_L and T_R are non-empty, then the maximum element according to the $(j \oplus k)$th-order in T_L is moved to nr

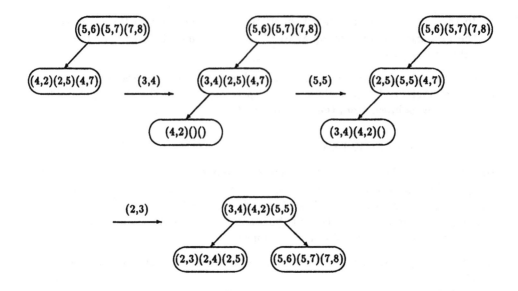

Fig. 4. A sequence of insertion in a gkd-tree with $t = \ell = 1$

(ii) If e is located in position ranging from $\ell + 2$ to $2\ell + 1$ in nr and T_R is not empty, then the minimum element according to the jth-order in T_R is moved to nr

(iii) If T_L is empty and e is located in position ranging from 1 to $\ell + 1$ in nr:
   1. If $\ell > 1$ and the elements in position $\ell+2$ and $\ell+3$ have different $(j \oplus k)$th coordinate values, the minimum element according to the $(j \oplus k)$th-order in T_R is moved to nr
   2. Otherwise, the subtree with nr as the root must be globally reorganized.

Deletion case (iii-2) is inherited from kd-trees that show the same weakness. However, gkd-trees will reduce the incidence of global reorganization since the probability that two adjacent positions in a node share the same coordinate value is smaller than the probability that any two elements have the same coordinate value, which is the case of standard kd-trees.

3 The Analysis of Gkd-trees

This section studies the average space and time performance of gkd-trees. Average space performance is measured by the ratio of the number of elements in the structure to the average space generated by the structure. Average time performance of the two search operations, total and partial match queries, is analyzed in terms of the number of binary comparisons between elements. The performance of insertions follows from that of total match queries plus a constant term due to node creations and fringe reorganizations.

The analysis will assume that gkd-trees are randomly generated. Random trees are built by the iterative process of inserting elements such that, at each step, the next element to be inserted may be placed into any of the available inter-element gaps with equal probability. The building process precludes deletions; otherwise, the probabilistic assumption cannot be enforced in the next insertion. In addition, it is assumed that, during insertions, if a fringe subtree grows to size $(2t + 1)(2\ell + 1)$ it is reorganized such that its left and right subtrees contain $t(2\ell + 1)$ elements each.

3.1 Fringe Analysis of Gkd-trees

Gkd-trees and gbs-trees share equivalent insertion and total match query operations. Therefore, gkd-tree performance for insertions, space and total match queries are asymptotically identical to those equivalent of gbs-trees.

To analyze these gkd-tree performances, we use *fringe analysis*, a special kind of Markov analysis, which models the dynamic changes in the fringe of the trees due to insertions [2, 3, 4]. The main difference between fringe analysis and a Markov chain is that transition probabilities in fringe analysis are not constant but dependent on the transition step. We omit the details and adapt the results of gbs-tree's analysis [4] to our proposed data structure.

For the sake of simplicity,

$$\Delta_{t,\ell}^{(i)} = H_{2(t+1)(2\ell+1)}^{(i)} - H_{(t+1)(2\ell+1)}^{(i)}$$

where $H_N^{(i)} = \sum_{j=1}^{N} 1/j^i$ denotes the Nth harmonic number of ith degree. The degree is omitted when its value is one. Note also that $H_N = H_N^{(1)} = \ln(N) + O(1)$ and $H_N^{(i)} = O(1)$ for $i > 0$ [7].

In addition to random generation, the analysis assumes that the insertions are such that the space is minimized at all times. This policy is highly expensive in practice; however, it simplifies the analysis whose results are close to those obtained from simulation [8].

Lemma 1. *If $F_\infty(t, \ell)$ denotes the asymptotic space performance of gkd-trees of node size $2\ell + 1$ and reorganization parameter t when the number of elements in the structure is very large, then*

$$F_\infty(t, \ell) = \frac{\Delta_{t,\ell}}{\sum_{j=t+1}^{2t+1} \frac{1}{j + \frac{1}{2t+1}}}.$$

A simple manipulation of the previous expression proves that, for a given reorganization parameter t, the asymptotic space utilization factor for very large nodes is

$$\lim_{l \to \infty} F_\infty(t, \ell) = \frac{\ln(2)}{H_{2t+1} - H_t}.$$

Thus, $F_\infty(t, \ell) \geq \ln(2) \simeq 69\%$ for any $\ell \geq 0$ and $t \geq 0$. This performance improves to above 83% if the simplest reorganization is applied; that is, $\lim_{\ell \to \infty} F_\infty(1, \ell) \geq 6\ln(2)/5 \simeq 83\%$. Furthermore, $\lim_{t \to \infty} \lim_{\ell \to \infty} F_\infty(t, \ell) = 1$. Space performance is larger than that measured by $F_\infty(t, \ell)$ since the ratio of pointers to N is smaller in gkd-trees than in kd-trees.

Lemma 2. *The average and the variance of the performance of total match queries (successful or unsuccessful) in a randomly generated gkd-tree with N elements are*

$$C_N = \frac{1}{\Delta_{t,\ell}} H_N + O(1)$$

$$V_N = \frac{\Delta_{t,\ell}^{(2)}}{\Delta_{t,\ell}^3} H_N + O(1)$$

respectively.

¿From Lemma 2, C_N approaches $\log_2(N)$ and V_N is almost constant—the performance of full balanced trees—when t or ℓ becomes large since $\lim_{t+\ell\to\infty} \Delta_{t,\ell} = \ln(2)$ and $\lim_{t+\ell\to\infty} \Delta_{t,\ell}^{(2)} = 0$. Table 1 tabulates the three coefficients for different values of t and ℓ. Observe that, the influence of ℓ on the time performance and the effect of t on the space efficiency.

Table 1. Numeric values of $F_\infty(t,\ell)$, $1/\Delta_{t,\ell}$ and $\Delta_{t,\ell}^{(2)}/\Delta_{t,\ell}^3$

	ℓ						ℓ			
t	0	1	2	10	100	t	0	1	2	10	100
0	1.0000	0.8222	0.7748	0.7138	0.6953	0	2.0000	1.6216	1.5489	1.4676	1.4453
1	1.0000	0.8966	0.8719	0.8417	0.8328	1	1.7143	1.5309	1.4953	1.4551	1.4440
2	1.0000	0.9274	0.9110	0.8912	0.8855	2	1.6216	1.5012	1.4776	1.4510	1.4436
3	1.0000	0.9442	0.9319	0.9173	0.9131	3	1.5760	1.4864	1.4689	1.4489	1.4433

$$F_\infty(t,\ell) \qquad\qquad 1/\Delta_{t,\ell}$$

	ℓ			
t	0	1	2	10	100
0	1.0000	0.3426	0.2067	0.0495	0.0052
1	0.5102	0.1725	0.1037	0.0248	0.0026
2	0.3426	0.1152	0.0692	0.0165	0.0017
3	0.2578	0.0865	0.0520	0.0124	0.0013

$$\Delta_{t,\ell}^{(2)}/\Delta_{t,\ell}^3$$

3.2 Analysis of Gkd-tree Partial Match Query

The technique used to analyze the partial match query operation of gkd-trees with $s < k$ specified coordinates is based on that of kdt-trees [5]. However, our analysis is more general and complete. Our technique is different from that of Flajolet and Puech [9], which only provides a proof for the average number of comparisons performed by partial match queries in kd-trees. The analysis will focus on the number of traversed nodes. The number of comparisons performed is roughly $(2\ell+1)F_\infty(t,\ell)$

times the number of traversed nodes. For practical purposes, $2\ell + 1$ can be taken as the upper-bound constant.

The framework of the analysis is:

(i) Definition of the random variable Y_N as the number of inter-element gaps in the left subtree of a randomly generated gkd-tree of N elements and $q_{r,N} = p[Y_N = r + 1]$

(ii) Derivation of a closed expression for $q_{r,N}$

(iii) Definition of the random variable $X_{t,\ell,u,N}$ as the number of traversed nodes when a partial match query with pattern u is executed in a randomly generated gkd-trees of N elements, node size $2\ell + 1$ and fringe reorganization parameter t. Furthermore, $p_{t,\ell,u,N,i} = p[X_{t,\ell,u,N} = i]$ defines the probability that the value of $X_{t,\ell,u,N}$ is i. The subscripts t and ℓ are omitted when they are understood

(iv) Formulation of the system of difference equations that describes the generating function over the sequence of $p_{i,N}$'s in terms of $q_{r,N}$

(v) From the previous equations, derivation of the expected value and the variance

(vi) Approximation of the system of difference equations that describe the expected value $E(X_N)$ and the variance $V(X_N)$ through continuously differentiable functions by an Eulerian system of differential equations

(vii) Proof of the statement which asserts that the solution of the Eulerian systems is $E(X_N) = C_1 N^\alpha (1 + o(1))$ where α is the unique root of the equation

$$\left((\lambda + \gamma + 1)^{\overline{\gamma+1}} \right)^{k-s} \left((\lambda + \gamma + 2)^{\overline{\gamma+1}} \right)^{s} = \left((\gamma + 2)^{\overline{\gamma+1}} \right)^{k}$$

such that $2\gamma + 1 = (2t + 1)(2\ell + 1)$

($viii$) Proof that the sequence $\{\alpha_{s,k,t,\ell}\}$ decreases in t and ℓ and converges to $1 - s/k$

(ix) Proof that $V(X_N) = C_2 N^{2\alpha} + C_3 N^\alpha (1 + o(1))$

(x) Proof that $\lim_{t+\ell \to \infty} C_2(t, \ell) = 0$, showing that the design criteria improve the variance considerably.

The Appendix provides a more formal version of this framework. Further details can be found in [8].

Table 2 shows how the combined effect of t and ℓ makes $\theta(s, k, t, \ell)$ converge to 0 rapidly ($\theta(s, k, t, \ell) = \alpha(s, k, t, \ell) - (1 - s/k)$). In all the four tables, $k = 4$. The first three tables consider different values of t and ℓ for each one of the three possible values of s. The fourth table fixes $\ell = 200$. Observe that θ is almost negligible even if $t = 0$.

Table 3 provides the exact average and variance time performance of randomly generated gkd-trees of size $N = 10000$ with $k = 3$. Note that the best improvement is obtained for values of t not greater than 2. For large values of ℓ, $t = 1$ seems to be the best choice.

Table 2. Numeric values of α and θ for $k = 4$

t	ℓ=0	1	2	3	4
0	0.7900	0.7753	0.7685	0.7645	0.7620
	0.0400	0.0253	0.0185	0.0145	0.0120
1		0.7620	0.7578	0.7558	0.7546
		0.0120	0.0078	0.0058	0.0046
2			0.7550	0.7536	0.7529
			0.0050	0.0036	0.0029
3				0.7526	0.7521
				0.0026	0.0021
4					0.7516
					0.0016

$$\alpha_{1,4,t,\ell}$$
$$\theta_{1,4,t,\ell}$$

t	ℓ=0	1	2	3	4
0	0.5616	0.5370	0.5263	0.5204	0.5167
	0.0616	0.0370	0.0263	0.0204	0.0167
1		0.5167	0.5107	0.5079	0.5063
		0.0167	0.0107	0.0079	0.0063
2			0.5067	0.5049	0.5039
			0.0067	0.0049	0.0039
3				0.5036	0.5028
				0.0036	0.0028
4					0.5022
					0.0022

$$\alpha_{2,4,t,\ell}$$
$$\theta_{2,4,t,\ell}$$

t	ℓ=0	1	2	3	4
0	0.3056	0.2809	0.2713	0.2662	0.2631
	0.0556	0.0309	0.0213	0.0162	0.0131
1		0.2631	0.2583	0.2561	0.2548
		0.0131	0.0083	0.0061	0.0048
2			0.2551	0.2537	0.2529
			0.0051	0.0037	0.0029
3				0.2527	0.2521
				0.0027	0.0021
4					0.2516
					0.0016

$$\alpha_{3,4,t,\ell}$$
$$\theta_{3,4,t,\ell}$$

t	s=1	2	3
0	0.7503	0.5004	0.2503
	0.0003	0.0004	0.0003
1	0.7501	0.5001	0.2501
	0.0001	0.0001	0.0001
2	0.7501	0.5001	0.2501
	0.0001	0.0001	0.0001
3	0.7500	0.5001	0.2500
	0.0000	0.0001	0.0000
4	0.7500	0.5000	0.2500
	0.0000	0.0000	0.0000

$$\alpha_{s,4,t,200}$$
$$\theta_{s,4,t,200}$$

Table 3. $E_{u,10000}$ and $V_{u,10000}$ for $1 \leq s \leq 3$, and different values of t and ℓ

		ℓ					ℓ		
t	0	1	2	3	t	0	1	2	3
0	1016.4	417.0	276.3	212.0	0	1184.4	501.7	337.0	260.7
	133192.9	16141.0	5562.4	2721.4		112227.1	13228.8	4537.2	2224.5
1	837.3	354.6	244.8	191.8	1	1007.3	437.0	306.6	247.2
	64072.8	6246.1	2168.7	972.4		52241.5	4989.8	1646.1	875.7
2	778.0	346.8	239.2	187.1	2	948.6	430.2	302.8	244.8
	42446.0	4137.9	1413.0	621.1		34197.0	3257.1	1063.6	607.1
3	751.9	345.0	239.5	186.1	3	923.4	428.4	300.7	246.0
	32147.5	3071.0	1178.5	509.0		25720.4	2585.7	815.3	472.6

$$u = S * *\qquad\qquad u = *S*$$

		ℓ					ℓ					ℓ		
t	0	1	2	3	t	0	1	2	3	t	0	1	2	3
0	1379.9	603.3	410.7	320.1	0	111.0	55.6	41.5	34.7	0	132.6	67.8	51.0	42.8
	96001.3	10987.0	3757.5	1871.9		1785.6	296.3	130.1	77.9		2002.5	340.4	151.4	92.1
1	1211.6	540.2	383.2	303.3	1	86.0	46.6	36.6	31.7	1	105.2	57.9	45.7	39.3
	43094.6	4069.1	1537.0	888.8		647.9	106.3	51.4	32.7		741.9	124.5	65.2	44.1
2	1156.5	536.9	380.7	298.6	2	79.1	45.2	35.9	31.2	2	97.5	56.4	45.0	38.6
	27769.3	2690.6	1074.0	632.7		393.1	72.5	37.3	25.1		454.3	85.8	50.0	34.8
3	1134.3	533.6	384.7	304.1	3	76.2	44.9	35.6	31.0	3	94.4	56.0	45.1	38.6
	20694.5	2030.0	860.7	572.7		288.3	60.1	33.0	22.3		334.4	68.5	44.9	32.8

$$u = * * S\qquad\qquad u = SS*\qquad\qquad u = S * S$$

		ℓ		
t	0	1	2	3
0	158.5	83.1	63.1	53.2
	2089.0	356.3	160.1	98.2
1	128.9	72.1	57.4	49.9
	769.2	131.3	67.9	50.2
2	120.6	70.5	56.8	49.6
	470.7	91.3	51.7	41.8
3	117.3	70.0	56.6	49.8
	346.9	76.8	45.0	39.1

$$u = *SS$$

4 Gkd-trees as External Multi-dimensional Data Structures

Gkd-trees can be adapted for secondary storage devices. The criterion of node size can be set to maximize the data block transfer of such devices. The data structure is very compact in space since indexing is avoided. In addition, its average space performance competes with efficient external data structures such as B-trees [10]. Partial match queries efficiently implement the intersection operation of subsets of elements when s, the number of specified coordinates, is a large proportion of k, the dimension of the data space. Preliminary simulation results, to be published [11], suggest that the structure performs well when the size of the domains D^j that conform the data space are large.

The space redundant indexed files [6] provide a more powerful query capability with set union and difference operations in addition to set intersection. Searches on each coordinate are logarithmic. However, the subsets obtained in each of the separate searches must be combined (intersected, merged, etc) in order to provide the final answer to the required query. This limitation restrict s to small values in practice.

Some variants of gkd-trees for external device data handling are currently under evaluation [11]. A special emphasis is placed on the use of gkd-trees as index trees that access multidimensional data stored in files organized into expansible buckets.

References

[1] Bentley, J.L.: Multidimensional Binary Search Trees Used for Associative Searching. Comm. ACM **18**, (1975) 509-517

[2] Knuth, D.E.: The Art of Computer Programming, Vol III. Sorting and Searching, Reading, MA.: Addison-Wesley 1973

[3] Poblete, P.V., Munro, J.I.: The Analysis of a Fringe Heuristic for Binary Search Trees. J. Algorithms **6**, (1985) 336-350

[4] Cunto, W., Gascón, J.L.: Improving Time and Space Efficiency in Generalized Binary Search Trees. Acta Inf. **24**, (1987) 583-594

[5] Cunto, W., Lau G., Flajolet P.: Analysis of Kdt-trees: Kd-trees Improved by Local Reorganizations. Proceedings of the Workshop on Algorithms and Data Structure, Ottawa, 1989, 24-38. Lecture Notes 382, F. Dehne, J.-R. Sack, N. Santoro Editors, Berlin: Springer Verlag 1989

[6] Wiederhold, G.: File Organization for Database Design. New York, N.Y.: McGraw-Hill 1987

[7] Knuth, D. E.: The Art Of Computer Programming, Vol. 1. Fundamental Algorithms, 2nd Edition, Reading, Ma: Addison-Wesley 1973

[8] Yriarte, V.: Arboles Kd Generalizados. Tesis de Maestría, Departamento de Computación, Universidad Simón Bolívar, Caracas 1992

[9] Flajolet, P., Puech, C.: Partial Match Retrieval of Multidimensional Data. J. ACM **33**, (1986) 371-407

[10] Bayer, R., McCreight, E.: Organization and Maintenance of Larger Ordered Indexes. Acta Inf. **14**, (1972) 173-189

[11] Cunto, W., Yriarte, V.: External Kd-trees. In preparation

[12] Spivak, M.: A Comprehensive Introduction To Differential Geometry, Vol. 1. Boston, Ma: Publish or Perish 1970

[13]   Elsgoltz, L.: Ecuaciones Diferenciales y Cálculo Variacional. Moscú, URSS: Editorial MIR 1977

[14]   Boole, G.: Calculus of Finite Differences. New York, N.Y.: Chelsea 1860, 5th Edition 1970

Appendix

A search pattern u is defined as a k-character word from the alphabet $\{S, *\}$ such that $u_j = S$ if the j-th coordinate value is specified, or $u_j = *$ otherwise. Looking only at the first coordinate, we will denote $u = Sv$ or $u = *v$ where v is the $(k-1)$-character suffix of u. Also, we represent shifts to the left of u by $u', u'', \ldots, u^{(k-1)}$ ($u = u^{(0)} = u^{(k)}$) and denote $\gamma = 2t\ell + t + \ell$.

Let $q_{r,N}$, with $0 \leq r \leq N - 2\ell - 1$, be the probability that the left subtree of a randomly generated gktree of node size $2\ell + 1$ and reorganization parameter t has r elements. The following lemma gives the closed formulas for $q_{r,N}$.

Lemma 3.

$$q_{r,N} = \delta_{r,0} \qquad\qquad 0 \leq N \leq 2l+1$$

$$q_{r,N} = \frac{\dbinom{r+\ell}{\ell}\dbinom{N-1-r-\ell}{\ell}}{\dbinom{N}{2\ell+1}} \qquad 2\ell+1 < N < 2\gamma+1$$

$$q_{r,N} = \frac{\dbinom{r+\ell}{\gamma}\dbinom{N-1-r-\ell}{\gamma}}{\dbinom{N}{2\gamma+1}} \qquad N \geq 2\gamma+1.$$

Proof. If the tree has at most $2\ell + 1$, its right subtree is empty. If the tree has more than $2\ell + 1$ elements, it can be easily proved that

$$q_{r,N} = q_{r,N-1}\frac{N-r-\ell-1}{N} + q_{r-1,N-1}\frac{r+\ell}{N}.$$

From this recursive equation, a simple induction on N for the cases when $2\ell + 1 < N < 2\gamma + 1$ and $N \geq 2\gamma + 1$ completes the proof.

Let $X_{t,\ell,u,N}$ denote the random variable equal to the number of traversed nodes during partial match query with a search pattern u in a randomly generated gkd-tree of size N with node size $2\ell + 1$ and reorganization parameter t. Taking into consideration that

$$p_{t,\ell,u,N,i} = P[X_{t,\ell,u,N} = i]$$
$$P_{t,\ell,u,N}(z) = \sum_{i \geq 0} p_{t,\ell,u,N,i} z^i$$
$$E_{t,\ell,u,N} = P'_{t,\ell,u,N}(1)$$
$$F_{t,\ell,u,N} = P''_{t,\ell,u,N}(1)$$

it follows that the average of the distribution is $E[X_{t,\ell,u,N}] = E_{t,\ell,u,N}$ and the variance is $V[X_{t,\ell,u,N}] = F_{t,\ell,u,N} - E^2_{t,\ell,u,N} + E_{t,\ell,u,N}$. To simplify the notation, subindices are omitted when they are clearly understood.

Lemma 4. *The probability generating function $P_{u,N}$ of the sequence of probabilities $\{p_{i,N}\}$ is:*

$$P_{u,N}(z) = 1 \qquad\qquad\qquad\qquad\qquad\qquad N = 0$$

$$P_{u,N}(z) = z \qquad\qquad\qquad\qquad\qquad\quad 0 < N \le 2\ell + 1$$

$$P_{u,N}(z) = \frac{2\ell}{N+1}z + 2z\sum_{r=0}^{N-2\ell-1} \frac{r+1}{N+1}q_{r,N}P_{u',r}(z) \qquad 2\ell+1 < N \le 2\gamma+1, u = Sv$$

$$P_{u,N}(z) = z\sum_{r=0}^{N-2\ell-1} q_{r,N}P_{u',r}(z)P_{u',N-2\ell-1-r}(z) \quad 2\ell+1 < N \le 2\gamma+1, u = *v$$

$$P_{u,N}(z) = \frac{2\ell}{N+1}z + 2z\sum_{r=\gamma-\ell}^{N-\gamma-\ell-1} \frac{r+1}{N+1}q_{r,N}P_{u',r}(z) \qquad N > 2\gamma+1, u = Sv$$

$$P_{u,N}(z) = z\sum_{r=\gamma-\ell}^{N-\gamma-\ell-1} q_{r,N}P_{u',r}(z)P_{u',N-2\ell-1-r}(z) \quad N > 2\gamma+1, u = *v.$$

Proof. $P_{u,0}(z) = 1$ since $p_{u,N,0} = \delta_{N,0}$ and $p_{u,0,i} = \delta_{0,i}$. If $0 < N \le 2\ell + 1$, the tree has only one node and the search performs exactly one visit; thus, $P_{u,N} = z$.

When $2\ell + 1 < N < 2\gamma + 1$, if the search pattern is of the form $u = Sv$, the search continues in only one of the subtrees with the search pattern u'. Thus,

$$p_{N,i} = \sum_{r=0}^{N-2\ell-1} q_{r,N}\left(\frac{r+1}{N+1}p_{u',r,i-1} + \frac{N-2\ell-r}{N+1}p_{u',N-2\ell-1-r,i-1}\right)$$

$$= 2\sum_{r=0}^{N-2\ell-1} \frac{r+1}{N+1}q_{r,N}p_{u',r,i-1}.$$

If $u = *v$, the search must recur in the left and right subtrees with the search pattern shifted. In this case

$$p_{u,N,i} = \sum_{r=0}^{N-2\ell-1} q_{r,N}\sum_{j=0}^{i-1} p_{u',r,j}p_{u',N-2\ell-1-r,i-1-j}.$$

In both cases, the probability generating functions follow immediately. The case when $n \ge 2\gamma + 1$ is similar to previous one.

Corollary 5.
$$E_{u,0} = 0$$
$$F_{u,0} = 0.$$

If $0 < N \le 2\ell + 1$,
$$E_{u,N} = 1$$
$$F_{u,N} = 0.$$

If $2\ell + 1 < N < (2t + 1)(2\ell + 1)$ and $u = Sv$ then

$$E_{u,N} = 1 + 2\sum_{r=0}^{N-2\ell-1} \frac{r+1}{N+1}q_{r,N}E_{u',r}$$
$$F_{u,N} = 2E_{u,N} - 2 + 2\sum_{r=0}^{N-2\ell-1} \frac{r+1}{N+1}q_{r,N}F_{u',r}.$$

If $N \geq (2t+1)(2\ell+1)$ and $u = Sv$ then

$$E_{u,N} = 1 + 2 \sum_{r=\gamma-\ell}^{N-\gamma-\ell-1} \frac{r+1}{N+1} q_{r,N} E_{u',r}$$
$$F_{u,N} = 2E_{u,N} - 2 + 2 \sum_{r=\gamma-\ell}^{N-\gamma-\ell-1} \frac{r+1}{N+1} q_{r,N} F_{u',r}.$$

If $2\ell + 1 < N < (2t+1)(2\ell+1)$ and $u = *v$ then

$$E_{u,N} = 1 + 2 \sum_{r=0}^{N-2\ell-1} q_{r,N} E_{u',r}$$
$$F_{u,N} = 2E_{u,N} - 2 + 2 \sum_{r=0}^{N-2\ell-1} q_{r,N} (E_{u',r} E_{u',N-2\ell-1-r} + F_{u',r}).$$

If $N \geq (2t+1)(2\ell+1)$ and $u = *v$ then

$$E_{u,N} = 1 + 2 \sum_{r=\gamma-\ell}^{N-\gamma-\ell-1} q_{r,N} E_{u',r}$$
$$F_{u,N} = 2E_{u,N} - 2 + 2 \sum_{r=\gamma-\ell}^{N-\gamma-\ell-1} (E_{u',r} E_{u',N-2\ell-1-r} + F_{u',r}).$$

Proof. Equations given in the previous lemma are differentiated and evaluated at $z = 1$.

Lemma 6. *Let $k, t, \ell \in \mathbf{N}$, $k \geq 2$, $t, \ell \geq 0$ and $u \in \{S, *\}^k$ be a fixed search pattern. To simplify the notation, $f_i(N) = E_{t,u^{(i)},N}$, $0 \leq i < k$, $f_k(N) = f_0(N)$, $N \geq 0$. Asymptotically, the functions f_i can be extended as real positive non-decreasing functions over $\mathbb{R}^+$ with infinite derivatives such that*

$$f_i(x) x^{2\gamma+1} = \frac{(\gamma+2)^{\overline{\gamma+1}}}{\gamma!} \left[\sum_{h=0}^{\gamma} \binom{\gamma}{h} (-1)^h x^{\gamma-h} \left(\int_0^x f_{i+1}(r) r^{\gamma+h} dr \right) \right] (1+o(1)), \quad u^{(i)} = *v$$

$$f_i(x) x^{2\gamma+2} = \frac{(\gamma+2)^{\overline{\gamma+1}}}{\gamma!} \left[\sum_{h=0}^{\gamma} \binom{\gamma}{h} (-1)^h x^{\gamma-h} \left(\int_0^x f_{i+1}(r) r^{\gamma+h+1} dr \right) \right] (1+o(1)), \quad u^{(i)} = Sv.$$

Proof. Clearly, the average performance of a partial match query is a non-decreasing function and it is bounded by the total match query average performance and the size of the tree. Hence, $\Omega(\ln N) \leq f_i(N) \leq N$, $0 \leq i < k$.

Suppose that $u^{(i)} = *v$ and $N \geq 2\gamma + 1$. By expanding the probability $q_{r,N}$ as polynomials and using properties of $o()$, the average performance becomes

$$f_i(N) N^{2\gamma+1}(1+o(1)) = \frac{2(2\gamma+1)!}{\gamma!^2} \left[\sum_{r=\gamma}^{N-\gamma-1} r^\gamma (N-r)^\gamma f_{i+1}(r) \right] (1+o(1)).$$

By expanding Newton's binomial and exchanging summations, we get

$$f_i(N) N^{2\gamma+1} = \frac{(\gamma+2)^{\overline{\gamma+1}}}{\gamma!} \left[\sum_{h=0}^{\gamma} \binom{\gamma}{h} (-1)^h N^{\gamma-h} \left(\sum_{r=0}^{N-2\gamma-1} f_{i+1}(r)(r+\ell)^{\gamma+h} \right) \right] (1+o(1)).$$

Each f_i can be extended as an $\mathbb{R}^+ \to \mathbb{R}^+$ function such that it remains non-decreasing and with an infinite number of derivatives [12]. Finally, by approximating the summation over the tree size by an integral, we obtain

$$f_i(N) N^{2\gamma+1} = \frac{(\gamma+2)^{\overline{\gamma+1}}}{\gamma!} \left[\sum_{h=0}^{\gamma} \binom{\gamma}{h} (-1)^h N^{\gamma-h} \left(\int_0^{N-2\gamma} f_{i+1}(r)(r+l)^{\gamma+h} dr \right) \right] (1+o(1)).$$

Further manipulation of the previous expression with lower order terms discarded, leads to the desired result. The case when $u = Sv$ is analogous.

Lemma 7. *The function $f_{i+1} = \Theta(f_i)$, $0 \leq i < k$.*

Proof. From the recurrent equations of Corollary 5 and for both cases when $u = Sv$ and $u = *v$, it is not difficult to prove that $f_i \leq 2f_{i+1}$ given the fact that the functions $f_i(x)$, $0 \leq i < k$ are non-decreasing, and $(r+1)/(n+1)$ and $q_{r,N}$ are smaller than 1. Inductively on i, $f_i \leq 2^j f_{(i+j)\oplus k}$. Therefore,

$$\frac{1}{2}f_i \leq f_{i+1} \leq 2^{k-1}f_i.$$

Theorem 8. *There exist constants A_i, $0 \leq i < k$, and α such that*

$$f_i(x) = A_i x^\alpha (1 + o(1))$$

where α is the unique positive real root of the equation

$$\left((\lambda+\gamma+1)^{\overline{\gamma+1}}\right)^{k-s}\left((\lambda+\gamma+2)^{\overline{\gamma+1}}\right)^s = \left((\gamma+2)^{\overline{\gamma+1}}\right)^k.$$

Furthermore, $0 < \alpha < 1$ and the coefficients A_i are such that

$$A_i = A_{i+1}\frac{(\gamma+2)^{\overline{\gamma+1}}}{(\alpha+\gamma+1)^{\overline{\gamma+1}}}, \; u^{(i)} = *v$$

$$A_i = A_{i+1}\frac{(\gamma+2)^{\overline{\gamma+1}}}{(\alpha+\gamma+2)^{\overline{\gamma+1}}}, \; u^{(i)} = Sv.$$

Finally,

$$E(X_{u,N}) = A_0 N^\alpha (1 + o(1)).$$

Proof. Let us consider the functions $f_i : \mathbb{R}^+ \to \mathbb{R}^+$, $0 \leq i < k$, as defined in Lemma 6, but discarding lower order terms, that is

$$f_i(x)x^{2\gamma+1} = \frac{(\gamma+2)^{\overline{\gamma+1}}}{\gamma!}\left[\sum_{h=0}^{\gamma}\binom{\gamma}{h}(-1)^h x^{\gamma-h}\left(\int_0^x f_{i+1}(r)r^{\gamma+h}dr\right)\right], \; u^{(i)} = *v$$

$$f_i(x)x^{2\gamma+2} = \frac{(\gamma+2)^{\overline{\gamma+1}}}{\gamma!}\left[\sum_{h=0}^{\gamma}\binom{\gamma}{h}(-1)^h x^{\gamma-h}\left(\int_0^x f_{i+1}(r)r^{\gamma+h+1}dr\right)\right], \; u^{(i)} = Sv.$$

These equations are differentiated $\gamma + 1$ times with respect to x obtaining a system of homogeneous differential equations of the form

$$\sum_{m=0}^{\gamma+1} \mathcal{A}_m x^m f_i^{(m)}(x) = \sum_{m=0}^{\gamma} \mathcal{B}_m x^m f_{i+1}^{(m)}(x), \; u^{(i)} = *v$$

$$\sum_{m=0}^{\gamma+1} \mathcal{C}_m x^m f_i^{(m)}(x) = \sum_{m=0}^{\gamma} \mathcal{D}_m x^m f_{i+1}^{(m)}(x), \; u^{(i)} = Sv$$

where $0 \leq i < k$ and, $\mathcal{A}_0,\ldots,\mathcal{A}_{\gamma+1}$, $\mathcal{B}_0,\ldots,\mathcal{B}_\gamma$, $\mathcal{C}_0,\ldots,\mathcal{C}_{\gamma+1}$ and $\mathcal{D}_0,\ldots,\mathcal{D}_\gamma$ are independent of x.

From Lemma 7, we can approximate f_{i+1} by $\zeta_i f_i$ and rewrite the previous system as follows

$$\sum_{m=0}^{\gamma+1}(\mathcal{A}_m - \zeta_1\mathcal{B}_m)x^m f_i^{(m)}(x) = 0, \; u^{(i)} = *v$$

$$\sum_{m=0}^{\gamma+1}(\mathcal{C}_m - \zeta_1\mathcal{D}_m)x^m f_i^{(m)}(x) = 0, \; u^{(i)} = Sv.$$

The solution of the previous system is an Eulerian system of differential equations with set of solutions

$$\left\{ \begin{pmatrix} A_{1,\alpha,j} \\ \vdots \\ A_{k,\alpha,j} \end{pmatrix} x^\alpha (\log x)^j \right\}$$

where $\alpha \in C$ is a root of the system characteristic equation, $0 \le j < \alpha$'s multiplicity [13](2.4).

To find the system characteristic equation, $f_i(x)$ is substituted by $A_i x^\alpha$ in the system of recurrent equations and by using [7](1.2.6-48), we get:

$$A_i = A_{i+1} \frac{(\gamma+2)^{\overline{\gamma+1}}}{(\alpha+\gamma+1)^{\overline{\gamma+1}}}, \ u^{(i)} = *v$$

$$A_i = A_{i+1} \frac{(\gamma+2)^{\overline{\gamma+1}}}{(\alpha+\gamma+2)^{\overline{\gamma+1}}}, \ u^{(i)} = Sv.$$

Since $A_0 = A_k$ and $A_i \ne 0$, $0 \le i < k$, manipulations of such coefficient relations lead to the system characteristic function

$$\mathcal{F}(\alpha) = \left((\alpha+\gamma+1)^{\overline{\gamma+1}} \right)^{k-s} \left((\alpha+\gamma+2)^{\overline{\gamma+1}} \right)^s - \left((\gamma+2)^{\overline{\gamma+1}} \right)^k.$$

This function has a unique real positive root in the interval $(0,1)$ which happens to be of multiplicity one ($\mathcal{F}(0) < 0$, $\mathcal{F}(1) > 0$ and $\mathcal{F}'(\alpha) > 0$ when $\alpha \ge 0$). Negative real roots and complex roots can be neglected since they produce negative or asymptotically null solutions.

Equations in Lemma 6 correspond to equations in this lemma with lower order terms discarded. Thus $f_i(N) = A_i N^\alpha (1 + o(1))$, $0 \le i < k$, are verified to be the solution of the original equations. The implication

$$f(t) = t^\alpha (1 + o(1)) \implies \int_0^t f(x)dx = (\alpha+1)^{-1} t^{\alpha+1} (1 + o(1))$$

shown in [14](7.3) is used in this process.

Corollary 9. *For a given domain dimension k and number of coordinates s, the minimum and the maximum asymptotic average performance of partial match queries correspond to the searching patterns $S^s *^{k-s}$ and $*^{k-s} S^s$ respectively.*

Proof. From the previous theorem $\alpha_\gamma \in (0,1)$, hence

$$\lim_{N\to\infty} \frac{E_{*,N}}{E_{*',N}} = \frac{(\gamma+2)^{\overline{\gamma+1}}}{(\alpha+\gamma+1)^{\overline{\gamma+1}}} > 1, u = *v$$

$$\lim_{N\to\infty} \frac{E_{*,n}}{E_{*',n}} = \frac{(\gamma+2)^{\overline{\gamma+1}}}{(\alpha+\gamma+2)^{\overline{\gamma+1}}} < 1, u = Sv.$$

Theorem 10. *Let $\{\alpha_{t,\ell,s,k}\}$ be the sequence obtained by solving the characteristic equation in Theorem 8 for different values of t, ℓ, s and k. The sequence decreases monotonically with limit $1 - s/k$ as t or ℓ increases.*

Proof. For each value of γ, let us define the function

$$\mathcal{F}_\gamma(x) = \left[(\gamma + 1 + x)^{\overline{\gamma+1}}\right]^{k-s} \left[(\gamma + 2 + x)^{\overline{\gamma+1}}\right]^s - \left[(\gamma + 2)^{\overline{\gamma+1}}\right]^k.$$

$\mathcal{F}_\gamma(x)$ is a non-decreasing and concave function in $(0, 1)$.

Let α be the root of $\mathcal{F}_\gamma(x)$ in $(0, 1)$. We will prove that $\mathcal{F}_{\gamma+1}(\alpha) > 0$ and, therefore, the root of $\mathcal{F}_{\gamma+1}(x)$ is lower than α. By developing $\mathcal{F}_{\gamma+1}(\alpha)$ and using the fact that $\mathcal{F}_\gamma(\alpha) = 0$, we conclude that,

$$\mathcal{F}_{\gamma+1}(\alpha) = \left[(\gamma + 2)^{\overline{\gamma+1}}\right] \frac{h(\alpha) - j(\alpha)}{(\gamma + 1 + \alpha)^{k-s}(\gamma + 2 + \alpha)^s},$$

where $h(x)$ and $j(x)$ are polynomial with positive coefficients such that $h(0) = j(0)$ and $h(1) > j(1)$. This implies that $\mathcal{F}_{\gamma+1}(\alpha) > 0$, thus proving that the sequence decreases monotonically.

On the other hand, the characteristic equation of Theorem 8 can be rewritten as

$$\left(\frac{(2\gamma + 2 + \alpha)^\alpha (1 + o(1))}{(\gamma + 1 + \alpha)^\alpha (1 + o(1))}\right)^s = \left(\frac{(2\gamma + 1 + \alpha)^{1-\alpha}(1 + o(1))}{(\gamma + \alpha)^{1-\alpha}(1 + o(1))}\right)^{k-s};$$

hence, the limit $1 - s/k$ can be obtained by solving the previous equation for $\gamma \to \infty$.

Theorem 11. *Let α_γ and $A_{\gamma,u}$ be as defined in Theorem 8. Therefore, there exist positive constants $B_{\gamma,u}$ and $C_{\gamma,u}$ such that*

$$V[X_{\gamma,u,N}] = \left(B_{\gamma,u} - A_{\gamma,u}^2\right) N^{2\alpha_\gamma} + (C_{\gamma,u} + A_{\gamma,u})N^{\alpha_\gamma}(1 + o(1)).$$

Proof. Let k, γ and u be fixed. The system of equations for $F_{\gamma,u,N}, F_{\gamma,u',N}, \ldots,$ $F_{\gamma,u^{(k-1)},N}$ given in Corollary 5 is linear. This means that its general solution is the summation of the general solution of its corresponding homogeneous system plus a particular solution [14](12.5). Theorem 8 supplies the general solution of the homogeneous system. We will prove that the order of the particular solution is expected to be of the order of the homogeneous solution squared.

Substituting $F_{t,u^{(i)},N} = B_i N^{2\alpha} + C_i N^\alpha(1 + o(1))$ into corresponding equations in Corollary 5 following a similar derivation to the one used in Theorem 8, we can determine the relation among the positive constants B_i and C_i. That is,

$$u^{(i)} = *v \quad \Longrightarrow \quad B_i = \frac{(\gamma+2)^{\overline{\gamma+1}}}{(2\alpha+\gamma+1)^{\overline{\gamma+1}}}B_{i+1} + \frac{(\gamma+2)^{\overline{\gamma+1}}(\Gamma(\alpha+\gamma+1))^2}{\gamma!\,\Gamma(2\alpha+2\gamma+2)}A_{i+1}^2$$

$$\text{and} \quad C_i = 2A_i + \frac{(\gamma+2)^{\overline{\gamma+1}}}{(\alpha+\gamma+1)^{\overline{\gamma+1}}}C_{i+1},$$

$$u^{(i)} = Sv \quad \Longrightarrow \quad B_i = \frac{(\gamma+2)^{\overline{\gamma+1}}}{(2\alpha+\gamma+2)^{\overline{\gamma+1}}}B_{i+1}$$

$$\text{and} \quad C_i = 2A_i + \frac{(\gamma+2)^{\overline{\gamma+1}}}{(\alpha+\gamma+2)^{\overline{\gamma+1}}}C_{i+1},$$

where $B_k = B_0$ and $C_k = C_0$.

The corresponding systems of equations for the B_i's and the C_i's have unique solutions, thus proving the theorem.

Theorem 12.

$$\lim_{\gamma \to \infty} V(X_{\gamma,u,N}) = O(N^{1-s/k}).$$

Proof. Since $D_{\gamma,u} = A^2_{\gamma,u}(B_{\gamma,u}/A^2_{\gamma,u} - 1)$ and $E_{\gamma,u,N} < N$, $A_{\gamma,u}$ is a bounded sequence, it suffices to prove that $\lim_{\gamma \to \infty} B_{\gamma,u}/A^2_{\gamma,u} = 1$.

Denoting $A_i = A_{\gamma,u^{(i)}}$, $B_i = B_{\gamma,u^{(i)}}$, by following Theorems 8 and 11,

$$u^{(i)} = *v \implies A_i = a_\gamma A_{i+1}$$
$$\text{and} \quad B_i = c_\gamma B_{i+1} + e_\gamma A^2_{i+1}$$
$$u^{(i)} = Sv \implies A_i = b_\gamma A_{i+1}$$
$$\text{and} \quad B_i = d_\gamma B_{i+1}$$

where

$$a_\gamma = \frac{(\gamma+2)^{\overline{\gamma+1}}}{(\alpha_\gamma + \gamma + 1)^{\overline{\gamma+1}}} \quad , \quad b_\gamma = \frac{(\gamma+2)^{\overline{\gamma+1}}}{(\alpha_\gamma + \gamma + 2)^{\overline{\gamma+1}}}$$

$$c_\gamma = \frac{(\gamma+2)^{\overline{\gamma+1}}}{(2\alpha_\gamma + \gamma + 1)^{\overline{\gamma+1}}} \quad , \quad d_\gamma = \frac{(\gamma+2)^{\overline{\gamma+1}}}{(2\alpha_\gamma + \gamma + 2)^{\overline{\gamma+1}}}$$

$$e_\gamma = \frac{(\gamma+2)^{\overline{\gamma+1}}(\Gamma(\alpha_\gamma + \gamma + 1))^2}{\gamma! \, \Gamma(2\alpha_\gamma + 2\gamma + 2)} .$$

But $A_{\gamma,u} = A_0$ and $B_{\gamma,u} = B_0$, thus we have to prove that $\lim_{\gamma \to \infty} B_0/A^2_0 = 1$. By using the Stirling approximation,

$$\lim_{\gamma \to \infty} a_\gamma = 2^{s/k} \quad , \quad \lim_{\gamma \to \infty} b_\gamma = 2^{(s/k)-1},$$

$$\lim_{\gamma \to \infty} c_\gamma = 2^{2s/k-1} \quad , \quad \lim_{\gamma \to \infty} d_\gamma = 2^{2(s/k-1)} \quad \text{and}$$

$$\lim_{\gamma \to \infty} e_\gamma = 2^{2s/k-1}.$$

The proof that $\lim_{\gamma \to \infty} B_0/A^2_0 = 1$ follows from the iterative application of previous equations in which coefficients have been substituted by their corresponding limit values.

This article was processed using the LaTeX macro package with LLNCS style

Fractional Cascading Simplified[*]

Sandeep Sen
Department of Computer Science and Engineering
Indian Institute of Technology, Delhi
New Delhi 110016, India

Abstract

We give an alternate implementation of the fractional cascading data-structure of Chazelle and Guibas to do iterative search for a key in multiple ordered lists. The construction of our data-structure uses randomization and simplifies the algorithm of Chazelle and Guibas vastly making it practical to implement. Although our bounds are asymptotically similar to the earlier ones, there are improvements in the constant factors. Our analysis is novel and captures some of the inherent difficulties associated with the fractional casading data structure. In particular, we use tools from branching process theory and derive some useful asymptotic bounds. The probability of deviation from the expected performance bounds decreases rapidly with number of keys.

1 Introduction

The problem of searching for a key in many ordered lists arises very frequently in computational geometry (see [2] for applications). Chazelle and Guibas [1] introduced fractional cascading as a general technique for solving this problem. Their work unified some earlier work in this area and gave a general strategy for improving upon the naive method of doing independent searches for the same key in separate lists. In brief, they devised a data-structure that would enable searching for the same key in n lists in time $O(\log M + n)$ where M is the size of the longest list. If N is the total size of all the lists then this data structure can be built in $O(N)$ preprocessing time and take $O(N)$ space. We shall give a more precise description of their data structure in the next section. Their solution, although elegant is difficult to implement and they leave open the question of simplifying it to be useful in practice.

---

[*]Part of this research was done when the author was a post-doctoral Member of Technical Staff at AT&T Bell Labs, Murray Hill, NJ during the year 1990-1991

In this paper, we give an alternate implementation of their data-structure that uses randomization. While retaining the salient features of their data-structure, we are able to simplify its construction considerably to an extent that is practical. The motivation of the new technique has been derived from the success of skip-lists (Pugh [5]). However, our method requires new analytical techniques which could have further applications. In particular we use tools from branching-process theory and derive some useful asymptotic bounds.

Our work still leaves open the issue of dynamic maintenance of fractional-casading data-structure, that attains optimal performance. However, it does simplify part of scheme which could be significant. The bounds for space and and preprocessing time that we obtain for the static case are (expected) worst case and hold with probability $1 - 2^{-O(N)}$. The search time holds with high probability, i.e. $1 - 1/N^k$ for any fixed $k > 0$. The following notation will be used in the paper. We say a randomized algorithm has resource (like time, space, etc.) bound $\tilde{O}(g(n))$ if there is a constant c such that the amount of resource used by the algorithm (on any input of size n) is no more than $c\alpha g(n)$ with probability $\geq 1 - 1/n^{\alpha}$, for any $\alpha \geq 1$.

2 Description of Fractional Cascading

In this section, we give a brief description of the problem setting and the approach taken by Chazelle and Guibas. Consider a fixed graph $G = (V, E)$ of $|V| = n$ vertices and $|E| = m$ edges. The graph G is undirected and connected and does not contain multiple edges. Each vertex v has a catalog C_v and associated with each edge e of G is a range R_e.

A catalog is an ordered collection of records where each record has an associated value in the set $\Re \cup \{\infty, -\infty\}$. The records are stored in a non-decreasing order of their values and more than one record can have the same value. A catalog is never empty (has at least a ∞ and a $-\infty$). A range is an interval of the form $[x, y], [-\infty, y], [x, \infty], [-\infty, \infty]$. The graph G together with the associated catalogs and ranges is called a catalog graph. This is the basic structure to which fractional cascading is applied.

For notational purposes, if the value k is an end-point of the range $R_{u,v}$, then k appears as the value of some record in both C_u and C_v. Moreover, if two ranges $R_{u,v}$ and $R_{v,w}$ have an end-point in common it appears twice in C_v. The space required to store a catalog graph is $N = \sum_{v \in V} |C_v|$. This includes the space to store the graph itself. A catalog graph G is said to be *locally bounded* by degree d if for each vertex v and each value of $x \in \Re$ the number of edges incident on v whose range includes x is bounded by d.

The input to a query is an arbitrary element k in the universe and a connected subtree $\Pi = (\bar{V}, \bar{E})$ such that $k \in R_e$ for all edges e in the subtree and $\bar{V} \subset V, \bar{E} \subset E$. The output of the query for each vertex $v \in \bar{V}$ is an element y such that $predecessor(y) < k \leq y$. We shall refer to such a pair of elements as *straddling pair* of k.

Theorem 1 (Chazelle-Guibas) *Let G be a catalog graph of size N and locally bounded degree d. In $O(N)$ space and $O(dN)$ time it is possible to construct a data-structure which allows multiple look-ups (query) in a subtree of size p in in time $O(pd + \log N)$. If d is fixed this is optimal. In addition, if the underlying catalog graph G is restructured to a constant degree graph, then the search time and the preprocessing time can be improved to $O(p \log d + \log N)$ and $O(N)$ respectively.*

3 Anatomy of the Data Structure

Our data-structure is very similar to [1] in the sense that we retain their idea of using augmented catalogs A_v for every vertex v such that $C_v \subset A_v$. But we shall use a different method for its construction. An augmented catalog A_v is also a linear list of records whose values form a sorted multiset. Augmented catalogs in neighbouring nodes of G will contain a number of records with common values. The corresponding records are linked together to correlate locations in the two catalogs. The objective is that given the location of a record in A_v, we would be able to find its location (the straddling pair) in the augmented catalog of a neighbour of v in constant additional time. More formally, for each node u and an edge e connecting u and v in G, we maintain a list of 'bridges' from u to v, B_{uv}, which is an ordered subset of records in A_v and lying in the range R_e. The end-points of R_e are the first and last records of B_{uv}. In node v, we maintain for each bridge in B_{uv} a companion bridge in B_{vu}. The value of a record is distinct from the record. Moreover each bridge is associated with a unique edge of G, implying that if a given value in A_u is chosen to be a bridge in both B_{uv} and B_{uw}, then it is duplicated and stored in different records of A_u.

A pair of consecutive bridges associated with the same edge $e = (u, v)$ defines a *gap*. Let a_u and b_u be two consecutive bridges in B_{uv} and a_v (respectively b_v) be the companion bridges in B_{vu}. If **value**$(a_u) <$ **value**(b_u), then the gap of b_u includes all elements of A_u positioned strictly between a_u and b_u and all elements of A_v between a_v and b_v (the bridges are not included). By definition, gap of b_v is the same as gap of b_u. One of the key strategy used by [1] is to maintain the invariant that no gap exceeds $6d$ -1 in size.

We now take a closer look at the information maintained with each record. Both C_v and A_v are maintained as linked lists. A record of C_v have the fields

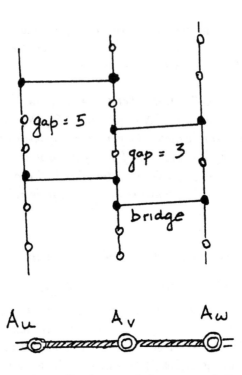

Figure 1: Gaps, Bridges and Augmented Catalogs

key and *up-pointer*. The key contains the value and the up-pointer is a pointer to the next record. A_v has several other fields :

(1) key: stores the value k of record r.
(2) C-pointer: holds a pointer $\nu(r)$, the successor of r in C_v.
(3) up-pointer, down-pointer : pointers to successor and predecessor in A_v.

In addition a bridge element also has pointers to its companion bridge and also the label of the edge for which it is a bridge (If r is a bridge in B_{uv} then it stores the label uv).

To answer a multiple look-up query (x, Π) where x is the key value and Π is the subtree, one begins by locating x in the first node of the path Π and then use the following properties:

Lemma 1 (CG) : *If we know the position of value x in A_v, we can compute the position of x in C_v in one step.*

This can be done by using the C-field.

Lemma 2 : *If we know the position of value x in A_v and $e = (v, w)$ is an edge of G such that $x \in R_e$, then we can compute the position of x in A_w in $O(|Gap_e(x)|)$ time. $Gap_e(x)$ is the set of elements in the gap (corresponding to edge e) that x belongs to.*

From the position of x in A_v follow up-pointers until a bridge is found which connects to A_w. Note that Chazelle-Guibas [1] maintained the invariant that all gap sizes are less than $6d$ which yields a search time of $O(\log N + d|\Pi|)$. This invariant was maintained during the construction of all the augmented catalogs which is done incrementally. Their algorithms start with empty catalog and then for each vertex v, the records of C_v are inserted in an increasing order into A_v. Between any two insertions, the gap invariants are restored. Note that a single insertion into A_v can alter the gaps leading to insertion of a new bridge which introduces new records and this could continue as a long chain of events.

We propose the following modification. Instead of explicitly maintaining gap invariants, we choose a newly inserted element r in A_v to be a bridge with probability p (we shall determine the exact value later). This is repeated for every edge incident on v and whose ranges cover r. If r is chosen to be a bridge for an edge (u, w), it leads to the insertion of a a new record in A_w and (possibly even in A_v if r is already a bridge). These new records are treated exactly the same way as described above (i.e. choose it with certain probability to be a bridge element). For each new record in the augmented catalog, we initialize the following fields:

(i) C-pointer: Can be determined from the predecessor and the successor elements in the augmented catalog. If this element came from C_v, then update the C-pointers for all the elements in A_v between this record and the previous element of C_v.
(ii) Initialize the edge field.

For each new record of C_v, this process is continued until there are no more bridges to insert.

4 Analysis

Let us first analyze the running time for a multiple look-up query. Given the position of x in A_v we follow the up-pointers until we find a bridge b_v which connects to A_u and then traverse the down-pointers until we locate the straddling pair. Since every element is chosen to be bridge element with probability p, the expected length of a gap is $2/p$. Thus from Lemma 1, the expected search time is $O(\log N + |\Pi|/p)$.

Moreover, if $|\Pi| \geq \log N$, we can show that the search time is $\tilde{O}(\log N + |\Pi|/p)$ using the observation that the search time is a sum of $O(\log N)$ independent random variables with a geometric distribution and parameter p (Sen [6]). There is however an oversight in the above reasoning since the query is a tree of size $|\Pi|$ which can assume various forms. In particular, Π can be any of the S positional trees of maximum degree d. It can be shown that $|S| < 2^{O(d|\Pi|)}$. If $1/p$ is proportional to d (which does turn out to be the case as we shall prove later) then the search time holds with high likelihood. Notice that there are $O(N)$ choices for the root of this tree and $O(N)$ combinatorially distinct search paths given the search tree (corresponding to the $O(N)$ intervals induced by all the key values).

Lemma 3 *The new scheme allows multiple query in a subtree Π in expected time $O(\log N + |\Pi| \cdot d)$. If the local degree d is bound by a constant then the query time is $\tilde{O}(\log N + |\Pi|)$, i.e. the bound holds with high probability.*

The more interesting aspect of the analysis is to bound the time and space during the data-structure construction. If we look more closely at the way the records are added to the augmented catalogs, the underlying stochastic process can be modelled by a branching process. The root corresponds to an inserted record from C_v and the number of children correspond to the bridges that are created by this record. Each bridge is created with probability p, which corresponds to two children. For each new record inserted from the C_v's the time and the space needed is proportional to the total progeny of this branching process. Each node can have upto 2d children where the number of children is a random variable which takes values $0, 2, 4 .. 2d$. The probability that this random variable takes value $2k$ is the same·as the probability that a binomial random variable with parameters (d, p) takes value k (i.e. there are k bridges created). The mean μ of this distribution is clearly $2pd$ and the generating function $G(s)$ can be worked out as $(q + ps^2)^d$. Here $q = 1 - p$.

From Feller [3], a branching process is finite if $\mu < 1$. Hence we choose p such that $2pd < 1$, that is $p < 1/2d$. This gives an expected gap length of greater than $4d$. From [3], if the generating function for the total progeny is denoted by $t(s)$, then $t = sG(t)$. In our case $G(t) = (q + pt^2)^d$. Moreover, the mean is $\frac{1}{1-\mu}$, which is $\frac{1}{1-2pd}$ in our case. If we choose $p = 1/3d$, this yields a mean progeny of 3. This in turn implies a total expected space bound of $O(\sum_{v \in V} C_v)$ which is $O(N)$.

We can get stronger bounds by estimating the probability of deviating from the mean value. The usual procedure is to use Chernoff bounds but in our case it is complicated by the fact that we cannot get an explicit generating function for $d > 2$ (since it involves solving equations of high degree). Instead we take an indirect approach. The total space and time for data-structure construction is the sum of N independent and identical random variables

$Y_i, 1 \leq i \leq N$, each of which is the total progeny of a branching process.
If $A = \sum_i X_i$, then from Chernoff Bounds,
Probability$[A \geq X] \leq s^{-X} t(s)^N$,
where $t(s)$ is the generating function for each Y_i. For $X > cN$, for some fixed
c, this can be rewritten as

$$Prob[A \geq cN] \leq \left(\frac{t(s)}{s^c}\right)^N$$

We shall prove that for some $s > 1$, there exists a constant c such that
$t(s)/s^c < 1$. Let $F(s,t) = t - s(q + pt^2)^d$. Then

$$F_t(s,t) = 1 - 2tpds(q + pt^2)^{d-1}$$

Hence $F_t(1,1) = 1 - 2pd$ and for $2pd < 1$, $F_t(1,1) \neq 0$. From *Implicit Function theorem*, (see Appendix) it follows that there exists a neighbourhood of $(s = 1, t = 1)$, such that

$$F(s,t) = F(1,1) \Leftrightarrow t = t(s)$$

Since $F(1,1) = 0$ this implies that there is a value $s > 1 + \epsilon$ for which $t(s) < 1 + \delta$ for some $\epsilon, \delta > 0$. By choosing c large enough, $t(s)/s^c$ can be made less than 1 and hence the probability of deviation from mean decreases as $1/2^{\Omega(N)}$.

For each new record in A_v, we need $O(d)$ time to determine if it will be a bridge with respect to any of the d (maximum) neighbours. Moreover inserting a new bridge takes time proportional to gap-size whose expected value is $O(d)$. So the total time for inserting $O(N)$ bridges is a sum of $O(N)$ independent geometric random variables with parameter p. (Unlike the argument for ensuring search time with high-probability, where we needed to consider all distinct query-trees, we simply bound this sum.) This is $O(Nd)$ with probability $1 - 2^{-\Omega(N)}$ using standard techniques like Chernoff bounds.

To complete the analysis for the time bound for building the data-structure, we have to ensure that the total number of C-pointer updates is also $O(N)$. For any new record inserted into A_v from C_v, the total number of C-pointer updates can be bound by the total number of records (i.e. the space bound). The records of C_v are inserted in an increasing order and so any record in A_v has its C-pointer updated at most once (not including when it is first created). Hence we state the following result

Theorem 2 *Let G be a catalog graph of size N and locally bounded degree d. Our algorithm constructs a data-structure for iterative search in $O(N)$ space and $O(dN)$ time to do iterative search in expected time $O(\log N + d|\Pi|)$. The bounds for preprocessing time and space hold with probability $1 - 2^{-\Omega(N)}$.*

Remarks

(1) The bounds for preprocessing time and search time can be improved to $O(N)$ and $\tilde{O}(\log N + \log d|\Pi|)$ respectively by using the same modifications as Chazelle-Guibas to restructure the catalog graph to a fixed degree graph.

(2) The constants associated with the expected bounds for preprocessing time and space are lower than the deterministic construction.

(3) Chazelle-Guibas also arrived at the figure $4d$ for minimum gap size from the observation that otherwise their analysis yields infinite time and space bound for construction. However, they give examples where the actual algorithm halts even when they use gap sizes of less than $4d$. Our analysis captures a more fundamental reason for this phenomenon. Although the mean $\mu < 1$, guarantees that the process is finite, the process dies out with probability x where $x = G(x)$ even when $\mu > 1$. So one can have a gap length of less than $4d$ and still terminate. The motivation for this is clearly a reduction in search time which is inversely proportional to the gap size.

(4) To allow insertions/deletions from the catalogs, our procedure for maintaining the augmented catalogs readily dynamize. The arguments for query-time and the space-bound remain identical to the static case. Unfortunately (as in the case of [1]), the bottleneck is maintaining the correspondence between the A_v and C_v. In particular, we are unable to analyze the number of C-pointer updates in the case of inserting or deleting a record from C_v. Using the priority queue of Fries et al. [4] to maintain this correspondence, we get similar bounds. Both the search time and update times are off by an $O(\log \log N)$ factor from the best possible.

5 Concluding remarks

Two of the outstanding problems posed by [1] still remain open, namely, the issue of high local-degree graphs and maintaining same asymptotic bounds in the dynamic case. Both in the Chazelle-Guibas and our schemes, the gap-size seems to be inherently tied with the local degree of the graph. In this respect, our analysis seems to provide a more concrete explanation, namely its connection with the convergence of a branching process. We conjecture that the gap-size can be made proportional to the local degree at each node instead of the maximum local degree of the graph - however analyzing such a branching process appears to be very complicated.

For the dynamization part, we believe that our algorithm will perform very well even without the complicated schemes of Fries et al. although at this point we are unable to do a tight analysis of the behaviour.

Acknowledgement: The author is very grateful to Paul Wright who pointed out the use Implicit Function theorem, to show existence of the generating function for $s > 1$.

References

[1] B. Chazelle and L. Guibas, "Fractional Cascading: I. A Data Structuring Technique," Algorithmica, Vol. 1, 1986, pages 133-162.

[2] B. Chazelle and L. Guibas, "Fractional Cascading: II. Applications," Algorithmica, Vol. 1, 1986, pages 163-191.

[3] W. Feller, "An Introduction to Probability Theory and Applications Volume I," *Pub* John Wiley, 1968.

[4] O. Fries, K. Mehlhorn and S. Näher, "Dynamization of geometric data structures," Proc. 1st ACM Computational Geometry Symposium, 1985, pages 168-176.

[5] W. Pugh, "Skip Lists: A Probabilistic Alternative to Balanced Trees," Communications of the ACM, Volume 33 Number 6, June 1990, pp. 668-676.

[6] S. Sen, "Some observations on skip lists," Information Processing Letters, 1991.

A   Implicit function theorem

We use the standard notation f_y to denote the partial derivative of f with respect to variable y.

Theorem 3 *Let $f(x,y)$ be continuously differentiable in D. Let (x_o, y_o)be any point in D such that $f_y(x_o, y_o) \neq 0$. Then there exist numbers $\delta > 0$ and $\epsilon > 0$ and a continuously differentiable function $g(x)$ defined for $|x - x_o| \leq \delta$ and $|y - y_o| \leq \epsilon$, then $f(x,y) = f(x_o, y_o) \Leftrightarrow y = g(x)$.*

Dynamic 2- and 3-Connectivity on Planar Graphs *

(Preliminary Version)

Dora Giammarresi[1] and Giuseppe F. Italiano[2]

[1] Dipartimento di Matematica e Applicazioni, Università di Palermo, Italy
[2] IBM T.J. Watson Research Center, Yorktown Heights, NY 10598

Abstract. We study the problem of maintaining the 2-edge-, 2-vertex-, and 3-edge-connected components of a dynamic planar graph subject to edge deletions. The 2-edge-connected components can be maintained in a total of $O(n \log n)$ time under any sequence of at most $O(n)$ deletions. This gives $O(\log n)$ amortized time per deletion. The 2-vertex- and 3-edge-connected components can be maintained in a total of $O(n \log^2 n)$ time. This gives $O(\log^2 n)$ amortized time per deletion. The space required by all our data structures is $O(n)$.

1 Introduction

In the last decade there has been a growing interest in dynamic problems on graphs, sparkled by the work of Even and Shiloach [5] who studied the problem of maintaining the connected components of an undirected graph during edge deletions. The goal of a dynamic algorithm is to update the solution of a problem after dynamic changes (faster than solving again the problem from scratch). We say that a problem is *fully dynamic* if the update operations include both insertions and deletions of edges. A problem is called *partially dynamic* if only one type of update, i.e., either insertions or deletions, is allowed. A partially dynamic problem dealing with insertions only is called *incremental*; if there are deletions only we call it *decremental*.

In this paper we study how to maintain information about edge and vertex connectivity on planar graphs in a decremental fashion. We recall that a graph $G = (V, E)$ is *planar* if it can be drawn in the plane without edge crossings (that is, no two edges intersect except at a vertex at which they are both incident). Such a drawing is called a *planar embedding* of the graph. A planar graph with fixed embedding is called *embedded planar* or in short *plane*. A graph $G = (V, E)$ is *k-vertex-connected* if removing any $(k - 1)$ vertices leaves G connected. G is *k-edge-connected* if removing any $(k - 1)$ edges leaves G connected. Vertex- and edge-connectivity problems arise naturally in many applications and have been extensively studied.

* Research partially supported by the ESPRIT II Basic Research Actions Program of the EC under Project ALCOM (contract No. 3075) and Project ASMICS. Work done while the first author was visiting Columbia University, partially supported by a CNR Fellowship and the second author was on leave from Università di Roma, Italy.

Although many dynamic algorithms have been proposed for several types of connectivity [2, 3, 4, 6, 7, 8, 9, 11, 12, 14, 16], in general the best that one can do when trying to solve a decremental problem is to use the corresponding fully dynamic algorithm. For instance, the best bound per operation known up to date both for fully dynamic and decremental connectivity is $O(n^{1/2} \log(m/n))$ for general graphs [3] and $O(\log n)$ for plane graphs [4]. For 2-edge connectivity is $O(n^{1/2} \log(m/n))$ for general graphs [3] and $O(\log^2 n)$ for plane graphs [11]. For 2-vertex connectivity is either $O(n \log(m/n))$ [3] or $O(m^{2/3})$ [15] for general graphs, $O(n^{2/3})$ [9] for planar graphs, and $O(n^{1/2} \log n)$ [15] for plane graphs. For 3-edge connectivity is $O(n^{2/3})$ and for 3-vertex connectivity is $O(n \log(m/n))$ [3]. This is somewhat surprising, since the algorithms for all the corresponding incremental problems are much faster, namely $O(\alpha(q,n))$ per operation [13, 14, 16]. Moreover, the previous fully dynamic algorithms are not able to maintain explicitly the 2- and 3-connected components, but only to answer queries about the connectivity of two given vertices.

In this paper, we start investigating efficient algorithms for decremental graph problems. As a first step, we study three such problems on planar graphs: 2-edge connectivity, 2-vertex connectivity, and 3-edge connectivity. Notice also that decremental problems on planar and plane graphs are actually the same. Indeed, when we have edge deletions only, the initial embedding of the planar graph has no need to change. We present algorithms and data structures that maintain the 2-edge-connected components of a planar graph under any arbitrary sequence of $O(n)$ deletions in $O(n \log n)$ worst-case time. This gives $O(\log n)$ amortized time per deletion. Then we show how to maintain the 2-vertex- and 3-edge-connected components under $O(n)$ edge deletions in $O(n \log^2 n)$ worst-case time. This gives $O(\log^2 n)$ amortized time per edge deletion. The space usage of all our data structures is $O(n)$. Our data structures are able to support an extensive repertoire of operations, including checking in constant time whether two given vertices are in the same 2-edge-, 2-vertex-, or 3-edge-connected component, or reporting efficiently a possible connectivity cut separating the two vertices, or listing all the vertices in the same connectivity component as a given vertex. The new time bound for 2-edge connectivity improves known bounds from $O(\log^2 n)$ to $O(\log n)$. For the other two problems the improvement is even sharper: from either $O(n^{2/3})$ or $O(n^{1/2} \log n)$ to $O(\log^2 n)$.

2 Preliminaries

In this section we introduce some graph-theoretical terminology. We assume that the reader is familiar with the basic terminology, as contained for instance in [1]. Let $G = (V, E)$ be an undirected graph with m edges and n vertices. An edge set $E' \subseteq E$ is an *edge-cut* for G if the removal of all the edges in E' disconnects G. An edge-cut E' for G a *min-edge-cut* if there is no other edge-cut E'' such that $|E''| < |E'|$. A graph G is said to be *k-edge-connected* if all its edge-cuts have cardinality at least k. A min-edge-cut of cardinality 1 is called a *bridge*. It is possible to show that k-edge connectivity defines an equivalence relationship that partitions the vertices of a graph G into equivalence classes called the *k-edge-connected components* of G. Similarly, a vertex set $V' \subset V$ is a *vertex-cut* for G if and only if the removal of all the vertices in V' disconnects G. A vertex-cut V' for G is a *min-vertex-cut* if there is no other vertex-cut V'' such that $|V''| < |V'|$. A graph G is *k-vertex-connected* if and only if all its vertex-cuts have cardinality at least k. A min-vertex-cut of cardinality one is called *articulation point*: a graph with no articulation points is said to be *biconnected*. The *biconnected components* of a graph G are the maximal subgraphs of G which are biconnected. The 2-edge-connected and the biconnected components of a graph can be naturally represented by tree-like data structures, called respectively *bridge-block tree* and *block tree*. These data structures capture all the information about the corresponding connectivity cuts in G.

The *bridge-block tree* (or *tree of 2-edge-connected components*) of G is a tree composed of square nodes (corresponding to vertices in G) and of round nodes (corresponding to 2-edge-connected components in G). Whenever a vertex belongs to a given 2-edge-connected component, there is a tree edge between the corresponding square and round nodes in the bridge-block tree. Furthermore, for each bridge $e = (u, v)$ we add a tree edge between the square nodes corresponding to u and v. We refer to the bridge-block tree of G as $T_{2E}(G)$. Note that there are only two types of tree edges in $T_{2E}(G)$: edges between square nodes (referred to as *dashed edges*), and edges between a round node and a square node (referred to as *solid edges*). Dashed edges of $T_{2E}(G)$ correspond to bridges of G. Since 2-edge connectivity is an equivalence relationship, a vertex of G belongs exactly to one 2-edge-connected component. This implies that there is exactly one solid edge incident to a square node of $T_{2E}(G)$. We associate to each round node ρ of $T_{2E}(G)$ the corresponding 2-edge-connected component, referred to as the *pertinent graph of ρ*. Given $T_{2E}(G)$ it is possible to reassemble the 2-edge-connected components into the original graph using the extra information stored in the pertinent graphs. If G has n vertices and m edges, the total size of $T_{2E}(G)$ is $O(m + n)$ (see Figure 1).

The *block tree* (or *tree of biconnected components*) of G is again a tree composed of square nodes (corresponding to vertices in G) and round nodes (corresponding to biconnected components in G): whenever a vertex belongs to a given biconnected component, there is a tree edge between the corresponding square and round nodes in the block tree. We refer to the block tree of G as $T_{2V}(G)$. Note that in a block tree every path alternates between square and round vertices. Furthermore, non-leaf square nodes of $T_{2V}(G)$ correspond to articulation points of G, and vice versa. Once again, we associate to each round node ρ of $T_{2V}(G)$ the corresponding biconnected component, referred to as the *pertinent graph of ρ*: given $T_{2V}(G)$ it is possible to reassemble the biconnected components into the original graph. The total size of $T_{2V}(G)$ is $O(m + n)$ (see Figure 1).

We now introduce some terminology describing the embedding of a planar graph, that will be used by our data structures. Most of this notation is borrowed from Guibas and Stolfi [10]. Next, we relate min-cut sets of a planar graph to some adjacency properties in its embedding.

A *subdivision S of the plane* is a connected set of *vertices* and *edges* that partition the plane in a finite collections of connected components called *faces*. A connected embedded planar graph $G = (V, E)$ corresponds to a subdivision S of the plane. If the graph is not connected, it will generate a collection of subdivisions. In the remainder of this section, we will restrict our attention to connected graphs only (if the graph G is not connected, we will consider each connected component of G separately).

Each undirected edge $e = (u, v)$ of the subdivision S can be directed in two ways. If e is the directed version of e originating in u and terminating in v, then $Sym(\mathbf{e})$ is the version of e directed from v to u. For each directed edge e, we can unambiguously define its origin and destination as the vertices that e leaves and enters respectively. We refer to these vertices as $Orig(\mathbf{e})$ and $Dest(\mathbf{e})$ respectively. Note that $Dest(\mathbf{e}) = Orig(Sym(\mathbf{e}))$. For each directed edge e we can unambiguously define its left and right faces referred to as $Left(\mathbf{e})$ and $Right(\mathbf{e})$ respectively. Note that $Right(\mathbf{e}) = Left(Sym(\mathbf{e}))$. Given a vertex v, the *edge ring of v* is defined as the circular list (ordered counterclockwise) of directed edges originating from v. We will refer to this list as $EdgeRing(v)$. For each vertex v, we define the *face ring of v* as the circular list (ordered counterclockwise) of faces that are on the right of edges in $EdgeRing(v)$. We refer to this list as $FaceRing(v)$. For each vertex v, we define the *vertex ring of v* as the circular list (ordered counterclockwise) of vertices that are destination of edges in $EdgeRing(v)$. We refer to this list as $VertexRing(v)$ (see Figure 2). Given a face f, the *edge contour of f* is defined as a counterclockwise ordered circular list of directed edges e such that $Left(\mathbf{e}) = f$. We will refer to this list as $EdgeContour(f)$. For each face f, we define the *face contour of f* as the circular list (ordered counterclockwise) of faces that are on the right side of edges in $EdgeContour(f)$. We refer to this list as $FaceContour(f)$. The *vertex contour of f* is defined as the circular list (ordered counterclockwise) of vertices that are destination of edges in $EdgeContour(f)$. We refer to this list as $VertexContour(f)$ (see Figure 2). We now characterize topological properties of min-cut-sets in the embedding of a planar graph.

Fact 1. Given an embedded planar graph G, an edge e of G is a bridge if and only if $Left(\mathbf{e}) = Right(\mathbf{e})$.

Fact 2. Given an embedded planar graph G, a vertex v is an articulation point if and and only if there is a face f that appears at least twice in $FaceRing(v)$.

All the concepts defined here can be naturally represented by the *quad-edge* data structure of Guibas and Stolfi [10]. Throughout this paper, we assume that the embedding of a planar graph is always given by a quad-edge data structure. We refer the interested reader to reference [10] for the details of this representation.

3   2-Edge Connectivity

In this section we present our algorithm for maintaining the 2-edge-connected components of a planar graph $G = (V, E)$ under an arbitrary sequence of edge deletions. We give an algorithm that operates on connected graphs. This is without loss of generality since if either G was disconnected at the beginning or G becomes disconnected because of an edge deletion, we can add *dummy edges* between its connected components to make it connected. Let $\widehat{G}$ be the connected graph obtained from G after introducing the dummy edges. We will keep the invariant that these dummy edges are bridges in $\widehat{G}$. This has two consequences. First, the 2-edge-connected components of $\widehat{G}$ are the same as the 2-edge-connected components of G. Second, if G has n vertices, $\widehat{G}$ has still $O(n)$ vertices and edges. Our algorithm will maintain $\widehat{G}$.

We maintain the 2-edge-connected components of $\widehat{G}$, by maintaining the bridge-block tree $T_{2E}(\widehat{G})$ (recall that the 2-edge-connected components of $\widehat{G}$ are explicitly maintained as the pertinent graphs of round nodes in $T_{2E}(\widehat{G})$). This allows us to perform several operations efficiently, such as testing whether two vertices are 2-edge-connected, or reporting a bridge separating two given vertices. For instance, to test whether two given vertices x and y are 2-edge-connected, we access the square nodes x and y in $T_{2E}(\widehat{G})$. By definition of bridge-block tree, x and y are in the same 2-edge-connected component of $\widehat{G}$ if and only if they are adjacent to the same round node of $T_{2E}(\widehat{G})$. As mentioned in Section 2, a square node u of a bridge-block tree is adjacent to exactly one round node, say ρ, and ρ can be found by following the unique solid edge leaving u. All this can be accomplished in $O(1)$ time.

The other data structures needed are the following. We maintain for each undirected edge $e = (u, v)$ in $\widehat{G}$ the two directed versions of this edge: e_u directed from u to v, and e_v directed from v to u. Given either one of the two directed edges, it is easy to locate the other in constant time since $e_u = Sym(e_v)$ and $e_v = Sym(e_u)$. For each directed edge e we maintain $Left(e)$ and $Right(e)$, respectively the left and right faces of e. By Fact 1, this information is crucial for 2-edge connectivity, since e is a bridge in $\widehat{G}$ if and only if both its directed versions are in the edge contour of the same face. As mentioned in Section 2, maintaining information about both $Left(e)$ and $Right(e)$ is redundant, since $Right(e)$ can be computed as $Left(Sym(e))$. However, this makes the description of our algorithms simpler. Furthermore, for each face f of $\widehat{G}$ we maintain the set $EdgeContour(f)$. We maintain also the cardinality of $EdgeContour(f)$, referred to as $|EdgeContour(f)|$. Our implementation of $EdgeContour(f)$ is a list L subject to the following three types of primitives. Primitive $rotate(i, L)$ cyclically permutes the order of the list L so that item i appears first. Primitive $pop(L)$ deletes the first item from a non-empty list L. Primitive $merge(L_1, L_2)$ combines two lists L_1 and L_2 into one. Using a doubly-linked implementation of a list, each $rotate$, pop and $merge$ can be supported in $O(1)$.

We now describe in detail how to delete an edge $e = (x, y)$. Let e and $Sym(e)$ be the two directed edges corresponding to undirected edge e. We pick e and check first whether $Left(e) = Right(e)$. If this is the case, then by Fact 1 e is a bridge in $\widehat{G}$. Consequently, the deletion of e does not affect the 2-edge-connected components of $\widehat{G}$. However, the removal of e would disconnect $\widehat{G}$. We do not remove bridge e but rather we mark it as a dummy edge.

Assume now that e is not a bridge, i.e., $Left(e) \neq Right(e)$. Let $f_l = Left(e)$ and $f_r = Right(e)$. Because of the deletion of e, the two faces f_l and f_r have to be merged into one. To do this, it suffices to rename either all occurrences of f_l in our data structure as f_r or all occurrences of f_r as f_l. We exploit the freedom implicit in renaming either face by renaming the smallest. Without loss of generality assume that $|EdgeContour(f_l)| \leq |EdgeContour(f_r)|$, otherwise interchange f_l and f_r in what follows. We now show how to rename face f_l as f_r and how to compute the new bridges caused by the deletion of (x, y). We initialize a set $NewBridges$ to $\emptyset$. For each e_i in $EdgeContour(f_l)$, we do the following. If $Right(e_i) = f_r$, then e_i was one of the edges between faces f_l and f_r. After deleting (x, y) and merging the two faces f_l and f_r, by Fact 1 the undirected edge e_i corresponding to e_i will become a bridge:

we add e_i to *NewBridges*. In any case, we set $Left(e_i)$ and $Right(Sym(e_i))$ to f_r. To compute the edge contour of the new face obtained by merging the two faces f_l and f_r, we first delete e and $Sym(e)$ from $EdgeContour(f_l)$ and $EdgeContour(f_r)$ respectively, and then merge the two edge contours. This can be done by executing a constant number of *rotate, pop* and *merge* operations.

At the end of this step all and only the new bridges caused by the deletion of (x,y) are in the set *NewBridges*. If *NewBridges* is empty, then deleting (x,y) does not change the 2-edge-connected components of $\widehat{G}$ and we stop. Assume that $NewBridges = \{e_1, e_2, \ldots, e_p\}$, $p \geq 1$. Note that all the new bridges must have been before in a same cycle. Consequently, the new bridges are splitting one existing 2-edge-connected component, say B_k. Let $B_{k_1}, B_{k_2}, \ldots, B_{k_{p+1}}$, $p \geq 1$, be the new 2-edge-connected components. Then, we have that the new bridge e_i is between B_{k_i} and $B_{k_{i+1}}$, $1 \leq i \leq p$, and the old 2-edge-connected component B_k is broken into a chain of new 2-edge-connected components (see Figure 3).

We start a visit of the old 2-edge-connected component B_k with the help of the new bridges e_i, $1 \leq i \leq p$. We interleave the traversals in the new 2-edge-connected components, as follows. As said before, let $e_1, e_2 \ldots e_p$ be the new bridges created by the deletion of (x,y) as encountered counterclockwise in the previous face f_l. Let u_i and v_i be the two endpoints of e_i, $1 \leq i \leq p$, $u_i \in B_{k_i}$ and $v_i \in B_{k_{i+1}}$. To visit the new 2-edge-connected component B_{k_i}, we visit the old 2-edge-connected component B_k starting from vertex u_i, $2 \leq i \leq p$, without traversing edge $e_{i-1} = (u_{i-1}, v_{i-1})$. To visit B_{k_1} and $B_{k_{p+1}}$ we start respectively from u_1 and v_p, without traversing edge (x,y). We carry out all these traversals simultaneously, in the sense that we alternate between them one step at the time. Whenever all of them but one are finished, we stop. Let B_{k_j}, for some $1 \leq j \leq p+1$, be the largest new 2-edge-connected component. At this point we have computed B_{k_i}, $1 \leq i \leq p+1$, $i \neq j$. B_{k_j} can be obtained by simply deleting from the old 2-edge-connected component B_k the new bridges plus all the vertices and edges of B_{k_i}, $1 \leq i \leq p+1$, $i \neq j$. This requires total time $O(\sum_{i \neq j} |B_{k_i}|)$.

Once the new 2-edge-connected components have been located, we update the bridge-block tree $T_{2E}(\widehat{G})$. Let ρ_k be the round node that corresponded to the old 2-edge-connected component B_k. Then ρ_k must be replaced in $T_{2E}(\widehat{G})$ with round nodes $\rho_{k_1}, \rho_{k_2}, \ldots, \rho_{k_{p+1}}$ corresponding to the new 2-edge-connected components, and dashed edges corresponding to the new bridges $e_1, e_2, \ldots, e_p$ must be inserted in the bridge-block tree. We refer to this operation as *breaking a round node*. We introduce p new round nodes ρ_{k_i}, such that ρ_{k_i} corresponds to the new 2-edge-connected component B_{k_i}, $1 \leq i \leq p+1$, $i \neq j$. For each B_{k_i}, $1 \leq i \leq p+1$, $i \neq j$, we do the following. We scan vertices of B_{k_i}, and for each vertex $w \in B_{k_i}$, we delete the solid edge (w, ρ_k) and insert a new solid edge (w, ρ_{k_i}) in $T_{2E}(\widehat{G})$. After all the vertices of B_{k_i}, $i \neq j$, have been moved, the square nodes incident to ρ_k correspond to all and only the vertices of B_{k_j}: we therefore change the name of this round node from ρ_k to ρ_{k_j}. At the end of this step, we have $(p+1)$ trees containing round nodes ρ_{k_i}, $1 \leq i \leq p+1$. We combine them into one tree by introducing dashed edges (u_i, v_i), $1 \leq i \leq p$, corresponding to the new bridges. Given the above implementation, the operation of breaking round node ρ_k can be accomplished in time $O(\sum_{i \neq j} |B_{k_i}|)$.

Theorem 3. *Let $G = (V, E)$ be a planar graph with n vertices. The 2-edge-connected components of G can be maintained in a total of $O(n \log n)$ time during an arbitrary sequence of edge deletions. The total space and preprocessing time required is $O(n)$.*

Proof. It is easy to see that the total space complexity and preprocessing required by our data structures is $O(n)$. Consider now the deletion of an edge $e = (x, y)$. First, we compute $f_\ell = Left(e)$ and $f_r = Right(e)$. If $f_\ell = f_r$ we only mark bridge e as dummy and stop. This clearly requires $O(1)$ time. If $f_\ell \neq f_r$, we first compute the possible new bridges created by the deletion of e. As shown before, this can be done in time proportional to the smaller face, namely in $O(|EdgeContour(f_\ell)|)$ if $|EdgeContour(f_\ell)| \leq |EdgeContour(f_r)|$. Then, we merge the two faces f_ℓ and f_r into one. This involves a constant number of *rotate*, *pop*, and *merge* primitives and therefore can be implemented in constant time. Last, we have to break the round node ρ_k of $T_{2E}(G)$. Let $T(n)$ be the total time spent in computing the new bridges and in breaking round nodes over a sequence of $O(n)$ edge deletions. Then any sequence of $O(n)$ edge deletions requires $O(n + T(n))$ time. $T(n)$ can be computed as follows. To bound the total time needed to find the new bridges, we have to bound the size of the edge contour of the face that gets renamed. Consider each time an edge e_i is moved from the edge contour of a face f_ℓ to the edge contour of a face f_r. Note that we do this only when $|EdgeContour(f_\ell)| \leq |EdgeContour(f_r)|$. As a result, after this move e_i finds itself in a new edge contour which is at least twice as larger as before. This gives at most $O(\log n)$ different moves for each edge, and therefore a total of $O(n \log n)$ time. Consider now the breaking of a round node into $(p + 1)$ round nodes. In our implementation, this requires time linear in the size of the p smallest new 2-edge-connected components. Let v be a vertex that belongs to one of the smallest p new 2-edge-connected components. After the update, v is in a new 2-edge-connected component which is at least twice as smaller as before. This yields $T(n) = O(n \log n)$.

4 2-Vertex and 3-Edge Connectivity

We now present our data structures and algorithms to maintain the biconnected components of a planar graph $G = (V, E)$ under an arbitrary sequence of edge deletions. Similarly to Section 3, we present an algorithm that works on connected graphs. Once again, let G be the original graph, and let $\widehat{G}$ be the augmented connected graph obtained after introducing dummy edges. This time our algorithm needs to be more careful in maintaining $\widehat{G}$, since the biconnected components of $\widehat{G}$ are not the same as the biconnected components of G. We recall that each dummy edge of $\widehat{G}$ is a bridge and therefore it is a biconnected component by itself. Since this is the only change in the biconnected components, two vertices are 2-vertex-connected in G if and only if they are 2-vertex-connected in $\widehat{G}$ but are not endpoints of the same dummy edge.

We maintain the biconnected components of $\widehat{G}$ by maintaining the block tree $T_{2V}(\widehat{G})$ of $\widehat{G}$. Since dummy edges are bridges, each dummy edge corresponds to a round node in $T_{2V}(\widehat{G})$. We mark such round nodes as *dummy*. Throughout the algorithm, we maintain the block tree $T_{2V}(\widehat{G})$ rooted arbitrarily at any round node.

Note that the biconnected components of $\widehat{G}$ are maintained explicitly as the pertinent graphs of the round nodes of $T_{2V}(\widehat{G})$. Maintaining $T_{2V}(\widehat{G})$ allows us answer efficiently questions on the 2-vertex connectivity of G. For instance, we can check in $O(1)$ time whether two vertices x and y are in the same biconnected component of G. We first check whether the two corresponding square nodes x and y are adjacent to a same round node as follows. We compute ρ_x and ρ_y, the parent (round) nodes of x and y respectively. We then compute p_x and p_y, respectively the parents of ρ_x and ρ_y. Square nodes x and y are adjacent to the same round node of $T_{2V}(\widehat{G})$ if and only if one of the following is true: (i) $\rho_x = \rho_y$; or (ii) p_x is defined and $p_x = y$; or (iii) p_y is defined and $p_y = x$. If x and y are not adjacent to the same round node, then x and y are not in the same biconnected component of G. Otherwise, let ρ be the round node between x and y: x and y are in the same biconnected component if and only if ρ is not dummy.

We maintain for $\widehat{G}$ the data structures defined in Section 3. That is, we maintain for each directed edge e $Left(e)$ and $Right(e)$, and for each face f the list $EdgeContour(f)$ subject to $rotate$, pop, and $merge$ primitives. In addition, we maintain $Dest(e)$ for each directed edge e. This allows us to compute for each face f $VertexContour(f)$ from $EdgeContour(f)$: namely we apply $Dest$ to each edge in $EdgeContour(f)$. Therefore, although our algorithm makes use of the vertex contours, it does not need to maintain them explicitly. Finally, we maintain for each vertex v $FaceRing(v)$. The primitives we need to support on $FaceRing(v)$ are lookups and updates (both inserts and deletes). We keep $FaceRing(v)$ as a balanced search tree, which allows us to support all these primitives in $O(\log n)$ time each.

We now describe in detail how to delete an edge $e = (x, y)$. The updates needed in the data structures are similar to the updates described in Section 3 for 2-edge connectivity. This time the details are somewhat more involved, however. If e is a bridge, then we mark e and the corresponding round node in the block tree as dummy. If e is not a bridge, then let $f_l = Left(e)$ and $f_r = Right(e)$. By Fact 1, $f_l \neq f_r$. As usual assume that $|EdgeContour(f_l)| \leq |EdgeContour(f_r)|$. Let v be a vertex of $\widehat{G}$. If the number of biconnected components of $\widehat{G}$ containing v increases after the deletion of edge $e = (x, y)$, we say that v is a *new articulation point*. The algorithm consists of four steps: (1) finding the new articulation points; (2) finding the new biconnected components; (3) updating the block tree $T_{2V}(\widehat{G})$; and (4) updating the graph $\widehat{G}$ by merging the two faces f_l and f_r together.

We begin by describing how to perform step (1). Because of the deletion of (x, y), the two faces f_l and f_r have to be merged into one face. Since this is the only change in the face set of $\widehat{G}$, by Fact 2, the new articulation points induced by the deletion of edge (x, y) are all and only the vertices that had previously both faces f_l and f_r in their face rings. We initialize a set $NewArtPoints$ to $\emptyset$. We then examine $VertexContour(f_l)$ (which can be computed by applying the operator $Dest$ to $EdgeContour(f_l)$). For each vertex v in $VertexContour(f_l)$ we check whether f_r belongs to $FaceRing(v)$. If $f_r \in FaceRing(v)$, then by Fact 2, v will be a new articulation point: we add v to $NewArtPoints$. At the end of this step, $NewArtPoints$ contains all and only the new articulation points caused by the deletion of edge (x, y).

We now describe step (2). If $NewArtPoints = \emptyset$, the deletion of (x, y) does not create new articulation points, and we stop. Assume that $NewArtPoints =$

$\{v_1, v_2, \ldots, v_p\}$, $p \geq 1$. Note that all the new articulation points must have been before in a same simple cycle containing the deleted edge (x, y). Consequently, the new articulation points v_i, $1 \leq i \leq p$, are splitting one existing biconnected component, say B_k. Let $B_{k_1}, B_{k_2}, \ldots, B_{k_{p+1}}$, $p \geq 1$, be the new biconnected components. Then, we have that the new articulation point v_i is between B_{k_i} and $B_{k_{i+1}}$, $1 \leq i \leq p$, and the old biconnected component B_k is broken into a chain of new biconnected components (see Figure 4). Define $v_0 = y$ and $v_{p+1} = x$. We compute the new biconnected components with the help of vertices v_i, $0 \leq i \leq p+1$. We distinguish among new biconnected components that consist of only one edge, referred to as *trivial biconnected components*, and *non-trivial new biconnected components*. The new trivial biconnected components can be easily computed as the new bridges caused by the deletion of (x, y). This task can be accomplished as shown in Section 3. We now show how to compute a non-trivial new biconnected component B_{k_i}, $1 \leq i \leq p+1$. We first pick a starting vertex s_i in B_{k_i} such that $s_i \neq v_i, v_{i+1}$ (such a vertex can always be computed in constant time). Next, we start a traversal of B_{k_i} starting from vertex s_i, similarly to what we did in Section 3.

Once the new biconnected components have been located, we perform step (3) by updating the block tree $T_{2V}(\widehat{G})$. Namely, the round node ρ_k corresponding to biconnected component B_k has to be replaced with a chain of round and square nodes corresponding to $B_{k_1}, v_1, B_{k_2}, v_2, \ldots, v_{p-1}, B_{k_p}, v_p, B_{k_{p+1}}$. The remaining square nodes that were previously incident to ρ_k become incident to the new biconnected component they belong to. This can be done by using a similar technique to the one shown in Section 3 to update the bridge-block tree $T_{2E}(\widehat{G})$. However, since $T_{2V}(\widehat{G})$ is maintained rooted, we have to direct properly all the new edges introduced in $T_{2V}(\widehat{G})$. We do this as follows. Denote by $p(\rho_k)$ the square node that was the parent of ρ_k in $T_{2V}(\widehat{G})$. We define $p(\rho_k)$ to be the *principal node* of ρ_k. If ρ_k was the root, $p(\rho_k)$ is undefined: in this case we define ρ_k to be the *principal node* of itself. Let w be an old vertex of B_k, $w \neq v_i$, $1 \leq i \leq p$, $w \neq p(\rho_k)$. If w is moved from ρ_k to some ρ_{k_i}, we keep the same direction of the edge: i.e., (w, ρ_k) was directed from w to ρ_k, and (w, ρ_{k_i}) will be directed from w to ρ_{k_i}. If $w = p(\rho_k)$, we keep again the same direction of the edge: (w, ρ_k) was directed from ρ_k to w, and (w, ρ_{k_i}) will be directed from ρ_{k_i} to w. At this point, we have still to re-direct edges in the chain $v_1, \rho_{k_1}, v_2, \ldots v_p, \rho_{k_{p+1}}$. We direct all the edges of the chain toward the principal node. This completes the update of the block tree. Note that this step can be accomplished in a total time that is linear in the size of the p smallest new biconnected components.

After the block tree has been updated, we perform step (4) and update $\widehat{G}$. The deletion of (x, y) causes the two faces f_l and f_r to be merged into one. The name of the new face will be the name of the larger face: f_r. The updates in $\widehat{G}$ are of the following two kinds. First, we have to compute the edge contour of the new face. Second, we have to change into f_r all the occurrences of f_l in the data structure. To compute the edge contour of the new face in $\widehat{G}$, we merge the two edge contours of f_l and f_r. This can be done by means of a constant number of *rotate*, *pop*, and *merge* primitives exactly as shown in Section 3. Updating to f_r all the previous occurrences of f_l in the data structure can be done as follows. The data structures used that store faces are the face rings and the fields *Left* and *Right* of each edge. The update in the fields *Left* and *Right* can be done exactly as shown in Section 3. The update of the face rings can be done as follows. By definition, the only vertices whose face rings contained f_ℓ were previously in the vertex contour of f_ℓ. For each vertex v in $VertexContour(f_\ell)$, we change f_ℓ to f_r in $FaceRing(v)$. Note that because of the actual deletion of edge (x, y), f_ℓ must be deleted (and not replaced) in $FaceRing(x)$ and $FaceRing(y)$. The above algorithm has the following time complexity.

Theorem 4. *Let $G = (V, E)$ be a planar graph with n vertices. The biconnected components of G can be maintained in a total of $O(n \log^2 n)$ time during an arbitrary sequence of edge deletions. The space usage is $O(n)$, while the time required for preprocessing is $O(n \log n)$.*

Proof. As in the proof of Theorem 3, the space required by the block tree (together with its pertinent graphs), all the edge contours and all the face rings of $\widehat{G}$ is $O(n)$. Again, all the data structures but the face rings can be initialized in $O(n)$ time. Since the face rings are balanced search trees, and there are at most $O(n)$ items in all the face rings, the time required for their initialization is $O(n \log n)$. We now bound the time required to process any sequence of $O(n)$ edge deletions. Consider the deletion of edge $e = (x, y)$. As usual denote by $f_\ell = Left(e)$ and $f_r = Right(e)$ the faces at the two sides of edge e. The algorithm consists of four steps: (1) finding the new articulation points $v_1, v_2, \ldots v_p$; (2) finding the new biconnected components $B_{k_1}, B_{k_2}, \ldots, B_{k_{p+1}}$; (3) updating the block tree $T_{2V}(\widehat{G})$; and (4) updating $\widehat{G}$. We first consider step (1). Let f_ℓ be the smaller face: $|EdgeContour(f_\ell)| \le |EdgeContour(f_r)|$. We compute $VertexContour(f_\ell)$ in linear time by using the *Dest* field of each directed edge in $EdgeContour(f_\ell)$. Note that $|VertexContour(f_\ell)| = |EdgeContour(f_\ell)|$. For each vertex v in $VertexContour(f_\ell)$ we check whether v is a new articulation point. This is done in $O(\log n)$ time by checking whether f_r is in $FaceRing(v)$. As a result, the total time required to compute the new articulation points caused by the deletion of edge (x, y) is $O(|EdgeContour(f_\ell)| \log n)$. Note that this bound differs from the bound achieved in Section 3 to compute the new bridges because of the extra $\log n$ factor. Applying now the same argument given in the proof of Theorem 3 yields a total of $O(n \log^2 n)$ time for step (1) during any sequence of edge deletion. As for steps (2) and (3), the same argument given in Theorem 3 shows that the overall time taken by these steps during any sequence of edge deletions is $O(n \log n)$. Finally we consider step (4), which consists of two subtasks: merging the two faces f_ℓ and f_r, and replacing the occurrences of f_ℓ with f_r in the data structure. Merging the two faces f_ℓ and f_r can be performed exactly in the same way as in the case of 2-edge connectivity, and therefore yields a total of $O(n \log n)$ over any sequence of edge deletions. The new part in the update of the data structure is changing to f_r all the previous occurrences of f_ℓ in the face rings. As said before, this can be done by either updating or deleting an item in the face ring of vertices of $VertexContour(f_\ell)$. Since each primitive on a face ring can be supported in $O(\log n)$, this can be accomplished in $O(|EdgeContour(f_\ell)| \log n)$ time while performing step (1). This yields a total of $O(n \log^2 n)$ time.

The same bounds of Theorem 4 hold for the problem of maintaining the 3-edge-connected components of a planar graph. The ideas and techniques underlying our algorithm have much the same flavor as the ones described for 2-edge and 2-vertex connectivity. However, this time the details are much more involved, and they will be given in the full paper.

Theorem 5. *Let $G = (V, E)$ be a planar graph with n vertices subject to edge deletions. The 3-edge-connected components of G can be maintained in $O(n \log^2 n)$ time during an arbitrary sequence of $O(n)$ edge deletions. The total space is $O(n)$ and the preprocessing time required is $O(n \log n)$.*

We mention that the bounds given in Theorems 4 and 5 can be reduced to $O(n \log n)$ worst-case time at the expense of $O(n^2)$ space, or to $O(n \log n)$ expected time with $O(n)$ space.

References

1. A. V. Aho, J. E. Hopcroft, and J. D. Ullman. *The Design and Analysis of Computer Algorithms.* Addison-Wesley, Reading, MA, 1974.
2. G. Di Battista and R. Tamassia. On-line graph algorithms with SPQR-trees. In *Proc. 17th ICALP*, 598–611, 1990.
3. D. Eppstein, Z. Galil, G. F. Italiano, and A. Nissenzweig. Sparsification – A technique for speeding up dynamic graph algorithms. Manuscript, 1992.
4. D. Eppstein, G. F. Italiano, R. Tamassia, R. E. Tarjan, J. Westbrook, and M. Yung. Maintenance of a minimum spanning forest in a dynamic plane graph. *J. Algorithms*, 13:33–54, 1992.
5. S. Even and Y. Shiloach. An on-line edge deletion problem. *J. Assoc. Comput. Mach.*, 28:1–4, 1981.
6. G. N. Frederickson. Data structures for on-line updating of minimum spanning trees. *SIAM J. Comput.*, 14:781–798, 1985.
7. G. N. Frederickson. Ambivalent data structures for dynamic 2-edge-connectivity and k smallest spanning trees. In *Proc. 32nd FOCS*, 632–641, 1991.
8. Z. Galil and G. F. Italiano. Fully dynamic algorithms for edge connectivity problems. In *Proc. 23rd STOC*, 317–327, 1991.
9. Z. Galil and G. F. Italiano. Maintaining biconnected components of dynamic planar graphs. In *Proc. 18th ICALP*, 339–350, 1991.
10. L. Guibas and J. Stolfi. Primitives for the manipulation of general subdivision and the computation of Voronoi diagrams. *ACM Trans. on Graphics*, 4:74–123, 1985.
11. J. Hershberger, M. Rauch, and S. Suri. Fully dynamic 2-connectivity on planar graphs. In *Proc. 3rd SWAT*, 1992.
12. A. Kanevsky, R. Tamassia, G. Di Battista, and J. Chen. On-line maintenance of the four-connected components of a graph. In *Proc. 32nd FOCS*, 793–801, 1991.
13. J. A. La Poutré. *Dynamic graph algorithms and data structures.* PhD Thesis, Department of Computer Science, Utrecht University, September 1991.
14. J. A. La Poutré. Maintainance of triconnected components of graphs. In *Proc. 19th ICALP*, 1992.
15. M. Rauch. Personal communication, 1992.
16. J. Westbrook and R. E. Tarjan. Maintaining bridge-connected and biconnected components on-line. *Algorithmica*, 7:433–464, 1992.

This article was processed using the LaTeX macro package with LLNCS style

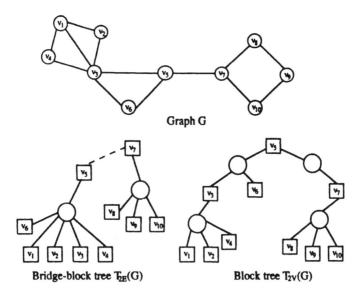

Graph G

Bridge-block tree $T_{2E}(G)$ Block tree $T_{2V}(G)$

Figure 1

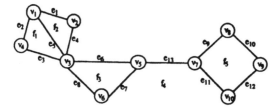

EdgeContour(f_2) = {e_1, e_5, e_4}   EdgeRing(v_3) = {e_8, e_6, e_4, e_5, e_3}

VertexContour(f_2) = {v_1, v_3, v_2}   VertexRing(v_3) = {v_6, v_5, v_2, v_1, v_4}

FaceContour(f_2) = {f_4, f_1, f_4}   FaceRing(v_3) = {f_4, f_3, f_4, f_2, f_1}

Figure 2

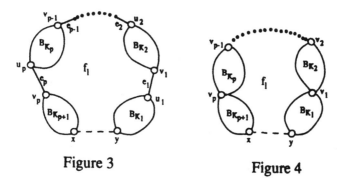

Figure 3   Figure 4

Fully Dynamic 2-Edge-Connectivity
in Planar Graphs

John Hershberger[1], Monika Rauch[2] and Subhash Suri[3]

[1] DEC Systems Research Center
[2] Princeton University
[3] Bell Communications Research

Abstract. We propose a data structure for maintaining 2-edge connectivity information dynamically in a planar graph. The data structure requires linear storage and preprocessing time for its construction, supports online updates (insertion and deletion of an edge) in $O(\log^2 n)$ time, and answers a query (whether two vertices are in the same 2-connected component) in $O(\log n)$ time. The previous best algorithm for this problem required $O(\log^3 n)$ time for updates.

1 Introduction

Connectivity in graphs is an important class of problems that has received considerable attention since the early work of Hopcroft and Tarjan, who designed linear-time algorithms for computing bi- and tri-connected components of a graph [1, 7, 10]. *NC* algorithms for two and three-connectivity have been proposed in [8, 11]. In recent years, attention has turned toward dynamic algorithms for graph connectivity. These algorithms maintain connectivity information as the underlying graph is modified by the insertion/deletion of edges or vertices. If only insertions or only deletions are permitted, then the algorithm is called *semi-dynamic*, and if both insertions and deletions are allowed, the algorithm is called *fully dynamic*. In this paper, we present a fully dynamic algorithm for maintaining 2-edge connectivity in embedded planar graphs. Our data structure requires linear space and preprocessing time, supports two-edge-connectivity queries in $O(\log n)$ time, and requires $O(\log^2 n)$ update time to insert or delete an edge.

There has been a considerable amount of work on dynamic graph algorithms in recent years. The 1-connectivity problem is to maintain the connected components of a graph under insertion/deletion of edges. If only insertions are allowed, then this problem reduces to the disjoint set union problem, and hence a sequence of n insertions and queries can be processed in $O(n\alpha(n))$ time, using the union-find data structure [9]. The problem becomes significantly harder if both insertions and deletions are allowed, and the best result to date for general graphs is an algorithm due to Frederickson that takes $O(\sqrt{m})$ time per update, where m is the number of edges in the graph. If the graph is planar and embedded, then a result of Eppstein et al. [2] achieves $O(\log n)$ time per operation for the 1-connectivity problem.

The semi-dynamic case of the 2-connectivity problem was considered by Westbrook and Tarjan [12], and they showed that a sequence of n insertions and queries can be processed in $O(n\alpha(n))$ time. In a recent breakthrough, Galil and Italiano [5] managed to obtain a sublinear time algorithm for fully dynamic 2-edge connectivity. For general graphs, their algorithm takes $O(m^{2/3})$ time per operation (update or query), where m is the current number of edges in the graph. For planar graphs, their time complexity improves to $O(\sqrt{n}\log\log n)$. Soon afterwards, Frederickson [4] improved the time bound in [5] to $O(\sqrt{m})$ per operation. Frederickson also presented a faster algorithm for planar embedded graphs, with query time $O(\log n)$ and update time $O(\log^3 n)$.

The main result of our paper improves the planar graph result of Frederickson by a factor of $\log n$. Our data structure supports online updates (insertion and deletion of an edge) in $O(\log^2 n)$ time, and answers a query whether two vertices are in the same 2-connected component in $O(\log n)$ time. The data structure can be built in linear time and requires linear storage.

Our algorithm introduces two new concepts, *edge bundles* and *coverage graph recipes*, which appear to be of general interest and may find applications in other planar graph algorithms. The former is a method for collapsing and manipulating edges that belong to the same equivalence class, for a given partitioning of vertices. The second is a method for compressing and uncompressing portions of a planar graph, for an efficient traversal of the topology tree. These concepts are the keys to our improved data structure.

2 Preliminaries

Let $G = (V, E)$ be an undirected planar graph, embedded in the plane. The initial embedding of G remains fixed throughout the course of the algorithm, and all the updates must respect the embedding. We let n denote the number of vertices of G and m the number of edges; the number of edges changes with updates, but planarity implies that $m \leq 3n - 6$.

We perform a standard transformation on G to convert it into a graph with maximum vertex-degree 3.[4] Suppose $v \in V$ is a vertex of degree $d > 3$, adjacent to vertices $u_1, u_2, \ldots, u_d$ in this cyclic order. In the transformed graph, the vertex v is replaced by a cycle $(v_1, v_2, \ldots, v_d)$, and the edge (v, u_j), for $1 \leq j \leq d$, is replaced by the edge (v_j, u_j). In order to maintain the vertex identity, we label exactly one vertex on the cycle to be v, and leave the others unlabeled. At v we also store an ordered list of all the other vertices on the cycle. Observe that the transformed graph has $O(m)$ vertices, and that it inherits the 2-edge-connectivity properties of the original graph: two vertices are 2-edge connected in G if and only if they are 2-edge connected in the transformed graph [6]. Thus, from now on, we assume that G has maximum vertex degree 3. Furthermore, since we will be primarily concerned with edge-connectivity, we use the term 2-connectivity instead of 2-edge connectivity.

Our data structure is based on a spanning tree of G. We use the notation T for the current spanning tree of G, and the notation $\pi(u, v)$ for the unique path in T between two vertices u and v. Our first lemma establishes the important connection between 2-connectivity and *edge-covering*. We say that a tree edge (x, y) is *covered*

---

[4] To maintain textual consistency and readability, we adopt the convention that vertex degrees are written as numerals ("1, 2, 3" rather than "one, two, three") in this paper.

by a *non-tree edge* (u, v) if (x, y) lies on $\pi(u, v)$. The edge (x, y) is *covered* if any non-tree edge covers it. A tree edge is called a *bridge* if it is not covered.

Lemma 1. *Two vertices u and v in G are 2-connected if and only if every edge of the tree path $\pi(u, v)$ is covered.*

3 Topology trees

We build a hierarchical representation of G based on the spanning tree T. The representation is a tree, called the *topology tree*, that has depth $O(\log n)$ [3, 4]. Each level of the topology tree partitions the vertices of G into connected subsets, called *clusters*. Two clusters are said to be *adjacent* if they are joined by an edge in the spanning tree T. The *external degree* of a cluster is the number of tree edges with exactly one endpoint inside the cluster. Our clusters will have maximum external degree 3. We now describe the rules for building the topology tree.

Every leaf of the topology tree corresponds to a cluster at level 0, which consists of a singleton vertex. An internal node of the topology tree corresponds to either a cluster of lower level, or the union of two adjacent lower level clusters subject to an external degree constraint. More precisely, a cluster at level i, for $i > 0$, is formed by either

1. the union of two *adjacent* clusters of level $i - 1$ such that the external degree of one cluster is 1 or the external degree of both is 2, or
2. one cluster of level $i - 1$, if the previous rule does not apply.

A cluster of level $i - 1$ belongs to exactly one cluster of level i, thus ensuring that the vertices of G are partitioned at each level.

In the topology tree, there is a node corresponding to each cluster that we form. The ancestor-descendant relationships in the tree encode the cluster containment information — the node corresponding to a cluster C at level i is the parent of the (one or two) clusters of level $i - 1$ whose union produced C. The greedy method of forming clusters reduces the number of clusters at each level by a constant fraction, thus ensuring that the height of the topology tree is $O(\log n)$.

Lemma 2 [4]. *The topology tree for the graph G has height $\Theta(\log n)$.*

When G is updated, the spanning tree T may change, and hence the topology tree may need restructuring. All restructuring operations follow a common routine: we break up all clusters containing some constant number of vertices, and then recombine the unbroken clusters, perhaps in different combinations. In terms of the topology tree, this corresponds to removing all the nodes along a constant number of root-to-leaf paths, transplanting certain subtrees, and rebalancing the tree. To facilitate our discussion, we introduce the following terminology. To *expand the topology tree* at a vertex u means that we break up all the clusters that contain u — observe that the clusters containing u correspond precisely to the nodes lying along the path between u and the root of the topology tree. We denote by c_i a cluster at level i and by $c_i(u)$ the cluster at level i that contains the vertex u. A level $i - 1$ cluster c_{i-1} contained in c_i is called a *sub-cluster* of c_i.

Let us consider the expansion of the topology tree at a vertex u. We start from the root, whose level is defined to be k, and walk down the tree, expanding the cluster $c_i(u)$ at each level. If $c_i(u)$ has only one child $c_{i-1}(u)$, we replace $c_i(u)$ by its child. Otherwise, $c_i(u)$ has two children, and we replace $c_i(u)$ by its two children and draw appropriate tree edges incident to the children from other clusters. In either case, we then recursively expand $c_{i-1}(u)$. The expansion stops when we reach the leaf-cluster representing the vertex u.

It is easy to see that, in the end, there is at most one cluster for each level i, where $0 < i \leq k$ — we create at most two clusters per level and recurse on one of them. We use the notation K_i for the cluster that exists at level i after this expansion; at the leaf-level, K_0 denotes the cluster containing the vertex other than u. See Figure 1.

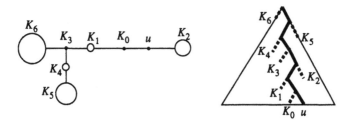

Fig. 1.: Expanding the topology tree at vertex u

The expansion procedure results in a tree of clusters linked by edges of T. If we wish to expand at two vertices, say, u and v, then after expanding at u, let K_j be the unexpanded cluster containing v. We repeat our expansion procedure on $K_j = c_j(v)$. In this way, we can expand the topology tree at any constant number of vertices in $O(\log n)$ time. The resulting partially-expanded tree has a constant number of clusters of each level. We prove that the sum of the level differences between adjacent clusters is only $O(\log n)$.

Lemma 3. *If we expand the topology tree at a constant number of vertices, the sum of the level differences between neighboring clusters is $O(\log n)$.*

To execute a query (u, v) or an update of an edge (u, v), we first expand the topology tree at u and then at v. In the case of a query, we answer the query and then merge the topology tree together in the same way as we expanded it. If we perform an update, edges and vertices are inserted and deleted (vertices are inserted or deleted to make sure that each vertex still has degree at most 3), and we may have to expand further at additional vertices. This changes the spanning tree and leads to a different topology tree, since only clusters that are neighbors in the spanning tree can be combined into larger clusters.

We merge back the clusters using a locally greedy heuristic — whenever possible, we merge two adjacent clusters of the lowest level, until the final topology tree is obtained; the merges respect the two clustering rules given earlier (see the full version of [4] for details). This process may force further expansion of some clusters. The number of expansions in each cluster is proportional to the sum of the differences between the level of the cluster and those of its neighbors. By Lemma 3, the total number of additional expansions is $O(\log n)$.

4 Edge Bundles

In the previous section, we described a hierarchical method for storing the spanning tree T. We now describe a data structure, called an *edge bundle*, for storing non-tree edges. Each edge bundle represents a set of "equivalent" edges incident to a particular cluster C. Edge equivalence is defined using the concept of edge *targets*; an edge's target is a cluster that contains its other endpoint (more about this shortly). Edges are equivalent iff they have one endpoint in C and belong to a contiguous group of edges with the same target.

We use two different interpretations for the targets of edge bundles in our data structure. The default target of an edge bundle is the lowest common ancestor in the topology tree (LCA) of the endpoints of the constituent edges. That is, the target cluster of the bundle is the lowest node v in the topology tree such that the edges in the bundle have both endpoints inside the cluster represented by v. The advantage of this targeting scheme is that the target of an edge bundle is independent of how the topology tree has been expanded. A possible disadvantage is that the other ends of the edges in a single edge bundle may be incident to multiple clusters. In Figure 2(a), three edge bundles are shown, one incident to each solid cluster. The target of all three bundles is the largest enclosing cluster, shown dashed.

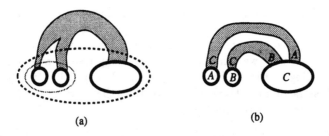

(a)   (b)

Fig. 2.: Different targeting schemes for edge bundles

During queries and restructuring of the topology tree, we use *precise targeting* for edge bundles. In this case, the targets are defined relative to a particular expansion of the topology tree. Each edge bundle consists of a set of edges connecting exactly two clusters, and the target stored at each bundle is just the other cluster to which the bundled edges are incident. Figure 2(b) shows the edge bundles for the clusters and edges of Figure 2(a), using precise targeting. The target of each of the four edge bundles is shown by a label on the bundle.

The data structure that represents an edge bundle is very simple: it is a record containing (1) a count of the number of edges in the bundle, and (2) the target cluster of the bundle.

5 Supernodes and Coverage Graphs

At each cluster that has external degree at most 2, we maintain one additional data structure, called a *coverage graph*. Suppose C is a cluster with external degree 2 and b_1, b_2 are its boundary vertices (the vertices incident to the tree edges that leave C). Furthermore, let $p = \pi(b_1, b_2)$ be the path in T between b_1 and b_2, and let T' be the subtree of T that connects the vertices in C. We build a data structure that records which parts of p are covered by edges incident to C.

The following scenario provides a motivation for this data structure. Let (a, b) be a non-tree edge, with $a \in C$ and $b \notin C$. Suppose that b lies in the part of T that is connected to C by b_1. Then (a, b) covers the whole path from a to b_1 in C. However, if an update changed T outside C, then b could suddenly be connected to C through b_2, in which case (a, b) would cover the path from a to b_2. If we had stored the information that (a, b) covers the path from a to b_1, we would have to change it, which would make update operations too expensive. Now suppose that there is a second edge (a', b'), with $a' \in C$ and $b' \notin C$, and further suppose that b and b' are in the same cluster when the topology tree is expanded at a. Then the tree path between a and a' is covered, no matter how b and b' are connected to C. Thus, for the purpose of computing coverage, we can collapse the path $p \cap \pi(a, a')$ to a single node, which we call a *supernode*. Instead of remembering the path p, we store only the ordered list of supernodes at C.

More precisely, we define the *tree path* $p(C)$ *of a cluster* C. If C has external degree 1, $p(C)$ consists of its boundary vertex; if C has external degree 2, $p(C)$ consists of the path in T between the boundary vertices of C. As described below, we store $p(C)$ at C in a condensed form. Assume (a, b) and (a', b') are the extreme edges of an edge bundle, such that a and a' are contained in C. We say that the *projection of a on p* is the closest vertex to a in T' that lies on p. We call the tree path from the projection of a to the projection of a' the *projection of the edge bundle on p*. A subpath of $p(C)$ is *internally covered* if it is covered by edges that have both endpoints in C.

A *supernode* of a cluster C is a maximal subpath $\pi(x, y)$ of $p(C)$ such that $\pi(x, y)$ intersects the projection of some edge bundle, and every edge of $\pi(x, y)$ either is covered by an edge internal to C or lies on the projection of some edge bundle on $p(C)$.

Since an edge bundle may be a single edge, a supernode may be just a vertex. The path from x to y is called the *path of the supernode*. The paths of two supernodes are disjoint; otherwise they would create a single supernode.

Instead of $p(C)$ we store at each C the list of supernodes in the order they appear along $p(C)$. Two consecutive supernodes are connected by a *superedge*. A superedge represents a subpath of $p(C)$ that is not completely internally covered. We call this representation of $p(C)$ by supernodes and superedges the *coverage graph of* C. If no supernode lies on $p(C)$, we introduce a single supernode to represent $p(C)$. This guarantees that if $p(C)$ has more than one supernode, then every supernode is incident to at least one edge bundle.

The data structure we use for a coverage graph of C is the following. If C has external degree 2, we have two doubly linked lists of edge bundles, one for each side of the tree path. Each list contains the edge bundles in the counterclockwise order of their embedding. The first edge bundle of each supernode in a counterclockwise traversal of the edge bundles is *marked*. The marker also keeps a skip count, that is, the number of consecutive supernodes following this supernode that have no edge bundles incident on this side. At the head of the list we keep an additional marker that indicates the number of consecutive supernodes with no edge bundles incident on this side before the first edge bundle. See Figure 3.

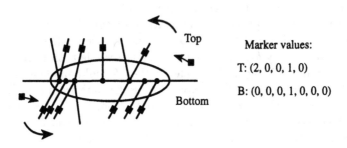

Marker values:

T: (2, 0, 0, 1, 0)

B: (0, 0, 0, 1, 0, 0, 0)

Fig. 3.: Storing the coverage graph

If C has external degree 1, the coverage graph is much simpler. It consists of a single doubly linked list of the edge bundles incident to C, in the counterclockwise order of their embedding. There is only one supernode in the coverage graph, so only one marker is needed in the list.

6   Recipes

We would like to store the coverage graph at each cluster, but that turns out to be too expensive, both in storage and time. Therefore we keep a *recipe* at each cluster C that explains how the coverage graphs of C's children can be computed from C's coverage graph. The coverage graph of the root of the topology tree is empty. We compute coverage graphs during a top-down traversal of the topology tree.

To compute the recipe for C, we build C's coverage graph by modifying the coverage graphs of its children, and then store at C a procedure that tells us how to reverse the construction. This procedure is the recipe. Whenever we expand the topology tree, we use the recipes to create the coverage graphs along the expanded path. Whenever we merge the topology tree, we first determine how to combine the coverage graphs of two clusters to create the coverage graph of their parent, and then we remember how to undo this operation in a recipe. Section 7 gives more details of the merging process.

Recipes use a specialized kind of pointer called a *location descriptor* to remember where the coverage graph of C has to be modified to create the coverage graphs of C's children. A location descriptor consists of a pointer to an edge bundle and an offset into the edge bundle (in terms of number of edges). It takes constant time to follow a location descriptor.

We now describe the structure of the recipes, which varies depending on the number of children of C and their external degrees. In this abstract we discuss only the most complex of the five cases; it illustrates all the main ingredients of the others.

Case 5: C has two children, both with external degree 2.

We will describe the process of going from children clusters to their parent. The recipe describes exactly the reversal of this process.

Let Y and Z be the children of C. On each side of the tree edge between Y and Z, there may be an edge bundle that connects Y and Z. To create the coverage graph for C, these edge bundles have to be removed. There may also be an edge bundle that starts in Y on one side of the spanning tree path, loops around the whole tree and ends on the other side of the spanning tree path in Z. Section 7 describes how we find such edges. We remove all the edge bundles connecting Y and Z and compress all the supernodes between the bundle endpoints into one supernode. To do this we concatenate the bundle lists of each side of the spanning tree path with each other and change the markers for the supernodes. We also merge newly adjacent edge bundles into a single edge bundle if they have the same target (see Figure 4).

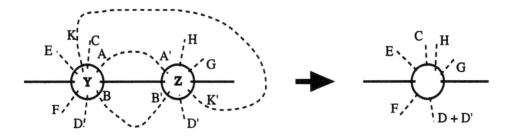

Fig. 4.: Combining coverage graphs; recipes for case 5

The recipe contains two location descriptors to indicate where we concatenated the edge bundle lists (and possibly merged adjacent bundles). We also have to store the edge bundles that were merged together and location descriptors pointing to the merged edge bundles. If there is an edge bundle that loops around the tree, we need two more location descriptors to mark its endpoints. Finally we have to store a location descriptor for each compressed supernode in Y or Z.

New edge bundles may be created during recipe evaluation. For each new edge bundle, the recipe stores a bundle record, preloaded with the count of bundle edges, and a location descriptor pointing to the place in the old edge bundle list where the new bundle is to be inserted. The target field of the bundle is easy to set: the LCA of the bundled edges is exactly the node at which the recipe is being evaluated. Each recipe evaluation creates $O(1)$ new bundles; this leads to the following lemma.

Lemma 4. *If the topology tree is expanded at a constant number of vertices and recipes are evaluated at the expanded clusters, the total number of edge bundles, supernodes, and superedges created is $O(\log n)$. The expansion takes $O(\log n)$ time.*

After we expand the topology tree at a constant number of vertices, we are left with an $O(\log n)$-size collection of clusters and edge bundles. The bundles use LCA targeting. To answer queries or perform updates, we need to transform the edge bundles to use precise targeting.

During the retargeting procedure, each edge bundle may be split into several smaller bundles, though all remain incident to the same supernode. The reason for this follows: When edge bundles are created, they identify a set of parallel edges connecting two clusters Y and Z. During the rest of the expansion process, Y and/or Z may be further subdivided into clusters, and hence the original set of parallel edges may connect up to $O(\log n)$ clusters. However, planarity ensures that the edges don't cross, and hence we can match the edge bundles originally incident to Y with those originally incident to Z.

Lemma 5. *Suppose that the topology tree has been partially expanded to get a collection of $O(\log n)$ clusters joined by tree edges and LCA-targeted edge bundles. Then we can compute a new set of precisely targeted edge bundles, incident to the same supernodes, in $O(\log n)$ time.*

For convenience, we refer to the graph formed by supernodes, superedges, and precisely targeted edge bundles as the *cluster graph*. The previous two lemmas imply the following theorem.

Theorem 6. *The cost of expanding the topology tree to get a cluster graph is $O(\log n)$.*

7 Queries and Updates

To answer the query whether u and v are 2-edge-connected, we want to test whether there is a bridge on $\pi(u, v)$, that is, whether an edge on $\pi(u, v)$ is uncovered. We expand the topology tree at u and v, which results in a cluster graph. By Lemmas 4 and 5, this graph has $O(\log n)$ supernodes, superedges and edge bundles. The vertices u and v are 2-edge-connected in the original graph iff $s(u)$ and $s(v)$ are 2-edge-connected in the cluster graph, where $s(u)$ and $s(v)$ are supernodes containing u and v, respectively. We check this in $O(\log n)$ time using a classical static algorithm. Then we merge the topology tree together in the same way we expanded it, leaving it as it was before the query.

Lemma 7. *Using the data structure described in Sections 3–6, we can determine whether two query vertices u and v are 2-edge-connected in $O(\log n)$ time.*

The general structure of an update is as follows. First we expand the topology tree at a constant number of vertices. Next we retarget the edge bundles to give each edge bundle a supernode as a target (see Lemma 5). Now a cluster graph has been created. On this graph we perform our changes, which may include insertion and deletion of edges and vertices, and making non-tree edges into spanning tree edges and vice versa. This part varies depending on what kind of update we are performing. Finally we merge the topology tree back together. In the previous sections we have explained how to create the cluster graph. In this section we explain the changes to the graph during updates and the merging of the topology tree. We support two kinds of updates, edge insertions and deletions. (It is straightforward to extend this to allow insertions/deletions of isolated vertices.)

Inserting an edge or deleting a non-tree edge. To insert an edge (u, v), we expand the topology tree at u and v, after first checking the degrees of u and v in the original graph. Let x stand for either u or v. If the degree of x before the insertion is 1 or 2, we expand the topology tree at x, create the cluster graph, and give x a new non-tree edge.

If the degree of x before the insertion is 3, we must create a four-vertex cycle for x and connect it appropriately. We expand the topology tree at x, then replace x by a four-vertex cycle. We connect the three original edges plus the newly inserted edge to the cycle in the proper order. We make three of the four cycle edges be tree edges; the last cycle edge and the newly inserted edge are non-tree edges.

If the degree of x becomes larger than 4, we already have a cycle for x. We insert a new vertex into the cycle and add the new edge in its proper place.

After making the appropriate modifications at u and v in the cluster graph, we merge the topology tree back together. (The details of this merge appear below.)

Deleting a non-tree edge (u, v) is essentially the inverse of an insertion, and we omit the details in this abstract.

Deleting a tree edge. When a tree edge (u, v) is deleted, the spanning tree is broken up into two parts. We expand the topology tree at the endpoints of the edge and create the cluster graph. Then we run along one of the two faces adjacent to (u, v) and find a group of parallel edges (represented by a pair of edge bundles) that connects the two parts of the spanning tree. This can be done in $O(\log n)$ time, by examining all the edge bundles in the cluster graph. (We are assuming here that the edge (u, v) was not a bridge edge of the graph G.) By repeatedly expanding the clusters incident to the pair of edge bundles, we can in $O(\log n)$ time identify a non-tree edge (u', v') that connects the two spanning tree components. We expand the topology tree at u' and v' and make (u', v') into a tree edge and (u, v) into a non-tree edge. Then we continue as for the deletion of a non-tree edge.

Merging the topology tree. We now describe how the topology tree is merged together after an edge insertion or deletion. The procedure has three steps. First, we compute the new topology tree for the updated cluster graph. Second, we compute the new edge bundles and their LCA targets based on the new topology tree. Third, we update recipes in all the clusters affected by the changed edge bundles. There are $O(\log^2 n)$ such clusters, and we spend amortized constant time apiece, for a total update time of $O(\log^2 n)$; all the other operations take $O(\log n)$ time.

We first determine the structure of the new topology tree for the modified cluster graph, as described in Section 3. This step is just preparatory for the full reconstruction of edge bundles, supernodes, and recipes, and does not involve any of those elements. We expand all the clusters affected by the restructuring algorithm (still only $O(\log n)$ clusters) to get a larger cluster graph.

The clusters of the cluster graph correspond to a *fringe* of nodes in the new topology tree: every node above them in the tree has been expanded, and no node below them has been expanded. To create recipes for the nodes above the fringe, we must compute the new edge bundles and their LCA targets based on the new topology tree. To do this we first compute precise targets for the edge bundles in the cluster graph. Now each edge bundle identifies one end of a group of parallel edges linking two clusters. For each pair of linked clusters, we compute their lowest common ancestor in the new topology tree and label each edge bundle with the LCA as its target. This takes $O(\log n)$ time altogether. Finally we merge any edge bundles that have the same target and are adjacent along the boundary of some cluster. This may involve merging supernodes as well.

There are two sets of topology tree nodes (clusters) for which recipes need to be computed. First, it is clear that the nodes above the fringe in the topology tree must have recipes created for them: these nodes may never have had recipes at all. Second, and less obviously, we must recompute recipes for certain nodes (clusters) below the fringe. The affected clusters are those that contain any of $O(\log n)$ vertices, namely the extreme vertices of edge bundles in the original or revised cluster graphs.

To prepare for recipe (re)creation, we expand the topology tree yet again at all $O(\log n)$ of the vertices identified above, then set the targets for edge bundles according to LCAs in the new topology tree. The $O(\log^2 n)$ expanded nodes below the fringe lie on $O(\log n)$ paths from the fringe to leaves. We process each fringe-to-leaf path in $O(\log n)$ time, then spend $O(\log n)$ time apiece on the $O(\log n)$ nodes above the fringe. (We could reduce the time spent above the fringe to $O(\log n)$ by being more careful, but it's not necessary.)

The recipes are computed by bottom-up merging (cf. Section 6): for each cluster we combine the coverage graphs of its sub-clusters, then record how to reverse the operation in a recipe. We now consider the process of merging two clusters. When we start, each cluster of the partially expanded topology tree has a coverage graph as described in Section 5.

We present a high level algorithm for merging two clusters; the subroutine details appear in the full paper. When we merge two clusters, we first identify the edge bundles incident to both clusters that will be internal to the merged cluster. Second, we identify the supernodes of the coverage graphs to which these bundles are incident. These supernodes and all the supernodes on the path connecting them will be coalesced. Coalescing takes time proportional to the number of supernodes being coalesced, and hence amortizes to $O(\log^2 n)$ overall. (This description is most applicable to the case of merging two degree-two nodes into one; there are minor differences when any of the nodes has degree one or three, but the basic ideas are the same.)

After we identify the bundles that will be internal to the merged node, we follow three steps: (1) delete these bundles; (2) merge adjacent bundles with the same target; and (3) coalesce supernodes. The recipe simply reverses this merging operation.

There are two subtasks to be explained more fully: (1) identifying bundles internal to the merged cluster, and (2) locating these bundles in the coverage graph. The full paper describes the data structures that implement these operations.

The following lemma summarizes the update performance of our data structure.

Lemma 8. *The 2-connectivity data structure can be updated in response to an edge insertion or deletion in $O(\log^2 n)$ time.*

We finish by restating the main theorem of our paper.

Theorem 9. *An embedded planar graph can be preprocessed into a linear-space data structure that supports insertion or deletion of an edge in $O(\log^2 n)$ time and answers a 2-connectivity query between two vertices in $O(\log n)$ time. If the graph has n vertices, then the preprocessing cost is $O(n)$.*

References

1. A. V. Aho, J. Hopcroft, and J. D. Ullman. *The design and analysis of computer algorithms.* Addison-Wesley, Reading, MA, 1974.
2. D. Eppstein, G. Italiano, R. Tamassia, R. E. Tarjan, J. Westbrook, and M. Yung. "Maintenance of a minimum spanning forest in a dynamic planar graph" *Proc. of 1st SODA*, 1990.
3. G. N. Frederickson. "Data structures for online updating of minimum spanning trees." *SIAM J. on Computing* (14), 1985, 781–798.
4. G. N. Frederickson. "Ambivalent data structures for dynamic 2-edge connectivity and k smallest spanning trees." *Proc. of 32nd FOCS*, 1991.
5. Z. Galil and G. Italiano. "Fully dynamic algorithms for edge connectivity problems." *Proc. of 23rd STOC*, 1991.
6. F. Harary. *Graph Theory.* Addison-Wesley, Reading, Massachusetts, 1969.
7. J. H. Hopcroft and R. E. Tarjan. "Dividing a graph into tri-connected components." *SIAM J. on Computing*, 1973.
8. G. L. Miller and V. Ramachandran. "A new graph triconnectivity algorithm and its parallelization." *Proc. 19th STOC*, 1987.
9. R. E. Tarjan. *Data Structures and Network Algorithms.* Society for Industrial and Applied Mathematics, Philadelphia, 1983.
10. R. E. Tarjan. "Depth-first search and linear graph algorithms." *SIAM J. on Computing*, 1972.
11. R. E. Tarjan and U. Vishkin. "An efficient parallel biconnectivity algorithm." *SIAM J. of Computing*, pp. 862–874, 1985.
12. J. Westbrook and R. E. Tarjan. "Maintaining bridge-connected and bi-connected components on-line." Tech Report, Princeton University, 1989.

This article was processed using the LaTeX macro package with LLNCS style

Non-Interfering Network Flows

C. McDiarmid
Corpus Christi College,
Oxford,
England.

B. Reed
Institute for
Discrete Mathematics,
Bonn, Germany.

A. Schrijver
Centre for Mathematics
and Computer Science,
Amsterdam,
The Netherlands.

B. Shepherd
Insitute for
Discrete Mathematics,
Bonn, Germany.

Abstract. We consider a generalization of the maximum flow problem where instead of bounding the amount of flow which passes through an arc, we bound the amount of flow passing "near" an arc. Nearness is specified by an extra distance parameter d. When $d = 0$ we get the usual network flow and $d = 1$ corresponds to bounding the flow through the nodes. A polynomial time algorithm is given to solve the max-flow and min-cost non-interfering flow problems for $d = 2$ and it is shown that the problems become NP-hard for $d \geq 3$. A polynomial time algorithm is outlined for arbitrary d when the underlying network is planar and how an integral flow can be obtained from a fractional one. Finally, we describe relationships with induced circuits and perfect graphs, VLSI chip design and the Hilbert basis problem.

1 Introduction

Historically, network (s,t) flows were studied as a model for the problem of shipping commodities through a network from a supply source to a destination where the arteries of the network had capacities on the amount of flow which could be shipped along them. Thus a solution to the *maximum (s,t) flow problem* would be interpreted in the graph $G = (V, E)$ representing the network as a maximum collection of $s - t$ paths such that for each edge, the number of paths containing that edge is no more than its capacity. (We presently restrict ourselves to undirected networks but we consider later some directed versions of the problem.) Ford and Fulkerson showed that it was enough to solve the problem

$$(1) \qquad max \; 1 \cdot y$$
$$y \cdot M \; \leq \; u$$
$$y \; \geq \; 0$$

where $u \geq 0$ is an integral vector of capacities on the edges and M is a matrix whose rows are indexed by the set of (s,t) paths paths and whose columns are indexed by the edges E. For a path P, its associated row in M is the incidence vector of its edge set. A feasible vector y for (1) is called a u-capacitated $s - t$ flow and its *value* is $1 \cdot y$.

The theory of network flows continues to be of interest in particular due to its relationship to wiring problems arising in the design of microchips. This imposes new constraints on the problem. In this paper we consider an extension which models the phenomenom of interference surrounding a wire, or path. Thus our aim is to solve (1) where we bound the amount of flow which passes within a specified distance of an edge. To be more precise, for an integer d and $S \subset E \cup V$, the *d-environment* of S, denoted by $E_d(S)$, consists of those elements $\beta \in E \cup V \setminus \{s,t\}$ for which there is a path P with at most d elements of $E \cup V$ such that (i) P is not incident with s or t and (ii) P connects β to an element of S (N.B. we count both edges and nodes in the path P). Note that the 1-environment of the edge set of an $s - t$ path P consists of the edges and internal nodes of that path; the 2-environment is the union of the 1-environment and the edges "hanging" from internal nodes of P.

Consider now the LP (1) where the rows of M are the incidence vectors of d-environments of·the edge sets of $s - t$ paths. Let u' be an integral vector of capacities on E. An *extension* u^* of u' is a vector of capacities on V where (i) the capacity of a node is at least as large as the minimum u' capacity of an edge incident to it and (ii) for each edge xy, one of u_x^* or u_y^* is bounded above by u'_{xy}. An integral capacity vector of the form $u = (u', u^*)$ where u^* is an extension of u', is called *tamed*. In the case that we have a tamed capacity vector and M is as described above, call the the resulting LP (1) the *maximum* (u, d)-*capacitated* (s,t) *flow* (or simply maximum (u, d) flow) *problem*. Denote this problem by $MF(G, s, t, u, d)$, or by $MF(u, d)$ when no confusion arises, and let $\mu(G, s, t, u, d)$ denote the value of a maximum (u, d) flow. For $d = 0$, this is just the usual maximum network flow problem. For $d = 1$, $MF(G, s, t, u, d)$ is equivalent to network flows with node capacities. This is because if u^* is an arbitrary vector of integral capacities on V, then there is an edge capacity vector of which u^* is an extension.

Note that when d is even, if a path weighting y is infeasible for (1), then there is a violated upper bound constraint which corresponds to an edge. This is because if some edge is in the $2k$-environment for a node $v \neq s, t$, then it is also in the environment of any edge α incident to v. Thus if v's constraint is violated, then so too is α's constraint. Similarly, if d is odd, then infeasibility of a path weighting y implies that some node constraint is violated. (N.B. We use here the fact that our capacity functions are tame.) Thus for d even (respectively odd) let M_d be a matrix whose rows are the incidence vectors of the edges (respectively nodes) in the d-environments of $s - t$ paths. A vector is called a *d-vector* if it assigns weights to edges or nodes when d is even or odd respectively. If we let u be an integral d-vector, then the above remarks together imply that $MF(u, d)$ is equivalent to:

$$(2) \qquad max \; 1 \cdot y$$
$$y \cdot M_d \leq u$$
$$y \geq 0.$$

We take this as our standard formulation of $MF(u, d)$. It is interesting to observe that this says that the "min-cuts" for (u, d) flows (i.e., dual solutions) will correspond to weighted sets of edges or nodes depending on whether d is even or odd. This provides a different perspective of the Menger edge and node cut theorems, (i.e., that they fit into this larger scheme which is based on some notion of parity).

In Section 2 we show how to solve $MF(u, 2)$ in polynomial time and show that $MF(u, d)$ is NP-hard for $d \geq 3$. We also consider the problem of computing a minimum cost $(u, 2)$ flow.

In Section 3 we consider the maximum (u, d)-capacitated *integral* (s, t) flow problem were we require an optimal integral solution to (2). We denote this problem by $MF'(G, s, t, u, d)$ and its optimal value by $\mu'(G, s, t, u, d)$. Note that $\mu'(G, s, t, \bar{1}, 2)$ is at least two if and only if s, t lie on a common circuit without chords, i.e., an *induced circuit*. The problem of determining whether two nodes lie on an induced circuit has been shown to be NP-complete by Fellows [2]. Thus computing integral $(u, 2)$ flows (and in fact integral (u, d) flows for $d \geq 2$) is NP-hard. On the other hand, we describe a polynomial time algorithm given in [9] which solves the integral (u, d) flow problems for planar networks. This work has several extensions and applications which we discuss in Section 4.

2   Computing (u, d) Flows

2.1   Maximum $(u, 2)$ Flows

Unlike the situations for $d = 0, 1$, we are no longer guaranteed that $\mu(G, s, t, u, 2)$ is an integer. Indeed we will see that the graph of Figure 1 is a graph for which $\mu(\bar{1}, 2) = \frac{8}{3}$. We describe two approaches to solving $MF(u, 2)$ in polynomial time.

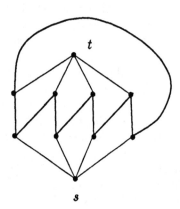

Figure 1:

Consider a directed graph D created from G by directing all edges incident with s and t in such a way that they become respectively source and sink. We then take two copies of each other edge and orient these in opposite directions. Finally we add the extra arc (t, s). We claim that $MF(u, 2)$ is computed by solving:

(3)
$$max\ x_{(t,s)}$$
$$A \cdot x = 0$$
$$x_a \geq 0 \qquad \text{for each arc } a,$$

$$x(\delta^-(v)) \le u_{sv} \qquad \text{for } sv \in E,$$
$$x(\delta^+(v)) \le u_{vt} \qquad \text{for } vt \in E,$$
$$x(\delta^-(x)) + x(\delta^-(y)) - x_{(x,y)} - x_{(y,x)} \le u_{xy} \qquad \text{for the remaining edges } xy \in E.$$

Here A is the node-arc incidence matrix of the digraph D and for $S \subset V$, $\delta^-(S)$ (respectively $\delta^+(S)$) denotes the set of arcs directed into (respectively out from) S. Note that except for the last three sets of *bounding* constraints this looks like the traditional LP formulation of the maximum (s,t) flow problem. Note also that if $u = \bar{1}$, then these constraints forbid assigning weight greater than $\frac{1}{2}$ to an $s-t$ paths with a chord. This is not an issue since such paths can always be short-cutted. We must use alternative methods, however, when we consider the directed problem in Section 4.1. Since (3) has a linear number of constraints, it can be solved by any polynomial algorithm for linear programming in time polynomially bounded by the input G, u, d. Moreover, the resulting solution (i.e., weights on the arcs of D) can be polynomially transformed to an optimal collection of paths.

The above approach is conceptually simple and varies little from the traditional solution to the max-flow problem. Unfortunately it does not extend to computing (u, d) flows in general. We need a different approach to solve the minimum cost and directed versions of the $(u, 2)$ flow problem. We outline this approach first for computing $\mu(u, 2)$ as it is also a means by which we can show the NP-hardness of $MF(u, d)$ for $d \ge 3$.

Note that the dual of (2) is:

(4)
$$\min u \cdot x$$
$$M_d \cdot x \ge 1$$
$$x \ge 0,$$

where x is a d-vector of variables.

Results of Grötschel, Lovasz and Schrijver [6], [8] imply that in order to solve (4) in polynomial time it is enough to have a polynomial time algorithm which determines whether a vector x satisfies the constraints of (4) (i.e., a separation algorithm). When $d = 2$, this separation problem can be reduced (in polynomial time) to a shortest path problem as follows. Given $x \in \mathbf{R}^E$ it is easy to check that $x \ge 0$. Thus determining whether x is feasible for (4) is equivalent to determining whether $x \cdot \chi^{C(P)} \ge 1$ for each induced (s,t)-path P (where $\chi^{C(P)}$ denotes the incident vector of edges in $E_2(E(P))$). We define weights w^x on E as follows: if uv is not incident to s or t, then

$$w_{uv}^x = \frac{1}{2}\left(\sum_{yu \in E : y \ne v} x_{yu} + \sum_{yv \in E : y \ne u} x_{yv} \right),$$

and

$$w_{us}^x = \frac{1}{2} \sum_{yu \in E} x_{uy}, \quad w_{ut}^x = \frac{1}{2} \sum_{yu \in E} x_{yu}.$$

We have defined w^x so that the weight of an induced path P, $\sum_{\alpha \in E(P)} w_\alpha^x$, is simply $x \cdot \chi^{C(P)}$. Thus determining whether $x \cdot \chi^{C(P)} \ge 1$ holds for each path P is equivalent to determining whether a smallest weight path has weight at least 1. Since $w^x \geq 0$ this latter problem is polynomially solvable. Thus the separation problem and hence the LP (4) is solvable in polynomial time. In fact, the Ellipsoid algorithm can be adapted (see [8]) to output an optimal dual solution in polynomial time. In our case this is a solution to the LP (2), i.e., a vector, y^*, of positive weights, each of which is associated with an induced (s,t)-path (in particular y^* can be represented in a polynomial number of bits).

One may now easily verify that Figure 1 depicts a graph where $\mu(G, s, t, \bar{1}, 2) = \frac{8}{3}$. This is done by exhibiting feasible solutions to (2) and (4). Namely, consider directing all edges in the graph "upwards" from s to t. Then assign weight $1/3$ to each of the 8 directed (s,t)-paths. Conversely, a feasible solution to (4) is obtained by setting each of the bold edges to a value $\frac{1}{3}$.

We close the section with a problem.

Problem 1: Give a combinatorial algorithm (not based on an algorithm which solves general linear programming) to compute $\mu(u, 2)$ in general graphs.

2.2 Minimum Cost $(u, 2)$ Flows

For the remainder of this section, c is a vector of non-negative costs on E. We define a vector c^* on the induced (s, t)-paths P such that $c_P^* = \sum_{\alpha \in E(P)} c_\alpha$. If the maximum in (2) is at least k, then the *minimum cost (u, d)-capacitated induced (s, t) flow of value k problem* $MC(G, s, t, u, d, c, k)$, or simply $MC(u, d, c, k)$, is:

$$(5) \qquad \begin{aligned} min \ & c^* \cdot y \\ yM \ & \leq \ u \\ 1 \cdot y \ & = \ k \\ y \ & \geq \ 0. \end{aligned}$$

The dual of (5) is

$$(6) \qquad \begin{aligned} max \ & (-u) \cdot x \ + \ k\beta \\ Mx \ + \ c^* \ & \geq \ \beta \cdot 1 \\ x \ & \geq \ 0. \end{aligned}$$

It is straightforward to check that (x, β) is a feasible solution to (6) if and only if the shortest (s, t)-path, with weights $(w^x + c)$ on the edges, is at least β. Hence, we may again use the Ellipsoid Algorithm to solve (5). Thus we have that

Theorem 2.1 $MC(u, d, k)$ *can be computed in polynomially bounded time.*

We raise the following question.

Problem 2: Find a combinatorial algorithm to solve $MC(G, s, t, u, d, c, k)$ for general or even for planar graphs.

2.3 Computing (u, d) Flows is Hard in General

We consider next the problems $MF(u, d)$ for $d > 2$ and we argue in a reverse manner to the previous section. Namely, if we can solve (2) for $d = 3$ for an arbitrary integral vector u, then we can solve the separation problem for its dual. The latter problem we call the *caterpillar problem* which is defined as: Given a graph G and nonnegative integral weights w on the nodes, find an $s - t$ path P for which $w(V(P) \cup N(V(P) - \{s,t\})$ is minimized. We show that the caterpillar problem is NP-hard for general graphs and hence so too is $MF(u,3)$.

We give a reduction of the node cover problem in general graphs to the caterpillar problem. Recall that the node cover problem is for a graph $G = (V, E)$ and integer k, to determine whether there is a subset S of size k such that $V - S$ is a stable set. We construct a graph G^* as follows. First take $|E|$ cycles of length four each of which is identified with an edge $e = uv \in E$ by labelling two of its non-adjacent nodes as e_u and e_v. The cycles are now concatenated on the unlabelled nodes to form a long chain. We then label the two extreme (unlabelled) nodes of degree two as s and t. We also add a new node n_v for each $v \in V$ and join it to e_v for every edge $e \in E$ incident to v. Each n_v is also joined to $|E|$ new degree one nodes to ensure that any minimum cardinality $s - t$ caterpillar will not use any of the nodes n_v. So suppose that P is an $s - t$ path not using any node n_v. Let the degree two nodes traversed by P be $E_P = \{s, e_{v_1}, e_{v_2}, \ldots, e_{v_{|E|}}, t\}$. Clearly $N_P = \{v \in V : e_v \in E_P \text{ for some } e \in E\}$ is a node cover of G. Moreover, the cardinality of P's caterpillar is simply $|E| + |N_P|$. Thus finding a minimum cardinality $s - t$ caterpillar in the new graph corresponds to finding a minimum size node cover for G.

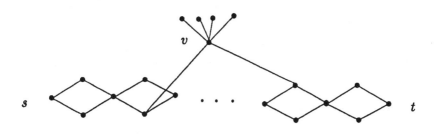

Figure 2:

The above ideas can be extended for $d > 3$ and so we have the following.

Theorem 2.2 $MF(u,d)$ *is NP-hard for* $d \geq 3$.

3 Planar Networks and Integral (u,d) Flows

3.1 The LP Approach to Integral $(u,2)$ Flows

We state the following result from [11].

Theorem 3.1 *For graphs G with $G - t$ planar, a $(u, 2)$-capacitated (s,t) flow y of value k can be transformed in polynomial time to an integral flow of value k. In particular, $\mu'(u, 2) = \lfloor \mu(u, 2) \rfloor$. Furthermore, the integral solution may be chosen to only use paths assigned positive weight in some optimal fractional solution.*

The above result is proved by uncrossing the paths assigned positive weights by y and then taking an appropriate subset of the resulting collection. Actually the latter statement of the theorem would be enough to solve $MF'(u, 2)$ in polynomial time since we could repeatedly delete nodes of G until we find a minimal subgraph $H \subseteq G$ such that $\mu(H, s, t, u_H, 2) \geq \lfloor \mu(G, s, t, u, 2) \rfloor$. At this point the graph H is precisely a collection of induced-disjoint $s - t$ paths. Thus the maximum $(u, 2)$ flow problem can be solved in polynomial time for planar graphs. In particular we have the following.

Corollary 3.2 *For graphs G such that $G-t$ is planar, seeing if s, t lie on an induced circuit can be checked in polynomial time.*

This is related to a well-known problem arising in the study of so-called perfect graphs. An *odd hole* is an odd induced circuit of size at least five. The problem is to determine the complexity of finding whether a graph contains an odd hole. Hsu [7] gives a polynomial algorithm for planar graphs. We propose the following generalization of Corollary 3.2 and Hsu's result.

Problem 3: Give an algorithm to determine whether two nodes in a planar graph lie on an odd induced circuit.

We state another result of [11] showing that minimum cost flows in planar graphs can also be transformed into integral flows.

Theorem 3.3 *If $G-t$ is planar and y is an optimal solution to $MC(G, s, t, u, d, c, k)$, then a $(u, 2)$-capacitated integral flow of value k with the same cost can be found in polynomial time.*

Specifically, this and Theorem 2.1 imply:

Corollary 3.4 *For planar graphs G and nonnegative integral costs on the edges, a minimum cost induced circuit through two specified nodes can be found in polynomial time.*

3.2 A Combinatorial Algorithm

We now consider a combinatorial approach to finding a maximum $(\bar{1}, d)$ integral flow in a planar network. The method yields a min-max theorem which resembles Menger's edge cut theorem. The method also extends to the problem $MF(u, 2)$, i.e., with general capacities. We must first introduce some concepts.

For a graph G and $d \geq 1$, a *d-path* is a simple path of length at most $d - 1$. A collection of $s - t$ paths is pairwise *d-separate* if there is no d-path in $G - \{s, t\}$ connecting internal nodes of distinct paths in the collection. Note that if $d \geq 1$, then an integral $(\bar{1}, d)$ flow in a graph is precisely a collection of d-separate paths.

We assume that G is embedded in the 2-sphere S_2. Let C be a closed curve in S_2, not traversing s or t. The *winding number* $w(C)$ of C is, roughly speaking, the number of times that C separates s and t. More precisely, consider any curve P from s to t, crossing C only a finite number of times. Let λ be the number of times C crosses P from left to right, and let ρ be the number of times C crosses P from o right to left (fixing some orientation of C, and orienting P from s to t). Then $w(C) = |\lambda - \rho|$. This number can be seen to be independent of the choice of P.

We call a closed curve C (with clockwise orientation relative to s) d-alternate if C does not traverse s or t, and there exists a sequence

$$(7) \quad (C_0, p_1, C_1, p_2, C_2, \ldots, p_l, C_l)$$

such that

(i) p_i is a d-path of $G \setminus \{s, t\}$ with endpoints s_i, t_i ($i = 1, \ldots, l$);

(ii) C_i is a (noncrossing) curve of positive length from t_{i-1} to s_i traversing a face of G ($i = 1, \ldots, l$ and $C_0 = C_l$);

(iii) C traverses the paths and curves given in (7) in the described order.

Here, by traversing p_i we mean the image of C "follows" p_i from one endpoint to the other.

Intuitively it makes sense that if G has a collection of k d-separate s-t paths, then any alternate curve C must satisfy $l \geq k w(C)$. Since any member of the collection must cross C $w(C)$ times and can only cross it on the paths p_i. Furthermore, no two paths may intersect the same p_i since the paths are d-separate, hence the number of p_i's is at least $k w(C)$. This argument is fine when $d \leq 2$, the problem for higher values of d is that there can exist a path which crosses C $w(C)$ times but does not intersect $w(C)$ of the p_i's. Fortunately, this can only happen when a maximum collection of d-separated paths is of size one! We now state the min-max theorem. A proof may be found in [9].

Theorem 3.5 *For a planar graph G:*

(i) *There exists an integral $(\bar{1}, d)$ flow of value k if and only if $l(C) \geq k \cdot w(C)$ for each d-alternate closed curve C.*

(ii) *A maximum value $(\bar{1}, d)$ flow from s to t can be found in polynomial time.*

(iii) *The curves C in (i) can be restricted to those with $w(C) < |V|$.*

Note that for the cases $d = 0, 1$ the max-flow min-cut theorems imply that we may restrict ourselves to alternate curves of winding number one. This is not the case for $d > 1$ otherwise $MF'(u, d)$ would always have an integral optimum as we have seen is not the case in Figure 1. It would be attractive if one could give a better bound on the required number of windings of an alternate curve. The proof of Theorem 3.5 is by means of a path shifting algorithm which also produces a collection of paths which is optimal for (2). This yields the following min-max formula for $MF(\bar{1}, 2)$ in planar graphs.

Theorem 3.6 *For a planar graph G,*
$\mu(G, s, t, \bar{1}, d) = min\{l(C)/w(C) : C \text{ is a } d\text{-alternate curve}\}.$

These methods can be applied to the case $d = 2$ with general capacities. In this case the *capacity* of a 2-alternate curve C, denoted by $cap(C)$, is defined as the sum of the capacities of the p_i's and if p_i is a single node, then it is the minimum capacity of an edge incident to it.

Theorem 3.7 *For a planar graph G and vector u of integers on the edges:*

(i) *There exists an integral $(u, 2)$ flow of value k if and only if $cap(C) \geq k \cdot w(C)$ for each 2-alternate closed curve C.*

(ii) $\mu(G, s, t, u, 2) = min\{cap(C)/w(C) : C \text{ is a 2-alternate curve.}\}$

Actually we do not know a counterexample to the obvious extension of this to $MF'(G, s, t, u, d)$ for $d > 2$. Of course one must change slightly the definition of capacity. The capacity of a d-path p_i becomes the minimum capacity of an element in its *center*. (The center of a path P are those element of $E(P) \cup V(P)$ whose d-environment contains both endpoints of P.) Vaguely speaking, the problem for higher values of d is that it is no longer obvious whether one can choose a noncrossing optimal collection of paths.

The algorithm can be extended using the techniques of [14] to find a maximum collection of 2-separated paths for graphs on a fixed surface. It gives a theorem resembling Menger's theorems. In particular, it states that for a graph embedded on a surface S of genus g either there are k 2-separated $s - t$ paths or there exist $f(g)$ alternate curves whose existence shows that there is no such collection. Here the function f corresponds to an upper bound on the number of homotopic possibilities there are for collections of k disjoint $s - t$ curves·in S (where we identify s and t with points on S). Note that in some sense the function f gives focus to why the induced circuit problem is computationally difficult in general graphs.

4   Extensions and Applications

4.1   Related Computational Problems

As far as solving for integral solutions, the combinatorial approach is generally preferable to the LP approach. The major advantage of the latter, of course is that it can be adapted to solve the minimum cost induced cycle problem. This in conjunction with Theorem 5 heightens the interest in resolving Problem 2. The shifting algorithms of [9],[10] generalize to solve the maximum (u, d) flow problem in planar graphs and can be extended (as we will note) to solve the maximum flow problems in directed planar networks. It also, as mentioned, yields a min-max theorem which is a closer analogy to the minimum cut conditions of Menger and Ford and Fulkerson. We now discuss a few more related computational problems.

1. Coloured_Paths
Input: A planar graph $G = (V, E)$ (possibly with multiple edges)

whose edges are coloured red and green. Source and sink s, t.

Output: A maximum collection of green-edge paths such that no red path joins internal nodes of distinct paths.

A similar characterization as to Theorem 3.5 holds for this problem although we have to change slightly the definition of an alternate curve: the faces it traverses are now faces of the green subgraph and paths p_i are simply red edge paths.

In this problem the red edges can be thought to represent forbidden or high penalty regions in the graph. This phenomenom occurs often in the design of VLSI chips where terminals are iteratively connected by wires. As the number of wires increases, certain regions of the chips become more congested and efforts should be made to avoid routing through these areas.

2. d-Separated Directed Paths

Input: A Planar digraph D with source s and sink t and integer d.

Output: A maximum collection of d-separated $s - t$ dipaths.

The method of Section 3.2 can also be extended to solve a directed version of the problem but we must modify slightly the definitions. (The proof is more involved than that of [10] and the details may be found in [9].) A collection of $s - t$ dipaths is pairwise d-separate if there is is no directed path of length less than d connecting internal nodes of distinct dipaths in the collection.

Here again, there is a min-max theorem where a d-alternate curve is defined as in (7) except that now the curves C_i are not required to only traverse a face:

(i) p_i is a d-dipath of $D \setminus \{s, t\}$ with endpoints s_i, t_i $(i = 1, \ldots, l)$;

(ii) C_i is a (noncrossing) curve of positive length from t_{i-1} to s_i and these are the only vertices of D that C_i intersects $(i = 1, \ldots, l$ and $C_0 = C_l)$;

(iii) C traverses the paths and curves given in (7) in the described order;

(iv) each C_i may cross arcs only from right to left (relative to the orientation derived from C) and *may not cross any arc in any d-path p_i*.

Here, the p_i's may be directed from s_i to t_i or conversely. and C may traverse p_i in either direction. Note that roughly speaking, condition (iv) requires that the arcs of D crossed by the C_i's point back towards s. It is interesting to note that in the directed case we do not require the paths in the collection to be induced, i.e., they may have backwards arcs. In fact, Fellows and Kratochvil [4] have recently shown that the problem of determining whether there is a single induced $s - t$ dipath in a planar digraph is NP-complete!

3. k-Star

Input: Planar graph G, node s and nodes $v_1, \ldots, v_k$.

Output: An induced subgraph consisting exactly of node-disjoint paths from s to each of the v_i's or proof that no such subgraph exists.

4. Internal Node

Input: Planar graph G and nodes u, v, w

Output: An induced $u - v$ path containing w or proof that none exists.

The fact that in Theorem 3.1 we need only that $G - t$ is planar implies that the above problems are also polynomially solvable.

5. Minium Weight Environment
Input: A planar graph G, integer d, prescribed nodes s, t and a d-vector w.
Output: An $s - t$ path whose d-environment has a minimum w-weight.

Finally, it would be interesting to extend the techniques of Section 3.5 to solve the capacitated version of integral d-separated paths. This would imply (see Section 2.3) the existence of polynomial time algorithms to solve the separation problems for the duals (4), that is the d-environment problem. In particular, this would imply the polynomial solvability of the caterpillar problem in planar graphs.

4.2   Hilbert Bases and a Polyhedral Description

Let G be a graph embedded on a cylinder. A path in G is *linear* if it joins the two ends of the cylinder. Let M_2 be a matrix whose rows are the incidence vectors of 2-environments of linear paths in G. Note that solving the resulting LP (2) is then equivalent to solving $MF(u, 2)$ where s and t are nodes corresponding to the two ends of the cylinder. We have from Theorem 3.1 that an integral optimum to (2) is obtained by rounding down the fractional optimum. Systems $M_2 \cdot x \geq 1$, $x \geq 0$ for which this holds for each integral u are said to have the *integer round-down property* (cf [13], [1]). Giles and Orlin have shown that this implies that the rows of the matrix

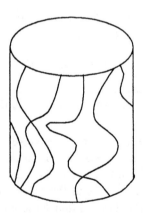

Figure 3:

$$
\begin{array}{cc}
M_2 & 1 \\
I & 0 \\
0 & -1
\end{array}
$$

form a Hilbert basis. Recall that a *Hilbert basis* is a set of vectors S such that each integral vector in the cone $\{x : x \cdot s \leq 0, \ \forall \ s \in S\}$ can be written as a nonnegative, integral combination of vectors in S. Thus we have the following special class of Hilbert bases, which we denote by $H_2(G, s, t)$. The *Hilbert basis problem* is to show that any integral vector in a $cone(S)$ can be expressed as the sum of at most $dim(cone(S))$ integral vectors in S.

Theorem 4.1 *For a planar graph G and nodes s, t, the associated Hilbert basis $H_2(G, s, t)$ satisfies the Hilbert basis conjecture.*

Proof: Suppose that (u, γ) is an integral vector in the cone generated by H_2. Then there exists $y_1, y_0 \geq 0$ and $\psi \geq 0$ such that $y_0 \cdot M_2 + y_1 \cdot I = u$ and $\gamma = 1 \cdot y_0 - \psi$. Thus Theorem 3.1 implies that $\mu(G, s, t, u, 2) \geq \lfloor 1 \cdot y_0 \rfloor$ and so there exists an integral vector $z_0 \geq 0$ such that $z_0 M_2 \leq u$ and $z_0 \cdot 1 \geq \lfloor 1 \cdot y_0 \rfloor \geq \gamma$. Thus $(u, \gamma) = z_0 \cdot M_2 + z_1 \cdot I - (0, 0, \ldots, 0, -\psi')$ for $\psi' = \lfloor 1 \cdot y_0 \rfloor - \gamma$ and an appropriate integral vector z_1. But for planar graphs Theorem 3.1 shows that z_0 may be chosen so that its number of positive components is at most the number of components where $z_0 \cdot M_2 = u$. Thus (u, γ) has a representation with at most $|E| + 1$ vectors from H_2, as desired. $\square$

We also obtain a polyhedral description for $P(G, s, t) = \mathbf{Q}^+ + conv(\{\chi^{C(P)} : P$ is an $s - t$ path$\})$ for s, t in a planar graph G. Results of Fulkerson [3] imply that it is enough to know the structure of the vertices of $\{x : M_2 \cdot x \geq 1, x \geq 0\}$. In fact Theorem 3.7 may be strengthened (see [11]) to show that each such vertex is $0 - \frac{1}{p}$ valued for some integer p.

Theorem 4.2 *For a planar graph G with nodes s, t, the polyhedron $P(G, s, t)$ is given by the nonegativity constraints and the rank inequalities.*

Acknowledgements: The authors are very grateful to Coelho de Pina Jr. for his careful reading and insightful comments on this work. We are also grateful to Andras Sebö for stimulating conversations. This paper stemmed from discussions at a workshop at Belairs Research Institute. We thank Wayne Hunt and his colleagues for their hospitality and providing an ideal working environment.

References

[1] S. Baum, L.E. Trotter Jr., Integer rounding for polymatroid and branching optimization problems, *SIAM Journal on Algebraic and Discrete Methods 2*, (1981), 416-425.

[2] M. R. Fellows, The Robertson-Seymour theorems: a survey of applications, *Contemporary Mathematics*, **89**, 1989.

[3] D.R. Fulkerson, Anti-blocking polyhedra, *Journal of Combinatorial Theory*, B **12**, (1972), 50-71.

[4] M. Fellows, J. Kratochvil, Personal communication, (1991).

[5] R. Giles, J.B. Orlin, Verifying total dual integrality, *manuscript* (1981).

[6] M. Grötschel, L. Lovász, and A. Schrijver (1981), *The ellipsoid method and its consequences in Combinatorial Optimization*, Combinatorica 1, 169-197.

[7] W.-L. Hsu, Recognizing planar perfect graphs, *J. of the A.C.M.* **34**, (1987), 255-288.

[8] M. Grötschel, L. Lovász, and A. Schrijver (1988), *Geometric algorithms and combinatorial optimization*, Springer.

[9] C. McDiarmid, B. Reed, A. Schrijver and B. Shepherd, Non-interfering dipaths in planar digraphs, *Centrum voor Wiskunde en Informatica Technical Report* (1991).

[10] C. McDiarmid, B. Reed, A. Schrijver and B. Shepherd, Induced circuits planar graphs, submitted to *Journal of Combinatorial Theory*.

[11] C. McDiarmid, B. Reed, A. Schrijver and B. Shepherd, Packing induced paths in planar graphs, *University of Waterloo Technical Report* (1990).

[12] W.S. Massey, Algebraic Topology: An Introduction, Graduate texts in mathematics, 56, *Springer-Verlag*, (1967).

[13] A. Schrijver, Theory of Integer and Linear Programming, *Wiley*, (1986).

[14] A. Schrijver, Disjoint circuits of prescribed homotopies in a graph on a compact surface, *J. Combinatorial Theory*, B **51** (1991), 127-159.

Triangulating Planar Graphs While Minimizing the Maximum Degree*

Goos Kant   Hans L. Bodlaender

Dept. of Computer Science, Utrecht University

P.O. Box 80.089, 3508 TB Utrecht, the Netherlands

Abstract

In this paper we study the problem of triangulating a planar graph G while minimizing the maximum degree $\Delta(G')$ of the resulting triangulated planar graph G'. It is shown that this problem is NP-complete. Worst-case lower bounds for $\Delta(G')$ with respect to $\Delta(G)$ are given. We describe a linear algorithm to triangulate planar graphs, for which the maximum degree of the triangulated graph is only a constant larger than the lower bounds. Finally we show that triangulating one face while minimizing the maximum degree can be achieved in polynomial time. We use this algorithm to obtain a polynomial exact algorithm to triangulate the interior faces of an outerplanar graph while minimizing the maximum degree.

1   Introduction

Planarity has been deeply investigated both in combinatorics and in graph algorithms research. Concerning undirected graphs, there are elegant characterizations of the graphs that have a planar representation and efficient algorithms for testing planarity (see, for instance, [1, 8, 6]). Also several algorithms are known to draw a planar graph in the plane with straight lines, leading to a so-called Fáry embedding. Elegant and simple algorithms for this problem are described in [14, 5, 16, 7], which all assume that the planar graph is triangulated. A triangulated planar graph has $3n - 6$ edges and adding any edge to it destroys the planarity. Every face is a triangle. In general, the restriction to triangulated planar graphs is no problem, since every planar graph can be triangulated in linear time and space, by the following modification of the algorithm in [14]: We construct an embedding of the biconnnected planar graph G. For every pair of neighbors u, w of v, we add (u, w), if there is no edge (u, w) *adjacent* to (u, v) in u and (w, v) in w yet. The resulting triangulated planar graph may have multiple edges, but after removing all multiple edges and adding the other edge in the involved faces of four vertices, this yields a correct triangulated planar graph. However, this algorithm may increase the degree of any vertex by $O(n)$, even if the maximum degree

---

*This work was supported by the ESPRIT Basic Research Actions of the EC under contract No. 3075 (project ALCOM).

of the given graph is bounded by a constant. As a concequence, the resulting picture may be significantly less readable, since a high degree implies small angles between adjacent edges.

The problem of drawing planar graphs readable in the plane gains a lot of interest for several other subclasses of planar graphs. Especially for biconnected and triconnected planar graphs some interesting and satisfying characterizations are known, leading to presentable drawing algorithms [17, 18, 2]. However, these algorithms, if they can be generalized to a more general subclass of planar graphs, will not always give compact and structured pictures for a more general subclass of planar graphs.

Recently, an investigation of augmenting a planar graph minimally by adding edges to admit these constraints has been presented in [13]. In this paper, Kant & Bodlaender present polynomial algorithms to make a planar graph biconnected or triconnected and still planar, working within 3/2 and 5/4 times optimal, respectively. Similar results with respect to outerplanar graphs are described in [11] and augmentation algorithms, without preserving planarity, are described in [3, 15, 9, 10].

In this paper we consider the problem of triangulating a planar graph while minimizing the maximum degree. This problem is very hard, as shown in the following theorem (using a reduction from the NP-hard planar graph 3-colorability problem):

Theorem 1.1 *Deciding whether a biconnected planar graph can be triangulated such that the maximum degree is $\leq K$ is NP-complete.*

Let $\Delta(G)$ be the maximum degree of a planar graph G and let $\Delta(G')$ be the maximum degree of some triangulation G' of G, then our goal is that $\Delta(G') \leq c_1\Delta(G) + c_2$, with c_1 (and c_2) as small as possible. We present worst-case lower bounds for c_1 and c_2, when G is a biconnected or triconnected planar graph. The added edges $\in G - G'$ are also called augmenting.

We present a simple linear algorithm to triangulate a planar graph G, such that the maximum degree in the resulting triangulated graph G' is only an additive constant larger than the worst-case lower bounds. To achieve this goal, a linear algorithm is presented to biconnect a planar graph while preserving planarity, increasing the degree of any vertex with at most 2. We also define an ordering for the vertices of a triconnected planar graph, which is interesting in its own sense. If only one face needs to be triangulated, then a simple dynamic programming approach can be applied to find an optimal solution. Using this technique, we present a polynomial exact algorithm to triangulate the interior faces of an outerplanar graph G while minimizing the maximum degree. This yields a maximal outerplanar graph if G is already biconnected.

This paper is organized as follows. In section 2 we give lower bounds for the maximum degree of the triangulated graphs. In section 3 we present a linear algorithm for triangulating planar graphs, working only an additive constant from the lower bounds. In section 4 we present an algorithm for triangulating one face. Section 5 contains some concluding remarks and some open problems. Omitted proofs can be found in the full paper.

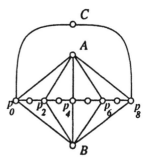

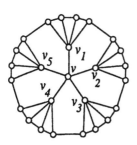

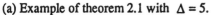

(a) Example of theorem 2.1 with $\Delta = 5$.   (b) Example of theorem 2.2 with $\Delta = 5$.

Figure 1: Examples of the lower bounds for the maximum degree.

2 Lower bounds for $\Delta(G')$

Regarding the problem of triangulating G into G' such that $\Delta(G') \leq c_1\Delta(G) + c_2$, we present here lower bounds for c_1 and c_2, when G is a biconnected or triconnected planar graph.

Theorem 2.1 *For every* $\Delta = \Delta(G) \geq 5$, *there exist triconnected planar graphs* G *such that for every triangulation* G' *of* G: $\Delta(G') \geq \Delta(G) + 3$.

Proof: See figure 1(a) for an example. (When Δ is larger, increase both the degree of v and the degrees of the vertices v_i by the same amount.) Case analysis shows that in every triangulation of this graph, at least one vertex receives three additional adjacent edges. □

Theorem 2.2 *For every* $\Delta > 1$ *there exists a biconnected planar graph* G *for which for every triangulation* $G', G \subseteq G'$: $\Delta(G') \geq \lceil \frac{3}{2}\Delta(G) \rceil$.

Proof: See figure 1(b) for an example. Observe that, as multiple edges are not allowed, each vertex p_i with i odd must have an edge to A or an edge to B. Hence either the degree of A or the degree of B increases by at least $\lceil \frac{\Delta}{2} \rceil$. □

So $c_1 \geq \frac{3}{2}$ for planar graphs in general. Our approximation algorithm, presented in section 3, matches these bounds up to an additive constant.

3 An Approximation Algorithm

3.1 Introduction

Our algorithm in this section will lead to a proof of the following theorem:

Theorem 3.1 *There is a linear algorithm to triangulate a planar graph* G *such that for the triangulation* G' *of* G, $\Delta(G') \leq \lceil \frac{3}{2}\Delta(G) \rceil + 21$.

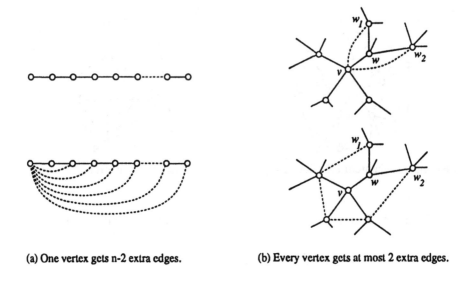

(a) One vertex gets n-2 extra edges. (b) Every vertex gets at most 2 extra edges.

Figure 2: Making different graphs biconnected.

A brief outline of the algorithm which obtains this result is as follows:

TRIANGULATE
    make the graph biconnected;
    while the entire graph is not triangulated **do**
        determine a triconnected component G' of G;
        compute the *canonical 3-ordering* for G';
        triangulate G';
        replace G' by an edge between the cutting pair;
    od

First we show how to biconnect G (keeping it planar) such that the degree of the vertices increases by not more than two. This degree-increase of two is sometimes necessary: for instance consider the tree $K_{1,3}$.

The problem of augmenting G such that it is biconnected and planar by adding $\leq K$ edges is NP-complete [13], but there is a linear algorithm to make G biconnected by adding edges while preserving planarity [14], which can be described as follows: if for any pair of consecutive neighbors u, w of v, u, w belong to different biconnected components (also called *blocks*), then the edge (u, w) is added. Unfortunately, this may increase the degree of a single node by $O(n)$, see figure 2(a). We modify the algorithm of [14] as follows: we inspecting the vertices in depth-first order, and test during the algorithm whether an added edge can be removed without destroying biconnectivity. The algorithm can be described more formally as follows:

BICONNECT
    determine an arbitrary embedding in the plane of G;
    number the cutvertices v_i of G in depth-first order;

```
for every cutvertex v_i (in increasing v_i-number) do
    for every two consecutive neighbors u, w of v_i in G do
        if u and w belong to different blocks then
            add an edge (u, w) to G
            if (v_i, u) or (v_i, w) was added to G then remove this edge;
    rof
rof
```

Lemma 3.2 *Algorithm* BICONNECT *gives a biconnected planar graph.*

Proof: For every cutvertex v_i, we add edges between the adjacent neighbors u, w of v_i, if they belong to different blocks, hence after this augmentation all neighbors of v_i belong to one common block. Thus v_i is not a cutvertex anymore. If (v_i, u) was added by BICONNECT then there was a path from v_i to u initially. But then there was initially a path from w to u. Adding (u, w) implies a cycle containing v_i, hence (u, v_i) can be removed without destroying the biconnectivity. Similar for (v_i, w). □

Lemma 3.3 *Every vertex receives at most 2 augmenting incident edges.*

Proof: Assume w.l.o.g. vertex v is the only element of a block. When visiting a neighbor w of v, v receives two incident augmenting edges, say (w_1, v) and (w_2, v). Then by the depth-first order v will be visited. If v is adjacent to at least one other block, then edges between neighbors of v are added and the edges (w_1, v) and (w_2, v) are removed. Hence v receives at most two incident augmenting edges (see figure 2(b)). □

Corollary 3.4 *There is a linear algorithm to augment a planar graph such that it is biconnected and planar and increases the degree of every vertex by at most 2.*

Hence we may now assume that G is biconnected. The main method of the triangulation is the following "zigzag"-method (see also figure 3):

ZIGZAG(F, v_1, v_p);
 (* F is a face of p vertices, numbered $v_1, \ldots, v_p$ around the ring; *)
 add edges $(v_p, v_2), (v_2, v_{p-1}), (v_{p-1}, v_3), (v_3, v_{p-2}), \ldots, (v_{\lfloor \frac{p}{2} \rfloor}, v_{\lfloor \frac{p}{2} \rfloor + 2})$;

Note that the degree of v_1 and $v_{\lfloor \frac{p}{2} \rfloor + 1}$ does not increase, the degree of v_p and $v_{\lfloor \frac{p}{2} \rfloor}$ (p even) or $v_{\lfloor \frac{p}{2} \rfloor + 2}$ (p odd) increases by 1, all other degrees increase by 2 (see figure 3(a)).

We cannot simply apply ZIGZAG to all faces: this could lead to a maximum degree of $3\Delta(G)$ and to multiple edges (see figure 3(b)). To apply a more subtle technique, we recall the *canonical ordering*, as introduced by De Fraysseix et al. [5] for *triangulated* planar graphs. We modify this ordering such that it holds for *triconnected* planar graphs. This ordering is then applied on the triconnected components of our biconnected planar graph G. We call this ordering the *canonical 3-ordering*.

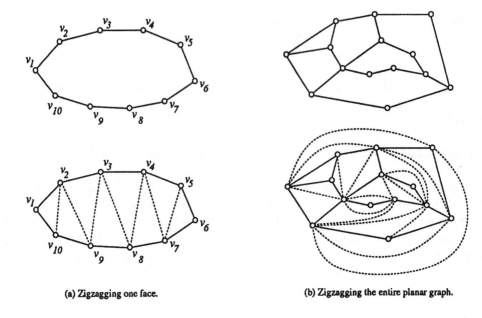

(a) Zigzagging one face.

(b) Zigzagging the entire planar graph.

Figure 3: Example of zigzagging a face and zigzagging a planar graph.

Theorem 3.5 *It is possible to order the vertices of a triconnected planar graph G in linear time in a sequence $v_1, \ldots, v_n$ such that in every step $k, k \geq 2$:*

1. *The subgraph G_k of G induced by the vertices $v_1, \ldots, v_k$ is biconnected and the boundary of its exterior face is a cycle C_k containing the edge (v_1, v_2).*

2. *either v_{k+1} is in the exterior face of G_{k+1} and has at least two neighbors in G_k, which are on C_k*

3. *or there exists an $l \geq 2$ such that $v_{k+1}, \ldots, v_{k+l}$ is a path in the exterior face of G_{k+l} and has exactly two neighbors in G_k, which are on C_k.*

Proof: (sketch) We use a similar reverse induction proof and a similar algorithm, as in [5]. We use counters (i) for each vertex v and face F, meaning that the edges already been visited and incident to v or F are composed of i intervals in the edge-list of v or F, where each edge-list is sorted in counterclockwise order with respect to some embedding. Any vertex or face with value (1) can be the next in the ordering. Updating these counters when traversing any edge or vertex can be done in constant time, implying a linear algorithm. $\square$

We use the canonical 3-ordering for triangulating the interior faces of each triconnected component of G. If deleting two vertices a, b of G disconnects G into one connect component V' and some components $G' - V'$, then we call V' a *2-subgraph*, and a, b is called the *cutting pair* of V'. If V' contains no other 2-subgraphs then V' is a triconnected component. Otherwise, if we replace all 2-subgraphs of V' by

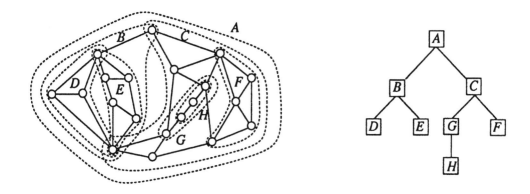

Figure 4: Example of a biconnected planar graph and the corresponding 2-subgraph-tree.

an edge between the cutting pair and eliminate the multiple edges, then V' is also a triconnected component. Notice that adding an edge between the cutting pair of a triconnected component V' makes V' triconnected. Hence with a small modification, we can apply the canonical 3-ordering on the 2-subgraphs of G.

Let T be a tree, where every node v' of T corresponds with a 2-subgraph V' of G. v' is an ancestor of w' in T, if $W' \subseteq V'$. The root r of T corresponds with the entire graph G. The leaves of T corresponds with the triconnected components of G. We call T the *2-subgraph-tree* (see figure 4). We start with triangulating the internal faces of the triconnected components of G and then the other 2-subgraphs V' of G, if all 2-subgraphs $W' \subseteq V'$ are already triangulated. To compute the canonical 3-ordering for V', we replace the 2-subgraphs $W' \subseteq V$ by an edge (a, b) between the cutting pair, and we *mark* this edge (a, b). When we triangulate V' and we visit edge (a, b), then we replace (a, b) by W'. We call this procedure *folding* and *unfolding*. Actually, this means that we triangulate V', if all 2-subgraphs W', for which v' is an ancestor of w', are already triangulated. We start with the leaves of T and work towards the root r of T.

We consider the two cases of the canonical 3-ordering: either a vertex v_{k+1} or a path $v_{k+1}, \ldots, v_{k+l}$ (implying a face F_{k+l}) will be added. When adding a vertex, several faces need to be triangulated, for which we use the algorithm for triangulating one face. We call in a step k the vertices $v_1, \ldots, v_k$ *old* and the vertices added in step $k+1$ *new*. Let v_a and v_b be the *leftmost* and *rightmost* old neighbor of the new added vertices in step $k+1$, then we call the old vertices $v \in C_k$ between v_a and v_b *internal*. Notice that every vertex is exactly once new and once internal, but it can be left- or rightmost more than once. Vertex v may receive in each phase (new, left- or rightmost, internal) incident augmenting edges. The upper bound for the increase of $deg(v)$ is the sum of the upper bounds for the increases in each of the three phases. The analysis for these bounds follows by inspecting the two cases (i) adding a face and (ii) adding a vertex, in the following subsections.

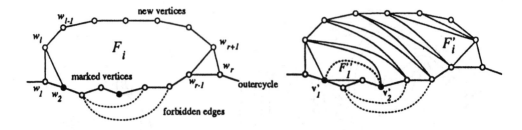

Figure 5: Example of triangulating one face.

3.2 Adding a face

When we add a face to G_k by adding the vertices of a path P, then we first change the marked edges (a, b) of P into the corresponding 2-subgraphs W' of G. Let F_{k+1} be the added face with all marked edges unfolded. We want to triangulate F_{k+1} such that the degree of the new and internal vertices increases by a constant, and the degree of the left- and rightmost vertex increases as little as possible.

Hereto we first assume that there is at least one internal vertex. Let $w_1, \ldots, w_r$ be the vertices of C_k such that w_1 is the leftmost and w_r the rightmost vertex, and $w_{r+1}, \ldots, w_l$ is the new added path (numbered counterclockwise around face F_{k+1}).

We add the edges (w_l, w_2) and (w_{r+1}, w_{r-1}), hence we need to triangulate the face with vertices $w_2, w_3, \ldots, w_{r-1}, w_{r+1}, \ldots, w_l$ and the degree of w_1 and w_r (the left- and rightmost vertex) does not increase. We call already existing edges (w_i, w_j) $2 \le i, j \le l$ forbidden, if $j > i+1$. For every forbidden edge (w_i, w_j) we label w_{i-1} and w_{j-1}, if they do not have incident forbidden edges. We renumber the labeled vertices by $v'_1, \ldots, v'_p$, in order of increasing w_i-number. We add the edges (v'_j, v'_{j+1}), $(1 \le j < p)$, thereby introducing several faces F''_j. Let F'_{k+1} be this face, containing all labeled vertices. We triangulate the faces as follows:

> **for** every face F''_j **do** ZIGZAG(F''_j, v'_j, v'_{j+1});
> ZIGZAG(F'_{k+1}, v'_1, v'_p);

In figure 5 an example of the triangulation is given.

Lemma 3.6 *The triangulation of F_{k+1} implies no multiple edges.*

Proof: Notice that if (w_i, w_j) is a forbidden edge, then there must be a vertex $w_x, i < x < j$, which is labeled. Let $v'_i = w_x$, then by adding the edges (v'_{i-1}, v'_i) and (v'_i, v'_{i+1}), there is no face F anymore with both $w_i, w_j \in F$. Hence the edge (w_i, w_j) will never be added by ZIGZAG to G. □

Lemma 3.7 *If there is an internal vertex, then the degree of the internal and new vertices increases by at most 5.*

Proof: Degrees of not labeled vertices increase by two by zigzagging F'_{k+1}. A labeled vertex v'_j gets two edges (v'_{j-1}, v'_j) and (v'_j, v'_{j+1}). Furthermore, v'_j gets two incident edges by zigzagging F'_{k+1}, one incident edge by ZIGZAG(F''_j, v'_j, v'_{j+1}), but no incident edges by ZIGZAG(F'''_j, v'_{j-1}, v'_j). Hence the degree of the labeled vertices increases by at most 5; the degree of the other vertices increases by at most 2. □

The problem arises when there is no internal vertex, because then at least one outgoing edge has to go to a left- or rightmost vertex, since they are neighbors on C_k. Let F_{k+1} be the added face with left- and rightmost vertex w_1 and w_2, respectively. w_1 and w_2 cannot be labeled, thus w_1 and w_2 are both element of F'_{k+1}. Calling ZIGZAG(F'_{k+1}, w_1, w_2) increases $deg(w_2)$ by 1 and does not increase $deg(w_1)$. We introduce a counter $extra(v)$ for every vertex v, counting the augmenting incident edges of v, added when v was left- or rightmost. When there is no internal vertex, we add an edge to the left- or rightmost vertex, according to the lowest $extra$-value, by ZIGZAG(F'_{k+1}, w_2, w_1) or ZIGZAG(F'_{k+1}, w_1, w_2), respectively. Initially $extra(v) = 0$ for all $v \in V$.

Lemma 3.8 *(i) For every four consecutive vertices v_1, v_2, v_3, v_4 on C_k, the following holds: If $extra(v_2) = 2$ then either $extra(v_1) = extra(v_3) = 0$ or $extra(v_1) = extra(v_4) = 0$ and $extra(v_3) = 1$. If $extra(v_2) = extra(v_3) = 1$ then $extra(v_1) = extra(v_4) = 0$. (ii) For every pair of consecutive vertices v_2, v_3 on C_k, $extra(v_2) + extra(v_3) \leq 3$. (iii) For every vertex v, $extra(v) \leq 2$.*

This implies that in every step when we add a face, the increase of $deg(v)$ is at most 5 when v is new or internal. Moreover, during all steps when adding a face, the increase of $deg(v)$, with v the left- or rightmost vertex, is totally at most 2. Hence in all phases of v, the increase of $deg(v)$ is bounded by a constant.

3.3 Adding a vertex

When we add a vertex v_{k+1} to G_k by the canonical 3-ordering, then we first replace the marked incident edges of v_{k+1} by the corresponding 2-subgraph. Assume there are t faces $F_1, \ldots, F_t$ to be triangulated, numbered from left to right. We do not want to increase the degree of the leftmost vertex $v \in F_1$ and the rightmost vertex $w \in F_t$. Therefore we add an edge between the two neighbors of v in F_1 and between the neighbors of w in F_t (only if F_1 and F_t are not triangles already). Due to these edges, $deg(v)$ and $deg(w)$ will not increase.

Moreover, to increase $deg(v_{k+1})$ as little as possible, we add in each face F_j an edge (w_1, w_2) between the two neighbors w_1 and w_2 of v_{k+1} in face F_j (only if F_j is not a triangle already). These newly added edges may imply multiple edges. In section 3.4 it is shown how this can happen and remedied. Remains now t reduced faces F_j to be triangulated. Hereto we apply the algorithm of section 3.2 to every face F_j. Doing this sequentially for $j = 1$ upto t, this will not imply multiple edges. It is easy to prove that an involved vertex v, belonging to two faces, F_j, F_{j+1}, can never be labeled, hence its degree will increase by at most 4. The degree of v_{k+1} will increase by at most 2 and the degree of the other vertices of F_j will increase by at most 5. Hence this increases the degree of the new vertex by at most 2 and the degree of the internal vertices by at most 5.

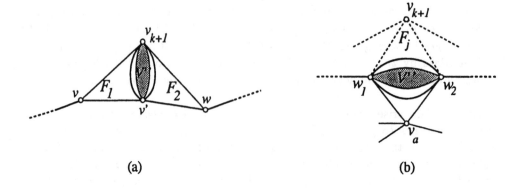

Figure 6: Reducing the faces may yield multiple edges.

3.4 Eliminating multiple edges

Applying the algorithms for triangulating one face in section 3.2 does not imply multiple edges, as proved in theorem 3.6. Hence applying this algorithm sequentially to the reduced faces $F_1, \ldots, F_t$ when adding a vertex does not imply multiple edges. However, reducing the faces $F_1, \ldots, F_t$ may imply multiple edges, as can be seen as follows:

1. We first add edges between the two neighbors of v and w in F_1 and F_t, respectively. If $t = 2$, then possibly these two neighbors of v and w are equal, say v'. Hence we have a multiple edge. To eliminate one of these two edges, an edge has to go to v or w. But when this occurs several times for v or w, the increase of $deg(v)$ or $deg(w)$ is not bounded by a constant.

2. Suppose there was already an edge between the neighbors w_1 and w_2 of v_{k+1} in F_j, hence adding such an edge implies a multiple edge. To eliminate this, an edge can go from v_{k+1} in F_j. But when this occurs for several faces F_j, the increase of $deg(v_{k+1})$ is not bounded by a constant.

To eliminate all multiple edges, obtained by reducing the faces $F_1, \ldots, F_t$, we construct a bipartite planar graph H. (One set of vertices in H corresponds to a subset of the vertices in G, one set corresponds to 2-subgraphs that must have an extra outgoing edge to remove one pair of multiple edges.) For case 1. and case 2. we do the following:

case 1. We represent the 2-subgraph V'' between v_{k+1} and v' by a vertex v'' in H. To eliminate the multiple edge, there has to go an edge from V'' to v or w. Hereto we add v and w to H (if not present already) with edges to v''.

case 2. Suppose we have the multiple edge (w_1, w_2) with the 2-subgraph V'' between these two edges. Let v_{k+1} and v_a be the other vertices, not part of V'', which have both w_1 and w_2 as neighbors (as in figure 6(b)). Then an edge has to go from v_a or v_{k+1} to V''. Hereto we represent V'' by a vertex v'' in H. Add the vertices v_a and v_{k+1} to H (if not present already) with edges to v''.

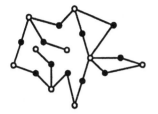 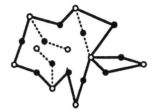 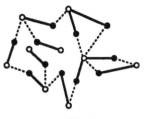

(a) example of H with white and black vertices.

(b) The elementary cycles and the disjoint paths.

(c) The set M of edges.

Figure 7: The construction of H and M.

H will be constructed during all steps of the triangulation of the entire graph G. H is planar and bipartite. Call vertices v'' representing the 2-subgraphs V'' *black* and the other vertices of H *white* (see figure 7(a)). We have to find a subset $M \subseteq E_H$ such that every black vertex v' has one incident edge $(v', v) \in M$ and the white vertices have as little as possible incident edges in M. These edges in M correspond with the edges, which will be added to eliminate the multiple edges. To obtain this, we do the following: using a simple modification of Eulers technique to find an Eulerian cycle in a graph, we can extract the elementary cycles C_{elem} from H. An elementary cycle is a cycle that uses each edge at most once. Thus $H - C_{elem}$ consists of paths P with disjoint begin- and endpoints (see figure 7(b)). From C_{elem} and every path P we add alternatingly one edge to M and one not. Recall that H is bipartite, and all white vertices have degree 2. Hence for every vertex in C_{elem} and every internal vertex of a path P, one incident edge e is in M and the other is not. But also for every black vertex, exactly one incident edge is in M, hence satisfying the constraints (see figure 7(c)). For every vertex $v \in H$ with $deg_H(v) \geq 2$, at most $\lceil deg_H(v)/2 \rceil$ neighbors are in M.

For each black vertex v'' exactly one incident edge, say (v'', v), is in M. We remove the corresponding multiple edge (w_1, w_2) or (v_{k+1}, v') in G' and we add the edge from a vertex in V'' v. This can easily be constructed in linear time. This removes all multiple edges and gives a triangulated planar graph G'.

Theorem 3.9 $\Delta(G') \leq \lceil \frac{3}{2}\Delta(G) \rceil + 21$.

Proof: Every vertex v receives at most 2 edges to admit biconnectivity. v receives at most 5 edges when it is new, at most 2 edges when it is left- or rightmost, and at most 5 edges when it is internal. Thus in G v receives at most 14 extra incident edges. In H from every vertex v at most $\lceil deg'(v)/2 \rceil$ incident edges are in M, implying $\lceil deg'(v)/2 \rceil$ augmenting edges in G to destroy the multiple edges. Since $deg'(v) \leq \Delta(G) + 14$, the theorem follows. □

Theorem 3.10 G' *can be constructed in linear time.*

We omit the proof of this result for space reasons. This completes the proof of theorem 3.1. In case G is already triconnected, a simpler variant of this algorithm can be used, as multiple edges cannot occur. One can show:

Theorem 3.11 *There is a linear algorithm to triangulate a triconnected planar graph G such that for the triangulation $G', \Delta(G') \leq \Delta(G) + 5$ holds.*

4 Triangulating one face

In this section we give a polynomial algorithm for the following problem: given a planar graph G with some embedding, triangulate one face F of G while minimizing the maximum degree.

Let the face F on G on n vertices $v_1, \ldots, v_n$ be given (numbered clockwise around the ring). We notice that in every triangulation the vertices v_1 and v_n have a common neighbor v_k ($2 \leq k \leq n-1$) inside face F, which splits the face F into two faces F_1 (with vertices $v_1, \ldots, v_k$) and F_2 (with vertices $v_k, \ldots, v_n$). If for some k, (v_1, v_k) or (v_k, v_n) is already present outside F, then for this value of k, this triangulation is not possible, since it would imply multiple edges. We can recursively triangulate the faces F_1 and F_2. Let F_1' and F_2' denote the triangulated faces, then the highest degree in F_1' and F_2' is important, and since F_1 and F_2 share v_k, the increase of $deg(v_k)$ in F_1' and F_2' must be added to $deg(v_k)$ in F. When we examine triangulations of a face F_{ij}, formed by vertices $v_i, v_{i+1}, \ldots, v_j$, we inspect the different values of increases of $deg(v_i), deg(v_j)$ and $deg(v_k)$ in F_1 and F_2. For avoiding case analysis when $k = 2$ or $k = n-1$, we delete the edges (v_i, v_{i+1}) ($1 \leq i < n$) and decrease $deg(v_i)$ by 2. Let $inc(v_i)$ denote the increase of $deg(v_i)$ when triangulating F_1 to F_1' (assuming $v_i \in F_1$). For a triangulation of a face F_{ij} with vertices $v_i, \ldots, v_j$ we have to store the different increases of $deg(v_i)$ and $deg(v_j)$ in a table. Let $D[i, j, i1, j1]$ be the maximum degree of F_{ij} by a triangulation such that $inc(v_i) = i1$ and $inc(v_j) = j1$. If such a triangulation does not exist, $D[i, j, i1, j1] = \infty$. A simple analysis shows the following recursive formulae if $i < j - 1$:

$$
D[i, j, i1, j1] := \min_{\substack{i < k < j \\ k1, k2 \\ (i,k) \text{ and } (j,k) \\ \text{not forbidden}}} \left\{ \max \left\{
\begin{array}{l}
D[i, k, i1 - 1, k1], D[k, j, k2, j1 - 1], \\
deg(v_i) + i1 + 1, \\
deg(v_j) + j1 + 1, \\
deg(v_k) + k1 + k2 + 2
\end{array}
\right\} \right\}
$$

If $i = j - 1$ then for all $i1, j1 \geq 0$: $D[i, j, i1, j1] = \max\{deg(v_i), deg(v_j)\}$. We want to compute $\min_{i1,j1}\{D[1, n, i1, j1], deg(v_1) + i1 + 1, deg(v_n) + j1 + 1\}$. We do this by using dynamic programming, based on the above formulaes, and some other ideas which help to decrease the running time of the algorithm, including binary search on the maximum degree. Details are omitted from this version.

Theorem 4.1 *There is an exact $O(n^3 \Delta(G) \log \Delta(G))$ algorithm to triangulate one face of n vertices of a graph G such that the maximum degree of the triangulation G' is minimized.*

The algorithm can be used in the following alternative for theorem 3.1: Use the canonical 3-ordering as described in section 3 and in each step we add a vertex or face to G_k and triangulate this face optimally. Though this seems to give a good approximation, there are inputs (see figure 1(b)), for which this algorithm will imply $\Delta(G') = 2\Delta + O(1)$. Similar to theorem 4.1, we can obtain:

Theorem 4.2 *There is an $O(n^3 \Delta(G) \log \Delta(G))$ algorithm to triangulate all interior faces of an outerplanar graph while minimizing the maximum degree.*

5 Final Remarks

In this paper we inspected the problem of triangulating a planar graph such that the maximum degree is minimized. It is shown that this problem is NP-complete for biconnected planar graphs. A linear approximation algorithm is presented, working only a constant from the presented lower bounds. This algorithm is heavily based on the canonical 3-ordering for triconnected planar graphs, which is a modification of the canonical ordering of [5]. This ordering can also be determined in linear time and leads to a good approximation algorithm. The main idea is: only information of vertices on the outerface is sufficient in every step. This technique also already lead to a linear implementation of the grid drawing algorithm of De Fraysseix et al. [12]. It is interesting to inspect related problems on planar graphs, for which the canonical 3-ordering also can lead to simple linear time approximations, by only maintaining local information in the vertices of the outerface.

This paper also gives more insight in the augmentation problems, which seems to be quite popular nowadays [4, 9, 10, 11, 13]. However, we were not able to prove that the algorithm in this paper works only an additive constant from optimal in the biconnected case. We conjecture that the algorithm in section 4 is only an additive constant worse than optimal, in case the input graph is biconnected. This conjecture is still open and interesting for further study. The NP-completeness result presented in this paper only holds for biconnected planar graphs and is open for triconnected planar graphs. It is interesting to inspect this problem, to come to a more combinatorial insight in planar graphs.

Acknowledgements

The authors wish to thank David Eppstein, for proposing the dynamic programming approach, described in section 4.

References

[1] Booth, K.S., and G.S. Lueker, Testing for the consecutive ones property, interval graphs and graph planarity testing using PQ-tree algorithms, *J. of Computer and System Sciences* 13 (1976), pp. 335–379.

[2] Chiba, N., T. Yamanouchi and Nishizeki, Linear algorithms for convex drawings of planar graphs, In: J.A. Bondy and U.S.R. Murty (Eds.), *Progress in Graph Theory*, Academic Press, Toronto, 1984, pp. 153–173.

[3] Eswaran, K.P., and R.E. Tarjan, Augmentation problems, *SIAM J. Comput.* 5 (1976), pp. 653–665.

[4] Frank, A., Augmenting graphs to meet edge-connectivity requirements, *Proc. 31th Annual IEEE Symp. on Found. on Comp. Science*, St. Louis, 1990, pp. 708–718.

[5] Fraysseix, H. de, J. Pach and R. Pollack, How to draw a planar graph on a grid, *Combinatorica* 10 (1990), pp. 41–51.

[6] Fraysseix, H. de, and P. Rosenstiehl, A depth first characterization of planarity, *Annals of Discrete Math.* 13 (1982), pp. 75–80.

[7] Haandel, F. van, *Straight Line Embeddings on the Grid*, Dept. of Comp. Science, Report no. INF/SCR-91-19, Utrecht University, 1991.

[8] Hopcroft, J., and R.E. Tarjan, Efficient planarity testing, *J. ACM* 21 (1974), pp. 549–568.

[9] Hsu, T., and V. Ramachandran, A linear time algorithm for triconnectivity augmentation, in: *Proc. 32th Annual IEEE Symp. on Found. on Comp. Science*, Porto Rico, 1991.

[10] Hsu, T., and V. Ramachandran, *On Finding a Smallest Augmentation to Biconnect a Graph*, Computer Science Dept., University of Texas at Austin, Texas, Tech. Rep. TR-91-12, 1991.

[11] Kant, G., *Optimal Linear Planar Augmentation Algorithms for Outerplanar Graphs*, Techn. Rep. RUU-CS-91-47, Dept. of Computer Science, Utrecht University, 1991.

[12] Kant, G., *A Linear Implementation of De Fraysseix' Grid Drawing Algorithm*, Manuscript, Dept. of Comp. Science, Utrecht University, 1988.

[13] Kant, G., and H.L. Bodlaender, Planar graph augmentation problems, Extended Abstract in: F. Dehne, J.-R. Sack and N. Santoro (Eds.), *Proc. 2nd Workshop on Data Structures and Algorithms*, Lecture Notes in Comp. Science 519, Springer-Verlag, Berlin/Heidelberg, 1991, pp. 286–298.

[14] Read, R.C., A new method for drawing a graph given the cyclic order of the edges at each vertex, *Congr. Numer.* 56 (1987), pp. 31–44.

[15] Rosenthal, A., and A. Goldner, Smallest augmentations to biconnect a graph, *SIAM J. Comput.* 6 (1977), pp. 55–66.

[16] Schnyder, W., Embedding planar graphs on the grid, in: *Proc. 1st Annual ACM-SIAM Symp. on Discr. Alg.*, San Francisco, 1990, pp. 138–147.

[17] Tutte, W.T., Convex representations of graphs, *Proc. London Math. Soc.*, vol. 10 (1960), pp. 304–320.

[18] Woods, D., *Drawing Planar Graphs*, Ph.D. Dissertation, Computer Science Dept., Stanford University, CA, Tech. Rep. STAN-CS-82-943, 1982.

How to Draw a Series-Parallel Digraph
(Extended Abstract)

*P. Bertolazzi**   *R. F. Cohen[†]*   *G. Di Battista[‡]*

R. Tamassia[†]   *I. G. Tollis[§]*

Abstract. Upward and dominance drawings of acyclic digraphs find important applications in the display of hierarchical structures such as PERT diagrams, subroutine-call charts, and is-a relationships. The combinatorial model underlying such hierarchical structures is often a series-parallel digraph. In this paper the problem of constructing upward and dominance drawings of series-parallel digraphs is investigated. We show that the area requirement of upward and dominance drawings of series-parallel digraphs crucially depends on the choice of planar embedding. Also, we present parallel and sequential drawing algorithms that are optimal with respect to both the time complexity and to the area achieved. Our results show that while series-parallel digraphs have a rather simple and well understood combinatorial structure, naive drawing strategies lead to drawings with exponential area, and clever algorithms are needed to achieve optimal area.

1 Introduction

In the field of graph drawing algorithms, special attention is devoted to planar graphs and to their planar representations. A *drawing* of a graph represents each vertex with a distinct point of the plane and each edge with a simple curve. In a *straight-line* drawing each edge is a straight-line segment. A *grid* drawing is such that vertices are points with integer coordinates. A *planar* drawing is such that no two edges intersect but (possibly) at common endpoints.

An *upward* drawing of an acyclic digraph is a planar straight-line drawing with the additional requirement that all the edges flow in the same direction, e.g., from bottom to top. See in Fig. 1.a an example of upward drawing. Upward drawings embody the notion of planarity for acyclic digraphs and are attracting an increasing theoretical (see, e.g., [6,7,9,20,4,12,1]) and practical (see, e.g., [15,5,16]) interest.

---

*IASI – CNR, Viale Manzoni 30, 00185 Rome, Italy.

[†]Department of Computer Science, Brown University, Providence, RI 02912–1910. Research supported in part by the National Science Foundation under grant CCR-9007851, by the U.S. Army Research Office under grant DAAL03-91-G-0035, by the Office of Naval Research and the Defense Advanced Research Projects Agency under contract N00014-91-J-4052, ARPA order 8225, and by Cadre Technologies, Inc.

[‡]Dipartimento di Informatica e Sistemistica, Università di Roma "La Sapienza", Via Salaria 113, 00198 Rome, Italy. Research supported in part by the ESPRIT II Basic Research Actions Program of the EC under Contract No. 3075 (project ALCOM), and by the Progetto Finalizzato Sistemi Informatici e Calcolo Parallelo of the Italian National Research Council.

[§]Department of Computer Science, The University of Texas at Dallas, Richardson, TX 75083–0688.

A planar straight-line drawing of an acyclic digraph is a *dominance drawing* [7] if for any two vertices u and v, there is a directed path from u to v if and only if $x(u) \leq x(v)$ and $y(u) \leq y(v)$. Notice that these two conditions cannot be simultaneously satisfied with equality since distinct vertices must be placed at distinct points. See in Fig. 1.b an example of a dominance drawing. Dominance drawings have the important feature of characterizing the transitive closure of a digraph by means of the geometric dominance relation among the vertices.

Planar drawings have been investigated from the point of view of their *area requirement*. The area of a drawing is defined as the area of the smallest covering rectangle with sides parallel to the axes, where we assume the existence of a resolution rule that prevents drawings to be arbitrarily scaled down (e.g., a minimum distance between vertices, or integer vertex coordinates). The existence of planar straight-line drawings with polynomial area for any planar undirected graph has been one of the most intriguing open problems in this field. This question has been settled by de Fraysseix, Pach, and Pollak [3] and, independently, by Schnyder [14], who show that every n-vertex planar graph has a planar straight-line grid drawing with $O(n^2)$ area.

Concerning acyclic digraphs, in [7] it is shown that there exists a family of planar acyclic digraphs that require exponential area in any upward drawing. This result sharply contrasts with the one given in [3,14] and leaves open the problem of identifying the structural properties of acyclic digraphs that affect the area requirement of upward and dominance drawings.

An *embedding* of a planar graph is a circular ordering of the edges incident on each vertex, as determined by a planar drawing. An *embedded* planar graph is a planar graph with a given embedding. Note that a planar graph may have an exponential number of embeddings.

From the point of view of the embeddings, the aforementioned results on drawing undirected planar graphs can be interpreted as follows: any embedded planar graph has an embedding-preserving planar straight-line drawing with $O(n^2)$ area. Thus, informally speaking, for producing straight-line drawings of planar undirected graphs any planar embedding is a "good" one.

What happens for acyclic digraphs? The answer given by [7] is that there exist

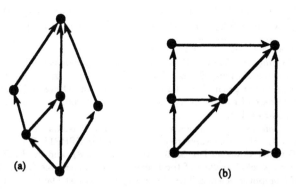

Figure 1: An upward drawing and a dominance drawing

"bad" planar embeddings such that their embedding-preserving upward drawings require exponential area. Moreover, there exist acyclic digraphs with only one planar embedding requiring exponential area.

In this paper we deal with the problem of constructing straight-line upward and dominance drawings of *series-parallel* digraphs. Such digraphs arise in a variety of problems such as scheduling, electrical networks, data-flow analysis, database logic programs, and circuit layout.

In Section 2, we review basic definitions and argue that in drawing series-parallel digraphs the choice of a "good" planar embedding is crucial. Namely, we show that there exists a family of embedded series-parallel digraphs that require exponential area in any embedding-preserving upward drawing.

In Section 3, we present an algorithm, called Δ-algorithm, that, given an embedded series-parallel digraph G with n vertices, constructs an upward grid drawing of G with $O(n^2)$ area by slightly modifying the embedding.

In Section 4, we show that for constructing dominance drawings the choice of an embedding must be more clever than the one for upward drawings. First, we show that the strategy used by the Δ-algorithm for restructuring the initial embedding yields dominance drawings that require exponential area. Second, we present a variation of the Δ-algorithm that constructs dominance grid drawings with $O(n^3)$ area and such that symmetries of the digraph are displayed.

In Section 5, we investigate the sequential and parallel complexity of the Δ-algorithm. We show that it can be implemented to run in $O(\log n)$ time on an EREW PRAM with $n/\log n$ processors, and hence in $O(n)$ sequential time.

2 A Lower Bound on the Area

A *series-parallel digraph* is recursively defined as follows: An edge joining two vertices is a series-parallel digraph. Let G_1 and G_2 be two series-parallel digraphs. Their series and parallel compositions defined below are also series-parallel digraphs: the series composition of G_1 and G_2 is the graph obtained identifying the sink of G_1 with the source of G_2; the parallel composition of G_1 and G_2 is the graph obtained by identifying the source of G_1 with the source of G_2 and the sink of G_1 with the sink of G_2.

An embedded series-parallel digraph G is naturally associated with a rooted binary tree T, which we call *decomposition tree* of G (also known as *parse tree*). The nodes of T are of three types: S-nodes, P-nodes, and Q-nodes. Tree T is defined recursively as follows: If G is a single edge, then T consists a single Q-node. If G is created by the parallel composition of series-parallel digraphs G' and G'', where G' is to the left of G'' in the embedding, let T' and T'' be the decomposition trees of G' and G'', respectively. The root of T is a *P-node* and has left subtree T' and right subtree T''. If G is created by the series composition of series-parallel digraphs G' and G'', where the sink of G' is identified with the source of G'', let T' and T'' be the decomposition trees of G' and G'', respectively. The root of T is an *S-node* and has left subtree T' and right subtree T'''. The leaves of T are Q-nodes. The internal nodes are either P-nodes or S-nodes. If G has n vertices, then T has $O(n)$ leaves and hence $O(n)$ nodes. Tree T can be constructed sequentially in $O(n)$ time using the recognition algorithm of [19].

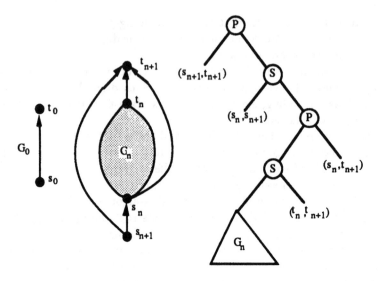

Figure 2: Recursive construction of the series-parallel digraph G_n

In the rest of this section we show that there exist embedded series-parallel digraphs such that any of their upward straight-line drawing that preserves the embedding requires exponential area.

We define a class of series-parallel digraphs with a planar embedding as follows (see Fig. 2): G_0 consists of two vertices s_0 and t_0 and an edge from s_0 to t_0. G_{n+1} is obtained from G_n by adding a new source s_{n+1} and a new sink t_{n+1}, an edge from t_n to t_{n+1}, an edge from s_{n+1} to s_n, an edge from s_n to t_{n+1} on the right of the embedding of G_n and an edge from s_{n+1} to t_{n+1} on the left of the embedding of G_n.

Theorem 1 *Any upward drawing of the $2n$-vertex embedded digraph G_n that preserves the embedding requires area $\Omega(4^n)$, under any resolution rule.*

Proof: We show that $Area(G_{n+1}) \geq 4 \cdot Area(G_n)$. Since any resolution rule guarantees a constant lower bound on the area of G_1, we have that $Area(G_n) = \Omega(4^n)$.

The drawing of G_n is bounded by a triangle Δ_n with vertices s_n, s_{n-1}, and t_n (see Fig. 3). Let σ and τ be the horizontal lines through s_n and t_n, respectively. Because of the upward requirement, vertex s_{n+1} must lie below σ, and vertex t_{n+1} must lie above τ. Let ρ be the line extending the edge (s_n, s_{n-1}). Since t_{n+1} must be connected to s_n from the right, it must lie to the right of ρ. Also, since t_{n+1} must be connected to s_{n+1} from the left, there exists a line λ through t_n intersecting σ and ρ, such that vertices s_{n+1} and t_{n+1} are to the left of λ. Hence, the drawing of G_{n+1} must contain the triangle Δ_{n+1} defined by lines σ, λ, and ρ. We will show that $Area(\Delta_{n+1}) \geq 4 \cdot Area(\Delta_n)$.

Consider the parallelogram Π defined by σ, τ, ρ, and the line through t_n parallel to ρ. Since edge (s_n, t_n) is a diagonal of Π, we have that $Area(\Pi) \geq 2 \cdot Area(\Delta_n)$. Triangle Δ_{n+1} is the union of Π and two similar triangles Δ' and Δ'' below and above τ, respectively. Without loss of generality, assume that $Area(\Delta') \geq Area(\Delta'')$ (otherwise a symmetric argument applies). Let p be the intersection of ρ and τ. The line through p parallel to λ partitions Π into a triangle and a quadrilateral, where

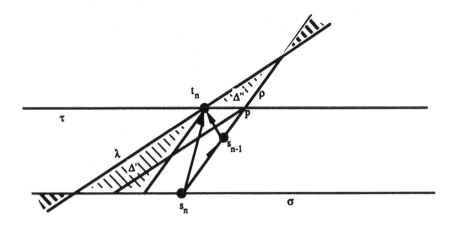

Figure 3: Illustration of the proof of Theorem 1

the triangle is congruent to Δ'', and the quadrilateral is congruent to a portion of Δ'. Hence, we conclude

$$Area(\Delta_{n+1}) = Area(\Pi) + Area(\Delta') + Area(\Delta'') \geq 2 \cdot Area(\Pi) \geq 4 \cdot Area(\Delta_n). \qquad \square$$

Notice that there exists a non-upward straight-line drawing of G_n that preserves the embedding with $O(n^2)$ area. It can be constructed with any of the algorithms presented in [3,14].

3 Upward Drawings

In this section, we show how to construct an upward drawing of a series-parallel digraph G with n vertices. We present a drawing algorithm that modifies the embedding of a given series-parallel digraph so that all the transitive edges are embedded on one side, say, the right side. We call such embedding *right-pushed*. A variation of this algorithm constructs upward grid drawings with $O(n^2)$ area.

The algorithm recursively produces a drawing Γ of G inside a *bounding* triangle $\Delta(\Gamma)$ that is isosceles and right-angled (see Fig. 4). Hereafter we denote the sides of $\Delta(\Gamma)$ with *base* (hypothenuse), *top side*, and *bottom side*. In a series composition, the subdrawings are placed one above the other. In a parallel composition, the subdrawings are placed one to the right of the other and are deformed in order to identify the end vertices, guaranteeing that their edges do not cross. The algorithm is described here at a level of detail that allows to deal with its correctness. It will be detailed further in Section 5 for the time complexity analysis.

Δ-algorithm
Input: a series-parallel digraph G and a decomposition tree of G;
Output: an upward drawing Γ of G.

- Modify the embedding of G into a right-pushed embedding and perform the corresponding modifications on T.
- If G consists of a single edge, it is drawn as a vertical segment of length 2, with bounding triangle having width 1 (see Fig. 4.a).

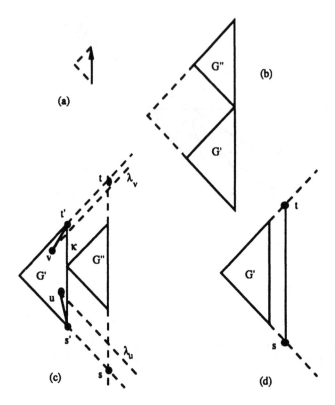

Figure 4: The behavior of the Δ-algorithm

- If G is the series composition of G' and G'', the two drawings Γ' and Γ'' of G' and G'' are first recursively produced. Then, Γ is drawn by translating Γ'' so that the sink of G' is identified with the source of G'' (see Fig. 4.b). The bounding triangle $\Delta(\Gamma)$ is obtained by extending the bottom side of $\Delta(\Gamma')$ and the top side of $\Delta(\Gamma'')$.

- Suppose that G is the parallel composition of G' and G''. The two drawings Γ' and Γ'' of G' and G'' are first produced. We consider the rightmost edges (s', u) and (v, t') incident on the source and sink of G', respectively (see Fig. 4.c). Let λ_u be the line through u that is parallel to the bottom side of the bounding triangle of G', and λ_v be the line through v that is parallel to the top side of the triangle of G. Also, let κ be the vertical line extending the base of the triangle of G'. We call *prescribed-region(Γ')* the region to the right of κ, λ_u, and λ_v. First, we translate Γ'' so that its triangle is anywhere inside prescribed-region(Γ'). Then, we identify the sources and sinks of G' and G'' by moving them to the intersections s and t of the base of $\Delta(G'')$ with the lines extending the top and bottom sides of $\Delta(G')$, respectively.

Note that in the parallel composition of G' with a single edge (the transitive edge (s, t) from the source to the sink of G), we move the source and sink of G' to the intersection of a vertical line (at least one unit) to the right of $\Delta(G')$, and the lines extending the top and bottom sides of $\Delta(G')$, respectively. See Fig. 4.d.

Theorem 2 *The Δ-algorithm correctly constructs an upward drawing Γ of a series-parallel digraph G.*

Proof: The proof of correctness is based on maintaining the following invariants:

1. The drawing is contained inside an isosceles right-angled triangle $\Delta(\Gamma)$, such that the base is vertical, and the other sides are to the left of the base.

2. The source and sink are placed at the bottom and top corner of $\Delta(\Gamma)$, respectively. The left corner of $\Delta(\Gamma)$ is not occupied by any vertex of G.

3. For any vertex v adjacent to the source s of G, the wedge formed at v by the rays with slopes $-\pi/2$ and $-\pi/4$ does not contain any vertex of G except s.

4. For any vertex v adjacent to the sink t of G, the wedge formed at v by the rays with slopes $\pi/2$ and $\pi/4$ does not contain any vertex of G except t.

Clearly, by the construction of the algorithm, Invariants 1 and 2 are always satisfied. Invariant 3 is immediately satisfied after a series composition since the relative position of the vertices of G' remains unchanged, and all the vertices of G'' are placed above the sink of G'. To continue the proof, we need the following lemma:

Lemma 1 *Let u' and u'' be neighbors of the source vertex s, such that edge (s, u') is to the left of edge (s, u''), and let $\lambda_{u'}$ and $\lambda_{u''}$ be the rays of slope $-\pi/4$ originating at u' and u'', respectively. If Invariant 3 holds, then $\lambda_{u'}$ is below $\lambda_{u''}$.*

In a parallel composition, G'' is represented to the right of G' and is above line λ_u, where (s, u) is the rightmost edge incident on the source s. By Lemma 1 no vertex of G'' is inside the angles associated with the neighbors of the source. Hence, Invariant 3 is satisfied for G if it was satisfied for G' and G''. Invariant 4 can be proved in a similar fashion as Invariant 3.

The above invariants show that every composition step yields a correct drawing provided the components are correctly drawn. A simple inductive argument completes the proof. □

As described in the algorithm, the series composition of two components exactly determines the relative positions of Γ' and Γ'' by identifying the source of G'' with the sink of G'. However, we have not described how to exactly place Γ'' with respect to Γ' in the parallel composition. We simply said that Γ'' has to be placed inside prescribed-region(Γ').

A possibility consists of translating Γ'' in the prescribed region so that the left corner of $\Delta(\Gamma'')$ is placed on the base of $\Delta(\Gamma')$. By Invariant 2 there is no vertex of G'' on this corner. Hence, the drawing is correct. This placement yields drawings with $O(n^2)$ area. In order to prove this bound we observe that the base of the resulting triangle is always equal to the sum of the bases of the triangles of Γ' and Γ''. Therefore the length of the base of $\Delta(\Gamma)$ is equal to $2 \cdot m$, where m is the number of edges of G. Hence the area of Γ is proportional to m^2. We conclude:

Theorem 3 *Let G be a series-parallel digraph with n vertices. A variation of the Δ-algorithm produces an upward grid drawing of G with $O(n^2)$ area.*

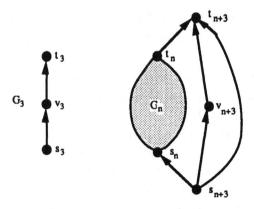

Figure 5: A class of series-parallel digraphs with a right-pushed embedding

4   Symmetric and Dominance Drawings

In this section we investigate dominance drawings of series-parallel digraphs.

First, we show that right-pushed embeddings may be "bad" with respect to the construction of dominance drawings. Namely, we exhibit a family of series-parallel digraphs with a right-pushed embedding requiring exponential area in any embedding-preserving dominance drawing.

Second, it is shown in [7] that every planar st-graph without transitive edges admits a dominance grid drawing with $O(n^2)$ area. Here we show that series-parallel digraphs admit planar straight-line dominance grid drawings with polynomial area, even in the presence of transitive edges. The drawings are obtained with a new strategy for modifying the embedding. We also show that the algorithm for producing dominance drawings constructs more symmetric drawings, with respect to the drawings of Theorem 3.

We define the following class of series-parallel digraphs with a right-pushed embedding (see Fig. 5). Digraph G_3 consists of vertices s_3, t_3, and v_3, and edges (s_3, v_3) and (v_3, t_3). For $n = 3k$ ($k \geq 1$) digraph G_{n+3} is constructed from G_n by adding vertices s_{n+3}, t_{n+3}, and v_{n+3}, and edges (s_{n+3}, s_n), (s_{n+3}, v_{n+3}), (t_n, t_{n+3}), (v_{n+3}, t_{n+3}), and (s_{n+3}, t_{n+3}), with the embedding shown in Fig. 5.

Theorem 4 *Any dominance drawing of the n-vertex embedded digraph G_n that preserves the embedding has area $\Omega(\sqrt[3]{4}^{n})$, under any resolution rule.*

Proof: Let A_n be the minimum area of a dominance drawing of G_n. We inductively prove that $A_{n+3} \geq 4 \cdot A_n$. Since $A_3 \geq c$, for some constant c depending on the resolution rule, this implies the claimed result.

Let Γ_{n+3} be a dominance drawing with minimum area A_{n+3}. By removing from Γ_{n+3} vertices s_{n+3}, t_{n+3}, and v_{n+3}, and their incident edges, we obtain a dominance drawing Γ_n of G_n. See Fig. 6. Let σ_1 and τ_1 be the horizontal lines through s_n and t_n, respectively. Let τ_2 and σ_2 be the vertical lines through s_n and t_n, respectively. Let λ_1 be the line through s_n and t_n. Because of the dominance hypothesis and the fact that (s_n, t_n) is the rightmost edge of Γ_n, we have that Γ_n is contained in the triangle Δ_n formed by lines τ_1, τ_2, and λ_1. Because of the dominance hypothesis we

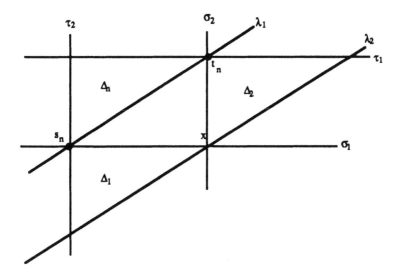

Figure 6: Illustration of the proof of Theorem 4

have also that:

1. s_{n+3} has to be placed to the left of τ_2 (included);
2. t_{n+3} has to be placed above τ_1 (included); and
3. v_{n+3} has to be placed below σ_1 (excluded) and to the right of σ_2 (excluded).

Let x be the intersection of σ_1 and σ_2; edge (s_{n+3}, t_{n+3}) has to be placed on the right of x and under x. Let λ_2 be the straight line through x and parallel to (s_{n+3}, t_{n+3}). Let Δ_1 be the triangle formed by lines σ_1, τ_2, and λ_2. Let Δ_2 be the triangle formed by lines σ_2, τ_1, and λ_2. We have that $A_{n+3} \geq A_n + Area(\Delta_n) + Area(\Delta_1) + Area(\Delta_2)$. But $Area(\Delta_1) + Area(\Delta_2) \geq 2 Area(\Delta_n)$ and $Area(\Delta_n) \geq A_n$. Hence, $A_{n+3} \geq 4 A_n$. □

In order to obtain dominance drawings with polynomial area we specialize the Δ-algorithm as follows.

- Modify the embedding in a right-pushed embedding with the additional requirement that in any parallel composition the smallest component is placed to the left of the largest one.

- In the parallel composition of G' and G'', place Γ'' so that the left corner of $\Delta(\Gamma'')$ is at the intersection of the rays with slopes $\pi/4$ and $-\pi/4$ originating from the source and sink of Γ', respectively.

- At the end, we rotate the entire drawing by $\pi/4$ clockwise.

Observe that, if Γ' and Γ'' are dominance drawings, our placement guarantees that their parallel composition is also a dominance drawing. Note also that a series composition of two dominance drawings is a dominance drawing since all the vertices of Γ'' are placed above and to the right of the sink of Γ'.

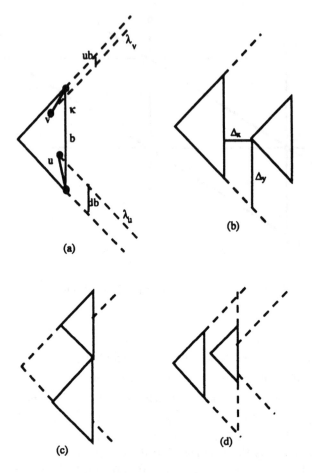

Figure 7: b, db, ub, Δ_x, and Δ_y

Theorem 5 *Let G be a series-parallel digraph with n vertices. A variation of the Δ-algorithm produces an $O(n^3)$-area dominance grid drawing of G that displays symmetries and isomorphisms of subgraphs of G.*

5 Complexity Analysis

In this section we first show how to implement the Δ-algorithm sequentially and then we discuss its parallelization.

Given a drawing Γ produced by the Δ-algorithm, we describe the bounding triangle $\Delta(\Gamma)$ by means of the length b of its base, and describe prescribed-region(Γ) by means of parameters db and ub, where db is the vertical distance between λ_u and the bottom corner of $\Delta(\Gamma)$, and db is the vertical distance between λ_v and the top corner of $\Delta(\Gamma)$. See Fig. 7.a.

The root of each subtree T of the decomposition tree of G is labelled by a procedure, called *Label*, with the values of $b(T)$, $db(T)$, and $ub(T)$ that describe the bounding triangle and the prescribed region of the upward drawing of the digraph whose decomposition tree is T.

In Procedure *Label* we suppose that in each parallel composition the displacement of $\Delta(\Gamma_2)$ with respect to the base of $\Delta(\Gamma_1)$ is denoted by means of Δ_x and Δ_y, where Δ_x is the horizontal distance between $\Delta(\Gamma_1)$ and $\Delta(\Gamma_2)$ and Δ_y is the vertical distance between the line extending the bottom side of $\Delta(\Gamma_1)$ and the left corner of $\Delta(\Gamma_2)$. See Fig. 7.b.

In order to place $\Delta(\Gamma_2)$ into the prescribed region of Γ_1 we must have that $\Delta_y \geq db(T_1)$, and $\Delta_y \leq b(T_1) - ub(T_1)$, where T_1 is the decomposition tree of G_1. Observe that in the variation of the Δ-algorithm of Theorem 3, $\Delta_x = 0$ and Δ_y can be choosen equal to $db(T_1)$.

procedure *Label* (T: decomposition tree of a series-parallel digraph G)
if $root(T)$ is a Q-node
then
    $b(T) = ub(T) = db(T) = 2$
else
    let T_1 and T_2 be the left and right subtrees of T, respectively
    for each $i = 1, 2$ **do** Label(T_i)
    if $root(T)$ is an S-node
    then (See Fig. 7.c)
        $b(T) = b(T_1) + b(T_2)$
        $db(T) = db(T_1)$
        $ub(T) = ub(T_2)$
    else (See Fig. 7.d)
        $b(T) = b(T_1) + b(T_2) + 2\Delta_x$
        $ub(T) = b(T_1) + 2\Delta_x - \Delta_y + ub(T_2)$
        $db(T) = db(T_2) + \Delta_y$

The labelled decomposition tree obtained in this way is an implicit representation of the drawing. In fact, it is easy to visit in pre-order the tree to obtain a drawing of the nested triangles. From that drawing it is then immediate to compute the coordinates of the vertices.

Theorem 6 *The Δ-algorithm can be implemented to run in $O(n)$ time and space on a series-parallel digraph with n vertices.*

Regarding the parallel implementation, we observe that the label procedure corresponds to evaluating arithmetic expressions over the decomposition tree. This can be done optimally in parallel using tree-contraction techniques, see e.g., [11]. To compute the final coordinates from the parameters of the bounding triangles we use a variation of the *Euler tour technique* [18].

Theorem 7 *The Δ-algorithm can be implemented to run in $O(\log n)$ time on an n vertex series-parallel digraph, using an EREW PRAM with $n/\log n$ processors.*

As shown in [2] the Δ-algorithm can also be efficiently dynamized.

References

[1] P. Bertolazzi and G. Di Battista, "On Upward Drawing Testing of Triconnected Digraphs," *Proc. 7th ACM Symp. on Computational Geometry*, 1991.

[2] R. Cohen, G. Di Battista, R. Tamassia, I.G. Tollis, and P. Bertolazzi, "A Framework for Dynamic Graph Drawing," *Proc. 8th ACM Symp. on Computational Geometry*, 1992.

[3] H. de Fraysseix, J. Pach, and R. Pollack, "Small Sets Supporting Fary Embeddings of Planar Graphs," *Proc. 20th ACM Symp. on Theory of Computing*, 1988.

[4] G. Di Battista, W.P. Liu, and I. Rival, "Bipartite Graphs, Upward Drawings, and Planarity," *Information Processing Letters*, 36, 317-322, 1990.

[5] G. Di Battista, E. Pietrosanti, R. Tamassia, and I.G. Tollis, "Automatic Layout of PERT Diagrams with X-PERT," *Proc. IEEE Workshop on Visual Languages*, 1989.

[6] G. Di Battista and R. Tamassia, "Algorithms for Plane Representations of Acyclic Digraphs," *Theoretical Computer Science*, 61, 175-198, 1988.

[7] G. Di Battista, R. Tamassia, and I.G. Tollis, "Area Requirement and Symmetry Display in Drawing Graphs," *Proc. 5th ACM Symp. on Computational Geometry*, 1989. To appear in *Discrete & Computational Geometry*.

[8] P. Eades and R. Tamassia, "Algorithms for Drawing Graphs: an Annotated Bibliograpy," Tech. Report CS-89-09, Brown Univ., 1989.

[9] P. Eades and L. Xuemin, "How to Draw a Directed Graph," *Proc. IEEE Workshop on Visual Languages*, 1989.

[10] S. Even, *Graph Algorithms*, Computer Science Press, Rockville, MD, 1979.

[11] A. Gibbons and W. Rytter, "An Optimal Parallel Algorithm for Dynamic Expression Evaluation and its Applications," *Proc. Symp. on Foundation of Software Technology and Theoretical Computer Science*, 1986.

[12] M.D. Hutton, A. Lubiw, "Upward Planar Drawing of Single Source Acyclic Digraphs," *Proc. 2nd ACM-SIAM Symp. on Discrete Algorithms*, 1990.

[13] T. Nishizeki and N. Chiba, *Planar Graphs: Theory and Algorithms*, Annals of Discrete Mathematics, North Holland, 1988.

[14] W. Schnyder, "Embedding Planar Graphs on the Grid," *Proc. 2nd ACM-SIAM Symp. on Discrete Algorithms*, 1990.

[15] K. Sugiyama, S. Tagawa, and M. Toda, "Methods for Visual Understanding of Hierarchical Systems," *IEEE Trans. on Systems, Man, and Cybernetics*, SMC-11, 109-125, 1981.

[16] R. Tamassia, G. Di Battista, and C. Batini, "Automatic Graph Drawing and Readability of Diagrams," *IEEE Trans. on Systems, Man, and Cybernetics*, SMC-18, 61-79, 1988.

[17] R. Tamassia and I.G. Tollis, "A Unified Approach to Visibility Representations of Planar Graphs," *Discrete & Comput. Geometry*, 1, 321-341, 1986.

[18] R.E. Tarjan and U. Vishkin, "Finding Biconnected Components and Computing Tree Functions in Logarithmic Parallel Time", *SIAM J. on Computing*, 1985.

[19] J. Valdes, R.E. Tarjan, and E.L. Lawler, "The Recognition of Series-Parallel Digraphs," *SIAM J. on Computing*, 11, 298-313, 1982.

[20] C. Thomassen, "Planar Acyclic Oriented Graphs," *Order*, 5, 349-361, 1989.

Coloring Random Graphs

Martin Fürer and C.R. Subramanian

Department of Computer Science
Pennsylvania State University
University Park, PA 16802
(E-mail: furer@cs.psu.edu)
(Fax: 814-865-3176)

Abstract. We present an algorithm for coloring random 3-chromatic graphs with edge probabilities below the $n^{-1/2}$ "barrier". Our (deterministic) algorithm succeeds with high probability to 3-color a random 3-chromatic graph produced by partitioning the vertex set into three almost equal sets and selecting an edge between two vertices of different sets with probability $p \geq n^{-3/5+\epsilon}$. The method is extended to k-chromatic graphs, succeeding with high probability for $p \geq n^{-\alpha+\epsilon}$ with $\alpha = 2k/((k-1)(k+2))$ and $\epsilon > 0$. The algorithms work also for Blum's balanced semi-random $\mathcal{G}_{SB}(n, p, k)$ model where an adversary chooses the edge probability up to a small additive noise p. In particular, our algorithm does not rely on any uniformity in the degree.

1 Introduction

The problem of 3-coloring a 3-chromatic graph is not only NP-complete, it also seems to be quite difficult to approximate. The best known approximation algorithm of Blum [1] is the result of a sequence of impressive improvements but still needs $\Omega(n^{3/8})$ colors.

These difficulties to handle the worst cases, sharply contrast the average case complexity. It is possible to k-color random k-chromatic graphs with high probability under suitable probability distributions, provided that the distributions favor graphs with reasonably high edge probability. A simple random model is $\mathcal{G}(n, p, k)$, where the n vertices are partitioned into k equally sized color classes, and then for each pair u, v of vertices of different colors, the edge $\{u,v\}$ is placed in the graph with probability p. Turner [4] has presented an algorithm which k-colors $\mathcal{G}(n, p, k)$ with high probability for all $p \geq n^{-1/k+\epsilon}$.

With high probability, graphs of $\mathcal{G}(n, p, k)$ have a very small variation in the degrees of their vertices. Some algorithms designed to handle average case graphs, depend on such properties. As pointed out by Blum [1], the graphs encountered in practical situations might neither be so "random", nor be close to worst cases. Fortunately some algorithms (like those of Turner and Blum) need very little randomness to succeed. The graphs could be produced by a semi-random source [3]. In the *balanced semi-random* graph model $\mathcal{G}_{SB}(n, p, k)$ of Blum, the random graph is created as follows. First, an adversary splits the n vertices into k color classes, each of size $\Theta(n)$. Then for each pair of vertices u, v, where u and v belong to different color classes, the adversary decides whether or not to include the edge $\{u, v\}$ in the graph. This decision is not final. Once the adversary has made a decision for a particular edge $\{u, v\}$, the decision is reversed with probability p. The later decisions of the adversary may depend on the outcomes of the earlier decisions. An alternative way to view this model is that in an order of its choosing, for each pair u, v of vertices belonging to different color classes, the adversary picks a bias p_{uv} between p and $1 - p$ of a coin which is flipped to determine whether the edge $\{u, v\}$ is placed in the graph. The probability p is called the noise rate of the source. With respect to randomness, this model lies between the random model $\mathcal{G}(n, p, k)$ and the worst case model.

Turner [4] and Blum [1] have given algorithms which succeed to 3-color random graphs supplied by $\mathcal{G}_{SB}(n, p, 3)$ with high probability. Turner's algorithm works for $p \geq n^{-1/3+\epsilon}$. Blum succeeds even for $p \geq n^{-1/2+\epsilon}$, and the analysis of his algorithm provides a clear explanation for this exponent. Blum's algorithm picks two vertices u and v, assuming they have different colors. All common neighbors have then the third color. As larger and larger such sets of vertices are colored, it is likely that their colors immediately enforce the coloring of more vertices. The main problem is to start this chain reaction. It is easy to see that there is a threshold probability $p = \Theta(n^{-1/2})$, corresponding to an average degree of $\Theta(\sqrt{n})$, where we start having a constant probability for two arbitrary vertices u and v to have any common neighbors. Therefore $p = \Theta(n^{-1/2})$, looks like a natural boundary for this kind of techniques.

It appears that Blum's analysis can be extended to k-coloring of graphs of $\mathcal{G}_{SB}(n, p, k)$ provided that $p \geq n^{-1/(k-1)+\epsilon}$.

In this paper, we show how to handle semi-random graphs with a noise rate p as low as $n^{-3/5+\epsilon}$ for 3-coloring, and $n^{-\alpha+\epsilon}$ with $\alpha = 2k/((k-1)(k+2))$ for k-coloring. A special feature of our analysis is a method of using matchings to moderate the exploitation of independence. The amount of independence between the appearance of different small subgraphs is considered a scarce resource, which should not be wasted. Furthermore, it is important to grow in parallel many small sets of identified vertices, rather than just producing one heap for each color class. In Section 2, we describe the algorithm introduce some basic notions and properties. In Section 3, we analyze the 3-chromatic case, which employs methods different from the k-chromatic case handled in Section 4.

2 Algorithm and Preliminaries

Let u, v be two vertices in a graph G. By *identifying* u and v, we mean removing the two vertices u and v from G and then adding a new vertex w which is adjacent to all the neighbors or u and v. Our algorithm is extremely simple. We just identify vertices when we know that they have the same color in any legal k-coloring. Showing that the algorithm is likely to succeed is not so easy, because quite often, we want to estimate the probability of several not quite independent events happening at the same time.

Algorithm 1 (Forced Identification)
```
while there are vertices u,v adjacent to the same (k − 1)-clique do
    identify u and v
```

Claim 1. *Let G be a k-chromatic graph. Then the graph resulting from applying the Forced Identification Algorithm to G, is independent of the order in which the vertices have been identified.*

For $k = 3$, we will show that the Forced Identification Algorithm reduces a graph given by $\mathcal{G}_{SB}(n, p, 3)$ to a triangle with high probability as long as $p \geq n^{-3/5+\epsilon}$. But first, we explain some notions used in the proof.

Definition 1 *Let G be a random graph supplied by $\mathcal{G}(n, p, k)$ or $\mathcal{G}_{SB}(n, p, k)$ (i.e., for every n we have a probability distribution consistent with some strategy of the adversary). We say that G has a property A with high probability, if $Prob\{G$ does not have property $A\} \leq e^{-n^\delta}$ for some constant $\delta > 0$.*

Lemma 1. *Let G be a random graph on n vertices and let $A_1, \ldots, A_{p(n)}$ be a polynomial number of graph properties. Suppose G has each of the properties A_i with high probability. Then, with high probability, G has all of these properties.*

Lemma 2. *Let A and B be two sets of vertices of size $\Omega(n^\beta)$ each. Let each edge between pairs of vertices from A and B appear with probability $p = \Omega(n^{-\alpha+\epsilon})$ independently. Then, with high probability, such a random bipartite graph contains a matching M of size $\Omega(n^\beta)$ if $\beta > \alpha - \epsilon$ or a matching M of size $\Omega(n^{2\beta-\alpha+\epsilon/2})$ if $\beta < \alpha - \epsilon$.*

3 The 3-Chromatic Case

Theorem 3. *If G is a random graph from $\mathcal{G}_{SB}(n, p, 3)$ with $p \geq n^{-3/5+\epsilon}$ for some $\epsilon > 0$, then with high probability, G is uniquely 3-colorable, and this unique 3-coloring can be obtained in polynomial time.*

Proof. First, let G be a random graph from $\mathcal{G}(n, p, 3)$ with the three color classes V_1, V_2, V_3, each of size $\Theta(n)$. We show that with high probability, V_3 collapses into a single vertex by a sequence of identifications.

For this proof, we view the process of identifying the vertices as follows. Initially, each vertex of V_3 is in a set by itself, forming a fine partition of V_3. In successive iterations, we form coarser partitions by taking unions. At each time we know that all vertices in the same set are forced to have the same color in each legal 3-coloring of G. Using Claim 1, we can assume that the sets forming a partition are all of roughly the same size. In successive iterations, we increase the size of the sets forming the partition until finally we reach the partition consisting of one set V_3.

In the following, we assume the probability p is at least $n^{-3/5+\epsilon}$ for a constant $\epsilon > 0$. Let W_1, W_2 be two subsets of V_1, V_2 of size $\Theta(n)$ each. For the analysis of the algorithm, we partition V_1 and V_2 into a constant number of such sets W_1 and W_2 to obtain several supplies of independent edges. Suppose at some intermediate point, the identified vertices of V_3 have already grown into subsets of size $\Theta(n^{t\epsilon})$ each. We have $\Theta(n^{1-t\epsilon})$ such sets in the partition. Let U be such a set. Then any given vertex v in V_1 or V_2 is adjacent to at least one vertex of this set U with probability at least $1 - (1-p)^{|U|} > 1 - e^{-p|U|}$. Hence, if $t\epsilon > 3/5 - \epsilon$, then with high probability, any vertex in W_1 or W_2 is adjacent to any set in the partition. Thus, if we take some edge between W_1 and W_2 (which exists with high probability), we can merge all the sets into one single set V_3 and stop the iteration.

On the other hand, if $t\varepsilon < 3/5 - \varepsilon$, then for the set U in the partition of V_3, each vertex in W_1 and W_2 will be adjacent to U with probability $q = \Theta(n^{-3/5+(t+1)\varepsilon})$. Hence Chernoff's bound implies that with high probability $\Theta(n^{2/5+(t+1)\varepsilon})$ vertices in W_1 will be adjacent of U. The situation for W_2 is similar. Let S_1 and S_2 be the neighbors of U in W_1 and W_2 respectively. We look for a large matching between S_1 and S_2. Using Lemma 2, we can find with high probability, a matching M of size $\Theta(n^{2/5+(t+1)\varepsilon})$ if $2/5 + (t+1)\varepsilon > 3/5 - \varepsilon$ or a matching M of size $\Theta(n^{4/5+2(t+1)\varepsilon-3/5+\varepsilon/2})$ if $2/5 + (t+1)\varepsilon < 3/5 - \varepsilon$.

In any case, a given set $U'(U' \neq U)$ in the partition of V_3 is adjacent to (both endpoints of) at least one edge of this matching M with probability at least $1 - (1-q^2)^{|M|} > 1 - e^{-q^2|M|}$. Hence, if $q^2|M| = \Omega(n^\delta)$ for some $\delta > 0$, then with high probability, every set can be merged with the set U, thereby resulting in a single set V_3 as the partition.

Otherwise, a given set U' in V_3 will be adjacent to some edge of this matching M with probability $q_1 = \Theta(\min(1, q^2|M|))$. Hence with high probability, we can merge $\Theta(n^{1-t\varepsilon}(\min(1, q^2|M|)))$ other sets with the set U. This new set contains $\Theta(nq_1)$ vertices. We can verify that this is either $\Omega(n^{1/5+3(t+1)\varepsilon})$ or $\Omega(n^{(4t+9/2)\varepsilon})$.

This holds true for any set U of the partition of V_3 with high probability. Hence using Lemma 1, with high probability, we can obtain a partition of V_3 into sets of size $\Omega(n^{3t\varepsilon})$ each. Furthermore, these sets are of size $\Omega(n^{3\varepsilon})$ for the first round, where we start with subsets of constant size, i.e., $t = 0$. This implies that if we start with a partition of singleton sets, then in $\Theta(\log(1/\varepsilon))$ iterations, we can merge all of V_3 into a single set. This requires only a constant number of iterations, and hence if we initially divide each of V_1 and V_2 into a constant number of subsets, we can identify all the vertices of V_3 into a single vertex.

Independently, similar arguments hold for V_1 and V_2 too. Hence, with high probability, the graph G is reduced to a triangle, thereby producing a 3-coloring of G.

The extension of the theorem from the distribution $\mathcal{G}(n,p,3)$ to $\mathcal{G}_{SB}(n,p,3)$ is straightforward, because the graphs of the second distribution can be viewed as graphs from the first distribution with additional "invisible" edges. Even though the algorithm cannot distinguish between the two kinds of edges, we can use Claim 1 repeatedly to obtain a computation which initially ignores the "invisible" edges. The preceding analysis shows that this computation already succeeds with high probability. $\qquad\Box$

4 The $(k+1)$-Chromatic Case $(k > 2)$

We assume G is a random graph given by $\mathcal{G}(n,p,k+1)$ with $p \geq n^{-\alpha+\varepsilon}$, where $\alpha = \frac{2(k+1)}{k(k+3)}$ and $\varepsilon > 0$ is a (arbitrary) constant. Let $V_1, V_2, \ldots, V_{k+1}$ be the $k+1$ color classes of G, each of size $\Theta(n)$. We will prove that with high probability, all the vertices of V_{k+1} are identified into a single vertex. Similar statements hold for the other color classes too, thereby yielding a $(k+1)$-clique as the resulting graph.

As in the 3-chromatic case, we start from a fine partition of V_{k+1} and move from one partition to another until we obtain the partition containing V_{k+1} as one single set. For the sake of brevity, we describe the analysis only for the initial partition consisting of singleton sets. Let W_i $(i = 1, \ldots, k)$ be a subset of V_i with size $\Theta(n)$ each. Let u be any vertex in V_{k+1}. Let S_i be the set of neighbors of u in W_i. With high probability we have $|S_i| = \Theta(n^{1-\alpha+\varepsilon})$.

Lemma 4. *Let α, β, and ε be positive real constants with $k\beta < \binom{k}{2}\alpha$. Let T be a random k-partite graph with parts T_i $(i = 1, \ldots, k)$ of size $\Theta(n^\beta)$ each. Let $n^{-\alpha+\varepsilon}$ be the probability of an edge between two vertices in different parts. Then the probability that T contains a k-clique is $\Theta(n^{k\beta - \binom{k}{2}(\alpha-\varepsilon)})$.*

Proof. First we note that $O(n^{k\beta - \binom{k}{2}(\alpha-\varepsilon)})$ is an upper bound on the probability of having at least one clique in T, because there are $\Theta(n^{k\beta})$ candidate k-tuples of vertices, and each of them defines a clique with probability $\Theta(n^{-\binom{k}{2}(\alpha-\varepsilon)})$. We claim that this upper bound is also a lower bound, thereby proving the lemma.

Consider an ordering of the vertices of each of the parts T_i. Based on this ordering, we can obtain a lexicographic ordering of k-tuples of vertices from $T_1, \ldots, T_k$. There are $\Theta(n^{k\beta})$ such k-tuples from $T_1, \ldots, T_k$. Let $m = \Theta(n^{k\beta})$ be the number of such k-tuples.

Let $\mathcal{G} = \{$ set of all possible k-partite graphs with parts $T_1, \ldots, T_k$, containing at least one k-clique $\}$. Partition $\mathcal{G}$ into $\mathcal{G}_1, \mathcal{G}_2, \ldots, \mathcal{G}_m$, where $\mathcal{G}_i = \{H : H \in \mathcal{G}$ and H does not have any k-clique given by the first $i-1$ k-tuples in the lexicographic ordering, but contains the k-clique of the ith k-tuple in the ordering $\}$. Clearly the sets $\mathcal{G}_1, \ldots, \mathcal{G}_m$ are disjoint. Hence

$$\text{prob}(\{H : H \in \mathcal{G} \wedge H \text{ has at least one } k\text{-clique}\})$$
$$= \text{prob}(\mathcal{G}_1 \cup \ldots \cup \mathcal{G}_m)$$
$$= \text{prob}(\mathcal{G}_1) + \ldots + \text{prob}(\mathcal{G}_m)$$

Let $(u_{i_1}, \ldots, u_{i_k})$ be the ith tuple in the lexicographic ordering. Now consider

$\mathcal{G}(1, i) = \{H :$ When all edges between the vertices $u_{i_1}, \ldots, u_{i_k}$ are

removed from H, then it does not contain any k-clique given by the first $(i-1)$ k-tuples in the lexicographic ordering$\}$.

$$\text{prob}(\mathcal{G}(1, i)) = \Theta(1)$$

$\mathcal{G}(2, i) \subseteq \mathcal{G}(1, i)$ is defined by

$$\mathcal{G}(2, i) = \{H : H \in \mathcal{G}(1, i) \text{ such that } H \text{ has the clique } (u_{i_1}, \ldots, u_{i_k})\}$$
$$\text{prob } (\mathcal{G}(2, i)) = p^l \text{prob}(\mathcal{G}(1, i)) = \Theta(p^l), \text{ where } l = \binom{k}{2}$$

Now partition $\mathcal{G}(2, i)$ into $\mathcal{G}(3, i)$ and $\mathcal{G}(4, i)$ where

$\mathcal{G}(2, i) = \mathcal{G}(3, i) \cup \mathcal{G}(4, i)$ and

$\mathcal{G}(3, i) = \{H : H \in \mathcal{G}(2, i) \wedge H$ does not have any k-clique formed by the k-tuples from the first $i - 1$ positions in the ordering$\}$

$\mathcal{G}(4, i) = \mathcal{G}(2, i) - \mathcal{G}(3, i)$

Clearly $\mathcal{G}(3, i) \subseteq \mathcal{G}_i$. Hence $\text{prob}(\mathcal{G}(3, i)) \leq \text{prob}(\mathcal{G}_i)$. Now we claim $\text{prob}(\mathcal{G}(4, i)) = o(p^l)$. As a result

$$\text{prob}(\mathcal{G}(3, i)) = \text{prob}(\mathcal{G}(2, i)) - \text{prob}(\mathcal{G}(4, i)) = \Theta(p^l) - o(p^l) = \Theta(p^l)$$

We prove that $\text{prob}(\mathcal{G}(4, i)) = o(p^l)$ as follows. We have $\text{prob}(\mathcal{G}(4, i)) \leq \text{prob}(\{H : H$ has the clique $(u_{i_1}, \ldots, u_{i_k})$ and also another clique which contains at least one edge from the clique $(u_{i_1}, \ldots, u_{i_k})\})$. Hence

$$\text{prob}(\mathcal{G}(4, i)) \leq \sum_{j=2}^{k-1} \binom{k}{j} \Theta(n^{(k-j)\beta} p^{\binom{k-j}{2}} p^{(k-j)j} p^{\binom{k}{2}})$$

$$= \sum_{j=2}^{k-1} \binom{k}{j} \Theta(n^{(k-j)\beta} n^{-(k-j)(k+j-1)/2(\alpha - \epsilon)} p^{\binom{k}{2}})$$

Since $k\beta < \binom{k}{2}\alpha$, the above sum turns out to be $\Theta(p^{\binom{k}{2}} n^{-\delta})$ for some constant $\delta > 0$. Thus $\text{prob}(\mathcal{G}(4, i)) = o(p^{\binom{k}{2}})$. Now it follows that $\text{prob}(\mathcal{G}_i) = \Theta(p^l)$ and hence

prob $(\{H : H \in \mathcal{G} \wedge H$ has at least one clique$\})$

$= \text{prob}(\mathcal{G}_1) + \ldots + \text{prob}(\mathcal{G}_m)$

$= \Theta(mp^l)$

$= \Theta(n^{k\beta} p^l)$

$= \Theta(n^{k\beta - \binom{k}{2}(\alpha - \epsilon)})$

$\square$

Let u, W_i, S_i be as defined before, i.e., let $u \in V_{k+1}$ be any fixed arbitrary vertex and S_i be the set of neighbors of u in W_i. With high probability, $|S_i| = \Theta(n^{1 - \alpha + \epsilon})$.

Let us estimate the probability that a given vertex $v \neq u$ in V_{k+1} is adjacent to a k-clique formed by the vertices of $S = \bigcup_{i=1}^{k} S_i$.

It is clear that for a given choice of edges between the sets S_i $(i = 1, \ldots, k)$, the quantity

prob$(\{v : v \in V_{k+1} \wedge v \neq u \wedge v$ is adjacent to a k-clique in $S\})$

depends only on the edges between v and the vertices in S. Hence for a given set E of edges between vertices in S, these probabilities for different vertices of V_{k+1} are independent of each other.

For $v \in V_{k+1}$ $(v \neq u)$, let $\mathcal{G}(S, v) = \{H : H$ contains a k-clique formed by vertices of the set S and v is adjacent to at least one such k-clique$\}$.

$$\text{prob}(\mathcal{G}(S,v)) = \text{prob}\{v \text{ is adjacent to } \Theta(n^{1-2\alpha+2\epsilon}) \text{ vertices in each } S_i \text{ and}$$
$$\text{the neighbors of } v \text{ contain a } k\text{-clique}\}$$
$$+ \text{prob}\{v \text{ is not adjacent to } \Theta(n^{1-2\alpha+2\epsilon}) \text{ vertices in}$$
$$\text{some } S_i \text{ and the neighbors of } v \text{ contain a } k\text{-clique}\}$$

The second part will be at most $\Theta(e^{-n^\delta})$ for some constant $\delta > 0$ since v is adjacent to $\Theta(n^{1-2\alpha+2\epsilon})$ vertices in each S_i with high probability.

For some ϵ and α, the first part can be evaluated easily using Lemma 4 as follows. Let $\alpha = \frac{2(k+1)}{k(k+3)}$. Then $k(1 - 2\alpha) < \binom{k}{2}\alpha$, and $k(1 - 2\alpha + 2\epsilon)$ is still less than $\binom{k}{2}\alpha$ for a sufficiently small constant $\epsilon > 0$. For such an ϵ, we can apply Lemma 4, but Theorem 5 will be true for all $\epsilon > 0$, because additional edges cannot do any harm (by Claim 1). Applying Lemma 4, we find the probability of the first part to be $(1-e^{-n^\delta})\Theta(n^{k(1-2\alpha+2\epsilon)-\binom{k}{2}(\alpha-\epsilon)})$. But this is equal to $\Theta(n^{k-(k+1)+\xi\epsilon}) = \Theta(n^{-1+\xi\epsilon})$ for some constant $\xi > 0$. Since these probabilities are independent for different vertices of V_{k+1}, we can identify $\Theta(n^{\xi\epsilon})$ vertices with the given vertex $u \in V_{k+1}$ with high probability.

Now using Lemma 1, we can conclude that, with high probability, each vertex $u \in V_{k+1}$ can be identified with $\Theta(n^{\xi\epsilon})$ vertices. Hence using Claim 1, V_{k+1} will be partitioned into sets of size $\Theta(n^{\xi\epsilon})$ vertices.

Now using an analysis similar to the one described above, we can prove the following:

Suppose V_{k+1} has already been partitioned into sets of size $\Theta(n^{t\epsilon})$ each. Suppose $W_i(i = 1, \dots, k)$ are subsets of V_i, each of size $\Theta(n)$. Then from this partition and the sets W_i, we can obtain with high probability, another partition in which V_{k+1} has been partitioned into sets of size $\Theta(n^{\eta t\epsilon})$ where $\eta > 1$ is a constant. This implies that, if we start with the initial partition of singleton sets, then in $\Theta(\log 1/\epsilon)$ iterations, we will obtain the partition consisting of one single set V_{k+1}. Similar statements hold true for the other color classes too.

The extension from the purely random distribution $\mathcal{G}(n, p, k + 1)$ to the distribution $\mathcal{G}_{SB}(n, p, k + 1)$ created by an adversary uses the same idea as in the case of 3-coloring.

Theorem 5. *If G is a random graph from $\mathcal{G}_{SB}(n, p, k + 1)$ with $p \geq n^{-\alpha+\epsilon}$, where $\alpha = \frac{2(k+1)}{k(k+3)}$ and $\epsilon > 0$ is an arbitrary constant, then with high probability, G is uniquely $(k + 1)$-colorable and this unique coloring can be found in polynomial time.*

5 Conclusions

We have given a simple algorithm for coloring random $(k + 1)$-chromatic graphs and also shown the probabilistic analysis. Using this algorithm, we are able to improve the previous bounds on the probability or the noise rate of the balanced semi-random model. Moreover, the bounds on the noise rate are optimal for the given way of analyzing the probabilities. This can easily be verified. However the optimal probability bounds for the algorithm may be lower than what we have given. It is an interesting open problem to estimate the optimal bounds for the given algorithm and prove that they are optimal for the given algorithm.

Another interesting problem is to modify the given algorithm to obtain another algorithm which colors every graph, and whose average time complexity is polynomial for graphs from $\mathcal{G}_{SB}(n, p, k+1)$. It is known that k-colorable graphs from the uniform distribution can be k-colored in polynomial time[2].

References

1. A. Blum, Some Tools for Approximate 3-Coloring, FOCS (1990), 554-562.
2. M.E. Dyer and A.M. Frieze, The Solution of Some Random NP-Hard Problems in Polynomial Expected Time, *J. Alg.* **10** (1989), 451-489.
3. M. Santha and U.V. Vazirani, Generating Quasi-random Sequences from Semi-random Sources, *J. Comp. Syst. Sci.* **33** (1986), 75-87.
4. J.S. Turner, Almost All k-colorable Graphs are Easy to Color, *J. Alg.* **9** (1988), 63-82.

This article was processed using the LaTeX macro package with LLNCS style

Testing superperfection of k-trees

T. Kloks [*]   H. Bodlaender [†]

Department of Computer Science, Utrecht University
P.O.Box 80.089, 3508 TB Utrecht, The Netherlands

Abstract

An interval coloring of a weighted graph with non-negative weights, maps each vertex onto an open interval on the real line with width equal to the weight of the vertex, such that adjacent vertices are mapped to disjoint intervals. The total width of an interval coloring is defined as the width of the union of the intervals. The interval chromatic number of a weighted graph is the least total width of an interval coloring. The weight of a subset of vertices is the sum of the weights of the vertices in the subset. The clique number of a weighted graph is the weight of the heaviest clique in the graph. A graph is called superperfect if, for every non-negative weight function, the clique number is equal to the interval chromatic number.

A k-tree is a graph which can be recursively defined as follows. A clique with $k + 1$ vertices is a k-tree. Given a k-tree with n vertices, a k-tree with $n + 1$ vertices can be obtained by making a new vertex adjacent to all vertices of a k-clique in the k-tree.

In this paper we present, for each constant k, a linear time algorithm to test if a k-tree is superperfect. We also give, for each constant k, a constant time algorithm to produce a complete characterization of superperfect k-trees. Finally we present a complete list of critical non-superperfect 2-trees. Answering a question of Golumbic ([11]), this shows the existence of triangulated graphs which are superperfect but not comparability graphs.

1 Introduction

Since the discovery of perfect graphs in 1960, much research has been devoted to special classes of perfect graphs, such as comparability graphs and triangulated graphs. The class of triangulated graphs contains well known graph classes such as interval graphs, split graphs, k-trees, and indifference graphs. The class of comparability graphs contains complements of interval graphs, permutation graphs, threshold graphs and P_4-free graphs (or cographs). Much work has been done in characterizing these graph classes and in finding relations between them. Interest has only increased since Lovász settled the perfect graph conjecture in 1972 ([13]). An explanation for this

---

[*]This author is supported by the foundation for Computer Science (S.I.O.N) of the Netherlands Organization for Scientific Research (N.W.O.), Email: ton@cs.ruu.nl.

[†]Email: hansb@cs.ruu.nl.

interest, from a theoretical point of view, might be the, as yet unsettled, strong perfect graph conjecture. From an algorithmic point of view, perfect graphs have become of great interest since the discovery of polynomial time algorithms (by Grötschel, Lovász and Schrijver) for NP-complete problems like Clique, Stable set and the Coloring problem, when restricted to perfect graphs (see [4]). From a practical point of view, special classes of perfect graphs have proven their importance by the large amount of applications (see for example [11] for applications in general and [6] for an overview of applications of interval graphs).

For computer science, the class of partial k-trees, which are subgraphs of k-trees, plays an increasingly important role. One reason for this is the existence of polynomial time algorithms for many NP-complete problems when restricted to partial k-trees for some constant k (see for example [2] and [1]). This has become even more interesting since the discovery of fast algorithms for the recognition of partial k-trees for constant k ([12], [3], [14], [17]). Of theoretical great importance is the work of Robertson and Seymour (see [18]), who showed that, for each k, there exists a finite set of forbidden minors for partial k-trees.

Most classes of perfect graphs can be recognized in polynomial time ([19], [5], [11], [16], [10]). An exception seems to be the class of superperfect graphs. Determining the interval chromatic number of a weighted interval graph with weights 1 and 2 is NP-complete. When restricted to weighted partial k-trees, for some constant k, and with weights bounded by some constant, it can be seen that the interval chromatic number can be determined in linear time, when the embedding of the graph in a k-tree is given. Until now we have not been able to find a polynomial algorithm to test superperfection on partial k-trees. In this paper we present our results for testing superperfection on k-trees.

The class of superperfect graphs contains that of the comparability graphs, but these classes are not equal as has been pointed out by Golumbic who showed the existence of an infinite class of superperfect graphs which are not comparability graphs (see [11]). However all these graphs are neither triangulated nor co-triangulated, and in [11] the question is therefore raised whether for triangulated graphs the classes of superperfect and comparability graphs coincide. For split graphs this equivalence has been shown. Our results show this is not the case in general. We show the existence of triangulated graphs which are superperfect but are not comparability graphs.

The results presented in this paper can be summarized as follows. We give a complete characterization, by means of forbidden induced subgraphs, of 2-trees which are superperfect. For each constant k we give a constant time algorithm which produces a complete characterization of superperfect k-trees, by means of forbidden configurations. With the aid of this characterization we find, for each constant k, a linear time algorithm to test superperfection of k-trees.

2 Preliminaries

We start with some definitions and easy lemmas. Most definitions and results in this section are taken from [11]. For further information on perfect graphs the reader is referred to this book or to [4].

Definition 2.1

An undirected graph $G = (V, E)$ is called a comparability graph, or a transitively orientable graph, if there exists an orientation (V, F) satisfying

$$F \cap F^{-1} = \emptyset \wedge F + F^{-1} = E \wedge F^2 \subseteq F$$

Such an orientation is called a *transitive* orientation.

So if F is a transitive orientation then $(a, b) \in F$ and $(b, c) \in F$ imply $(a, c) \in F$.

It is easily checked that if a graph G is a comparability graph, then this also holds for every induced subgraph of G. A comparability graph can not contain an induced odd cycle of length at least 5, or the induced complement of a cycle with length at least 5. This last part can be seen as follows: consider the complement of a cycle with length at least 6 and assume there is a transitive orientation. Since the orientation is acyclic, there must exist at least one sink node s. Consider the square which contains exactly two neighbors of s but not s itself. The subgraph induced by the square and s can not be transitively oriented such that s is a sink node. In [11] it is shown that comparability graphs are *perfect* (i.e. for every induced subgraph the chromatic number is equal to the maximum size of a clique) and can be recognized in polynomial time (see also [19]). Comparability graphs share all these properties with *triangulated* graphs.

Definition 2.2

A graph is triangulated if it contains no chordless cycle of length greater than three.

Definition 2.3

Let $G = (V, E)$ be a graph. A *simplicial vertex* of G is a vertex of which the neighborhood forms a clique. An ordering of the vertices $\sigma = [v_1, \ldots, v_n]$ is called a *perfect elimination scheme* if each v_i is a simplicial vertex in $G[\{v_i, \ldots, v_n\}]$, which is the subgraph induced by $\{v_i, \ldots, v_n\}$.

Fulkerson and Gross ([8]) characterized triangulated graphs by means of a perfect elimination scheme.

Lemma 2.1 *A graph is triangulated if and only if there exists a perfect elimination scheme.*

A special type of triangulated graphs are k-*trees*.

Definition 2.4

A k-tree is defined recursively as follows: A clique with $k+1$ vertices is a k-tree; given a k-tree T_n with n vertices, a k-tree with $n+1$ vertices is constructed by making a new vertex x_{n+1} adjacent to a k-clique of T_n, and nonadjacent to the $n - k$ other vertices of T_n.

A triangulated graph is a k-tree if it is connected, every maximal clique contains $k + 1$ vertices and every minimal vertex separator is a k-clique. It is clear that for k-trees there exists a perfect elimination scheme $\sigma = [v_1, \ldots, v_n]$ such that for each $1 \le i \le n - k$, the neighborhood of v_i is a clique with k vertices in the subgraph induced by $\{v_{i+1}, \ldots, v_n\}$. Notice that for $k = 1$, k-trees are just ordinary trees. So k-trees are a natural generalization of trees.

A *weighted* graph is a pair (G, w), where G is a graph and w is a weight function which associates to every vertex x a non-negative weight $w(x)$. For a subset S of the vertices we define the weight of S, $w(S)$ as the sum of the weights of the vertices in S.

Definition 2.5

An *interval coloring* of a weighted graph (G, w) maps each vertex x to an open interval I_x on the real line, of width $w(x)$, such that adjacent vertices are mapped to disjoint intervals. The *total width* of an interval coloring is defined to be $|\bigcup_x I_x|$. The *interval chromatic number* $\chi(G, w)$ is the least total width needed to color the vertices with intervals.

Determining whether $\chi(G, w) \leq r$ is an NP-complete problem, even if w is restricted to values 1 and 2 and G is an interval graph, this has been shown by L. Stockmeyer as reported in [11]. In this paper we shall only use the following alternative definition of the interval chromatic number (see [11]).

Theorem 2.1 *If (G, w) is a weighted undirected graph, then*

$$\chi(G, w) = \min_F (\max_\mu w(\mu))$$

where F is an acyclic orientation of G and μ is a path in F.

If w is a weighting and F is an acyclic orientation, then we say that F is a *superperfect orientation* with respect to w if the weight of the heaviest path in F does not exceed the weight of the heaviest clique.

Definition 2.6

The *clique number* $\Omega(G, w)$ of a weighted graph (G, w) is defined as the maximum weight of a clique in G.

We use the capital Ω to avoid confusion with the weighting w. It is easy to see that $\Omega(G, w) \leq \chi(G, w)$ holds for all weighted graphs, since for any acyclic orientation and for every clique there exists a path in the orientation which contains all vertices of the clique.

Definition 2.7

A graph G is called *superperfect* if for *every* non-negative weight function w, $\Omega(G, w) = \chi(G, w)$.

Notice that each induced subgraph of a superperfect graph is itself superperfect, and also that every superperfect graph is perfect. If G is a comparability graph, then there exists an orientation such that every path is contained in a clique. This proves the following theorem (see also [11]).

Theorem 2.2 *A comparability graph is superperfect.*

The converse of this theorem is not true. In [11] an infinite class of superperfect graphs is given that are not comparability graphs. However, none of these graphs is triangulated. In [11] (page 214) the question is raised if the converse of the theorem holds for triangulated graphs; is it true or false that, for *triangulated* graphs, G is a comparability graphs if and only if G is superperfect? In the next section we answer this question in the negative, and we give a complete characterization of superperfect 2-trees.

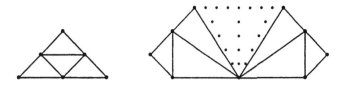

Figure 1: 3-sun (left) and wing (right)

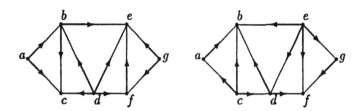

Figure 2: two orientations of the wing with seven vertices

3   2-trees and superperfection

In this section we give a characterization of 2-trees that are superperfect by means of forbidden subgraphs. In 1967 Gallai, [9], published a complete list of critical non-comparability graphs (this list can also be found in [4] page 78). Extracting from this list the triangulated graphs which are subgraphs of 2-trees (or: have *treewidth* at most two), we find a characterization of 2-trees which are comparability graphs. We find two types of forbidden induced subgraphs, which we call the 3-*sun* and the *odd wing*. They are illustrated in figure 1. Notice that a 3-sun and a wing are 2-trees, and that a wing has at least seven vertices. We call a wing odd (even) if the total number of vertices is odd (even). The following lemma is easy to check.

Lemma 3.1 *A wing is a comparability graph if and only if it is even.*

We thus find the following characterization of 2-trees that are comparability graphs.

Theorem 3.1 *A 2-tree is a comparability graph if and only if it does not contain a 3-sun or an odd wing.*

The next theorem shows that the smallest odd wing, with seven vertices, (which is not a comparability graph) is superperfect. As we shall see later, this is in fact the only odd wing that is superperfect.
Remark. Notice that in [11] (page 212, figure 9.9) this graph is mistakenly placed in the position of a non-superperfect graph. See also [15] and [7]; the result of [15] is wrong: A wing is an interval graph.

Theorem 3.2 *The odd wing with seven vertices is superperfect.*

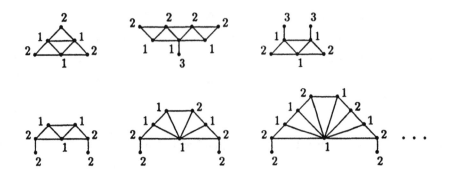

Figure 3: critical non-superperfect graphs

Proof. Label the vertices of the graph as in figure 2. We consider two orientations of this wing as illustrated in figure 2 and we show that for every weighting one of these orientations is superperfect. Notice that both orientations are such that there is exactly one path not contained in a triangle. In the first orientation this is the path $\{a, b, e\}$ and in the second orientation the path $\{c, d, f\}$. Consider a non-negative weighting w of the vertices. Suppose the orientation of the first type is *not* superperfect with respect to w. Then the path $\{a, b, e\}$ must be heavier then every triangle. Since $\{a, b, c\}$ is a triangle, this implies that $w(e) > w(c)$. But then $w(\{c, d, f\}) < w(\{e, d, f\})$, and since $\{e, d, f\}$ is a triangle, the second orientation is superperfect with respect to w.
□

In the last part of this section we give a complete characterization of the superperfect 2-trees. In figure 3 we give an (infinite) list of forbidden induced subgraphs. The following lemma can be easily checked.

Lemma 3.2 *The graphs illustrated in figure 3 are not superperfect. The weight function that is shown is such that for any acyclic orientation, there exists a path which is heavier than the heaviest clique.*

Theorem 3.3 *A 2-tree is superperfect if and only if it does not contain an induced subgraph isomorphic to one of the graphs shown in figure 3.*

Proof. For brevity the proof is omitted. □

Notice that the list of forbidden induced subgraphs is infinite. In the next section we show that for each k there is a finite characterization of superperfect k-trees, by means of forbidden configurations. Furthermore we give for each k a constant time algorithm to find this characterization. As a consequence we find a linear time algorithm to check if a k-tree is superperfect.

4 k-trees and superperfection

In this section, let k be some constant, and let G be a k-tree. We start by showing that we can restrict the set of orientations to test if G is superperfect.

Definition 4.1
A *coloring* of a triangulated graph $G(V, E)$ with $k + 1$ colors is a function $C : V \rightarrow \{1, \ldots, k + 1\}$, such that $C(x) \neq C(y)$ whenever x and y are adjacent.

In this paper we only use colorings with $k + 1$ colors; we do not always mention the number of colors. Notice that a coloring of a k-tree is unique up to a permutation of the colors:

Lemma 4.1 *If C and C' are two colorings of a k-tree $G = (V, E)$ then there exists a permutation π of the colors $\{1, \ldots, k + 1\}$ such that for every vertex x, $C(x) = \pi(C'(x))$.*

Since a coloring of a k-tree is unique up to a permutation of the colors, the following set of orientations is uniquely defined for each k-tree.

Definition 4.2
Let G be a graph and let C be a coloring of G with $k + 1$ colors. For each permutation π of the colors we define an orientation F_π as follows. Direct the edge (x, y) from x to y if $\pi(C(x)) < \pi(C(y))$. We define $\mathcal{F}_c(G)$ as the set of orientations obtained in this way.

The following lemma follows immediately from definition 4.2.

Lemma 4.2 *If G is a k-tree then:*

1. $|\mathcal{F}_c| = (k + 1)!$.

2. *Each $F \in \mathcal{F}_c$ is acyclic.*

3. *If $F \in \mathcal{F}_c$ then each path μ in F has at most $k + 1$ vertices.*

Definition 4.3
Let G be a k-tree. We define $\mathcal{F}^*$ as the set of acyclic orientations of G, of which every path contains at most $k + 1$ vertices.

Notice that $\mathcal{F}_c \subset \mathcal{F}^*$.

Lemma 4.3 $\mathcal{F}_c = \mathcal{F}^*$.

Proof. We proof that $|\mathcal{F}^*| = (k + 1)!$. Let $F \in \mathcal{F}^*$. Let S be a k-clique in G, and let x and y be two vertices which are adjacent to all vertices of S. Since S is a clique and F is acyclic, there is a unique ordering of the vertices of S, say $s_1, s_2, \ldots, s_k$, such that $s_i \rightarrow s_j$ if and only if $i > j$. Since x is adjacent to all vertices of S, there exists an index $0 \leq t_x \leq k$ such that $x \rightarrow s_i$ for all $1 \leq i \leq t_x$ and $s_j \rightarrow x$ for all $t_x < j \leq k$. The same holds for y with index t_y. Consider the case $t_x < t_y$. Then F has a path of length $k + 2$:
$$(s_k, s_{k-1}, \ldots, s_{t_y+1}, y, s_{t_y}, \ldots, s_{t_x+1}, x, s_{t_x}, \ldots, s_1)$$
Since $F \in \mathcal{F}^*$, we find that $t_x = t_y$. Now consider the recursive construction of G as a k-tree. Start with an acyclic orientation of a $(k + 1)$-clique. This can be done in $(k + 1)!$ manners. If we add a new vertex v and make it adjacent to a k-clique, by the argument above, the orientations of the edges incident with v are determined. Hence $|\mathcal{F}^*| = (k + 1)!$. $\qquad\square$

Definition 4.4

Let F be an acyclic orientation. A path μ in F is *contained* in a path μ', if all vertices of μ are also vertices of μ'.

Lemma 4.4 *Let $F \in \mathcal{F}_c$. Then any path μ in F is contained in a path μ' with $k+1$ vertices.*

Proof. Let C be a coloring and let $F = F_\pi$. The colors in the path μ must appear in the same order as in the permutation. Assume there is a gap between adjacent colors c_1 and c_2 in the path (i.e. there is a color in the permutation between c_1 and c_2). Since the edge of the path with colors c_1 and c_2 is contained in a $(k+1)$-clique, the missing colors can be put between c_1 and c_2. Thus we can make a longer path μ' containing μ. $\qquad\square$

Theorem 4.1 *Let G be a k-tree. Then G is superperfect if and only if*

$$\forall_w [\min_{F \in \mathcal{F}_c} \max_\mu w(\mu) = \Omega(G, w)]$$

Proof. Assume G is superperfect. Let w be a non-negative weighting. There is an orientation F, such that $\max_\mu w(\mu) = \Omega(G, w)$. If every path in F has at most $k+1$ vertices, we are done. Assume F has a path with more than $k+1$ vertices. Now increase all weights with some constant $L > \Omega(G, w)$. Let w' be this new weighting. Notice that $\Omega(G, w') = \Omega(G, w) + (k+1)L$. Since G is superperfect, there must be an orientation F' for this new weighting w'. Suppose F' also has a path μ with more than $k+1$ vertices. Then (with $|\mu|$ the number of vertices of μ):

$$
\begin{aligned}
\Omega(G, w') = \Omega(G, w) + (k+1)L \;&\geq\; w'(\mu) \\
&=\; w(\mu) + |\mu|L \\
&\geq\; w(\mu) + (k+2)L \\
&\geq\; (k+2)L
\end{aligned}
$$

Since $L > \Omega(G, w)$, this is a contradiction. We may conclude that $F' \in \mathcal{F}^* = \mathcal{F}_c$. We show that F' is also a good orientation for the weighting w. Let ν be a path in F'. By lemma 4.4, ν is contained in a path ν^* with $k+1$ vertices. Hence

$$w(\nu) \leq w(\nu^*) = w'(\nu^*) - (k+1)L \leq \Omega(G, w') - (k+1)L = \Omega(G, w)$$

The converse is trivial. $\qquad\square$

Definition 4.5

Let G be a triangulated graph and let C be a coloring of G with $k+1$ colors. For each permutation π of the colors, let $\mathcal{P}(\pi)$ be the set of paths in F_π, which are *not* contained in a clique, and which have $k+1$ vertices. If Q is a set of paths in G, we say that Q forms a *cover* if, for every permutation π, there is a path $\mu \in Q$ which can be oriented such that it is in $\mathcal{P}(\pi)$. A cover is called *minimal* if it contains $\frac{1}{2}(k+1)!$ paths.

Lemma 4.5 *Let G be a k-tree. If for some permutation π, $\mathcal{P}(\pi) = \emptyset$, then G is a comparability graph (hence superperfect).*

Proof. Suppose $\mathcal{P}(\pi) = \emptyset$. Consider the orientation F_π. If there is a path in F_π which is not contained in a clique then, by lemma 4.4, $\mathcal{P}(\pi)$ can not be empty. Hence, every path in F_π is contained in a clique. Since F_π is acyclic, the lemma follows. $\square$

Definition 4.6
Let G be a k-tree and let C be a coloring of G. Let S be a maximal clique of G, and let Q be a minimal cover. Define $LP(G, S, Q)$ as the following set of inequalities:

1. For each vertex x: $w(x) \geq 0$.

2. For each maximal clique $S' \neq S$: $w(S') \leq w(S)$.

3. For each path μ of Q: $w(\mu) > w(S)$.

We call the second type of inequalities, the *clique inequalities*. The inequalities of the third type are called the *path inequalities*.

Lemma 4.6 *There are $n - k - 1$ clique inequalities and $\frac{1}{2}(k + 1)!$ path inequalities.*

Proof. Notice that a k-tree has $n - k$ cliques with $k + 1$ vertices. $\square$

Theorem 4.2 *Let G be a k-tree with a coloring C. G is not superperfect if and only if there is a maximal clique S and a minimal cover Q such that $LP(G, S, Q)$ has a solution.*

Proof. Suppose $LP(G, S, Q)$ has a solution. Take this solution as a weighting. Then, clearly, for any orientation F_π there is a path (in Q and in $\mathcal{P}(\pi)$) which is heavier than the heaviest clique S. By theorem 4.1 G is not superperfect. On the other hand, if G is not superperfect, there exists a weighting w such that for every orientation F_π there is a path which is heavier than the heaviest clique (hence it can not be contained in a clique). By lemma 4.4 we may assume this path has $k + 1$ vertices, hence it is in $\mathcal{P}(\pi)$. Take S to be the heaviest clique and let Q be a minimal cover for these paths. $\square$

Notice that we could use theorem 4.2 to make a polynomial time algorithm to test superperfection on k-trees: There are at most n^{k+1} different paths of length k, hence the number of minimal covers is at most $(n^{k+1})^{\frac{1}{2}(k+1)!}$. Since the number of maximal cliques of G is at most $n - k$, we only have to check for a polynomial number of sets of inequalities if it has a solution. This checking can be done in polynomial time, e.g. by the ellipsoid method. We now show that there also exists a linear time algorithm.

Consider a set of inequalities $LP(G, S, Q)$ which has a solution. Notice that if some vertex y does *not* appear in the path inequalities then we can set the weight $w(y) = 0$. This new weighting is also a solution. Hence we can transform the set of inequalities as follows:

Definition 4.7

Let G be a k-tree and let C be a coloring of G. Let S be a maximal clique, and let Q be a cover. Let H be the subgraph of G induced by the vertices of S and of all paths in Q. Define $LP'(H, S, Q)$ as the following set of inequalities:

1. For each vertex x of H: $w(x) \geq 0$.

2. For each maximal clique $S' \neq S$ of H: $w(S') \leq w(S)$.

3. For each path μ in Q: $w(\mu) > w(S)$.

The following lemma follows directly from the definitions 4.6 and 4.7.

Lemma 4.7 *$LP(G, S, Q)$ has a solution if and only if $LP'(H, S, Q)$ has a solution.*

Lemma 4.8 *The number of inequalities of $LP'(H, S, Q)$ is bounded by a constant.*

Proof. There are $\frac{1}{2}(k+1)!$ paths in Q, each involving $k+1$ variables. The clique S has also $k+1$ vertices, hence it follows that the subgraph H, has at most $(\frac{1}{2}(k+1)!+1)(k+1)$ vertices. Since H is triangulated, the number of maximal cliques in H is bounded by the number of vertices. Hence the number of clique inequalities is bounded by $(\frac{1}{2}(k+1)!+1)(k+1)$. $\quad\square$

Notice that we now have the following algorithm to test superperfection of k-trees.

Algorithm to check superperfection of G

- Generate all $(k+1)$-colored triangulated graphs H, with at most $(1 + \frac{1}{2}(k+1)!)(k+1)$ vertices, for which there exists:

  1. A maximal clique S with $k+1$ vertices
  2. A set Q of $\frac{1}{2}(k+1)!$ paths, which is a minimal cover,

  such that $LP'(H, S, Q)$ has a solution.

- Make a coloring of G (with $k+1$ colors).

- Check if a graph H from this list is an induced subgraph of G (preserving colors). If G does have a subgraph from the list, G is not superperfect, otherwise it is.

Notice that generating the list takes constant time (if k is a constant). Since the subgraphs have constant size, we can check if such a subgraph is an induced subgraph of G in linear time, using standard techniques for (partial) k-trees (see [1]).

Theorem 4.3 *The algorithm correctly determines if G is superperfect, and does so in linear time.*

Proof. Assume G is not superperfect. Let C be a coloring of G. By theorem 4.2, $LP(G, S, Q)$ has a solution for some maximal clique S and some minimal cover Q. Take H the colored subgraph induced by vertices of S and of paths in Q. By lemma 4.7, $LP'(H, S, Q)$ has a solution, so the subgraph H is in the list. Conversely, suppose the colored graph G has a colored induced subgraph H from the list. Then H has a clique

S with $k + 1$ vertices and a minimal cover Q such that $LP'(H, S, Q)$ has a solution. Since H is an induced subgraph of G preserving colors, the clique S is also a maximal clique in G and the cover Q is also a cover for G. Since $LP(G, S, Q)$ has a solution, G can not be superperfect. □

Notice that the list only has to contain those subgraphs H, of which every vertex is either in the maximal clique S or on some path in Q (with S and Q as defined in the algorithm). For reasons of simplicity, we left this detail out of the algorithm.

5 Conclusions

In this paper we presented a linear time algorithm to test superperfection of k-trees, for some constant k. We also showed there exists a list of constant size, of forbidden colored triangulated graphs, such that G is superperfect if and only if the colored graph G does not have an induced subgraph (preserving colors) from this list. Finally, we completely characterized the 2-trees which are superperfect, by means of forbidden induced subgraphs. To test superperfection on *partial* k-trees remains an open problem.

6 Acknowledgements

We like to thank Prof. D. Seese for valuable suggestions.

References

[1] S. Arnborg, J. Lagergren and D. Seese, Easy problems for tree-decomposable graphs, *J. Algorithms*, **12**, 308-340, 1991.

[2] S. Arnborg and A. Proskurowski, Linear time algorithms for NP-hard problems restricted to partial k-trees. *Disc. Appl. Math.*, **23**, 11-24, 1989.

[3] H.L. Bodlaender and T. Kloks, Better algorithms for the pathwidth and treewidth of graphs, *Proceedings of the 18th International colloquium on Automata, Languages and Programming*, 544-555, Springer Verlag, Lecture Notes in Computer Science, vol. 510, 1991.

[4] C. Berge and C. Chvatal, *Topics on perfect graphs*, Annals of Discrete Mathematics **21** 1984.

[5] K.S. Booth and G.S. Lueker, Testing for the consecutive ones property, interval graphs, and graph planarity using PQ-tree algorithms, *Journal of Computer and System Sciences* **13**, 335 − 379, 1976.

[6] J.E. Cohen, J. Komlós and T. Mueller, The probability of an interval graph, and why it matters, *Proc. of Symposia in Pure Math.* **34**, 97 − 115, 1979.

[7] P.C. Fishburn, An interval graph is not a comparability graph, *J. Combin. Theory* **8**, 442 − 443, 1970.

[8] D.R. Fulkerson and O.A. Gross, Incidence matrices and interval graphs, *Pacific J. Math.* **15**, 835 − 855, 1965.

[9] T. Gallai, Transitiv orientierbaren Graphen, *Acta Math. Sci. Hung.* **18**, 25 − 66, 1967.

[10] P.C. Gilmore and A.J. Hoffman, A characterization of comparability graphs and of interval graphs, *Canad. J. Math.* **16**, 539 − 548, 1964.

[11] M.C. Golumbic, *Algorithmic Graph Theory and Perfect Graphs*, Academic Press, New York, 1980.

[12] J. Lagergren and S. Arnborg, Finding minimal forbidden minors using a finite congruence, *Proceedings of the 18th International colloquium on Automata, Languages and Programming*, 532-543, Springer Verlag, Lecture Notes in Computer Science, vol. 510, 1991.

[13] L. Lovász, Normal hypergraphs and the perfect graph conjecture, *Discrete Math.*, **2**, 253 − 267, 1972.

[14] J. Matoušek and R. Thomas, Algorithms finding tree-decompositions of graphs, *Journal of Algorithms* **12**, 1-22, 1991.

[15] M. Jean, An interval graph is a comparability graph, *J. Combin. Theory* **7**, 189 − 190, 1969.

[16] A. Pnuelli, A. Lempel, and S. Even, Transitive orientation of graphs and identification of permutation graphs, *Canad. J. Math.* **23**, 160 − 175, 1971.

[17] B.A. Reed, Finding approximate separators and computing treewidth quickly, To appear in STOC'92.

[18] N. Robertson and P.D. Seymour, Graph minors — a survey. In I. Anderson, editor, *Surveys in Combinatorics* 153-171. Cambridge Univ. Press 1985.

[19] J. Spinrad, On comparability and permutation graphs, *SIAM J. Comp.* **14**, No. 3, August 1985.

Parametric Problems on Graphs of Bounded Tree-width

David Fernández-Baca* and Giora Slutzki

Department of Computer Science, Iowa State University, Ames, IA 50011

Abstract. We consider optimization problems on weighted graphs where vertex and edge weights are polynomial functions of a parameter λ. We show that, if a problem satisfies certain regularity properties and the underlying graph has bounded tree-width, the number of changes in the optimum solution is polynomially bounded. We also show that the description of the sequence of optimum solutions can be constructed in polynomial time and that certain parametric search problems can be solved in $O(n \log n)$ time, where n is the number of vertices in the graph.

1 Introduction

We shall consider parametric optimization problems whose nonparametric version takes the following familiar form. Given a graph G with real-valued vertex and edge weight functions $w_V : V(G) \rightarrow \mathbf{R}$ and $w_E : E(G) \rightarrow \mathbf{R}$, respectively, find an optimum subgraph H satisfying a property P. Well-known examples of such problems are minimum-weight dominating set and the traveling salesman problem. Let us write $\mathsf{val}_G(H)$ to denote $\sum_{v \in V(H)} w_V(v) + \sum_{e \in E(H)} w_E(e)$, where H is a subgraph of G. We can express all optimum subgraph problems as

$$z_G^P = \mathrm{opt}\{\mathsf{val}_G(H) : H \text{ a subgraph of } G \text{ satisfying } P\}, \tag{1}$$

where "opt" is either "min" or "max", depending on the problem.

It is well known that many optimum subgraph problems that are NP-hard in general are polynomially solvable for restricted classes of graphs [GaJo79]. Recently, a long line of work has culminated in the development of various methodologies for devising polynomial-time (and, indeed, often *linear*-time) algorithms for graphs of *bounded tree-width* [ALS91, ArPr89, Bod87, BPT88, BLW87, Cou90, Wim87]. In essence, all these approaches deal with subgraph problems that have certain "regularity" properties that make them amenable to dynamic programming solutions. The class of regular problems is broad, and includes the subgraph problems mentioned above, as well as many others, such as the maximum cut problem and the Steiner tree problem (see, e.g., [ALS91, BPT88, BLW87]). Here we shall study the implications that regularity properties have on the parametric versions of these problems.

Parametric optimization problems arise in *sensitivity analysis* [Gus83], *minimum-ratio optimization* [Meg79, Meg83], *Lagrangian relaxation* [Fis81], and, in general, in environments where the data evolves continuously with time. We will focus on parametric optimum subgraph problems where vertex and edge weights are functions of

---

⋆ Supported in part by the National Science Foundation under grant No. CCR-8909626.

a real-valued parameter λ, $W_V : V(G) \times \mathbf{R} \to \mathbf{R}$ and $W_E : E(G) \times \mathbf{R} \to \mathbf{R}$, respectively. Let us write $\mathsf{Val}_G(H, \lambda)$ to denote $\sum_{v \in V(H)} W_V(v, \lambda) + \sum_{e \in E(H)} W_E(e, \lambda)$, where H is a subgraph of G. The function of interest to us is

$$Z_G^P(\lambda) = \mathrm{opt}\{\mathsf{Val}_G(H, \lambda) : H \text{ a subgraph of } G \text{ satisfying } P\}. \tag{2}$$

While, for concreteness, we shall often only deal with minimization problems, all our results easily extend to maximization problems. In our subsequent discussions we will fix P, while G may vary. Thus, we shall often write Z_G instead of Z_G^P. Similarly, we shall often write z_G instead of z_G^P.

For tractability, we shall consider the case where weights are polynomial functions of λ. Thus, for any subgraph H of G, $\mathsf{Val}_G(H, \lambda)$ is a polynomial in λ. By (2), Z_G is the lower envelope of a set of polynomials $\mathsf{Val}_G(H, \lambda)$, implying that Z_G is a piecewise-polynomial function of λ. Z_G partitions the λ-axis into a sequence of intervals, where each interval is the maximal set of λ-values for which $Z_G(\lambda) = \mathsf{Val}_G(H, \lambda)$ for some particular subgraph H satisfying P. The boundary points between intervals are called *breakpoints*. We can represent Z_G by (1) listing these intervals in order, from left to right and (2) for each interval providing the associated optimum subgraph H. Clearly, such a representation is finite. In the special case where weights are linear functions of λ, Z_G will be a piecewise-linear concave function [Gus83].

We shall concentrate on two kinds of issues: construction and search. In construction problems we shall be interested in computing the entire representation of the function Z_G. Search problems involve finding a value of λ at which a particular event occurs. Our two main results deal with *regular* optimum subgraph problems (in the sense of Bern et al. [BLW87] and Borie et al. [BPT88]) on graphs of bounded tree-width. First, we show that for every regular graph property P, the number of breakpoints of Z_G^P is polynomially-bounded in $n = |V(G)|$. As a byproduct of the proof we obtain a polynomial-time algorithm to construct Z_G^P. The second result is a proof that, for every regular property P, there exist $O(n \log n)$ algorithms for certain kinds of parametric search problems.

Related Work. Several researchers have considered parametric versions of combinatorial optimization problems (see, e.g., [Mur80, Car83, GGT89, vHKRW89]). The algorithmic approach followed in this paper was first used in [FeSl89] to analyze various optimization problems on trees. It has subsequently been used to analyze the parametric maximum independent set problem on outerplanar graphs [ZhGo91] and parametric nonserial dynamic programming problems on partial k-trees [FeMe90]. Many of the ideas used here are based on the work of Megiddo [Meg79, Meg83].

Organization of the paper. Section 2 reviews the notions of tree-width and regularity. In section 3 we present some results relating parse trees and tree-decompositions and review the dynamic programming approach to solving optimum subgraph problems. In section 4 we study the properties of Z_G^P. In section 5 we present our parametric search algorithms. Finally, section 6 discusses related results and open problems.

2 Preliminaries

2.1 Tree-width

The following definition is due to Robertson and Seymour [RoSe86].

Definition 1 Let G be an undirected graph. A *tree decomposition* of G is a labeled tree (T, χ), where χ is the labeling function for T, such that for all $i \in V(T)$, $\chi(i) = \chi_i \subseteq V(G)$, and such that the following conditions hold.

1. $\bigcup_{i \in V(T)} \chi_i = V(G)$.
2. For every $(u, v) \in E(G)$, $\{v, u\} \subseteq \chi_i$ for some $i \in V(T)$.
3. If j lies on the path of T from i to k, then $\chi_i \cap \chi_k \subseteq \chi_j$.

The *width* of a tree-decomposition is $\max_{i \in V(T)}(|\chi_i| - 1)$. The *tree-width* of a graph G is the minimum over all tree-decompositions (T, χ) of G of the width of (T, χ).

We denote the set of all graphs of tree-width at most w by Γ_w. Many important classes of graphs have bounded tree-width; see, e.g., [ArPr89, vLe90]. It is known that a graph G is a partial k-tree if and only if $G \in \Gamma_k$ [vLe90]. We say that sets $Y, Y' \subseteq V(G)$ are *separated* by $S \subseteq V(G)$ if every path in G from Y to Y' goes through S. A set $S \subseteq G$ is a *separator* of G if $G - S$ is not connected. The next result follows easily from Theorem 2.5 of [RoSe86].

Theorem 2.1 Let $G \in \Gamma_w$ and suppose $Q \subseteq V(G)$. Then, there exists a partition (B_1, B_2, B_3, S) of $V(G)$ with $|S| \leq w + 1$ such that for $i, j \in \{1, 2, 3\}$, $i \neq j$, B_i and B_j are separated by S, and $|B_i \cap Q| \leq |Q - S|/2$ for $i = 1, 2, 3$.

2.2 Regular Graph Properties

The various subgraph problems that are amenable to dynamic programming on graphs of bounded tree-width share two key properties. First, the space of potential solutions to these problems can be partitioned into a finite number of equivalence classes [ALS91, ArPr89, Bod87, BPT88, BLW87, Cou90]. Second, there is a finite set of rules whereby partial solutions, computed for portions of the input graph, can be combined into larger partial solutions. Rules are expressed in tables which are fixed for each problem and each family of graphs.

Let $\mathcal{G}$ be the set of all finite graphs. A graph $G \in \mathcal{G}$ is a *k-terminal graph* if it is given together with a list $\mathbf{terms}(G) = \langle t_1, \ldots, t_s \rangle$, $1 \leq s \leq k$, of distinct vertices of G called *terminals*. The set of all k-terminal graphs will be denoted by $\mathcal{G}_k$. A *k-terminal graph composition operator* is a (partial) function $\varphi : \mathcal{G}_k^r \to \mathcal{G}_k$, where $r = r(\varphi) \geq 0$ is the *arity* of φ. The resulting graph $G = \varphi(G_1, \ldots, G_r) \in \mathcal{G}_k$ is obtained by identifying the terminals of $G_1, \ldots, G_r$ in some precisely prescribed way. The terminals of G are obtained from the terminals of the composing graphs.

In the remainder of this section, $\mathcal{R}$ will denote a finite set of k-terminal composition operations and $\mathcal{R}_i \subseteq \mathcal{R}$ is the subset of operations of arity i. Note that $\mathcal{R}_0$ is a subset of $\mathcal{G}_k$. The graphs in $\mathcal{R}_0$ are called the *base* or *primitive* graphs. We will always assume that $\mathcal{R}_0 \neq \emptyset$. A *family of decomposable graphs* over $\mathcal{R}$ is the smallest family of graphs $\mathcal{R}^* \subseteq \mathcal{G}_k$ defined recursively as follows: (1) $\mathcal{R}_0 \subseteq \mathcal{R}^*$; (2) for all $r \geq 1$, for each $\varphi \in \mathcal{R}_r$, and for all $G_1, \ldots, G_r \in \mathcal{R}^*$, if $\varphi(G_1, \ldots, G_r)$ is defined, then $\varphi(G_1, \ldots, G_r) \in \mathcal{R}^*$. The equality $G = \varphi(G_1, \ldots, G_r)$ is called a *decomposition of G with respect to $\mathcal{R}$*. Let (T, δ) be a rooted, ordered tree with labeling function $\delta : V(T) \to \mathcal{R}$ satisfying the following compatibility requirement: If $v \in V(T)$ has $r \geq 0$ children in T, then $\delta(v) \in \mathcal{R}_r$. Given such a tree (T, δ), we define an induced partial function $\gamma_T : V(T) \to \mathcal{R}^*$ as follows:

(R1) If $v \in V(T)$ is a leaf (i.e., $\delta(v) \in \mathcal{R}_0$), then $\gamma_T(v) = \delta(v)$.
(R2) If $v \in V(T)$ is an internal node with children $v_1, \ldots, v_r$, then

$$\gamma_T(v) = \delta(v)(\gamma_T(v_1), \ldots, \gamma_T(v_r)).$$

We define the set $\mathcal{T_R}$ of *trees over $\mathcal{R}$* as the set of all trees (T, δ) for which the induced function γ_T is total. Let $(T, \delta) \in \mathcal{T_R}$. A tree $(T, \delta) \in \mathcal{T_R}$ is a *parse tree* of $G \in \mathcal{R}^*$ if $G = \gamma_T(v)$, where v is the root of T.

When dealing with optimization problems on graphs, we will be interested in the set of all graph × subgraph pairs $\hat{\mathcal{G}} = \{(G, H) : G \in \mathcal{G}, H \text{ a subgraph of } G\}$ and its k-terminal version, $\hat{\mathcal{G}}_k$. For $D = (G, H) \in \hat{\mathcal{G}}_k$, the *signature* of D is the list $\mathbf{terms}(G)$, together with information as to which nodes in this list are in H. We extend the k-terminal operations $\varphi \in \mathcal{R}$ to $\hat{\mathcal{G}}_k$ by letting $\hat{\mathcal{R}} = \bigcup_{r \geq 0} \hat{\mathcal{R}}_r$, where $\hat{\mathcal{R}}_0 = \{(\varphi, H) : \varphi \in \mathcal{R}_0, H \text{ a subgraph of } \varphi\}$ and, for $r \geq 1$, $\hat{\mathcal{R}}_r = \{\hat{\varphi} : \varphi \in \mathcal{R}_r\}$ with $\hat{\varphi}$ defined as follows. If $\varphi(G_1, \ldots, G_r)$ is defined, then $\hat{\varphi}((G_1, H_1), \ldots, (G_r, H_r)) = (G, H)$, where $G = \varphi(G_1, \ldots, G_r)$ and H is specified by $V(H) = \bigcup_{i=1}^r V(H_i)$, and $E(H) = \bigcup_{i=1}^r E(H_i)$, modulo vertex identification resulting from application of φ. Analogously with the definition of $\mathcal{R}^*$, we can define a family $\hat{\mathcal{R}}^* \subseteq \hat{\mathcal{G}}_k$ as the closure of the set $\hat{\mathcal{R}}$.

Of great importance here will be a special kind of predicates on $\hat{\mathcal{G}}$, called *regular predicates*. We shall define them in an algebraic framework. Let $\mathcal{D} = \hat{\mathcal{R}}^*$. Then $\mathcal{D} = (\mathcal{D}, \hat{\mathcal{R}})$ is an algebra, with domain (carrier) $\mathcal{D}$ and operations $\hat{\mathcal{R}}$. Suppose $\mathcal{C} = (\mathcal{C}, \mathcal{Q})$ is an algebra such that there is an arity-preserving one-to-one correspondence between $\hat{\mathcal{R}}$ and $\mathcal{Q}$ — i.e. $\mathcal{C}$ is *similar* to $\mathcal{D}$. Let P be a predicate on $\hat{\mathcal{G}}$ and let $h : \mathcal{D} \to \mathcal{C}$ be a mapping. We say that h *respects* P if for every $D_1, D_2 \in \mathcal{D}$,

(H1) $h(D_1) = h(D_2) \implies P(D_1) = P(D_2)$.

Note that the value of P for D_i is independent of the terminals associated with D_i. The mapping h is a *homomorphism with respect to $\mathcal{D}$ (or $\mathcal{R}$)* if for every $\hat{\varphi} \in \hat{\mathcal{R}}$, $r = r(\hat{\varphi})$, and for every $D_1, \ldots, D_r \in \mathcal{D}$ such that $\hat{\varphi}(D_1, \ldots, D_r)$ is defined,

(H2) $h(\hat{\varphi}(D_1, \ldots, D_r)) = \bar{\varphi}(h(D_1), \ldots, h(D_r))$

where $\bar{\varphi} \in \mathcal{Q}$ is the unique operation corresponding to $\hat{\varphi} \in \hat{\mathcal{R}}$.

We say that a predicate P is *regular with respect to $\mathcal{D}$ (or $\mathcal{R}$)* if there exists a finite algebra $\mathcal{C}$ similar to $\mathcal{D}$ and a mapping $h : \mathcal{D} \to \mathcal{C}$ that satisfies (H1) and (H2). Note that h defines an equivalence relation $\sim_h$ on $\mathcal{D}$, where $D_1 \sim_h D_2$ if and only if $h(D_1) = h(D_2)$, and that $\sim_h$ has at most $|\mathcal{C}|$ equivalence classes.

From now on, rather than referring to the algebra $\mathcal{C}$ we will let $\mathcal{C} = \{C_1, \ldots, C_N\}$ be the set of equivalence classes of $\mathcal{D}$ with respect to $\sim_h$. Every $\hat{\varphi} \in \hat{\mathcal{R}}_r$ is thus "lifted" to an operation $\bar{\varphi} : C^r \to C$ defined by $\bar{\varphi}([D_1], \ldots, [D_r]) = [\bar{\varphi}(D_1, \ldots, D_r)]$, where $[D_i]$ is the equivalence class containing D_i. Without loss of generality, we will assume that each equivalence class uniquely determines the signature of its elements. We will refer to C as the *set of equivalence classes of D with respect to predicate P*. A class C_i is said to be *accepting* if there exists a pair $D \in C_i$ such that $P(D)$ holds. Several well-known problems — including dominating set, maximum cut, Steiner tree, traveling salesman, and independent set — have been shown to be regular on *any* family of decomposable graphs [BPT88].

We will need one further piece of notation. For $\varphi \in \mathcal{R}_r$, $\mathsf{MT}(\varphi, i)$ will denote the set of all ordered r-tuples $(i_1, \ldots, i_r)$ such that $\bar{\varphi}(C_{i_1}, \ldots, C_{i_r}) = C_i$.

3 The Basic Algorithm

In this section we shall describe the scheme underlying our subsequent results. We first show the existence of parse trees with certain useful properties for every $G \in \Gamma_w$. Next, we review the dynamic programming algorithm for nonparametric optimum subgraph problems of Bern et al. [BLW87].

3.1 Separators and Parse Trees

The close relationship between graphs of bounded tree-width and decomposable graphs has been shown by Wimer [Wim87]. For our purposes, we shall need a variant of one of his results. Let $\mathcal{R}$ be a family of k-terminal composition operations. A decomposition $G = \varphi(G_1, \ldots, G_r)$ with respect to $\mathcal{R}$ is said to be *balanced* if $|V(G_i)| \le |V(G)|/2 + c$, where c is a constant that depends only on $\mathcal{R}$. A parse tree (T, δ) of an n-vertex graph G is said to be *balanced* if (1) $|V(T)| = O(n)$, (2) the height of T is $O(\log n)$, and (3) for every internal node v of T with children $v_1, \ldots, v_r$, the decomposition $\gamma_T(v) = \delta(v)(\gamma_T(v_1), \ldots, \gamma_T(v_r))$ is balanced with respect to $\mathcal{R}$.

Let $\mathcal{U}[k, r]$ denote the set of all k-terminal graph composition operators $\varphi : \mathcal{G}_k^s \to \mathcal{G}_k$ where $0 \le s \le r$, and where, for each φ of arity zero, $|V(\varphi)| \le k$. Note that for every fixed k and r the size of $\mathcal{U}[k, r]$ is bounded by a constant. The proof of the next result relies on a simple observation. Let G be a graph and, for $X \subseteq V(G)$, let $G[X]$ denote the subgraph of G induced by X. Suppose $(A_1, \ldots, A_l, S)$ is a partition of $V(G)$ such that, for $i \ne j$, A_i and A_j are separated by S. Then, one can view G as the composition of $|S|$-terminal graphs $G_1, \ldots, G_l$, where $G_i = G[A_i \cup S]$ and $\text{terms}(G_i)$ consists of the vertices of S in some arbitrary (but fixed) order. Obviously, the corresponding composition operator is an element of $\mathcal{U}[|S|, l]$.

Theorem 3.1 *For each fixed w, every $G \in \mathcal{G}_{4w+4} \cap \Gamma_w$ has a balanced decomposition with respect to $\mathcal{U}[4w + 4, 5]$.*

Proof sketch. The following procedure, based on ideas from [Lag91], produces the desired decomposition. We write $\text{terms}(G)$ to denote the set of vertices in $\text{terms}(G)$.

Procedure DECOMPOSE
Input: $G \in \mathcal{G}_{4w+4} \cap \Gamma_w$, where $\text{terms}(G) = \langle t_1, \ldots, t_k \rangle$.
Output: Balanced decomposition $G = \varphi(G_1, \ldots, G_r)$, with $\varphi \in \mathcal{U}[4w + 4, 5]$.

Step 1. If $|V(G)| \leq 4w + 4$, return the decomposition $G = G$.

Step 2. Find a partition (A_1, A_2, A_3, S_1) of $V(G)$ satisfying Theorem 2.1 with $Q = V(G)$. Let $H_i = G[A_i]$, for $1 \leq i \leq 3$. Assume w.l.o.g. that $|A_1 \cap \text{terms}(G)| \geq |A_2 \cap \text{terms}(G)| \geq |A_3 \cap \text{terms}(G)|$.

Step 3. If $|A_1 \cap \text{terms}(G)| \leq k/2$, return the decomposition $G = \varphi(G_1, G_2, G_3)$, where, for $1 \leq i \leq 3$, $G_i = G[A_i \cup S_1]$ and $\text{terms}(G_i)$ consists of S_1, together with the vertices in $\text{terms}(G) \cap A_i$, in any order, and φ is an appropriate composition operator in $\mathcal{U}[4w + 4, 5]$.

Step 4. If $|A_1 \cap \text{terms}(G)| > k/2$, find a partition (B_1, B_2, B_3, S_2) of H_1 satisfying Theorem 2.1 with $Q = \text{terms}(G) \cap A_1$. For $i = 1, 2, 3$, let $G_i = G[B_i \cup S_1 \cup S_2]$ and let $\text{terms}(G_i)$ consist of the vertices in S_1, S_2, and $\text{terms}(G) \cap B_i$, in any order. For $i = 4, 5$, let $G_i = G[A_{i-2}]$ and let $\text{terms}(G_i)$ consist of S_1, together with the vertices in $\text{terms}(G) \cap A_{i-2}$, in any order. Return the decomposition $G = \varphi(G_1, \ldots, G_5)$, where φ is an appropriate composition operator in $\mathcal{U}[4w + 4, 5]$.

□

Corollary 3.2 *For every fixed w there exists $\mathcal{R} \subseteq \mathcal{U}[4w + 4, 5]$ such that (i) $\Gamma_w \subseteq \mathcal{R}^*$, and (ii) every $G \in \mathcal{G}_{4w+4} \cap \Gamma_w$, has a balanced parse tree (T, δ) in $T_\mathcal{R}$.*

Proof. A balanced parse tree (T, δ) of any $G \in \mathcal{G}_{4w+4} \cap \Gamma_w$ can be produced as follows. If $n = |V(G)| \leq 4w + 4$, (T, δ) consists of a single node v where $\delta(v) = G$. Otherwise, use procedure DECOMPOSE to find a balanced decomposition $G = \varphi(G_1, \ldots, G_r)$. Next, recursively construct a parse (T_i, δ_i) for each G_i. The parse tree of G will consist of a root v where $\delta(v) = \varphi$, and subtrees $(T_1, \delta_1), \ldots, (T_r, \delta_r)$. Clearly, for every $v \in V(T)$, $\delta(v) \in \mathcal{U}[4w + 4, 5]$ and (T, δ) is balanced. □

Let $w \geq 1$ be an integer. We shall say that a family $\mathcal{R}$ of composition operators is *w-adequate* if $\mathcal{R} \subseteq \mathcal{U}[4w + 4, 5]$ and every $G \in \mathcal{G}_{4w+4} \cap \Gamma_w$ has a balanced parse tree in $T_\mathcal{R}$. It follows from the proof of Corollary 3.2 that there exists a w-adequate family of composition operators for every $w \geq 1$.

3.2 Dynamic Programming on Decomposable Graphs

The following result was proved in [BLW87].

Theorem 3.3 *Suppose that $\mathcal{F} = \mathcal{R}^*$ is a class of decomposable graphs and P is a property that is regular with respect to $\mathcal{R}$. Then there exists a linear-time algorithm that, given a linear-size parse tree of $G \in \mathcal{F}$, finds an optimum-weight subgraph of G satisfying P.*

We sketch the proof of this theorem, because the underlying approach serves as a basis for our subsequent results. Let $\mathcal{C} = \{C_1, \ldots, C_N\}$ denote the set of equivalence classes of $\hat{\mathcal{R}}^*$ with respect to P. Let G be in $\mathcal{F}$. We define $z_G^{(i)}$ and z_G to be

$$z_G^{(i)} = \min\left\{\{+\infty\} \cup \{\text{val}_G(H) : (G, H) \in C_i\}\right\} \tag{3}$$

and

$$z_G = \min \left\{ \{+\infty\} \cup \{z_G^{(i)} : C_i \text{ is an accepting class}\} \right\}. \tag{4}$$

Note that the value of z_G is equal to $\mathrm{val}_G(H)$, where H is an optimum subgraph of G satisfying P (in the sense of equation (1)). If G is a primitive graph, then $z_G^{(i)}$ can be computed directly from equation (3) — i.e., by exhaustive enumeration. Otherwise, suppose $G = \varphi(G_1, \ldots, G_r)$ is a decomposition of G with respect to $\mathcal{R}$. Then, $z_G^{(i)}$ can be computed using the following equation

$$z_G^{(i)} = \min \left\{ \sum_{j=1}^{r} z_{G_j}^{(i_j)} - \mathrm{sv}[\varphi; i] : i = (i_1, \ldots, i_r) \in \mathrm{MT}(\varphi, i) \right\}, \tag{5}$$

where $\mathrm{sv}[\varphi; i]$ is the sum of the weights of shared vertices; i.e., those vertices that have been accounted for more than once in $\sum_{j=1}^{r} z_{G_j}^{(i_j)}$. Observe that, given the $z_{G_j}^{(i_j)}$'s, equation (5) can be evaluated in $O(1)$ time. As argued in [BLW87], these facts and the existence of linear-size parse trees can be used to obtain linear-time algorithms to compute z_G.

4 Parametric Problems

We shall now study the properties of Z_G^P when P is a regular graph property and G has bounded tree-width. Let d be a nonnegative integer. The term *d-th degree polynomial* will be used to refer to any polynomial of degree at most d. A function $f : \mathbf{R} \to \mathbf{R}$ is a *d-th degree piecewise polynomial function (d-ppf)* if it is the lower envelope of some finite set of d-th degree polynomials in λ; we write $b(f)$ to denote the number of breakpoints of f. The following result gives upper bounds for the number of breakpoints of the sum and lower envelope of d-ppfs.

Lemma 4.1 *Let $f_1, \ldots, f_m$ be d-ppf's. Then*

(i) $b\left(\sum_{j=1}^{m} f_j\right) \leq \sum_{j=1}^{m} b(f_j)$, *and*

(ii) $b\left(\min_{1 \leq j \leq m} f_j\right) \leq s(m, d)\left(\sum_{j=1}^{m} b(f_j) + 1\right) - 1$,

where $s(m, d)$ is a function that depends only on m and d. $\square$

The main result of this section is the following theorem. In its proof, we assume a model of computation where finding the roots of a d-th degree polynomial function is a primitive operation.

Theorem 4.2 *Let $w \geq 1$ and let P be a predicate that is regular with respect to some w-adequate set of composition operators $\mathcal{R}$. Then, for any $G \in \Gamma_w$ whose vertex and edge weights are d-th degree polynomial functions of λ, $b(Z_G^P)$ is polynomially bounded in $|V(G)|$. Furthermore, given a linear-size parse tree (T, δ) of G, Z_G^P can be constructed in polynomial time.*

Proof sketch. The main idea behind our proof is *algorithm simulation*, a technique inspired, in part, by [Meg79], and used in [FeSl89]. Let us refer to the dynamic programming algorithm described in section 3.2 as *algorithm $\mathcal{A}$*. Note that we can use $\mathcal{A}$ to compute $Z_G(\lambda)$ for *any* fixed λ. Based on this observation, we can obtain an algorithm $\mathcal{A}_C$ that constructs Z_G in its entirety by simulating the behavior of algorithm $\mathcal{A}$ for all possible parameter values at once. Wherever algorithm $\mathcal{A}$ adds or subtracts real numbers, $\mathcal{A}_C$ adds or subtracts d-ppf's and wherever algorithm $\mathcal{A}$ finds the maximum or minimum of real numbers, $\mathcal{A}_C$ computes upper or lower envelopes of d-ppf's. In order to specify algorithm $\mathcal{A}_C$ more precisely, we define a function $Z_G^{(i)} : \mathbf{R} \to \mathbf{R}$ that is the parametric analog of $z_G^{(i)}$, for $1 \leq i \leq N$:

$$Z_G^{(i)}(\lambda) = \min \left\{ \{+\infty\} \cup \{\mathsf{Val}_G(H, \lambda) : (G, H) \in C_i\} \right\}, \tag{6}$$

and a function $Z_G : \mathbf{R} \to \mathbf{R}$ that is the parametric analog of z_G:

$$Z_G(\lambda) = \min \left\{ \{+\infty\} \cup \{Z_G^{(i)}(\lambda) : C_i \text{ is an accepting class}\} \right\}. \tag{7}$$

Note that Z_G and $Z_G^{(i)}$ are d-ppf's. Now, if G is a primitive graph, then, for $i = 1, \ldots, N$, $\mathcal{A}_C$ computes $Z_G^{(i)}$ in $O(1)$ time directly from equation (6). Clearly, $b(Z_G^{(i)}) = O(1)$. If G is not a primitive graph, let $G = \varphi(G_1, \ldots, G_r)$ be a decomposition of G with respect to $\mathcal{R}$. Then, $\mathcal{A}_C$ computes $Z_G^{(i)}$ using:

$$Z_G^{(i)}(\lambda) = \min \left\{ \sum_{j=1}^{r} Z_{G_j}^{(i_j)}(\lambda) - \mathsf{SV}[\varphi; \mathbf{i}](\lambda) : \mathbf{i} = (i_1, \ldots, i_r) \in \mathsf{MT}(\varphi, i) \right\}, \tag{8}$$

where, as in equation (5), $\mathsf{SV}[\varphi; \mathbf{i}]$ is the sum of the weight functions of the vertices that contribute more than once to $\sum_{j=1}^{r} Z_{G_j}^{(i_j)}(\lambda)$. Note that $\mathsf{SV}[\varphi; \mathbf{i}]$ is a d-th degree polynomial for all φ and $\mathbf{i}$. Thus, $Z_G^{(i)}$ is the lower envelope of certain functions that are sums of d-ppf's. If these d-ppf's have a polynomial number of breakpoints, then Lemma 4.1 implies that so will $Z_G^{(i)}$; moreover, it follows that $Z_G^{(i)}$ is computable in polynomial time. These observations will form the basis of our proof. Let us define

$$\beta(n) = \max\{b(Z_G) : G \in \Gamma_w, |V(G)| \leq n\}$$

and $\quad \beta^{(i)}(n) = \max\{b(Z_G^{(i)}) : G \in \Gamma_w, |V(G)| \leq n\}.$

Using equation (7) and Lemma 4.1, we can derive

$$\beta(n) \leq c_1 \sum \{\beta^{(i)}(n) : C_i \text{ is an accepting class}\} + c_1 - 1, \tag{9}$$

where c_1 is a constant. Thus, polynomial bounds on the $\beta^{(i)}(n)$'s imply polynomial bounds on $\beta(n)$. Using (8) and the w-adequacy of $\mathcal{R}$, we can show that there exist constants α, c_3, and $a_j^{(i)}$, $1 \leq i, j \leq N$, such that

$$\beta^{(i)}(n) \leq \sum_{j=1}^{N} a_j^{(i)} \beta^{(j)}(\alpha n) + c_3 - 1, \tag{10}$$

for $i = 1, \ldots, N$. For lack of space, we omit the details of this derivation. This system of inequalities can be expressed as $\bar{\beta}(n) \leq A\bar{\beta}(\alpha n) + \bar{c}$, where $\bar{\beta}(n) = (\beta^{(1)}(n), \ldots, \beta^{(N)}(n))$, $A = \left(a_j^{(i)}\right)$ is an $N \times N$ matrix and $\bar{c}$ is a constant N-vector. Clearly, each $\beta^{(j)}$ is polynomially-bounded. Thus, by (9), $\beta(n)$ is polynomially bounded. Because of this, we can conclude that we can use the parse tree (T, δ) and equations (6), (7), and (8) to calculate Z_G in a bottom-up fashion in polynomial time. We omit the details. $\square$

5 Parametric Search Problems

We are interested in search problems where we must find the λ-value at which a particular event occurs in Z_G^P. We shall restrict our attention to minimization problems where weights are *linear* functions of λ. The following three problems are instances of parametric search:

(P1) Given a value λ_1 and a subgraph H which is optimum at λ_1, find the largest $\lambda^* \geq \lambda_1$ such that $Z_G^P(\lambda) = \mathrm{Val}_G(H, \lambda)$ for all $\lambda \in [\lambda_1, \lambda^*]$.
(P2) Given $t \in \mathbf{R}$, find $\lambda^* \in \mathbf{R}$ such that $Z_G^P(\lambda^*) = t$.
(P3) Find λ^* such that $Z_G^P(\lambda^*) = \max_\lambda Z_G(\lambda)$.

Each of the above problems has important applications [Gus83, Meg79, Fis81]. The main result of this section is the following theorem.

Theorem 5.1 *Let $w \geq 1$ and let P be a predicate that is regular with respect to some w-adequate set of composition operators $\mathcal{R}$. Then, given any weighted n-vertex graph $G \in \Gamma_w$, together with a balanced parse tree $(T, \delta) \in T_\mathcal{R}$ of G, problems (P1), (P2), and (P3) can be solved in $O(n \log n)$ time.*

We shall make use of the following lemma, whose proof is omitted.

Lemma 5.2 *Suppose that $\mathcal{F} = \mathcal{R}^*$ is a class of decomposable graphs and that P is property that is regular with respect to $\mathcal{R}$. Then, for each of (P1), (P2), and (P3), there is an oracle which, given a linear-size parse tree of $G \in \mathcal{F}$, answers the following question in linear time: Given $\lambda_0 \in \mathbf{R}$, determine whether or not $\lambda^* \geq \lambda_0$.*

We will also need one further result, due to Cole [Cole87]. A *combinational circuit* is a directed acyclic graph $\mathcal{B}$ whose nodes are *combinational elements*, and where an edge from element e_1 to element e_2 implies that the output of e_1 is an input to e_2 [CLR90]. Combinational elements are computational units that have a constant number of inputs and outputs and that perform well-defined operations. An element is said to be *active* if all its inputs are known, but the associated operation has not been carried out yet. We shall say that an element has been *resolved* if the associated operation has been carried out. Suppose we have a weight function $\omega : V(\mathcal{B}) \to \mathbf{R}$. The *active weight*, W, of $\mathcal{B}$ is the sum of the weights of its active elements. An *oracle with respect to ω* is a procedure that is guaranteed to resolve a set of active elements whose total weight is at least $W/2$.

Lemma 5.3 *[Cole87] Let B be a combinational circuit of size M and depth $f(M)$. Let $d_{\min} = \min\{d_I, d_O\}$, where d_I (d_O) denotes the maximum fan-in (fan-out) of an element of B. Then, there exists a weight function ω such that B can be evaluated with $O(f(M)\log d_{\min} + \log M)$ calls to an oracle with respect to ω.*

Proof of Theorem 5.1. We shall use ideas from [Meg83, Cole87]. As in section 4, let Z_G and $Z_G^{(i)}$ be the parametric analogs of z_G and $z_G^{(i)}$. The proof proceeds by first showing how to obtain a combinational circuit B_0 whose inputs are the vertex and edge weights at λ and whose output is $Z_G(\lambda)$. Next, we use B_0 to construct a circuit B, whose inputs are the vertex and edge weight functions, and whose output is the function Z_G. Finally, we devise a search algorithm $\mathcal{A}_S$ that guides the execution of B to determine the behavior of Z_G in the neighborhood of λ^*.

We obtain B_0 from a balanced parse tree (T, δ) of G and from the multiplication tables for the operators in $\mathcal{R}$ and the equivalence classes $C_1, \ldots, C_N$ as follows. For $v \in V(T)$, let G_v denote $\gamma_T(v)$. Note that for every $v \in V(T)$ and every $i, 1 \leq i \leq N$, there exists a $O(1)$-size circuit $\mathsf{CT}(\varphi, i)$, where $\varphi = \delta(v)$, that computes $z_{G_v}^{(i)}$, given the appropriate inputs. These inputs will be edge and vertex weights of φ if $\varphi \in \mathcal{R}_0$ or they will be the $z_{G_j}^{(i_j)}$'s for graphs corresponding to children of v. In either case, the structure of $\mathsf{CT}(\varphi, i)$ depends only on $\mathsf{MT}(\varphi, i)$. There is also a $O(1)$-size circuit that computes the value of z_G from the values of the $z_{G_u}^{(i)}$'s, where u is the root of T. All of these circuits can be constructed entirely out of min gates, adders, and subtractors, and can be assembled into a circuit B_0 of size $O(n)$ and depth $O(\log n)$ by following the structure of T.

To construct the parametric circuit B, we modify circuit B_0 by replacing its elements, which manipulate real numbers, with elements that manipulate functions of λ and by replacing its inputs with the corresponding functions of λ. Thus, the min gates will compute lower envelopes of their inputs and the adders will construct the sum of their input functions. The size and depth of B are identical to those of B_0. For $v \in V(B)$, we write $S_v(\lambda)$ to denote the output of v at λ. Clearly, the inputs and outputs of every circuit element are piecewise linear concave functions, and $Z_G = S_u$, where u is the output node of B.

The search algorithm $\mathcal{A}_S$ simulates the execution of B to find the structure of Z_G within a certain closed interval I^* such that (1) $\lambda^* \in I^*$ and (2) Z_G is a line in I^*. At all times, $\mathcal{A}_S$ maintains an interval $I = [\lambda_L, \lambda_R]$ such that $\lambda^* \in I$; initially $\lambda_L = -\infty$ and $\lambda_R = +\infty$. We shall say that $\mathcal{A}_S$ has *resolved* an element v of B if S_v is a line in the interval I and the equation of this line has been computed. An unresolved element v is *active* if v is an input, or all elements u such that $(u, v) \in E(B)$ have been resolved. At each stage, algorithm $\mathcal{A}_S$ chooses which elements to resolve according to a weight function ω defined on $V(B)$ and the following procedure.

Procedure RESOLVE(λ_L, λ_R)

Step 1. Let A be the set of active nodes of B. Let A_1 be the set of adders and subtractors in A and let A_2 be the set of min gates in A. Resolve each $v \in A_1$ by constructing S_v in its entirety in the interval $I = [\lambda_L, \lambda_R]$.

Step 2. For each $v \in A_2$, construct S_v in its entirety in the interval I, by taking the lower envelope of the inputs S_{u_1} and S_{u_2}. Let λ_v denote the single breakpoint of this function in the interval (λ_L, λ_R); if S_v has no breakpoints in this open interval, set $\lambda_v = +\infty$. Let $U = \{v \in A_2 : \lambda_v \in (\lambda_L, \lambda_R)\}$.

Step 3. Resolve each $v \in A_2 - U$. If U is empty, stop. Otherwise, compute the weighted median λ_{v_m} of the set $\{\lambda_v : v \in U\}$, where the weight of λ_v is the weight $\omega(v)$ of the corresponding element of B. Use the oracle of Lemma 5.2 to determine whether or not $\lambda^* \geq \lambda_{v_m}$. If the answer to this call is "yes", then set $\lambda_L = \lambda_{v_m}$ and resolve all elements $v \in U$ such that $\lambda_v \leq \lambda_{v_m}$. Otherwise, set $\lambda_R = \lambda_{v_m}$ and resolve all elements $v \in U$ such that $\lambda_v \geq \lambda_{v_m}$.

We omit the proof that RESOLVE is a $O(n)$-time oracle with respect to any weight function ω. At the end of the simulation, I will be an an interval containing λ^* within which Z_G is a line whose equation is known. At this point, computing λ^* will be straightforward. Since the fan-in of any element is at most 2, Lemma 5.3 implies that A_S will run in $O(n \log n)$ time. □

6   Discussion

The main results of this paper, Theorems 4.2 and 5.1, rely on the existence of balanced decompositions and parse trees. The key to constructing these is an algorithm delivering a partition satisfying Theorem 2.1. A linear-time separator algorithm is easily obtainable from what is known as the *embedding w-tree* of the graph [ArPr89] (see, e.g., [FeMe90]). Since such embeddings can be constructed in linear time for graphs of tree-width 1, 2, or 3 [MaTh91], we can construct partitions of such graphs in $O(n)$ time, and balanced parse trees in $O(n \log n)$ time. Things are more complicated for graphs of tree-width $w > 3$. It is easy to see that all of our results are valid if instead of a partition satisfying Theorem 2.1, we have a partition (B_1, B_2, B_3, S) satisfying all the conditions of this theorem, except that we only guarantee that $|B_i \cap Q| \leq (1 - d_0(w))|Q - S|$, for some function $d_0(w)$ such that $0 < d_0(w) \leq 1/2$ for $w > 3$. We refer to the separator S in this partition as an *approximate separator*. Lagergren [Lag91] has shown how an approximate separator can be produced in $O(n \log n)$ time (see also [Reed92]). The approximate separator algorithm can be used to construct parse trees of size $O(n)$ and height $O(\log n)$ in $O(n \log^2 n)$ time.

Theorem 4.2 implies that if vertex and edge weights are bounded-degree polynomials in λ, then problems (P1)–(P3) can be solved in polynomial time by simply constructing Z_G. Unfortunately, we do not see any easy way to extend the results of section 5 to problems where costs are polynomial functions of λ. There are, however, two search problems for which we have $O(n)$-time algorithms.

(P4) Given $\lambda_0 \in \mathbf{R}$, find a $\lambda^* > \lambda_0$ such that Z_G^P has no breakpoints in (λ_0, λ^*).

(P5) Find λ_∞ such that Z_G^P has no breakpoints in $(\lambda_\infty, +\infty)$.

Problem (P4) is a simplified version of the sensitivity analysis problem, while problem (P5) is the *steady state problem* [Ata85, FeSl89]. Observe that, given a solution to (P5), we can, in $O(n \log n)$ time, find the first or last breakpoint of Z_G for the case where weights are linear.

In the next theorem we assume, as in section 4, a model of computation where computing the roots of a d-th degree polynomial is a primitive operation.

Theorem 6.1 *Let $w \geq 1$ and let P be a predicate that is regular with respect to some w-adequate set of composition operators $\mathcal{R}$. Then, given any weighted n-vertex graph $G \in \Gamma_w$, together with a $O(n)$-size parse tree $(T, \delta) \in \mathcal{T}_\mathcal{R}$ of G, problems (P4) and (P5) can be solved in $O(n)$ time.* $\square$

The restriction to graphs of bounded tree-width seems to be important in achieving our bounds. Without it, some problems do indeed have an exponential number of breakpoints in the worst case [Car83]. However, the bound of Theorem 4.2 can probably be sharpened considerably in certain special cases. A natural candidate for further study is the maximum independent set problem. Improving the running times of the algorithms for the search problems described in Section 5 is another intriguing problem. We see no obvious reason why $\Omega(n \log n)$ should be a lower bound for the solution of these problems

Acknowledgements

The first author thanks Bruno Courcelle for his hospitality in Bordeaux, and Hans Bodlaender, Jens Lagergren, and Andrzej Proskurowski for valuable discussions during the early stages of this work.

References

[ArPr89] S. Arnborg and A. Proskurowski. Linear time algorithms for NP-hard problems restricted to partial k-trees. *Discr. Appl. Math.*, 23:11–24 (1989).

[ALS91] S. Arnborg, J. Lagergren, and D. Seese. Easy problems for tree-decomposable graphs. *J. Algorithms*, 12:308–340.

[Ata85] M. Atallah. Dynamic computational geometry. *Comp. & Maths. with Appls.*, 11(12):1171–1181, 1985.

[BLW87] M.W. Bern, E.L. Lawler, and A.L. Wong. Linear time computation of optimal subgraphs of decomposable graphs. *J. Algorithms*, 8:216–235, 1987.

[BPT88] R.B. Borie, R.G. Parker, and C.A. Tovey. Automatic generation of linear-time algorithms from predicate-calculus descriptions of problems on recursively-constructed graph families. Manuscript, 1988. To appear in *Algorithmica*.

[Bod87] H.L. Bodlaender. Dynamic programming on graphs with bounded tree-width. Technical Report RUU-CS-88-4, University of Utrecht, 1988. Extended Abstract in *Proceedings of ICALP88*.

[Car83] P. Carstensen. Complexity of some parametric integer and network programming problems. *Math. Programming*, 26:64–75, 1983.

[Cole87] R. Cole. Slowing down sorting networks to obtain faster sorting algorithms. *J. Assoc. Comput. Mach.*, 34(1):200–208, 1987.

[Cou90] B. Courcelle. The monadic second-order logic of graphs I: Recognizable sets of finite graphs. *Information and Computation*, 85:12–75, 1990.

[CLR90] T.H. Cormen, C.E. Leiserson, and R.L. Rivest. *Introduction to Algorithms*. MIT Press, Cambridge, Mass., 1990.

[FeSl89] D. Fernández-Baca and G. Slutzki. Solving parametric problems on trees. *J. Algorithms*, 10:381–402 (1989).

[FeMe90] D. Fernández-Baca and A. Medepalli. Parametric module allocation on partial k-trees. Technical Report 90-25, Department of Computer Science, Iowa State University, December 1990.

[Fis81] M.L. Fisher. The Lagrangian relaxation method for solving integer programming problems. *Management Science*, 27:1–18 (1981).

[GGT89] G. Gallo, M.D. Grigoriades, and R.E. Tarjan. A fast parametric maximum flow algorithm and its applications. *SIAM J. Computing*, 18(1):30–55, 1989.

[Gus83] D. Gusfield. Parametric combinatorial computing and a problem in program module allocation. *J. Assoc. Comput. Mach.*, 30(3):551–563, July 1983.

[vHKRW89] C.P.M. van Hoesel, A.W.J. Kolen, A.H.G. Rinooy and A.P.M. Wagelmans, *Sensitivity analysis in combinatorial optimization: a bibliography.* Report 8944/A, Econometric Institute, Erasmus University Rotterdam, 1989.

[GaJo79] M. Garey and D. Johnson. *Computers and Intractability: A Guide to the theory of NP-Completeness.* Freeman, San Francisco, 1979.

[Lag91] J. Lagergren. Algorithms and minimal forbidden minors for tree-decomposable graphs. (Doctoral Dissertation) Technical Report TRITA-NA-9104, Royal Institute of Technology, Sweden, March 1991.

[MaTh91] J. Matoušek and R. Thomas. Algorithms finding tree-decompositions of graphs. *J. Algorithms*, 12:1–22 (1991).

[Meg79] N. Megiddo. Combinatorial optimization with rational objective functions. *Math. Oper. Res.*, 4:414–424 (1979).

[Meg83] N. Megiddo. Applying parallel computation algorithms in the design of serial algorithms. *J. Assoc. Comput. Mach.*, 30(4):852–865, 1983.

[Mur80] K. Murty. Computational complexity of parametric linear programming. *Math. Programming*, 19:213–219, 1980.

[Reed92] B.A. Reed. Finding approximate separators and computing tree width quickly. To appear in *STOC 92.*

[RoSe86] N. Robertson and P.D. Seymour. Graph minors II: Algorithmic aspects of tree-width. *J. Algorithms*, 7:309–322, 1986.

[RoSe] N. Robertson and P.D. Seymour. Graph minors XIII: The disjoint paths problem. *To appear.*

[vLe90] J. van Leeuwen. Graph Algorithms. In J. van Leeuwen (ed.) *Handbook of Theoretical Computer Science*, MIT Press, Cambridge, Mass., 1990.

[Wim87] T.V. Wimer. Linear algorithms on k-terminal graphs. Ph.D. Thesis, Report No. URI-030, Clemson University (1987).

[ZhGo91] B. Zhu and W. Goddard. An algorithm for outerplanar graphs with parameter. *J. Algorithms*, 12:6657–662 (1991).

This article was processed using the LaTeX macro package with LLNCS style

Efficient two-dimensional searching

by

Gaston H. Gonnet

Informatik, E.T.H.
Zurich, Switzerland.

PAT trees, PAT arrays (or suffix arrays) resolve the problem of searching strings in $O(\log n)$ expected time, independently of the size of the answer. Prefixes, ranges of prefixes, longest repetitions of text, longest repetitions with a particular prefix and repetitions longer than a certain length can also be found very efficiently.

The structure lends itself to run automata, and hence provides the first known algorithm to search for a general regular expression in sublinear (in the size of the text) time. It also allows the approximate matching of a string against a database, (using the dynamic programming algorithm for string similarity) to be run in sublinear time.

Here we describe the construction of a spiraled PAT tree which allows picture (two-dimensional) searching, in $O(\log n)$ time, at a cost of $O(n)$ storage (n being the number of pixels in the picture database), and independent on the size of the answer. Although designed originally for pixels, the algorithm also works for text, and gives an algorithm which does 2-dimensional searching in logarithmic searching cost, once the index is built. Most of the abilities of one-dimensional PAT structures can be mapped to two-dimensional structures. Prefix searching is quite interesting while it appears to be useful. Approximate 2-D string matching using dynamic programming is also interesting and useful.

The spiraled PAT trees and arrays depend on a geometrical parameter which defines the index. We explore several mechanisms to cope with searching geometrical areas which are different from the index one.

Improvements on Geometric Pattern Matching Problems[*]

L. Paul Chew[1] and Klara Kedem[2]

[1] Department of Computer Science, Cornell University, Ithaca, NY 14853, USA
[2] Department of Computer Science, Tel-Aviv University, Tel-Aviv 69978, Israel

Abstract. We consider the following geometric pattern matching problem: find the minimum Hausdorff distance between two point sets under translation with L_1 or L_∞ as the underlying metric. Huttenlocher, Kedem, and Sharir have shown that this minimum distance can be found by constructing the upper envelope of certain Voronoi surfaces. Further, they show that if the two sets are each of cardinality n then the complexity of the upper envelope of such surfaces is $\Omega(n^3)$. We examine the question of whether one can get around this cubic lower bound, and show that under the L_1 and L_∞ metrics, the time to compute the minimum Hausdorff distance between two point sets is $O(n^2 \log^2 n)$.

1 Introduction

A central problem in pattern recognition, computer vision, and robotics, is the question of whether two point sets A and B *resemble* each other. One approach to this problem, first used by Huttenlocher and Kedem [5] and further developed by Huttenlocher, Kedem and Sharir [7], is based on the minimum Hausdorff distance between point sets in the plane under translation. The Hausdorff distance between two point sets A and B is defined as

$$H(A, B) = \max(h(A, B), h(B, A))$$

where

$$h(A, B) = \max_{a \in A} \min_{b \in B} d(a, b) \ .$$

Here, $d(\cdot, \cdot)$ represents a more familiar metric on points; for instance, the standard Euclidean metric (the L_2 metric) or the L_1 or L_∞ metrics. Huttenlocher and Kedem observe that the Hausdorff distance remains a metric even when minimized with respect to all possible translations of the point sets. Thus, it can be used to define a metric on point sets (and more general shapes) that is independent of translation. Intuitively, it measures the maximum mismatch between two point sets after the sets have been translated to minimize this mismatch.

---

[*] The first author was supported by AFOSR Grant AFOSR-91-0328, ONR Grant N00014-89-J-1946, NSF Grant IRI-9006137, and DARPA under ONR contract N00014-88-K-0591. The second author was supported by the Eshkol grant 04601-90 from the Israeli Ministry of Science and Technology.

Another approach to the basic problem of when two point sets resemble each other is that of Alt, Mehlhorn, Wagener and Welzl [1]. They state the problem as follows: given two sets A and B of n points each in the plane, a transformation group G, and some tolerance ε, A and B are *approximately congruent* if there is a bijection $\ell : B \to A$ and a transformation $M \in G$, such that the distance between $M(b)$ and $\ell(b)$ is less than or equal to ε for all $b \in B$.

These two approaches reflect different views of the qualitative problem of pattern matching. In [1], the assumption is that there is a congruence between the point sets, but their location in the plane is not precisely measured (due to sensor or representation inaccuracy); hence, the sets A and B are of equal cardinality but the corresponding points cannot be matched exactly. In [5, 7], the authors assume that the data achieved by sensors is not precise, in the sense that one point might be split into two points or some cluster of nearby points might be merged into one; hence, the sets A and B can be sets of different cardinality. The different approaches are illustrated in Fig. 1 and Fig. 2. Each figure shows two different point-resemblance problem instances, (a) and (b). In each instance, one set is represented by circles and the other by crosses. The two approaches provide different solutions for instance (a) and the same solution for instance (b).

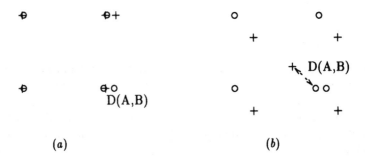

(a)　　　　　　　　　　　　　　　　(b)

Fig. 1. The minimum Hausdorff distance under translation.

In this paper we address the problem of finding the minimum Hausdorff distance between two sets in the plane. We show that this problem can be solved by examining the intersection of many sets where each set is the union of translates of a convex shape. Convex shapes of particular interest here are the circle, corresponding to the L_2 or standard Euclidean metric, and the square, corresponding to both the L_1 and L_∞ metrics. Our approach enables us to improve on the algorithms of [5, 7] for the L_1 and L_∞ metrics.

We show that the arrangement that results from our intersection of unions can have complexity $\Omega(n^3)$. We then prove the surprising result that sweeping using Bentley's segment tree (see, for instance, [11, 10]) combined with a type of parametric search [3, 4] can be used to work around this $\Omega(n^3)$ bound, allowing the minimum Hausdorff distance under translation to be computed in time $O(n^2 \log^2 n)$ when the underlying metric is the L_1 (or L_∞) metric. When the sets being compared have

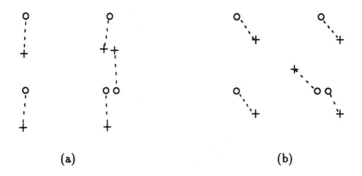

Fig. 2. The best approximate congruence under translation.

different numbers of points (say m and n) then this time bound is $O(mn \log^2 mn)$, a significant improvement over the best previous time bound of $O(mn(m+n) \log mn)$ presented in [7]. In addition, this answers an open problem from [7] where it is asked whether there are faster methods for computing the minimum Hausdorff distance than their technique of computing an entire upper envelope of Voronoi surfaces.

Using our 2D result, we provide an $O((mn)^2 \log^2 mn)$ algorithm to solve the minimum Hausdorff distance in three dimensions when the underlying metric is the L_∞ metric. In general, we show that each higher dimension multiplies this bound by mn, giving a time bound of $O((mn)^{(d-1)} \log^2 mn)$ for the L_∞ metric in dimension d. The closest-related previous result is the $O((mn)^2(m+n)\alpha(mn) \log^2 mn)$ algorithm presented in [7] for finding the minimum Hausdorff distance in 3-space when the underlying metric is the L_2 metric.

2 Preliminaries

Let A and B be two sets of points in the plane, and let ε be a positive real number. Take set A and put a circle of radius ε about each point of A, creating set A_ε. Note that the "circle" depends on the underlying distance metric: it is a standard circle for the L_2 metric, a square for the L_∞ metric, and a square tipped at 45 degrees for the L_1 metric. A_ε can be written as $A \oplus C(\varepsilon)$ where $C(\varepsilon)$ is the "circle" of size ε and $\oplus$ represents the Minkowski sum. Consider the set $A_\varepsilon \oplus -b$ where $-b$ represents the reflection of the point b through the origin. This set can be thought of as the set of translations that map b into A_ε. The set of translations that map all points $b \in B$ into A_ε is then $\cap_{b \in B}(A_\varepsilon \oplus -b)$; we denote this set by $S(A, \varepsilon, B)$. In [2], the authors use this kind of construction in their analysis of the time bound for computing a minimum Hausdorff distance for sets of segments under the L_2 metric.

We define the *Hausdorff decision problem* for a given ε to be the question of whether the minimum Hausdorff distance under translation is less than some given ε. We say that the Hausdorff decision problem for sets A and B and for ε is *true* if there exists a translation t such that the Hausdorff distance between A and $B \oplus t$ is less than or equal to ε. The following lemma is an immediate consequence of these definitions.

Lemma 1. *The Hausdorff decision problem for point sets A and B and for ε is true iff* $S(A, \varepsilon, B) \cap -S(B, \varepsilon, A) \neq \emptyset$

3 A Lower Bound

In this section, we show that $S(A, \varepsilon, B)$ can have $\Omega(n^3)$ disjoint regions, where n is the number of points in each set A and B, by using a construction similar to that used in [5, 7] to show that an upper envlope of Voronoi surfaces can have complexity $\Omega(n^3)$. The construction applies to many types of metrics that are convex distance functions, including the L_1, L_2, and L_∞ metrics. For clarity, we present our example here in terms of the L_2 metric – pictures of overlapping circles are much easier to interpret than pictures of overlapping squares.

To start our example, consider the set (A_ε) consisting of a pair of circles; the pair consists of one circle centered at the origin and one centered at $(1, 0)$. Now consider the intersection of k copies of this pair of circles (equivalent to shifting by k vectors $-b \in B$); the first copy is untranslated, the second is translated in the positive x direction by δ, where $\delta < \frac{1}{k}$, the third is translated by 2δ, etc. When ε, the radius of our circles, is near $1/2$, the resulting intersection is a set of tall, thin regions as illustrated in Fig. 3.

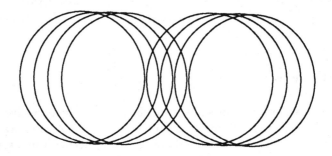

Fig. 3. The intersection of k copies of a pair of circles.

This intersection has k regions, but we can get k^2 regions by using a slightly more complicated construction. First note that if we were to rotate our initial pair of circles by 90 degrees then the resulting intersection would consist of long and thin horizontal regions. We start with three circles, one at the origin, one at $(1, 0)$ and one at $(0, 1)$. By using appropriate translations (i.e., $(i\delta, \frac{k\delta}{2})$ and $(\frac{k\delta}{2}, i\delta)$ for $i = 0 \ldots k - 1$) we get two sets of skinny regions, one horizontal and one vertical; these intersect to form $\Omega(k^2)$ separate regions.

To get up to $\Omega(n^3)$ regions in $S(A, \varepsilon, B)$ we let A consist of $\frac{n}{3}$ copies of the 3 unit circles used above, where the copies are separated widely enough that they do not interfere with one another. The set of translations above corresponds to shifts by $-b$ for all $b \in B$. If $|B| = k$ we get $\Omega(nk^2)$ regions. If $k = n$ then the intersection is of complexity $\Omega(n^3)$.

4 Minimum Hausdorff Distance for the L_1 (L_∞) Metric

Surprisingly, under the L_1 and L_∞ metrics the time to compute the minimum Hausdorff distance between point sets A and B in the plane is $O(n^2 \log^2 n)$ ($O(mn \log^2 mn)$ when the point sets are of differing cardinality). This is surprising because the construction in the previous section shows that the bound of $\Theta(n^3)$ for the complexity of the structure used to calculate the minimum one-way Hausdorff distance is tight for L_1 and L_∞. We work around this bound by observing that we can solve a Hausdorff decision problem without having to build or visit the entire structure. We take advantage of properties of the segment tree, reformulating our problem in such a way that we can use a sweeping segment tree to detect whether the Hausdorff decision problem for a given ε is true.

Theorem 2. *Given $\varepsilon > 0$ and two point sets A and B in the plane, and using L_∞ (L_1) as the underlying metric, we can determine whether the minimum Hausdorff distance between A and B is less than ε in time $O(mn \log mn)$, where $m = |A|$ and $n = |B|$.*

Proof. We place a square (an L_∞ circle) of size ε about each point of A to form A_ε. By Lemma 1, the minimum Hausdorff distance between A and B is less than ε iff $S(A, \varepsilon, B) \cap -S(B, \varepsilon, A) \neq \emptyset$. For clarity, we look only at the problem of determining whether $S(A, \varepsilon, B)$ is empty; the intersection of this set with $-S(B, \varepsilon, A)$ is straightforward.

We determine whether $S(A, \varepsilon, B)$ is empty or nonempty by sweeping a segment tree across the arrangement of translates of A_ε and counting how many different translates of A_ε cover each point in the plane – the idea is that the intersection is nonempty iff we can find a point on the plane that is covered by at least one square from *each* copy of A_ε. Since the squares within a single copy of A_ε can overlap, simply counting the squares at each point is not enough.

The trick is to convert A_ε into a structure without any overlap. Then counting is enough – a point that is covered to depth n has been covered by something from each of the n copies of A_ε. We observe that the set A_ε can be subdivided into $O(m)$ nonoverlapping rectangles. This follows from the fact that the complexity of the boundary of translates of squares (A_ε) is $O(m)$ [9].

Taking A_ε as a set of nonoverlapping rectangles, we form n translates of these rectangles, producing an arrangement of $O(mn)$ rectangles. Since the rectangles within a given copy of A_ε are nonoverlapping, we can easily detect whether the intersection of all the translates is nonempty: the intersection is nonempty iff there is some point on the plane which is covered by exactly n rectangles.

A segment tree [10, 11] can be used to sweep across a set of k rectangles, reporting the maximum covering depth encountered, in time $O(k \log k)$. Since we have $O(mn)$ rectangles, the total time to detect whether the intersection is nonempty is $O(mn \log mn)$. $\qquad\square$

We can use the technique of the above theorem to solve the minimum Hausdorff distance problem at the cost of an additional log factor in the time bound.

Theorem 3. *Given point sets A and B in the plane and using the L_∞ (L_1) metric, the minimum Hausdorff distance between A and B can be determined in time $O(mn \log^2 mn)$, where $m = |A|$ and $n = |B|$.*

Proof. The minimum Hausdorff distance problem is an optimization problem for which we can use the Hausdorff decision problem as a subroutine. In the proof of Theorem 2 we place a square of size ε about each point of A and of B in order to find whether the intersection $S(A, \varepsilon, B) \cap -S(B, \varepsilon, A)$ is empty or not. Now we want to determine the minimum ε where the intersection is still non-empty. It is easy to see that the desired minimum value is achieved at some ε_0 for which a certain (initially unknown) pair of squares touch each other.

We need to search among all possible values of ε where two squares touch at their boundaries. (Note that we actually have to search twice, once for touches at vertical boundaries and once for touches at horizontal boundaries.)

On the whole there are mn squares, therefore $O((mn)^2)$ values. If we try to use simple binary search on these values then we waste too much time first computing all the values and then sorting them. Fortunately, these values can all be represented implicitly in $O(mn)$ space as a sorted matrix (i.e., each row and each column are in sorted order). We do not compute the matrix of ε's explicitly; entries within the table are calculated just when needed. If the squares are in sorted order then the resulting implicit table is automatically sorted.

We use a result due to Frederickson and Johnson [3, 4] to solve this kind of optimization problem: Given a sorted matrix of size N by N it takes time $O(N + D \log N)$ to solve the optimization problem, where D is the time it takes for the decision problem. For our problem this time bound becomes $O(mn \log^2 mn)$. $\qquad\square$

5 Higher Dimensions

Theorem 4. *Given point sets A and B in 3-space and using the L_∞ metric, the minimum Hausdorff distance between A and B can be determined in time $O((mn)^2 \log^2 mn)$, where $m = |A|$ and $n = |B|$. In dimension d the time bound is $O((mn)^{(d-1)} \log^2 mn)$.*

Proof. Briefly, in 3D we can split the arrangement of cubes into $O(mn)$ slices. Within each slice, we can run the 2D algorithm. An extra factor of mn occurs for each additional dimension. $\qquad\square$

6 Conclusions and Further Research

The minimum Hausdorff distance problem for point sets under translation can be viewed as an intersection problem involving the intersection of translates of a union of "circles". Using this viewpoint, we have presented improved algorithms for the minimum Hausdorff distance problem for the L_1 and L_∞ metrics in the plane and for the L_∞ metric in higher dimensions. The time bound of $O(n^2 \log^2 n)$ for the 2-dimensional problem is unexpected in the sense that the problem's solution requires examination of a structure with complexity $\Omega(n^3)$.

In terms of practical implementation for the problem of point-set resemblance, the minimum Hausdorff distance can be determined more efficiently than an approximate congruence [1], taking time $O(n^2 \log^2 n)$ versus time $O(n^6 \log n)$ for the L_∞ versions of the problems. Of course, these two algorithms answer different questions (see Sect. 1), so direct comparisons are not entirely meaningful. Other methods that may perform well in practice include the *approximate* approximate congruence methods of Heffernan and Schirra [8] and the rasterized version of the minimum Hausdorff method [6].

We showed that for one type of Hausdorff problem – the minimum Hausdorff distance problem for points in the plane using the L_∞ (L_1) metric – we can break the cubic bound by using segment trees and a form of parametric search. Are there other natural metrics for which we can break the cubic bound? In particular, can the bound be broken for the more-familiar L_2 metric?

7 Acknowledgements

The authors wish to thank Paul Heffernan and a member of the SWAT committee for their helpful comments.

References

1. H. Alt, K. Mehlhorn, H. Wagener, and E. Welzl, "Congruence, similarity, and symmetries of geometric objects", *Discrete and Computational Geometry*, 3(1988), pp. 237–256.
2. P.K. Agarwal, M. Sharir, and S. Toledo, "Applications of parametric searching in geometric optimization", to appear in Proc. Third ACM-SIAM Symposium on Discrete Algorithms, 1992.
3. G.N. Frederickson, "Optimal algorithms for tree partitioning", *Proceedings of the Second Annual ACM-SIAM Symposium on Discrete Algorithms*, 1991, pp 168–177.
4. G.N. Frederickson and D.B. Johnson, "Finding kth paths and p-centers by generating and searching good data structures", *Journal of Algorithms*, 4(1983), pp 61-80.
5. D.P. Huttenlocher and K. Kedem, "Efficiently computing the Hausdorff distance for point sets under translation", *Proceedings of the Sixth ACM Symposium on Computational Geometry*, 1990, pp. 340–349.
6. D.P. Huttenlocher, G.A. Klanderman, and W.J. Rucklidge, "Comparing Images Using the Hausdorff Distance Under Translation", Cornell Computer Science Department, TR-91-1211, 1991.
7. D.P. Huttenlocher, K. Kedem, and M. Sharir, "The upper envelope of Voronoi surfaces and its applications", *Proceedings of the Seventh ACM Symposium on Computational Geometry*, 1991, pp 194–203.
8. P.J. Heffernan and S. Schirra, "Approximate decision algorithms for point set congruence", Saarbrucken Computer Science Department, Germany, Technical Report MPI-I-91-110, 1991.
9. K. Kedem, R. Livne, J. Pach, and M. Sharir, "On the union of Jordan regions and collision-free translational motion amidst polygonal obstacles", *Discrete and Computational Geometry*, 1(1), 1986, pp. 59–72.

10. K. Mehlhorn, *Data Structures and Algorithms 3: Multi-Dimensional Searching and Computational Geometry*, Springer-Verlag, 1984.
11. F.P. Preparata and M.I. Shamos, *Computational Geometry*, Springer-Verlag, New York, 1985.

This article was processed using the LaTeX macro package with LLNCS style

Determining DNA Sequence Similarity Using Maximum Independent Set Algorithms for Interval Graphs

Deborah Joseph* Joao Meidanis** Prasoon Tiwari***

Computer Sciences Department, University of Wisconsin-Madison, Madison WI 53705, USA

Abstract. Motivated by the problem of finding similarities in DNA and amino acid sequences, we study a particular class of two dimensional interval graphs and present an algorithm that finds a maximum weight "increasing" independent set for this class. Our class of interval graphs is a subclass of the graphs with interval number 2. The algorithm we present runs in $O(n \log n)$ time, where n is the number of nodes, and its implementation provides· a *practical* solution to a common problem in genetic sequence comparison.

1 Introduction

The work presented in this paper is motivated by a common problem in genetic sequence comparison. The problem is to determine the functionality of a DNA or amino acid sequence by comparing the sequence to a database of sequences whose functionality is known. This problem arises in large scale genome sequencing projects. The sequence comparison methods used are frequently syntactic. However, the matches found are rarely exact (unless the query sequence has previously been sequenced and included in the database). In fact, it is very difficult to put a quantitative bound on the number of mismatches that is acceptable – usually a geneticist wants to see the *most similar* sequences from the database.

In recent years, many sequence comparison algorithms have been developed and implemented for solving problems such as the one described above (see [14], [17], [12], [15], [13], [2], [5], and [8], among others). These algorithms can have either a *global* or a *local* character. Global methods seek an alignment of two given sequences that maximizes a certain *similarity measure*. These methods are normally allowed to insert gaps into the sequences in order to improve the alignment. The *similarity measure* then rewards matching characters in the alignment and penalizes mismatches and gaps. Several such measures have been proposed, and some are currently being closely investigated (see [18] and [9]). In general, global algorithms output a single answer, or a small set of *best* answers.

---

* Supported by NSF Presidential Young Investigator Grant DCR-8451387.

** Permanent Affiliation: Computer Science Dept., State University of Campinas, Cx. Postal 6065, 13081 – Campinas – SP, Brazil. Partially supported by FAPESP, Brazil, under grant 87/0197-2.

*** Supported by Wisconsin Alumini Research Foundation and by National Science Foundation under grant CCR-9024516

In contrast, local comparison methods find similar subregions within the two sequences. In these methods, inserting gaps in the subregions is either not allowed, or heavily penalized. As a result, these methods frequently yield a large, perhaps overlapping, collection of pairs of matching subregions. Associated with each matching pair of subregions is a *score*. The score indicates the quality and length of the match. Most of the existing local comparison implementations can be configured to produce all the matches with scores above a given threshold. If the database is very large and diverse, as is the case of several publicly available ones, most database entries will yield no matched subregions, indicating no detectable similarity to the query. Some entries will provide a few matched regions with relatively high score; these can be inspected manually by experts for significant similarities. However, some entries will generate a large number of overlapping matched regions. These large collections of matched subregions are very hard to further analyze manually.[4]

Nevertheless, for several biological reasons local comparison methods are appropriate for determining sequence similarity when the goal is to predict functionality. First, similar functionality between genetic sequences can often be accurately predicted by observing that relatively short subregions[5] are conserved ([7]). Second, in most higher organisms[6] genes are composed of regions that are expressed, that is, translated into protein sequences (these regions are called exons), and also regions that are removed from the gene before it is translated (these regions are called introns). When comparing DNA sequences, with the goal of predicting functionality, we are interested in the similarity between exon regions, not intron regions. However, if we simply have a DNA sequence obtained from a large sequencing project, it is difficult, if not impossible, to tell which regions represent introns, and which are exons.[7]

Finally, local similarity methods are particularly useful for dealing with *frameshift errors*. DNA sequencing is an imperfect process and even with repeated sequencing there will very often be errors present in long query sequences. When trying to determine functionality, the type of error that causes the most problem is the erroneous insertion, or deletion, of a nucleotide base. This is because of the biological mechanisms[8] in which information in a codon (3 adjacent nucleotide bases in a DNA sequence) is translated to form a single amino acid in a protein sequences. To overcome the problems caused by inserted, or deleted, nucleotide bases systems such as FIND-IT and NUTSS (see [16], [7]) begin by translating the given DNA sequence into 3 possible protein sequences; one for each of the possible ways of grouping bases into triplets. They then use local comparison methods to compare the translated DNA sequences to sequences in databases of protein sequences. Thus, the matching subregions reported by these algorithms come from one or more of the 3 possible translations.

---

[4] These observations are based on experience using the FIND-IT system ([16]) to analyze a 2 minute region (a DNA sequence of more than 91,000 nucleotide bases) of the E. coli genome sequenced in F. Blattner's laboratory at the University of Wisconsin.

[5] These are often called motifs.

[6] Eukaryotes.

[7] Codon preference methods have been proposed to aid in distinguishing between introns and exons, but currently these methods don't solve the problem.

[8] Transcription and translation.

Thus, despite the fact that local comparison methods may produce large numbers of matched subregions which are difficult to analyze manually, for this type of sequence comparison problem they are superior to global methods for many biological reasons. One method for circumventing the problem of too many matched subregions is to group matches with distinct, non overlapping regions into a small number of "families", and analyze these families as a whole. This is the approach taken by systems such as FIND-IT ([16]), and the setting in which our algorithm will be presented.

In the context of sequence comparison, sometimes the maximum score match is used as a measure of the similarity between two sequences (see [2]). An alternative approach is to generate a number of high scoring matches, group them into families, and use the highest scoring family as the measure of similarity. The score of a family is the sum of the scores of its members.

For instance, suppose that sequence q is compared to proteins i_1 and i_2 seeking all matches with score ≥ 1, and three matches result: one between q and i_1 with score equal to 5, and two between q and i_2 with scores equal to 3 and 4, located as in Figure 1. The highest score match measure will declare i_1 as the item closest to the query (5 vs. 4), whereas the highest score family measure will give preference to i_2 (7 vs. 5).

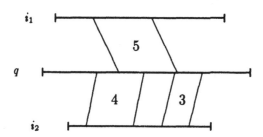

Fig. 1. Resulting matches after sequence q is compared to both i_1 and i_2. Scores appear inside matches. Sequence i_1 wins if maximum match scores are used, i_2 wins if family scores used.

In this paper, we propose a natural way of constructing families by grouping individual matches together. Under this formulation, a family of matches becomes an "increasing" independent set in a certain weighted graph, with each match corresponding to a node. The resulting graph can be viewed as a two dimensional version of the classical interval graphs called *double interval graphs* (see [4], [19]). Biological applications of interval graphs have a long history (see [20] and references there).

We present an algorithm to find an increasing independent set with maximum total weight using a sweep technique very common in interval graph algorithms (see [6], [3]). The algorithm runs in $O(n \log n)$ time and uses $O(n)$ space on graphs with n nodes.

The rest of the paper is organized as follows. In Section 2 we use the "graphic matrix" plot introduced in [11] to rephrase the problem of combining segment pairs in terms of rectangle graphs. The algorithm itself is presented in Section 3, together with a proof of correctness and analysis of its computational resource requirements. Finally, in Section 4 we discuss some directions for future research.

2 Modeling the Problem

Recall that we have a database of sequences, and a query sequence q. In general, several sequences from the database, each with additional associated information, are returned in response to the query. If s is a response sequence, then its associated information identifies a collection of segments on s, and for each such segment, the corresponding matching segment on q (Figure 2). In addition, each matching segment of s is accompanied by a nonnegative *score*.

For a response sequence s, let $s_1, s_2, \ldots, s_l$ be the segments of s returned in response to the query q. Suppose that s_i matches the segment q_i of q, and that the score associated with this match is is w_i. A *family* of segment pairs is a sequence $(s_{i_1}, q_{i_1}), (s_{i_2}, q_{i_2}), \ldots, (s_{i_k}, q_{i_k})$ such that s_{i_j} and q_{i_j} are strictly to the left of $s_{i_{j+1}}$ and $q_{i_{j+1}}$, respectively, in sequences s and q, respectively. The *weight* of this family is $\sum_{j=1}^{k} w_{i_j}$. Our objective is to determine a maximum weight family.

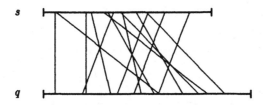

Fig. 2. A collection of seven matches between a query sequence q and a response sequence s. Each match is indicated by two parallel lines.

The following two-dimensional representation of the (input) data is convenient for expressing our algorithm for determining the maximum weight family. Lay down the sequences q and s along the X and the Y axes, respectively. Align the sequences such that the ith character of q and the jth character of s are identified with points $(i, 0)$ and $(0, j)$, respectively. Then, each segment pair (s_{i_j}, q_{i_j}) naturally corresponds with a rectangle in the first quadrant; we will associate weight w_{i_j} with this rectangle.

This construction is analogous to the "graphic matrix" plot mentioned in [11], [10], and [15]. The only differences are that the rectangles are actually squares in the cited references, and only their main diagonals are used to represent them.

Observe that a set of rectangles represents a family if and only if: (a) no pair of rectangles intersect when projected on either coordinate axis; and (b) the rectangles form a sequence such that every rectangle, except the first one, is strictly below, and to the left of the next rectangle in the sequence.

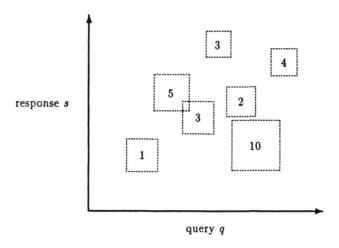

Fig. 3. Rectangle representation of the matches in Figure 2. Numbers inside rectangles indicate scores.

There is an analogy between our setting and the notion of an independent set in an interval graph. Notice that if we consider the projection of the rectangles in either axis, we obtain a set of intervals, and the implied interval graphs. The set of nodes (rectangles) corresponding to segment pairs in a family projects to independent sets in the interval graphs on each axes. (The converse is not necessarily true.) Figure 4 shows the graph representation for the collection of matches in Figures 2 and 3. Two nodes (matches) are adjacent if and only if their projections intersect in one of the coordinate axes. These graphs are special cases of the graphs with interval number 2 defined by Trotter and Harary (see [19]).

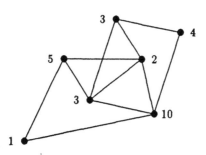

Fig. 4. Weighted graph for the rectangles in Figure 3

The "below-and-left" condition, however, is not captured by the independent set abstraction alone. Therefore, we need to introduce the notion of an "increasing" independent set which depends upon the planar representation at hand. An independent set is *increasing* when there is an ordering $r_1, r_2, \ldots, r_m$ of its rectangles such that r_i is located completely below and to the left of r_{i+1}, for $1 \leq i \leq m - 1$. With these definitions, the problem of finding the maximum weight family is equivalent to finding the maximum weight increasing independent set in a rectangle graph.

Formally, we have the following definitions. A two dimensional interval or *rectangle* is a subset of the plane that can be written as a product of two finite closed intervals, one from each coordinate axis. Given a rectangle r, we define

$$Xmin(r) = \text{minimum } X\text{-coordinate of any point in } r$$
$$Xmax(r) = \text{maximum } X\text{-coordinate of any point in } r$$
$$Ymin(r) = \text{minimum } Y\text{-coordinate of any point in } r$$
$$Ymax(r) = \text{maximum } Y\text{-coordinate of any point in } r$$

so that for any rectangle r we have

$$r = [Xmin(r), Xmax(r)] \times [Ymin(r), Ymax(r)].$$

In what follows, we will also assume that we are given a collection of rectangles and each rectangle r has a nonnegative weight $Weight(r)$ associated with it.

For two rectangles r and s, we say that r *precedes* s (notation: $r \to s$) when $Xmax(r) < Xmin(s)$ and $Ymax(r) < Ymin(s)$. In other words, r is located strictly below and to the left of s in the coordinate plane. We also say that s *dominates* r.

An *increasing independent set* (IIS) is an ordered set $A = \{r_1, r_2, \ldots, r_m\}$ of rectangles such that $r_j \to r_{j+1}$ for $1 \leq j \leq m-1$. If A is nonempty, then the rectangle r_1 is called the *head* of A, and it is denoted by $Head(A)$. The empty IIS has no head. For an IIS A, $Weight(A)$ denotes the sum of the weights of all rectangles in A, and $Xmin(A)$ denotes $Xmin(Head(A))$ (with similar notations for $Xmax(Head(A))$, etc.)

3   An Algorithm for Determining Maximum Weight IIS

We use a dynamic program to determine a maximum weight IIS starting at every rectangle. Observe that, if we know the maximum weight IIS starting with every rectangle which lies completely to the right of a rectangle r, then we can easily determine the maximum weight IIS starting at r as follows. For all rectangles t that dominate r, append r to the beginning of the maximum weight IIS starting with t. From these candidate sequences, pick out the one that has maximum weight. This gives an $O(n^2)$ algorithm for solving the maximum weight IIS problem. The rest of this section is devoted to describing an $O(n \log n)$ algorithm for the problem.

3.1   Outline of the Algorithm

We will begin by giving a careful description of our algorithm for finding a maximum weight IIS. In the next subsection, we will discuss the data structure necessary for this algorithm to run in $O(n \log n)$ time. The core of the algorithm is a sweep along the X-axis from right to left. (A sweep along the Y-axis will work equally well.) During this sweep, we maintain a collection D of IISs. This collection satisfies the properties listed below.

Let D be the collection of IISs when we arrive at x during the scan. Let S be any IIS that lies completely to the right of x. The IISs in D can be ordered, say $A_1, A_2, \ldots, A_k$, such that:

P1. $x < Xmin(A_i)$, i.e., all IISs in D lie completely to the right of x.
P2. $Ymin(A_i) < Ymin(A_{i+1})$ and $Weight(A_i) > Weight(A_{i+1})$.
P3. $Ymin(S) \leq Ymin(A_k)$.
P4. If $Ymin(A_i) < Ymin(S) \leq Ymin(A_{i+1})$ then $Weight(A_{i+1}) \geq Weight(S)$.
(Assume that $Ymin(A_0) = -\infty$.)

Notice that these properties imply that A_1 is a maximum weight IIS lying completely to the right of x, and that A_i is the maximum weight IIS lying completely to the right of x and on/above $Ymin(A_i)$. The collection D is used to quickly locate the largest weight IIS lying completely above and to the right of a given point (x, y) in the plane.

Next, we describe a simple method of maintaining properties P1-P4 during the sweep. The collection D will (possibly) change only if we encounter the left end point, $Xmin(r)$, for some rectangle r. In addition, we will need to do some bookkeeping when we encounter the right end point, $Xmax(r)$. The algorithm is outlined below (see also Figure 5).

Let $r_1, r_2, \ldots, r_n$ be the set of input rectangles with weights $w_1, w_2, \ldots, w_n$, respectively. Start by sorting the set $\{Xmin(r_i), Xmax(r_i) \mid 1 \leq i \leq n\}$ of left and right end points of the input rectangles. Let $x_1 \leq x_2 \leq \ldots \leq x_{2n}$ be the sorted sequence. Set $x_0 = -\infty$. The collection D is initialized to be the empty collection. We start at the point $x = x_{2n}$, and scan points $x_{2n-1}, x_{2n-2}, \ldots x_1$ in that order. The scan ends when it arrives at x_0. For each i, $1 \leq i \leq n$, initialize $Next(i)$ to be empty; this will be used for bookkeeping purposes.

In the following discussion, we will assume that all the x_i's are distinct. The case when x_i's are not distinct is a straightforward extension. It will also be convenient to define $Ymin(A_{k+1}) = \infty$, and $Weight(A_{k+1}) = 0$.

Suppose that x_i is the next point to be scanned, and that the current collection D, denoted by $D_i = \{A_1, A_2, \ldots, A_k\}$, satisfies properties P1-P4. Technically, we should have two subscripts on the A's, and we should define $D_i = \{A_{i1}, A_{i2}, \ldots, A_{ik_i}\}$. Instead, we will use only one subscript; we will either use the phrase "A_l-from-D_i" or a particular D_i will be implied by the context.

The following two cases may arise:

Case 1: x_i is the right end point of a rectangle r_j. ($x_i = Xmax(r_j)$): Define $D_{i-1} = D_i$. We need to update $Next(j)$ as follows. If $Ymax(r_j) > Ymin(A_k)$, then set $Next(j) = (i, null, null)$. Otherwise, find the least l such that $Ymax(r_j) < Ymin(A_l)$, and set $Next(j) = (i, l, Head(A_l))$. Notice that A_l is the largest weight IIS in D_i such that $r_j \rightarrow Head(A_l)$.

As D_i is unchanged, and the set of IISs lying completely to the right of x_i and x_{i-1}, respectively, is the same, it is easy to verify that properties P1-P4 hold when the sweep moves to point x_{i-1}.

Case 2: x_i is the left end point of a rectangle r_j. ($x_i = Xmin(r_j)$): Suppose that $Next(j) = (l, m, \acute{r})$. If $m = null$, then define B to be an IIS consisting of the rectangle r_j alone. Otherwise, define B by appending r_j to the front of A_l-from-D_m.

Input: Rectangles $r_1, r_2, \ldots, r_n$ with weights $w_1, w_2, \ldots w_n$, respectively.
Output: Maximum weight IIS.
Sort the set $\{Xmin(r_i), Xmax(r_i) \mid 1 \leq i \leq n\}$.
Let $x_1, x_2, \ldots, x_{2n}$ be the sorted sequence.
Initialize $D = \emptyset$
for $j \leftarrow 2n$ downto 1 do
      if x_i is a right end point, i.e., $x_i = Xmax(r_j)$ then

> Find an IIS A in D such that $Ymin(A)$ is the least possible, but $Ymin(A) > Ymax(r_j)$. If no such IIS exists, then define $succ(j) = null$, $Total\_weight(r_j) = w_j$, and $R_j = \{r_j\}$. Otherwise, let $succ(r_j) = Head(A)$, $Total\_weight(r_j) = Total\_weight(succ(r_j)) + w_j$, and construct an IIS R_j by appending r_j to the front of A.

      else x_i is a left end point. Let $x_i = Xmin(r_j)$.

> Find an IIS A in D such that $Ymin(A)$ is the least possible, but $Ymin(A) > Ymax(r_j)$. If such an A exists, then add R_j to D, and delete all IISs C from D such that $Ymin(C) \leq Ymin(R_j)$ and $Total\_weight(Head(C)) \leq Total\_weight(r_j)$.

    endif
endfor
Output the IIS A in D with least $Ymin(A)$. (This can be constructed using the $succ()$ links.)

Fig. 5. The algorithm

Next, in D_i, locate the least v such that $Ymin(r_j) < Ymin(A_v)$. If $Weight(B) \leq Weight(A_v)$, then define D_{i-1} to be the same as D_i.

Otherwise, if $v = 1$, then define $D_{i-1} = D_i \cup \{B\}$. On the other hand, if $v > 1$, then determine maximum u such that $Weight(A_{v-p}) \leq Weight(B)$ for $1 \leq p \leq u$. Define $D_{i-1} = (D_i \cup \{B\}) - \{A_{v-p} \mid 1 \leq p \leq u\}$.

Now we verify that properties P1-P4 hold at point x_{i-1}. These are easy to verify in case $D_{i-1} = D_i$. Therefore, consider the case when $D_{i-1} \neq D_i$. (In the following, the A_t's are from the set D_i.) Then

$$D_{i-1} = \{A_1, A_2, \ldots, A_{v-u-1}, B, A_v, A_{v+1}, \ldots, A_k\},$$

and P1 and P3 are easily seen to hold.

Recall that $Head(B) = r_j$, and P2 at x_i implies that, if $v - p > 0$ and $p > u$, then $Weight(A_{v-p}) > Weight(B)$. Therefore, $Ymin(A_{v-u-1}) < Ymin(B)$ and $Weight(A_{v-u-1}) > Weight(B)$. By definition of v, and the fact that B has been added to D_i, it follows that $Ymin(B) < Ymin(A_v)$ and $Weight(B) > Weight(A_v)$. Therefore, P2 holds at x_{i-1}.

For establishing P4, we only need to consider S such that $Ymin(A_{v-u-1}) < Ymin(S) \leq Ymin(B)$. For S such that $Ymin(A_{v-u-1}) \leq Ymin(S) \leq Ymin(A_{v-1})$, P4 at x_i implies that $Weight(S) \leq Weight(A_t)$ for some t in the range $v - u - 1 < t < v$. But then, $Weight(A_t) \leq Weight(B)$.

Next consider S such that $Ymin(A_{v-1}) < Ymin(S) < Ymin(B)$. For any such S that does not begin with r_j (i.e., $Head(S) \neq Head(B)$), P4 at x_{i-1} is implied by P4 at x_i, and the fact that $Weight(B) > Weight(A_v)$. On the other hand, if $Head(S) = Head(B)$, then we can establish $Weight(S) \leq Weight(B)$ using the definition of $Next(j)$ and the fact that P1-P4 held before the point x_l was scanned.

Hence, we have established that properties P1-P4 hold through the process.

Finally, observe that we can obtain a list of the rectangles in A_l by following the third component of the triplet $Next()$.

3.2 Efficient Implementation of the Set D

We use the *concatenable queues* described in [1] to store the collection D. These structures are 2-3 search trees, and permit the execution of several operations in $O(\log n)$ time, where n is the number of items in the structure. For our purposes, we only need operations for search, insertion, concatenation, and splitting, all of which can be implemented in $O(\log n)$. Furthermore, the amount of storage needed never exceeds $O(n)$.

There are two keys associated with each IIS A in D : $Weight(A)$, and $Ymin(A)$. In a concatenable queue, elements are stored in increasing (or decreasing) order of keys. Since we have two keys, this is apparently a problem. Fortunately, property P2 implies that the order of ascending $Weight()$ is the same as the order of descending $Ymin()$. Hence, if we keep the keys sorted based on one of the keys, they will automatically be sorted with respect to the other key as well.

Let's now explain how to implement the required operations on this data structure.

1. Determining A_l in Case 1 is essentially a search on the $Ymin()$ key.
2. Determining u in Case 2 is a search on the $Weight()$ key.
3. Deletion of $A_{v-u}, A_{v-u+1}, \ldots, A_{v-1}$ and insertion of B can be accomplished by two splits, a concatenation, and an insertion (see Figure 6).

Since all these operations can be performed in $O(\log n)$ steps, the total time required by the sweep is $O(n \log n)$. As the sorting can also be performed in $O(n \log n)$ steps, the algorithm can be executed in $O(n \log n)$ steps.

4 Conclusions

In this paper we have taken a problem in computational genetics, found a formal model for it in the theory of interval graphs, and presented an efficient solution to the problem using fairly simple algorithmic techniques. We believe that our paper makes important contributions in two ways. First, we have not simplified the biological problem that needed to be solved – ours is an efficient solution for one of the *real* problems that systems such as FIND-IT ([16]) face. Second, although the interval

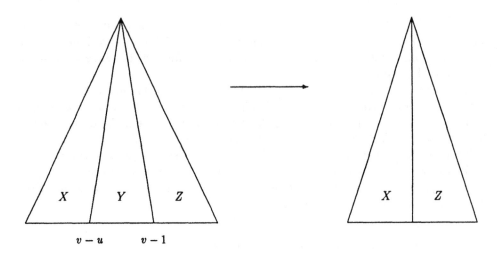

Fig. 6. Removing a range from D: split at $v - u$ and $v - 1$, then concatenate X and Z.

graph model that we form for the biological problem is important for the algorithmic solution, most important for the efficient solution is the observation that the interval graph does not have to be explicitly represented using adjacency lists or matrices. Our implicit representation makes an $O(n \log n)$ algorithm possible.

Nevertheless, several interesting questions remain unanswered. The algorithm that we presented finds the family that represents the best match (w.r.t. combined weight of individual matches) between the query sequence and the database sequence. In practice one would like to find all very good matches. Many of the very good matches may be overlapping. We would like an algorithm that efficiently finds and presents all very good matches. Emphasis should be placed on developing a good method of presentation of the output.

Two questions having to do with the requirements for combining segment pairs are also interesting.

The first question relates to the "below-and-to-the-left" condition. For biological reasons we required that either $r \to s$ or $s \to r$ for distinct rectangles r, s belonging to the same family. There are situations in which we can relax this condition. An appropriate model might allow a fixed number of rectangle pairs that violate the condition provided that the sequences forming the rectangles are nonoverlapping. What is the best algorithm for the maximum weight family under the new requirements?

A second question relates to our definition of adjacency in rectangle graphs. It can be argued that the requirement that projections do not intersect at all is too strong, making us miss very good families just because of a few bases meeting at the extremities. In this respect, a two dimensional version of the tolerance graphs seems to be the appropriate model. An efficient algorithm for the maximum weight increasing independent set for this class would be interesting. However, to be practical from the biological standpoint the measure of weight used for the independent set must be modified to reflect the overlap.

Acknowledgements

We would like to thank Jude Shavlik for bringing to our attention the combinatorial problems faced by systems such as FIND-IT, and for many helpful discussions. We also thank Fred Blattner and his group in the Genetics Department at the University of Wisconsin-Madison with whom we have been actively collaborating on problems in computational genetics.

References

1. A. V. Aho, J. E. Hopcroft, and J. D. Ullman. *The Design and Analysis of Computer Algorithms.* Addison-Wesley, Reading, MA, 1974.
2. Stephen F. Altschul, Warren Gish, Webb Miller, Eugene W. Myers, and David J. Lipman. A basic local alignment search tool. *J. Mol. Biol.*, 215, 1990.
3. Alan A. Bertossi and Alessandro Gori. Total domination and irredundance in weighted interval graphs. *SIAM J. Disc. Math.*, 1(3):317–327, 1988.
4. Martin Charles Golumbic. *Algorithmic Graph Theory and Perfect Graphs.* Academic Press, 1980.
5. Osamu Gotoh. Optimal sequence alignment allowing for long gaps. *Bull. Math. Biol.*, 52(3):359–373, 1990.
6. U. I. Gupta, D. T. Lee, and J. Y.-T. Leung. Efficient algorithms for interval graphs and circular-arc graphs. *Networks*, 12:459–467, 1982.
7. Steven Henikoff, James C. Wallace, and Joseph P. Brown. Finding protein similarities with nucleotide sequence databases. In Russell F. Doolittle, editor, *Molecular Evolution: Computer Analysis of Protein and Nucleic Acid Sequences*, volume 183 of *Methods in Enzymology*, pages 111–132. Academic Press, 1990.
8. Xiaoqiu Huang, Ross C. Hardison, and Webb Miller. A space-efficient algorithm for local similarities. *Comput. Applic. Biosci.*, 6(4):373–381, 1990.
9. Eric Lander, Jill P. Mesirov, and Washington Taylor. Study of protein sequence comparison metrics on the connection machine CM-2. *J. Supercomp.*, pages 255–269, 1989.
10. D. J. Lipman and W. R. Pearson. Rapid and sensitive protein similarity search. *Science*, 227:1435–1441, 1985.
11. J. Maizel and R. Lenk. Enhanced graphic matrix analysis of nucleic acid and protein sequences. *Proc. Nat. Acad. Sci. USA*, 78:7665–7669, 1981.
12. Hugo M. Martinez. An efficient method for finding repeats in molecular sequences. *Nucleic Acids Research*, 11(13):4629–4634, 1983.
13. Webb Miller and Eugene W. Myers. Sequence comparison with concave weighting functions. *Bull. Math. Biol.*, 50(2):97–120, 1988.
14. Saul B. Needleman and Christian D. Wunsch. A general method applicable to the search for similarities in the amino acid sequence of two proteins. *J. Mol. Biol.*, 48:443–453, 1970.
15. William R. Pearson and David J. Lipman. Improved tools for biological sequence comparison. *Proc. Nat. Acad. Sci. USA*, 85:2444–2448, 1988.
16. Jude Shavlik. Finding genes by case-based reasoning in the presence of noisy case boundaries. In *Proc. DARPA Cased-Based Reasoning Workshop*, pages 327–338, Washington, DC, 1991.

17. T. F. Smith and M. S. Waterman. Identification of common molecular subsequences. *J. Mol. Biol.*, 147:195–197, 1981.

18. T. F. Smith, M. S. Waterman, and W. M. Fitch. Comparative biosequence metrics. *J. Molec. Evol.*, 18:38–46, 1981.

19. William T. Trotter, Jr. and Frank Harary. On double and multiple interval graphs. *J. Graph Theory*, 3:205–211, 1979.

20. M. S. Waterman and J. R. Griggs. Interval graphs and maps of DNA. *Bull. Math. Biol.*, 48(2):189–195, 1986.

This article was processed using the LaTeX macro package with LLNCS style

New Results on Linear Programming and Related Problems

EMO WELZL

Institut für Informatik
Freie Universität Berlin
Arnimallee 2-6, W 1000 Berlin 33, Germany
e-mail: emo@tcs.fu-berlin.de

We consider the problem of minimizing some linear function in d variables subject to n linear inequalities. Geometrically, this corresponds to finding a point extremal in some direction inside a polyhedron defined as the intersection of n halfspaces in d-space, see e.g. [Shr]. There is vivid interest in this problem, starting with Dantzig's simplex method (which was shown to require exponential time on certain inputs, but works very good in practice).

Recent years have brought some progress on the complexity of this problem in the *unit cost model*, and the best result known at this point is a randomized 'combinatorial' algorithm which computes the solution with expected $O(d^2 n + e^{O(\sqrt{d \log d})} \log n)$ arithmetic operations. The bound relies on two algorithms by Clarkson, [Cla], and the subexponential algorithms due to Kalai, [Kal], and Matoušek, Sharir, and Welzl, [MSW].

Note that the expectation is with respect to the internal 'coin flips' of the algorithm, and it holds for any input – in contrast to the polynomial average bounds (first proved by Borgwardt) for the simplex method, which assume that the input is random with respect to some distribution. The polynomial bounds (due to Khachiyan and Karmarkar) in the bit-model are also in terms of the size of the coefficients of the input (not only in d and n).

One nice feature of the combinatorial algorithms is that they can be formulated in a quite general abstract framework, and so they are applicable to a number of nonlinear optimization problems, as e.g. computing the smallest ball (ellipsoid) enclosing n points in d-space, computing the largest ellipsoid in a convex d-polytope with n facets, computing the distance between two convex d-polytopes with n facets or with n vertices, etc. Similar bounds as the one claimed above hold for all these problems, provided one can solve basic problems of small sizes efficiently (like (1) computing the smallest ball enclosing $d + 1$ points in d-space, (2) computing the distance between two d-simplices). Recently, Gärtner, [Gär], established subexponential bounds for the two basic cases (1) and (2) just mentioned. Other cases (like those involving ellipsoids) are still open.

(References to the more classical results mentioned above can be found in [Shr].)

References

[Cla] K. L. Clarkson, Las Vegas algorithms for linear and integer programming when the dimension is small, manuscript, 1989.

[Gär]   B. Gärtner, A subexponential algorithm for abstract optimization problems, manuscript, 1992.

[Kal]   G. Kalai, A subexponential randomized simplex algorithm, to appear in *Proc. 24th ACM Symposium on Theory of Computing*, 1992.

[MSW]   J. Matoušek, M. Sharir, E. Welzl, A subexponential bound for linear programming, to appear in *Proc. 8th Annual ACM Symposium on Computational Geometry*, 1992.

[Shr]   A. Schrijver, *Theory of Linear and Integer Programming*, John Wiley & Sons, 1986.

Dynamic Closest Pairs - A Probabilistic Approach

Mordecai J. Golin*

The dynamic closest pair problem is to find the closest pair among a set of points that is continuously being changed by insertions and deletions. In this paper we present a simple, robust, easily coded heuristic for solving the planar closest pair problem. We prove that this heuristic uses only $O(\log n)$ expected time to perform an insertion or deletion when the input points are chosen from a very wide class of distributions in the plane.

1   Introduction

The *closest pair* problem, that of finding the closest pair in a set of n points, is one of the oldest and most basic problems in computational geometry and as such has been extensively studied. In two dimensions it is known that the problem requires at least $n \log n$ time to solve [9]. There are algorithms that find the closest pair in matching $O(n \log n)$ worst case time [12] [7]. In addition there are randomized algorithms that run in $O(n)$ expected time [10] [15]. and another algorithm which has been shown to run in $O(n)$ expected time where the average is taken over all inputs of n points chosen independently, identically, distributed from the uniform distribution over the unit square [3].

More recently work has concentrated on the *dynamic* closest pair problem. In this problem we modify the initial point set by inserting and deleting points to and from it. After every change the closest point is recomputed and reported. Overmars [8] presented an algorithm which had an update time of $O(n)$. Smid [13] then developed an algorithm to update the closest pair in $O(n^{2/3} \log n)$ (later $O(\sqrt{n} \log n)$) time. Recently Smid has developed another algorithm [14], one which performs updates in $O(\log^2 n \log \log n)$ *amortized* time. This seems to be the current state of the art. (If deletions are prohibited then there is a new algorithm which performs updates in $O(\log n)$ time [11].)

In this paper we take a different approach to the dynamic closest pair problem. Our approach is probabilistic in the same sense that Bentley, Weide and Yao's [3] static algorithm was probabilistic. We assume that the input points are chosen from a given distribution. We present a relatively simple heuristic which performs updates in expected $O(\log n)$ time per update where the average is taken over all inputs. Our algorithm is robust and easily coded and thus, in a practical sense, is "useful".

The heuristic is extremely intuitive and easy to code. It is based on the idea of comparing points whose projection on the x-axis is "close." This idea has been used before

*INRIA Rocquencourt, 78153 Le Chesnay, France. This work was supported by a Chateaubriand fellowship from the French Ministère des Affaires Étrangères, by the European Community, Esprit II Basic Research Action Number ?075 (ALCOM) and by NSF grant CCR-8918152

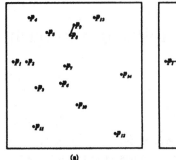 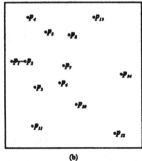 

| (a) | (b) | (c) |

Figure 1: The closest pairs in each of the illustrated point sets are connected by a line. Figure (a) is the base set $S_i = \{p_1, \ldots, p_{14}\}$ for which $\delta(S_i) = d(p_8, p_9)$. Figure (b) illustrates that $delete(p_9)$ will return p_1, p_2 and the distance between them. Figure (c) illustrates that $insert(p_{15})$ will return p_{10}, p_{15} and the distance between them.

in a similar milieu by Bentley and Papidimitriou [2] in their algorithm for solving the post office problem. The major technical problem encountered in our analysis is that the random variables to be analyzed are strangely conditional upon each other. We sidestep this problem by introducing a separation principle which *almost* allows us to consider the random variables as being independent. Our results are valid for a wide variety of distributions, specifically, when the input points are chosen from the uniform distribution over any region whose boundary is rectifiable (has length).

In section 2 we present formal definitions of the dynamic closest pair problem and of the stochastic process that we use to model its inputs. We then describe our heuristic and state its probabilistic running time. In section 3 we present the details of the probabilistic analysis. We conclude in section 4.

2 The Problem, the Model, and the Algorithm

2.1 The Problem

Let p, q be two points in the plane. We use $d(p, q)$ to denote the distance between them. This distance can be any L_p (or the L_∞) metric but, for the sake of simplicity, we will always assume

$$d(p, q) = \sqrt{(p.x - q.x)^2 + (p.y - q.y)^2}$$

i.e. the Euclidean (L_2) metric. The results in this paper, will however, remain valid in any L_p (or the L_∞) metric.

Suppose $S = \{p_1, \ldots, p_n\}$ is a set of n points. The *closest distance* in S is defined by $\delta(S) := \min_{p_i \neq p_j} d(p_i, p_j)$. A *closest pair* in S is a pair of points $q, q' \in S$ such that $d(q, q', = \delta(S)$. The dynamic closest pair problem is to keep track of the closest pair in the set while the set is being continuously modified by the insertion and the deletion of points, e.g, implementing the following two operations (see figure 1):

Insert(p): $S_i := S_{i-1} \cup \{p\}$; return$(q, q', \delta(S_i))$ where $q, q' \in S_i$, $\delta(q, q') = \delta(S_i)$;

Delete(p): $S_i := S_{i-1} \setminus \{p\}$; return($q$, q', $\delta(S_i)$) where $q, q' \in S_i$, $\delta(q, q') = \delta(S_i)$;

where S_i is the set of points after the i'th update. That is, at time i a point is either inserted into or deleted from the set S_{i-1} and the closest pair in the new set S_i is returned.

Set $n = |S_{i-1}|$, the number of points in S_{i-1}. The cost of an insertion or deletion at time i is usually expressed as a function of n. If we say that the expected cost of an insertion is $O(\log n)$ our intended meaning is that it is $O(\log |S_{i-1}|)$.

2.2 The Model

Let C be some fixed planar set possessing finite non-zero measure. There is a very natural stochastic process that models an input sequence to the dynamic closest pair problem. Set $S_0 = \emptyset$. The input process is defined inductively. Suppose $S_{i-1} = \{p_1, \ldots, p_n\} \subseteq C$. Using any arbitrary rule decide if the update to be performed at time i will be an insertion or a deletion. If the update is an an insertion then choose the new point p to be inserted from the uniform distribution over C with p independent of S_{i-1}. If the update is a deletion then choose a point p_i randomly, i.e. with probability $1/n$, from S_{i-1} and delete it. In other words, at each step either a "random" point is inserted or a "random" point is deleted.

An immediate consequence of these definitions is that, for each i, the points in S_i have the same distribution as points chosen independently identically distributed (I.I.D.) from the uniform distribution over C. This fact will be central to our analyses.

In the rest of this paper we assume that the input sequence to the dynamic closest pair problem is the stochastic process described above. We further assume that C is closed, bounded (but not necessarily connected) and has a rectifiable boundary. By rectifiable we mean that ∂C, the *boundary* of C, is composed of a finite number of curves each of which has finite length. This is a very general condition. All convex regions have rectifiable boundaries as do all polygons. Similarly, the intersection and/or union of two regions with rectifiable boundaries have rectifiable boundaries.

2.3 The Heuristic And Its Running Time

We now come to the main purpose of this paper, the introduction of a fast expected time heuristic for solving the dynamic pair problem. Henceforth we will assume that at time i the value of $\delta(S_{i-1})$ and the identity of a closest pair q, q' with $d(q, q') = \delta(S_{i-1})$ have been stored after being computed in the last update. We will also assume the existence of a data structure $\mathcal{P}$ that stores the points of S_{i-1} sorted by x coordinate (breaking ties arbitrarily) and permits fast range searching. Formally $\mathcal{P}$ supports the following operations in the given deterministic times where $n = |S_{i-1}|$:

Insert_point(p) Inserts point p into $\mathcal{P}$ in $O(\log n)$ time.

Delete_point(p) Deletes point p from $\mathcal{P}$ in $O(\log n)$ time.

Range(X,d) Return all points in $\{p \in S_{i-1} : |p.x - X| \le d\}$ in time $O(\log n + K)$ where K is the number of points returned.

Pred(p) Returns the predecessor (by x-coordinate) of p in $O(1)$ time.

Succ(p) Returns the successor (by x-coordinate) of p in $O(1)$ time.

Insert(p): /* Set $S_i = S_{i-1} \cup \{p\}$; return($q, q', \delta(S_i)$) */
for all points $p' \in Range(p.x, \delta)$ **do**
   if $d(p, p') < \delta$ **then**
      begin $\delta = \delta(p, p')$; $q := p$; $q' = p'$ **end**
insert_point(p);
return(q, q', δ;)

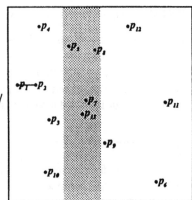

Figure 2: Pseudo-code for $insert(p)$ along with an illustrated example. In the example the initial set was $S_i = \{p_1, \ldots, p_{12}\}$ with $\delta(S_i) = d(p_1, p_2)$. An $insert(p_{13})$ was then done and p_{13} was compared to all points whose x-coordinate was within $d(p_1, p_2)$ of p_{13}'s x-coordinate i.e. to all of the points in the shaded region. The new closest pair p_{13}, p_7 was found and reported.

The first two operations are named $insert\_point$ and $delete\_point$ to distinguish them from the $insert$ and $delete$ defined in 2.1. Data structures which permit the above operations are well known and can easily be constructed from off-the-shelf algorithmic components. For example, a 2-3 tree [1] that sorts the points by x-coordinate, keeps all of the points stored in its leaves, and stores the leaves as a doubly linked list fulfills all of the requirements.

We start with the insertion heuristic. It uses the concept, introduced in Bentley and Papadimitriou [2] for solving the post office problem, of comparing points whose projections on the x-axis are close. Let $p = (p.x, p.y)$ be the point to be inserted at time i into set S_{i-1} to form $S_i = S_{i-1} \cup \{p\}$. Notice that $\delta(S_i) \neq \delta(S_{i-1})$ only if there is a point $p' \in S_{i-1}$ such that $d(p, p') < \delta$. If this occurs then $|p.x - p'.x| < \delta$ because of the triangle inequality. Thus, to implement $input(p)$, we need only compute the distances between p and each of the the points returned by $range(p.x, \delta(S_{i-1}))$ and check if any of these distances is less than δ. If one is less then we update δ and the corresponding closest pair q, q' accordingly, otherwise we do nothing. We conclude by adding p to the data structure. (Figure 2.)

The running time of this procedure is obviously dependent on the set S_{i-1} and the point p. Let $K = |\{s \in S_{i-1} : |p.x - s.x| \leq \delta(S_{i-1})\}|$ be the number of points returned by the $range$ command. Then the cost of the algorithm will be $O(\log n + K)$ for the $range$ command and $O(\log n)$ for the $insert\_point$ command. Since in the worst case $K = n$ this algorithm has a bad worst case running time. In the next section we will prove:

Theorem 1 *Let S_{i-1} and p be defined as in the probabilistic model. Let K be the number of points returned by $Range(p.x, \delta(S_{i-1}))$. Then $E(K) = O(1)$.*

Consequentially the $insert(p)$ operation will take only $O(\log n)$ deterministic time $+ O(1)$ expected time. We turn now to the deletion heuristic. It is essentially a modification of the sweep-line technique [5][7] for finding closest pairs in the static case.

We start by calling $delete\_point(p)$ and then checking if $p = q$ or $p = q'$. If p is not one of these points then nothing further needs to be done because $\delta(S_i) = \delta(S_{i-1})$. Otherwise take

Delete(p): /* Set $S_i = S_{i-1} \setminus \{p\}$; return($q, q', \delta(S_i)$) */
delete_point(p);
if $p = q$ **or** $p = q'$ **then**
  begin
       Let s be the point with the smallest x-coordinate;
       $t := succ(s)$; $\delta = d(s,t)$;
       repeat
            $t = succ(t)$;
            $s = pred(t)$;
            while $s.x > t.x - \delta$ **do**
            begin
                if $d(s,t) < \delta$ **then**
                    begin $q = s$; $q' = t$; $\delta = d(s,t)$; **end**
                $s = pred(s)$;
            end
       until t is the point with the largest x-coordinate
  end
return(q, q', δ;)

Figure 3: Pseudo-code for sweep version of $delete(p)$ and a worked example.

the $n-1$ points in S_i, sort them by their x-coordinates, and relabel them as $p_{(1)}, \ldots, p_{(n-1)}$ (with $p_{(j)} = (p_{(j)}.x, p_{(j)}.y)$) such that $p_{(1)}.x \leq p_{(2)}.x \leq p_{(3)}.x \leq \cdots \leq p_{(n-1)}.x$. Let δ_j be the closest-pair distance among the first j sorted points i.e. $\delta_j = \delta(\{p_{(1)}, \ldots, p_{(j)}\})$ If $\delta_j \neq \delta_{j-1}$ then δ_j must be the distance between $p_{(j)}$ and some point to $p_{(j)}$'s left:

$$\delta_j = \min\left(\delta_{j-1}, \min_{k<j} d(p_{(j)}, p_{(k)})\right).$$

The triangle inequality tells us that if $|p_{(j)}.x - p_{(k)}.x| > \delta_{j-1}$ then $d(p_{(j)}, p_{(k)}) > \delta_{j-1}$. Thus, given δ_{j-1}, we can calculate δ_j by simply calculating the distance between $p_{(j)}$ and each of the points in the δ_{j-1} width strip to the left of $p_{(j)}$:

$$\{p_{(k)} \in S_i : k < j \text{ and } p_{(j)}.x - p_{(k)}.x \leq \delta_{j-1}\}.$$

The algorithm uses this fact to successively calculate $\delta_2, \delta_3, \delta_4, \ldots$ and concludes when it has calculated $\delta_{n-1} = \delta(S_i)$.

    The way that the points in S_i are stored in the data structure makes it very easy to implement this algorithm. We use the $succ$ operation to sweep a vertical line through the $p_{(1)}, \ldots, p_{(n-1)}$ in sorted x order. By the time the sweep-line encounters point $p_{(j)}$ we have already calculated δ_{j-1}. Using the $pred$ operation we can then calculate δ_j by comparing $p_{(j)}$ to all of the points in the δ_{j-1} strip to its left. Figure 3 provides pseudo-code and a worked example. In the example the base set is $S_i = \{p_1, \ldots, p_{15}\}$ where $\delta(S_i) = d(p_{14}, p_{15})$. The next update is $delete(p_{15})$ so the new closest pair in $S_i \setminus \{p_{15}\}$ must be computed using the sweep. In the example points are labelled by increasing x-coordinate so that

$p_{(i)} = p_i$. The picture illustrates the stage when all points p_i, $i < 10$, have been examined and their closest pair p_1, p_2 has been found. The point p_{10} is then being compared to all of the points within distance $d(p_1, p_2)$ to its left (the shaded region): p_7, p_8, p_9. Making these comparisons shows that the closest pair among the first 10 points is p_9, p_{10}.

How fast is this routine? The $delete\_point(p)$ requires $O(\log n)$ time. The remainder of the routine, the sweep, is only called when $p = q$ or q', an event having probability $2/n$. We define N_j to be the number of points in the δ_{j-1} strip to the left of point $p_{(j)}$: $N_j = |\{s \in S_i : 0 \le p_{(j)}.x - s.x \le \delta_{j-1}\}|$. In this notation the sweep uses $O(\sum N_j)$ time. In section 3 we shall prove the following:

Theorem 2 *Let S_{i-1}, p and $S_i = S_{i-1} \setminus \{p\}$ be defined as in §2.2. Furthermore, suppose that p is one of a closest pair q, q' in S_{i-1}. Let N_j be defined as above. Then, for $j > n^{1/3} \log n$,*

$$E(N_j) = O\left(\sqrt{\frac{n}{j}}\right).$$

By definition, $N_j < j$. Furthermore it is well known that $\sum_{j<n} \sqrt{1/j} = (\sqrt{n})$. Summing over j and applying the theorem we find

$$E\left(\sum_{j<n} N_j\right) = \sum_{j<n^{1/3}\log n} E(N_j) = O(n)$$

and the expected cost of the second part of the routine (assuming that $p = q$ or $p = q'$) is $O(n)$.

We find that the sweep part of the algorithm only takes $O(\frac{2}{n}n) = O(1)$ expected time. Therefore this implementation of $delete(p)$ requires only $\log n$ deterministic $+ O(1)$ expected time where most of the work actually occurs during the $delete\_point(p)$ call.

3 Proofs of the Theorems

Before proving Theorems 1 and 2 we introduce some useful notation and prove two utility lemmas. Let C be a bounded measurable region. We define $c = \text{Area}(C)$ and

$$h_C = \max\{|y - y'| : \exists x, y, y' \text{ such that } (x, y) \in C \text{ and } (x, y') \in C.\}$$

where h_C is the *height* of C. We also define the α-neighborhood of the boundary of C to be

$$N_\alpha^C = \{p \in C : \exists q \in \partial C \text{ such that } d(p, q) \le \alpha\},$$

the region in C which is within distance α of the boundary ∂C. e.g if C is the disk $\bigcirc$ then N_α^C is the annulus $\bigcirc$ (the shaded region).

We will also need to be able to describe the part of the C to the left of a vertical line. For this we introduce the notation

$$X(q) = Area\left(\{p \in C : p.x \le q.x\}\right) \quad \text{and} \quad C(s) = \{q \in C : X(q) \le s\} \quad (1)$$

where q is a point and s a scaler. The region $C(s)$ is defined so that $Area(C(s)) = s$: if a point q is chosen from the uniform distribution over C then $Pr(q \in C(s)) = \frac{s}{c}$. Informally $C(s)$ can te found by moving a vertical line from $-\infty$ to the right until it has swept over s of C's area. We will need the following lemma:

Lemma 1 *Let C be a closed bounded planar region whose boundary ∂C is rectifiable and let $s \geq 0$. Then there is some constant b_C such that the α-neighborhood of the boundary of $C(s)$ satisfies Area $\left(N_\alpha^{C(s)} \right) \leq b_C \alpha$ for $\alpha \leq 1$.*

Proof: Suppose ∂C is composed of k component curves of lengths $l_1, l_2, \ldots, l_k$ where $\sum_{i \leq k} l_i = l = length(\partial C)$. Let $t_i : [0, l_i] \to \partial C$ be the equations of the k component curves parameterized so that the oriented arclength distance between $t_i(a)$ and $t_i(b)$ along curve t_i is $b - a$. Place balls of radius 2α along each curve so that their centers are 2α arclength apart along the curve, i.e. at $t_i(0), t_i(2\alpha), t_i(4\alpha) \ldots$. Any point which is within α of a point on the curve t_i will be contained within one of these balls. Stated formally

$$N_\alpha^C \subseteq \bigcup_{i=1}^{k} \left(\bigcup_{j=0}^{\lfloor l_i/2\alpha \rfloor} B\left(t_i(2j\alpha), 2\alpha \right) \right)$$

where $B(q, r) = \{ p : d(p, q) \leq r \}$. Thus $Area(N_\alpha^C) \leq \sum_{i \leq k} (4\pi\alpha^2(1 + \lfloor l_k/2\alpha \rfloor)) \leq 4k\pi\alpha^2 + 2\pi l\alpha$. Since any point in $N_\alpha^{C(s)}$ must either be in N_α^C or within distance α of the vertical line which is the right boundary of $C(s)$ we find that $Area\left(N_\alpha^{C(s)} \right) \leq 4k\pi\alpha^2 + 2\pi l\alpha + h_C\alpha$. Setting $b_C = 4k\pi + 2\pi l + h_C$ proves the lemma. ∎

The next lemma provides bounds on the distribution function of δ.

Lemma 2 *Let C be a closed bounded rectifiable region. Let $s \in [0, c]$ and $C(s)$ be as defined in (1). Choose n points I.I.D. from the uniform distribution over $C(s)$.*
(a) If $\alpha \leq s/4b_C$ where b_C is the constant defined in the previous lemma then

$$\Pr\left(\delta(\{p_1, \ldots, p_n\}) \geq \alpha \right) \leq e^{-n^2\alpha^2/16s} \left[1 + O\left(\frac{n\alpha^2}{s} + \frac{n^3\alpha^4}{s^2} \right) \right] + e^{-n/18}. \quad (2)$$

(b) Let b_C be the constant defined in Lemma 1. If $s \geq 16b_C^2 \log^2 n/n^2$ then $\mathrm{E}(\delta(\{p_1, \ldots, p_n\})) = O(\sqrt{s}/n)$.
(c) Let $p_1, \ldots, p_n$ be chosen I.I.D. from C. Then $\mathrm{E}(\delta(\{p_1, \ldots, p_n\})) = O(1/n)$.

Proof: (a) Let $S = \{p_1, \ldots, p_n\}$. We will upper-bound $\Pr(\delta(S) \geq \alpha)$ by gridding the plane with squares of side $\alpha/\sqrt{2}$ and upper-bounding the probability that two of the p_i fall into the same square. Define the grid squares as

$$G(i.j) = \left\{ (x, y) : \frac{i\alpha}{\sqrt{2}} < x \leq \frac{(i+1)\alpha}{\sqrt{2}}, \frac{j\alpha}{\sqrt{2}} < y \leq \frac{(j+1)\alpha}{\sqrt{2}} \right\}$$

for all integral pairs (i, j). We will also use the alternative notation $G(p)$ to denote the grid square $G(i, j)$ that point p is located in, i.e. $p \in G(p)$. Let

$$\mathcal{G} = \{(i, j) : G(i, j) \subseteq C\} \quad \text{and} \quad S' = S \cap \left(\bigcup_{(i,j) \in \mathcal{G}} G(i, j) \right)$$

be, respectively, the set of indices of the grid squares that are totally contained in C and the set of the points in S that fall within those grid squares. If two points are located in the same grid square then they are less than distance α apart so

$$\Pr(\delta(S) \geq \alpha) \leq \Pr(\delta(S') \geq \alpha) \leq \Pr\left(\nexists p, q \in S' \text{ such that } G(p) = G(q) \right). \quad (3)$$

Let $g = |\mathcal{G}|$ be the number of grid squares totally contained in C and $n' = |S'|$ be the number of points that fall within those squares. Write $S' = \{q_1, \ldots, q_{n'}\}$. The points in S' have the same distribution as n' points I.I.D. uniformly distributed in $\bigcup_{(i,j)\in\mathcal{G}} G(i,j)$ so the probability that no two points in S' fall into the same box is

$$
\prod_{i=2}^{n'} \Pr(q_i \notin \cup_{j<i} G(q_j)) = \prod_{i<n'} \left(1 - \frac{i}{g}\right) = \exp\left(\sum_{i<n'} \ln\left(1 - \frac{i}{g}\right)\right)
$$

$$
= \exp\left(-\sum_{i<n'} \frac{i}{g}\right) \cdot exp\left(O\left(-\sum_{i<n'} \frac{i^2}{g^2}\right)\right)
$$

$$
= e^{-\frac{n'^2}{2g}} \left[1 + O\left(\frac{n'}{g} + \frac{n'^3}{g^2}\right)\right]
$$

where $\exp(x) = e^x$. Combining this result with (3) yields

$$
\Pr(\delta(S) \geq \alpha) \leq e^{-\frac{n'^2}{2g}} \left[1 + O\left(\frac{n'}{g} + \frac{n'^3}{g^2}\right)\right]. \tag{4}
$$

To complete the proof of part (a) we must bound g and n'. First note that

$$
\frac{g\alpha^2}{2} = Area\left(\bigcup_{(i,j)\in\mathcal{G}} G(i,j)\right) \leq Area(C(s)) = s \tag{5}
$$

To bound g in the other direction note that if $G(i,j) \cap C(s) \neq \emptyset$ but $G(i,j) \not\subseteq C(s)$ then $G(i,j)$ must intersect the boundary $\partial C(S)$ somewhere so every point in $G(i,j)$ is within α of $\partial C(S)$; thus $G(i,j) \subseteq N_\alpha^{C(s)}$. Using the assumption $\alpha \leq s/4b_C$ and applying Lemma 1 we find that

$$
\frac{g\alpha^2}{2} = Area\left(\bigcup_{(i,j)\in\mathcal{G}} G(i,j)\right) \geq Area(C(s)) - Area\left(N_\alpha^{C(s)}\right) \geq 3s/4 \tag{6}
$$

Combining this with the upper bound (5) gives $\alpha^2/2s \leq 1/g \leq 2\alpha^2/3s$. We bound n' by noting that from equation (6) $\gamma = Area\left(\bigcup_{(i,j)\in\mathcal{G}} G(i,j)\right) / Area(C(s)) \geq 3/4$. Since n' is a binomial random variable with parameters n and γ (i.e. $\Pr(n' = i) = \binom{n}{i}\gamma^i(1-\gamma)^{n-i}$) standard Chernoff bounds [6, eq. (7)] tell us that

$$
\Pr(n' \leq n/2) \leq \Pr(n' \leq 2\gamma n/3) \leq e^{-n/18}.
$$

Thus with probability at least $1 - e^{-n/18}$ we may assume that $n' > n/2$. Combining this fact with $\alpha^2/2s \leq 1/g \leq 2\alpha^2/3s$ and the trivial bound $n' \leq n$ we find that with probability at least $1 - e^{-n/18}$

$$
\frac{n'^2}{2g} \geq \frac{n\alpha^2}{8s}, \quad \frac{n'}{g} = O\left(\frac{n\alpha^2}{s}\right), \quad \frac{n'^3}{g^2} = O\left(\frac{n^3\alpha^4}{s^2}\right).
$$

Inserting these inequalities back back into (4) proves part (a) of the lemma.

(b) We are given that C is bounded. Therefore there is some constant a dependent upon C such that $\delta \leq a$. Use the fact that $\mathrm{E}(\delta) = \int_0^a \Pr(\delta \geq \alpha)\, d\alpha$ to derive

$$\mathrm{E}(\delta) = \int_0^{\frac{\sqrt{s}\log n}{n}} \Pr(\delta \geq \alpha)\, d\alpha + O\left(a\Pr\left(\alpha \geq \frac{\sqrt{s}\log n}{n}\right)\right). \tag{7}$$

Equation (2) gives a formula for $\Pr(\delta(S) \geq \alpha)$ for all $\alpha < s/4b_C$. The thing to note here is our assumption that $s \geq \frac{16b_C^2 \log^2 n}{n^2}$. This forces $\frac{\sqrt{s}\log n}{n} < s/4b_C$ and thus for all $\alpha \leq \frac{\sqrt{s}\log n}{n}$ we may evaluate $\Pr(\delta(S) \geq \alpha)$ using equation (2). The proof of the lemma follows by evaluating and integrating.

(c) This is part (b) with $s = Area(C)$. ∎

We are now ready to prove Theorems 1 and 2.

Proof of Theorem 1:

Set $S = S_{i-1} = \{p_1, \ldots, p_n\}$ and $\delta = \delta(S_{i-1})$. It is given that the n points in S are chosen I.I.D. from C. Let X and d be arbitrary reals. We define $L_S(X, d)$ and $R_S(X, d)$ to be the number of points in S that are in the strips of width d to the left and right of the line $x = X$:

$$L_S(X, d) = |\{q \in S : 0 \leq X - q.x \leq d\}|, \quad R_S(X, d) = |\{q \in S : 0 \leq q.x - X \leq d\}|. \tag{8}$$

Then $K \leq L_S(p.x, \delta) + R_S(p.x, \delta)$. We will prove that $\mathrm{E}(L_S(p.x, \delta)) = O(1)$. A symmetric argument will show that $\mathrm{E}(R_S(p.x, \delta)) = O(1)$ and complete the proof.

Note that if X and d are fixed constants then $L_S(X, d)$ is a binomial random variable with parameters n and $\gamma = Area(\{q \in C : 0 \leq X - q.x \leq d\})/c \leq dh_C/c$ where h_C is the height of C and $c = Area(C)$. Thus

$$\mathrm{E}(L_S(X, d)) = |S|\gamma \leq \frac{ndh_C}{c}. \tag{9}$$

This inequality does not involve X so, leaving d fixed and taking X to be the random variable $p.x$, we find $\mathrm{E}(L_S(p.x, d)) \leq \frac{ndh_C}{c}$.

Suppose for a moment that d is a random variable *independent* of S and $p.x$. Then the above reasoning gives $\mathrm{E}(L_S(p.x, d)) \leq n\mathrm{E}(d)h_C/c$. Unfortunately, in our case, $d = \delta$ which is very dependent on S. In fact, after conditioning on $d = \delta$ the points in S are no longer I.I.D. from the *uniform* distribution over C because *they are constrained to be at least δ apart from each other.* Thus $L_S(p.x, \delta)$ is not a binomial random variable. The trick here is that it will be possible to (almost) separate S and δ so that we no longer worry about their dependence. We define two new sets

$$S' = \{p_1, p_2, \ldots, p_{\lfloor n/2 \rfloor}\}, \quad S'' = \{p_{\lfloor n/2 \rfloor + 1}, p_{\lfloor n/2 \rfloor + 2}, \ldots, p_n\}.$$

These two sets partition S. Fix $\delta' = \delta(S')$ and $\delta'' = \delta(S'')$. Then

$$\mathrm{E}(L_S(p.x, \delta)) \leq \mathrm{E}(L_{S'}(p.x, \delta'')) + \mathrm{E}(L_{S''}(p.x, \delta')) \leq |S'|\mathrm{E}(\delta'')h_C/c + |S''|\mathrm{E}(\delta')h_C/c. \tag{10}$$

The first inequality comes from the fact that $\delta \leq \min(\delta', \delta'')$; the second from the fact that S' and δ'' are independent of each other as are δ' and S''.

Lemma 2 (c) tells us that $E(\delta') = O\left(1/|S'|\right)$ and $E(\delta'') = O\left(1/|S''|\right)$. Inserting these equalities back into equation (10) yields $E(L_S(p.x, \delta)) = O\left(\frac{|S'|}{|S''|} + \frac{|S''|}{|S'|}\right) = O(1)$ and finishes the proof of Theorem 1. ∎

We now prove the second theorem.

Proof of Theorem 2: We will assume without loss of generality that $c = Area(C) = 1$. Otherwise scale the x and y coordinates by the same factor so that $c = 1$.

We must prove that $E(N_j) = O\left(\sqrt{n/j}\right)$. The technical difficulty in proving this theorem arises from the fact that the N_j are random variables of the set S_i and the distribution of S_i is not very well understood because the $n - 1$ points in S_i are not I.I.D. *uniformly* distributed in the square. This is because $S_i = S_{i-1} \setminus \{p\}$ where p *is specified to be a member of a closest pair in* S_{i-1}. To sidestep this problem we introduce new random variables M_j, $j = n^{1/3} \log n, \ldots, n$ that are functions of S_{i-1}, a set whose distribution we do know. These M_j will satisfy (i) $N_j \leq \max(M_j, M_{j+1})$ and (ii) $E(M_j) = O\left(\sqrt{n/j}\right)$. The proof that $E(N_j) = O\left(\sqrt{n/j}\right)$ will follow accordingly.

Let $q_1, \ldots, q_n$ be the points in S_{i-1}. Sort them by x-coordinate and relabel them as $q_{(1)}, \ldots, q_{(n)}$ such that $q_{(1)}.x \leq q_{(2)}.x \leq q_{(3)}.x \leq \cdots \leq q_{(n)}.x$. Fix j and set $S_{(j)} = \{q_{(1)}, \ldots, q_{(j)}\}$ to be the set containing the first j sorted points. Now randomly select $\lfloor j/2 \rfloor$ of the points in $S_{(j)}$ and put them into a new set S'. Take the remaining $\lceil j/2 \rceil$ points and put them into another new set S''.

We define $d_j = \max(\delta(S'), \delta(S''))$ and $M_j = |\{q \in S_i : 0 \leq p_{(j)}.x - q.x \leq d_{j-1}\}|$. Note that if $q \in S_{i-1}$ and $0 \leq p_{(j)}.x - q.x \leq d_{j-1}$ then $q \in S_{(j)}$. Note also that $d_j \geq \delta(S_{(j)})$. Thus

$$|\{q \in S_{(j)} : 0 \leq p_{(j)}.x - q.x \leq \delta(S_{(j-1)})\}| \leq M_j. \tag{11}$$

M_j satisfies an even stronger property. Let u be an arbitrary point in $S_{(j-1)}$. Then

$$|\{q \in S_{(j-1)} \setminus \{u\} : 0 \leq p_{(j)}.x - q.x \leq \delta\left(S_{(j-1)} \setminus \{u\}\right)\}| \leq M_j. \tag{12}$$

With these facts we can now show that $N_j \leq \max(M_j, M_{j+1})$.

Recall that $S_i = S_{i-1} \setminus \{p\}$ for some $p \in S_i$. Suppose that $p = q_{(k)}$. Then, for all $j' < k$, we have $p_{(j')} = q_{(j')}$ and thus, from (11), if $j < k$ then $N_j \leq M_j$. For $j' > k$ we have $p_{(j')} = q_{(j'+1)}$ and thus, from (12), if $j > k$ then $N_j \leq M_{j+1}$. Combining the last two sentences yields $N_j \leq \max(M_j, M_{j+1})$.

It remains to prove that $E(M_{j+1}) = O\left(\sqrt{n/j}\right)$. From the definitions of M_j, d_j and L_S we find that

$$\begin{aligned} M_{j+1} &= L_{S_{(j)}}(q_{(j+1)}.x, d_j) = L_{S'}(q_{(j+1)}.x, d_j) + L_{S''}(q_{(j+1)}.x, d_j) \\ &\leq L_{S'}(q_{(j+1)}.x, \delta(S')) + L_{S'}(q_{(j+1)}.x, \delta(S'')) \\ &\quad + L_{S''}(q_{(j+1)}.x, \delta(S')) + L_{S''}(q_{(j+1)}.x, \delta(S'')). \end{aligned}$$

To prove that $E(M_{j+1}) = O\left(\sqrt{n/j}\right)$ it will therefore suffice to prove that both $L_{S'}(q_{(j+1)}.x, \delta(S')) = O\left(\sqrt{n/j}\right)$ and $L_{S'}(q_{(j+1)}.x, \delta(S''))O\left(\sqrt{n/j}\right)$. (the other two terms will follow by symmetry arguments).

Before doing this we present a quick lemma that allows us to restrict the range of $q_{(j+1)}.x$. Recall the definitions of $X()$ and $C()$ from (1).

Lemma 3

$$\Pr\left(X(q_{(j+1)}) \notin \left[\frac{1}{2}\frac{(j+1)}{(n+1)}, \frac{3}{2}\frac{(j+1)}{(n+1)}\right]\right) \leq \frac{4}{(j+1)}.$$

Proof: Recall our assumption that $c = Area(C) = 1$ If $s \in [0,1]$. then

$$\Pr(X(q) \leq s) = \Pr\{q \in C(s)\} = Area(C(s))/c = s/c = s.$$

Thus $X(q_{(j+1)})$ has the same distribution as the random variable Z which is constructed as follows. Choose n real numbers I.I.D. uniformly from the unit interval $[0,1]$. Sort them and let Z be the $j+1$'st sorted point. Z is known as *the $j+1$'st order statistic* and it's mean and variance are well known (Z has a Beta distribution):

$$\mathrm{E}(Z) = \frac{j+1}{n+1}, \quad \mathrm{Var}(Z) = \frac{(j+1)(n+j)}{(n+1)^2(n+2)}$$

[4, p. 17], [5, p. 85]. The lemma follows from Chebyshov's inequality. ∎

Until otherwise indicated we will henceforth assume that $X(q_{(j+1)}) = s = dj/n$ where $d \in [1/3, 2]$ is some constant. Conditioned on $X(q_{(j+1)}) = s$, we have that the set $S_{(j)}$ of points to the left of $p_{(j)}$ has the same distribution as j points chosen I.I.D. uniformly from the region $C(s)$. It follows that the points in S'' have the same distribution as $|S''| = \lceil j/2 \rceil$ points chosen I.I.D. uniformly from $C(s)$. Since $j > n^{1/3} \log n$ and $s = dj/n$ we have that, for large enough n, $s = \frac{dj}{n} \geq \frac{16b_C^2 \log^2 |S''|}{|S''|^2}$ so we can use Lemma 2 (b) to find that $\mathrm{E}(\delta(S'')) = O(\sqrt{s}/j) = O\left(\sqrt{d/nj}\right)$.

The points in S' have the same distribution as $|S'| = \lfloor j/2 \rfloor$ points chosen I.I.D. uniformly from $C(s)$. Following the line of reasoning originally developed in the proof of Theorem 1 we find that if α is a random variable independent of S' then $\mathrm{E}\left(L_{S'}(q_{j+1}.x, \alpha)\right) \leq h_C|S'|\mathrm{E}(\alpha)/s$. By virtue of their construction S' and S'' are independent of each other so substituting $\alpha = \delta(S')$ yields

$$\mathrm{E}\left(L_{S'}(q_{j+1}.x, \delta(S''))\right) = O\left(\frac{|S'|\mathrm{E}(\delta(S''))}{s}\right) = O\left(\frac{j\sqrt{d/nj}}{dj/n}\right) = O\left(\sqrt{\frac{n}{j}}\right). \quad (13)$$

We dispense with d in the last equality because we assume that $X(q_{(j+1)}) = s = dj/n$ where d is bounded in the range $d \in [1/3, 2]$.

To evaluate $\mathrm{E}\left(L_{S'}(q_{j+1}.x, \delta(S'))\right)$ we randomly select $\lfloor |S'|/2 \rfloor$ of the points in S' and put them into a new set V. We then take the remaining $\lceil |S'|/2 \rceil$ points and put them into another new set V'. The exact same line of reasoning that led us to (10) also yields

$$\mathrm{E}(L_{S'_{(j)}}(q_{j+1}.x, \delta(S'))) \leq |V| \cdot \frac{\mathrm{E}(\delta(V'))}{s} + |V'| \cdot \frac{\mathrm{E}(\delta(V))}{s} = 2O\left(j\frac{\sqrt{d/nj}}{dj/n}\right) = O\left(\sqrt{\frac{n}{j}}\right). \quad (14)$$

We again dispense with d in the last equality because we are still assuming that $X(q_{(j+1)}) = s = dj/n$ where d is bounded in the range $d \in [1/3, 2]$. From the discussion following (12) we see that (13) and (14) together imply that $\mathrm{E}\left(M_{j+1} | d \in [1/3, 2]\right) = O\left(\sqrt{n/j}\right)$ where we are conditioning over the event $q_{(j+1)}.x = dj/n$. To complete the evaluation of $\mathrm{E}(M_{j+1})$ note that Lemma 3 states that $\Pr(d \notin [1/3, 2]) \leq 4/j$. Since $M_{j+1} \leq j$ we derive $\mathrm{E}(M_{j+1} | d \notin [1/3, 2]) \leq j \cdot 4/j = 4$. We have therefore proven that $\mathrm{E}(M_{j+1}) = O\left(\sqrt{n/j}\right)$. This last fact taken together with the previously proven $N_j \leq \max(M_j, M_{j+1})$ proves that $\mathrm{E}(N_j) = O\left(\sqrt{n/j}\right)$ and we are done. ∎

Acknowledgements: The author would like to thank Jon Bentley for conversations concerning an earlier version of this paper. He would also like to thank Michiel Smid for introducing him to the "state of the art" in the dynamic closest pair problem.

References

[1] A. Aho, J. Hopcroft and J. D. Ullman, *The Design and Analysis of Computer Algorithms*, Addison-Wesley, Reading, Mass. (1974)

[2] J. L. Bentley and C. H. Papadimitriou, "A Worst-Case Analysis of Nearest-Neighbor Searching by Projection," *7th Int. Conf. on Automata, Languages and Programming*, (August 1980) 470-482.

[3] J.L. Bentley, B.W. Weide, and A.C. Yao, "Optimal Expected-Time Algorithms for Closest Point Problems," *ACM Trans. on Mathematical Software*, 6(4) (Dec. 1980) 563-580.

[4] Luc Devroye, *Non-Uniform Random Variate Generation*, Springer-Verlag, New York. (1986).

[5] M. J. Golin *Probabilistic Analysis of Geometric Algorithms (Thesis)*, Princeton University Technical Report CS-TR-266-90. June 1990.

[6] Torben Hagerup and Christine Rub, "A Guided Tour of Chernoff Bounds," *Information Processing Letters*, 33 (1989/90) 305-308.

[7] Klaus Hinrichs, Jurg Nievergelt, and Peter Schorn, "Plane-Sweep Solves the Closest Pair Problem Elegantly," *Information Processing Letters*, 26 (January 11, 1988) 255-261.

[8] M..H. Overmars, *The Design of Dynamic Data Structures*, Lecture Notes in Computer Science, volume 156, Springer-Verlag, Berlin. (1983).

[9] F. P. Preparata and M. I. Shamos *Computational Geometry: An Introduction*, Springer-Verlag, New York. (1985).

[10] M.O. Rabin, "Probabilistic Algorithms," *Algorithms and Complexity: New Directions and Recent Results (J.F. Traub ed.)*, (1976) 21-39.

[11] C. Schwarz, M. Smid, and J. Snoeyink, "An Optimal Algorithm for the On-Line Closest Pair Problem," *8'th ACM Symposium on Computational Geometry*, 1992.

[12] Michael Ian Shamos, "Computational Geometry," *Thesis (Yale)*, (1978).

[13] Michiel Smid, " Maintaining the Minimal Distance of a Point Set in less than Linear time Algorithms Review," 2(1) (May 1991) 33-44.

[14] Michiel Smid, "Maintaining the Minimal Distance of a Point Set in Polylogarithmic Time (revised version)" Technical report MOI-I-91-103, Max Planck Institut, Saarbrucken, April 1991

[15] Bruce W. Weide, "Statistical Methods in Algorithm Design and Analysis," *Thesis (Carnige-Mellon University). CMU-CS-78-142*, (August 1978).

Two- and Three-Dimensional Point Location in Rectangular Subdivisions*
(Extended Abstract)

Mark de Berg
Department of Computer Science
Utrecht University

Marc van Kreveld
Department of Computer Science
Utrecht University

Jack Snoeyink
Department of Computer Science
University of British Columbia

Abstract

We apply van Emde Boas-type stratified trees to point location problems in rectangular subdivisions in 2 and 3 dimensions. In a subdivision with n rectangles having integer coordinates from $[1, U]$, we locate an integer query point in $O((\log \log U)^d)$ query time using $O(n)$ space when $d \leq 2$ or $O(n \log \log U)$ space when $d = 3$. Applications and extensions of this "fixed universe" approach include spatial point location using logarithmic time and linear space in rectilinear subdivisions having arbitrary coordinates, point location in c-oriented polygons or fat triangles in the plane, point location in subdivisions of space into "fat prisms," and vertical ray shooting among horizontal "fat objects." Like other results on stratified trees, our algorithms run on a RAM model and make use of perfect hashing.

1 Introduction

The point location problem—which seeks to preprocess a set of disjoint geometric objects to be able to determine quickly which object contains a query point—is an important and well-studied problem in computational geometry. The usual goal of such study is logarithmic-time algorithms and linear-space structures, since this is the lower bound for one-dimensional search in a comparison-based model. In two dimensions, researchers have developed several solutions that attain these bounds; see Preparata [13] for a survey. In three dimensions, these bounds have not yet been attained, even though recent work on dynamic planar point location has lead to advances in spatial point location. Goodrich and Tamassia's [8] method, which achieves $O(\log^2 n)$ query time using $O(n \log n)$ space, is the current best.

---

*This research was supported by the ESPRIT Basic Research Action No. 3075 (project ALCOM). The first author was also supported by the Dutch Organization for Scientific Research (N. W. O.).

We consider the special case of rectangular subdivisions. For our purposes, a *rectangle* in d dimensions is the Cartesian product of d intervals that are closed on the left and open on the right. A *rectangular subdivision* is a partition of a rectangle R into disjoint rectangles $R_1, R_2, \ldots, R_n$ whose union covers R; the *size* of this subdivision is n. The problem of point location in a subdivision is to report the rectangle R_i that contains a query point $q \in R$. Edelsbrunner, Haring, and Hilbert [5] extended a planar point location method of Edelsbrunner and Maurer [6] to solve point location in a d-dimensional rectangular subdivision in $O(\log^{d-1} n)$ query time. Their algorithm handles arbitrary coordinates and runs on a pointer machine.

We use a stronger model of computation, the random access machine (RAM), to support the perfect hashing of Fredman, Komlós and Szemerédi [7]. Furthermore, for the first half of this paper, we require that the rectangle corners and query points lie in a fixed size integer grid $[1, U]^d$. Stratified trees, a data structure introduced by van Emde Boas [14] and extended by him and others [9, 11, 16, 17], exploit the power of a RAM on a fixed universe. They have been used for log-logarithmic time queries in one-dimensional point location, more commonly known as searching a list for the successor of a query point. Müller [12] used a type of stratified tree as a two-dimensional point location structure, to answer queries in a rectangular subdivision of size n using $O((\log \log U)^2)$ time and $O(n \log U)$ space.

We give a new type of stratified tree that emphasizes a tradeoff between space and query time. In two dimensions, we can preserve Müller's $O((\log \log U)^2)$ query time using only $O(n \log \log U)$ space or reduce space to linear and increase the query time to $O(\log^\epsilon U)$, where ϵ is a postive constant. We can also achieve $O((\log \log U)^2)$ query time with linear space by extending van Emde Boas' pruning technique [15] to two dimensions. In three dimensions, we can extend the point location method but not the pruning: We achieve $O((\log \log U)^3)$ query time using $O(n \log \log U)$ space or $O(\log^\epsilon U)$ time using linear space, for any constant $\epsilon > 0$. Section 2 describes the data structure and subsections 2.1, 2.2 and 2.3 describe integer point location in rectangular subdivisions of one-, two-, and three-dimensional integer grids.

In Section 3 we apply the point location method to other problems that are not initially defined on fixed integer grids. In subsection 3.1, we normalize a three-dimensional rectangular subdivision having arbitrary coordinates to allow point location using linear space and logarithmic time. Point location in k rectangular subdivisions of $d \leq 3$ dimensions that have total size n takes $O(\log n + k(\log \log n)^d)$ time. In subsection 3.2, we perform point location among c-oriented polygons or fat triangles in the plane in $O((\log \log n)^2)$ time after a constant number of normalizations. This allows point location in a subdivision of 3-space into c-oriented or fat prisms in logarithmic time and linear space. In subsection 3.3, we perform vertical ray shooting queries among n horizontal objects in 3-space using $O(\log n (\log \log n)^2)$ time. If the objects are rectangles or c-oriented triangles, the space is $O(n \log n)$; and if the objects are fat triangles, the space is $O(n \log n \log \log n)$.

2 Stratified trees and point location

In this section we describe our variant of stratified trees and show how they can be used to solve point location problems efficiently on a RAM. Conceptually, a stratified tree is an interval tree T built on the universe $[1, U]$ with a search tree built on the levels of T. The actual implementation depends upon perfect hashing to reduce storage space. First we describe how we think of stratified trees and then how they are implemented. We assume that the universe size U is a power of 2 and take all logarithms as base 2.

An *interval tree* T on $[1, U]$ is a complete binary tree that stores intervals of $[1, U]$— our definition will be slightly different from that in Edelsbrunner [4]. Number the levels of the interval tree T from the root, level 0, to the leaves, level $\log U$. Number the leaves of T from left to right with integers from 1 to U. With the jth leaf we associate the range $\rho(j) = [j - 1/2, j + 1/2]$; an internal node τ of T is associated with the range $\rho(\tau)$ that is the union of the ranges of the leaves of the subtree rooted at τ. From Figure 1 you can see that the set of ranges on level ℓ partition the integers of $[1, U]$ into 2^ℓ equal-size subsets.

An interval $I = [i_{\min}, i_{\max})$ with integer bounds from $[1, U]$ *spans* a tree node $\tau \in T$ if I contains the range $\rho(\tau)$. Interval I is *contained in* τ if I is contained in $\rho(\tau)$. Interval I *cuts* node τ if $\rho(\tau)$ contains exactly one of the endpoints of I. We can further distinguish whether I cuts τ *on the left*, meaning that I contains the lower bound of $\rho(\tau)$, or whether I cuts τ *on the right*, meaning that I contains the upper bound of $\rho(\tau)$. As Figure 1 illustrates, the interval I is contained in the root and in one node per level down to some level $\ell_I - 1$. Then I cuts two nodes per level, one on the right and one on the left, from ℓ_I down to the leaves and spans any nodes between them.

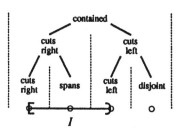

Figure 1: Interval tree T on $[1, 4]$ and interval $I = [1, 3)$

We next form a *level-search tree*, a balanced k-ary search tree on the levels of T. Figure 2 shows an interval tree with a ternary level-search tree. The level-search tree is formed by assigning $k - 1$ evenly spaced levels to the root and recursively constructing k subtrees for the levels in between. Thus, its height is $h = \Theta(\log \log U / \log k)$.

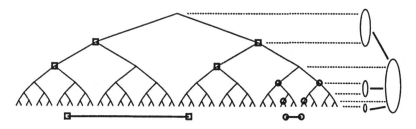

Figure 2: A stratified tree storing two intervals

At a level-search tree node, there is a natural ordering of the associated levels and children.

We can say that an associated level is *directly below* a child if it is one level deeper than the deepest level in the subtree rooted at the child. Now, given a subdivision of $[1, U]$ into n intervals with integer bounds, we store these intervals in the stratified tree as follows. Give interval I to the nodes that it cuts on level ℓ_I, which are the highest level cut nodes, and also to the nodes that it cuts on all other levels directly below the path in the level-search tree from ℓ_I to the root, as depicted in Figure 2. Notice that each interval is given to at most $2h$ interval tree nodes. By way of contrast, each interval is stored in all cut nodes in a van Emde Boas tree.

Recall that this was the conceptual view of stratified trees. If $n \ll U$, then most of the interval tree nodes do not receive intervals; to actually store these empty nodes would be wasteful. To implement stratified trees, we create only the level-search tree and store only the *full* nodes of each level, that is, the nodes that receive at least one interval.

We label the nodes at level ℓ from 1 to 2^ℓ; the label of the node that contains the integer $q \in [1, U]$ is one greater than the binary number represented by the first ℓ bits of q. We store the labels of full nodes and pointers to their intervals in a table using the perfect hashing scheme of Fredman, Komlós and Szemerédi [7]. (See also Mehlhorn and Näher [11].) This scheme stores m full nodes in $O(m)$ space and locates a stored node in $O(1)$ time. The deterministic preprocessing time is the minimum of $O(mU)$ and $O(m^3 \log U)$; the expected randomized preprocessing time is $O(m)$. Thus we have:

Theorem 2.1 *To store n intervals that partition $[1, U]$ in a stratified tree with a level-search tree of height h requires $O(nh + \log U)$ space and expected preprocessing time.*

Proof: The level-search tree structure takes $O(\log U)$ space, neglecting storage for associated levels. These levels store nodes containing $2hn$ intervals, thus, the maximum number of nodes and amount of storage is $O(nh)$ for all levels. The preprocessing is dominated by computing perfect hash tables; it is easy to assign intervals to levels and nodes in $O(nh)$ total time. ∎

Remark: The dynamic perfect hashing technique of Dietzfelbinger et al. [3] can be used to make these stratified trees dynamic. The amortized expected time to delete j intervals and replace them by k intervals that have the same union is $O((j + k)h)$. The space to store a level remains linear and the time to lookup whether a node is stored remains constant.

In the next subsections, we show how stratified trees answer point location queries in fixed universes.

2.1 One-dimensional point location

As a warm-up exercise for higher-dimensional point location, we show how to answer point location queries in one dimension using our variant of stratified trees. We prove Theorem 2.2.

Theorem 2.2 *Using a stratified tree on $[1, U]$ with a level-search tree of height $1 \leq h \leq \log \log U$, one can perform point location in an interval subdivision $I_1, I_2, \ldots, I_n$ using $O(nh)$ space and expected preprocessing time and $O(h(\log U)^{1/h})$ query time. By pruning, one can achieve $O(n)$ space and preprocessing time and $O(\log \log U)$ query time.*

When we have very few intervals, say $n = O(\log U)$, we punt the stratified trees and simply use a balanced binary search tree on the interval endpoints. This gives $O(\log \log U)$ query time using $O(n)$ space. Otherwise, we build a stratified tree using $O(nh)$ space according to Theorem 2.1.

Consider a stratified tree node τ and the (at most two) intervals it receives. If τ receives an interval I_j that cuts τ on the left (right), store I_j's upper (lower) bound. If τ receives no intervals, that is, if τ is empty, then some interval I_j spans $\rho(\tau)$.

Since an interval I_j is always given to two adjacent nodes at level ℓ_{I_j}, every point in I_j is in $I_j \cap \rho(\tau')$ for some full node τ'.

Suppose, just for one paragraph, that we had given each interval I_j to every node that it cuts—this is precisely what is done in forming a van Emde Boas tree. We could then determine if the interval containing a query integer q was stored above or below a level ℓ of the interval tree by the following procedure: Take the label of q, which is one greater than the number determined by the first ℓ bits of q, and, by hashing in constant time, determine if the node $\tau \in T$ with that label is empty or is stored at level ℓ. If τ is empty, then q is inside an interval I_j that spans τ and, therefore, also spans τ's descendents—we need not search deeper in the interval tree. Otherwise, test q against the intervals cutting τ on the left and right. If either interval contains q, stop and report it; otherwise q is inside an interval I_j contained in τ and, therefore, also contained in τ's ancestors—we need not search higher in the tree.

We have not given each interval to every cut node, however, so we must remember intervals that we have seen as we move down the level-search tree. To answer a point location query for an integer q, we begin at the root of the level-search tree and set the interval $\mathcal{I} = [1, U]$.

With a node ν of the level-search tree are associated $k - 1$ levels of the interval tree, $\ell_1, \ell_2, \ldots, \ell_{k-1}$. When the search reaches ν, we use the hash table for each level ℓ_j to determine if the interval tree node τ_j that contains q on level ℓ_j is empty or stores one or two intervals. For each full node τ_j, we can check in constant time whether an interval stored with τ_j contains q and if one does, we stop and report it. Otherwise, we use $\mathcal{I}$ to help decide in which child of ν to continue the search in the level-search tree. First, we shrink $\mathcal{I}$ by the closest interval boundaries found to the right and left of q. Second, we determine which node τ_j contains the interval $\mathcal{I}$. If none does, then we continue the search in the highest child of ν, otherwise we continue in the child of ν that is directly below level ℓ_j. Lemma 2.3 proves the correctness of this procedure.

Lemma 2.3 *Let ν be a level-search tree node. Let $\mathcal{I}$ be the largest interval containing the query q that is disjoint from all intervals found in levels associated with nodes on the path to and including ν. The interval I_j that contains q can be found in the subtree of the highest child of ν below all associated levels having a node that contains $\mathcal{I}$.*

Proof: Think of adding the root and leaf levels to those associated with ν; then we can find two levels ℓ and ℓ', with one child between them, such that $\mathcal{I}$ is contained in a node τ at level ℓ and not contained in a node at level ℓ'.

We know that any interval stored in the stratified tree that intersects $\mathcal{I}$ is contained in $\mathcal{I}$—including the interval that contains q. Since $\mathcal{I}$ is contained in τ, we need not search higher than level ℓ.

Now, consider the node $\tau' \in T$ that contains q at level ℓ'. The interval $\mathcal{I}$ either cuts or spans τ'—we shall prove that it spans τ'. Suppose, instead, that $\mathcal{I}$ cuts τ. Then $\mathcal{I}$ contains some interval I' that is stored in the stratified tree and cuts τ. But the highest level node that I' cuts must then be between ℓ and ℓ', so I' would be stored in node τ and would be found to contain q or to shorten the interval $\mathcal{I}$. Thus, $\mathcal{I}$ spans τ and also spans all descendents of τ—we need not search lower than ℓ'. This establishes the lemma. ∎

The proof of Theorem 2.2 is almost complete. For each of the h levels of the level-search tree, a query examines $k - 1$ levels of the interval tree in constant time apiece. Query time is $O(hk)$ and space is $O(nh)$, where $h = \log\log U / \log k$. Varying the height parameter h gives a space/query time tradeoff: Choosing k a constant gives $O(\log\log U)$ query time and $O(n \log\log U)$ space. Choosing h a constant gives $O((\log U)^{1/h})$ query time and $O(n)$ space.

The tradeoff afforded by h is unnecessary for one-dimensional point location. Instead, one can use van Emde Boas' technique of *pruning* [15] to reduce the space to linear and increase the query time by only a constant factor. Choose every $\log\log U$th interval boundary to form $n / \log\log U$ super-intervals and store these super-intervals in a stratified tree using $O(n)$ space. Given a query, find the containing super-interval in the stratified tree, then use linear search to find the actual interval. The query time remains $O(\log\log U)$. This completes the proof of Theorem 2.2.

Because the union of rectangles is not a rectangle, pruning in higher dimensions is more difficult. At the end of the next section, we use the planar separator theorem to show that pruning is still possible in two dimensions. For three dimensions, however, we need the space/time tradeoff to attain linear space.

2.2 Two-dimensional point location

In this section, we show how to perform point location in a rectangular subdivision of the plane by using two layers of stratified trees. Theorem 2.4 improves a theorem of Müller [12].

Theorem 2.4 *Using stratified trees on $[1, U]^2$, one can perform 2-d point location in a rectangular subdivision $R_1, R_2, \ldots, R_n$ in $O(h(\log U)^{1/h} \log\log U)$ query time using $O(nh)$ space and expected preprocessing time, for any integer $1 \le h \le \log\log U$. By pruning, one can achieve $O(n)$ space after $O(n \log n)$ deterministic and $O(n)$ expected preprocessing time.*

To perform point location in a rectangular subdivision of two dimensions, we form a stratified tree on the intervals of the x-axis in much the same way as in the previous section. We give each rectangle to the highest level nodes that its x-interval cuts and to the nodes cut on all other levels directly below the the path to the root of the level-search tree.

Consider a node τ in the interval tree T: it has range $\rho(\tau) = [x_{\min}, x_{\max}]$ and receives a set of rectangles $\mathcal{R}$ that cut it. (See Figure 3.) The intersection of the line $x = x_{\min}$ or $x = x_{\max}$ with $\mathcal{R}$ is a set of intervals—it is not a subdivision because there are gaps left by rectangles that span τ or that are stored elsewhere in the stratified tree. If we fill in these gaps, however, we can use one-dimensional point location to find the projection of a query point q onto the the lines $x = x_{\min}$ and $x = x_{\max}$. Since filling in the gaps at most doubles the number of intervals, we can locate both projections in $O(\log \log U)$ time using space proportional to the number of rectangles received by τ. This proves that the total space required is $O(nh)$.

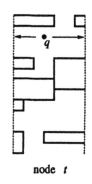

node t

Figure 3: Rectangles given to τ

If the projection of q lies in a y-interval of a rectangle $R \in \mathcal{R}$, then we can check in constant time if q also lies in the x-interval of R. Thus, to locate the rectangle containing a query point q, we begin at the root of the level-search tree and set the line segment $\mathcal{I}$ to the portion of the horizontal line through q with x coordinates in $[1, U]$. At a level-search tree node ν, we use hashing to obtain the node containing q at each of the $k - 1$ levels associated with ν and use one-dimensional point location to check for rectangles containing q. If none is found, we shrink the segment $\mathcal{I}$ to lie between closest rectangles intersecting $\mathcal{I}$ to the right and left of q. We continue the search in the child of ν directly below the lowest associated level that has a node containing $\mathcal{I}$. Again, Lemma 2.3 proves the correctness of this procedure.

For each of the h levels of the level-search tree, a query examines $k - 1$ levels of the tree in $O(\log \log U)$ time apiece. Thus, query time is $O(hk \log \log U)$, where $h = \log \log U / \log k$. Except for the pruning, this establishes Theorem 2.4. Choosing k a constant gives a $O((\log \log U)^2)$ query time algorithm using $O(n \log \log U)$ space. Choosing h a constant gives a $O((\log U)^{1/h} \log \log U)$ algorithm with linear space.

Remark: If we use dynamic perfect hashing [3] in the primary structure, then an operation that replaces j rectangles by k rectangles that have the same union induces $O((j + k)h)$ changes in 1-dimensional structures. Each of these takes $O(1)$ expected amortized time, if dynamic, pruned stratified trees are used for the 1-dimensional subproblems. Thus, the entire space/query time tradeoff can be achieved. Unfortunately, the 2-dimensional pruning described below cannot be used in a dynamic setting.

For the remainder of this section, we develop a two-dimensional analogue of van Emde Boas' pruning technique to reduce the space to linear while increasing the query time by only a constant factor. Specifically, we prove in the full paper that we can collect the rectangles $R_1, R_2, \ldots, R_n$ into groups of size $O((\log \log U)^2)$ and cover these groups by a new subdivision into $m = O(n / \log \log U)$ rectangles $R'_1, R'_2, \ldots, R'_m$ such that every new rectangle R'_i intersects only one group. We can store the new rectangles $R'_1, R'_2, \ldots, R'_m$ in a point location structure that uses $O(n)$ space and find the rectangle R'_i containing a query point in $O((\log \log U)^2)$ time. Rectangle R'_i tells us a unique group of $O((\log \log U)^2)$

rectangles that we can search exhaustively.

Lemma 2.5 *Let rectangles $R_1, R_2, \ldots, R_n$ form a rectangular subdivision. In $O(n \log n)$ time, one can partition these n rectangles into sets $\mathcal{R}_j$ of size $O((\log \log U)^2)$ and find a new subdivision $R'_1, R'_2, \ldots, R'_m$ of at most $m = O(n/\log \log U)$ rectangles such that each new rectangle R'_i intersects the rectangles of only one set $\mathcal{R}_j$.*

> **Proof:** We repeatedly apply Lipton and Tarjan's planar separator Theorem [10] to form connected sets of $O((\log \log U)^2)$ rectangles and show that the number of rectangles removed is $O(n/\log \log U)$. ∎

Remark: On integer grids, this data structure can improve many algorithms that use point location as a subroutine. For a simple example, reporting the k horizontal segments that intersect a vertical query segment can be performed in $O((\log \log n)^2 + k)$ time by preprocessing Chazelle's hive graph [2].

2.3 Three-dimensional point location

By now, the method for three-dimensional point location should come as no surprise: use a stratified tree with two-dimensional point location as secondary structures at each node. That the method breaks down in four dimensions may come as more of a surprise. For the abstract, we merely state Theorem 2.6.

Theorem 2.6 *Using stratified trees on $[1, U]$, one can perform 3-dimensional point location in a rectangular subdivision $R_1, R_2, \ldots, R_n$ in $O(h(\log U)^{1/h}(\log \log U)^2)$ query time using $O(nh)$ space, for any integer $1 \leq h \leq \log \log U$.*

Varying the height parameter h gives a space/query time tradeoff: Choosing h a constant gives linear space and $O((\log U)^{1/h}(\log \log U)^2)$ query time. Choosing $h = \lceil \log \log U \rceil$ gives $O(\log \log U)$ query time and $O(n(\log \log U)^2)$ space.

This method cannot be extended to higher dimensions because we can no longer "fill the gaps" using linear space. There are sets of m boxes in 3-dimensions that are contained only in rectangular subdivisions of size $\Omega(m^{3/2})$.

3 Application to other domains

The previous section developed data structures for point location problems in fixed universes; the problem domains were fixed size integer grids. In this section, we look at some easy applications of these data structures to problem domains that are not (initially) fixed grids. These include a logarithmic-time linear-space point location structure for rectangular subdivisions of three dimensions using arbitrary coordinates, locating a single point in several subdivisions, point location among c-oriented polygons or fat triangles in the plane and among prisms with c-oriented or fat bases, and vertical ray shooting among horizontal rectangles or c-oriented or fat triangles. Our approach is two-pronged: First, to extract one or more grids from a problem and preprocess them for point location. Second, to normalize a query point to these grids by binary search and then perform point locations. We gain by having only one search for the normalization of a query point.

3.1 Point location in rectangular subdivisions of real 3-space

We begin with two simple examples of normalization. The first is the problem that motivated this research.

Theorem 3.1 *Given a three-dimensional rectangular subdivision of size n having arbitrary real coordinates, we can answer point location queries in $O(\log n)$ time on a RAM using linear space and $O(n \log n)$ expected preprocessing time.*

Proof: Every rectangular box is a product of intervals from the three coordinate directions. For each of the three coordinate directions, form a sorted and ranked list of the bounds of these intervals and replace each real interval $[a, b)$ with the integer interval $[rank(a), rank(b))$. This takes $O(n \log n)$ preprocessing.

The rectangular subdivision can now be considered as a subdivision of the integer grid whose maximum coordinate is n. We can therefore apply Theorem 2.6 and compute a linear space point location structure that reports the rectangle containing a query grid point in $O(\log n)$ time. (If the height of level-search trees in all dimensions are taken to be $h = 3$, for example, then these bounds are attained without pruning and the expected time to build the data structures is $O(n)$.)

To answer a query for a real point q, we normalize q to a grid point: we replace each coordinate of q by its rank, which we determine by a binary search in the list of bounds for that coordinate. Then we report the box that contains this grid point. Both normalization and grid-point location take $O(\log n)$ time. ∎

A related (and trivial) example deals with locating a point in several subdivisions of total size n. All bounds of rectangle intervals from a given axis can be collected into one sorted and ranked list and all normalizations can performed on this list.

Theorem 3.2 *A query point can be located in k rectangular subdivisions of $d \leq 3$ dimensions that have total size n in $O(\log n + k(\log \log n)^d)$ time.*

Remark: By collecting the planar subproblems that arise in the skewer trees of Edelsbrunner, Haring, and Hilbert [5] and applying Theorem 3.2, we obtain an $O((\log n)^{d-2}(\log \log n)^2)$ point location method for rectangular subdivisions of $d > 2$ dimensions.

3.2 Point location among c-oriented polygons and fat triangles

In this section we explore a method to extend the rectangular point location scheme to a set $\mathcal{P}$ of polygons in the plane that have disjoint interiors. Suppose we can find a rectangular subdivision of the plane such that each rectangle in the subdivision intersects only a constant number of polygons of $\mathcal{P}$; we call such a subdivision a *sparse rectangularization $SR(\mathcal{P})$* of the set $\mathcal{P}$ of polygons. Then we could answer a point location in $\mathcal{P}$ by locating the query point q in $SR(\mathcal{P})$ and comparing q to the polygons of $\mathcal{P}$ that intersect the rectangle containing q.

Not every set $\mathcal{P}$ admits a sparse rectangularization: if, for example, the set contains a vertex with more than a constant number of incident polygons then no sparse rectangularization of $\mathcal{P}$ exists. Therefore we study the restricted class of c-oriented polygons; polygons whose edges are parallel to a fixed set of c orientations, for some constant c. Furthermore, rather than looking for a single sparse rectangularization, we partition a set of c-oriented polygons into a constant number of sets of quadrilaterals, each of which admits a sparse rectangularization. The size of all rectangularizations (the total number of rectangles) will be linear in the number of polygon edges. Different rectangularizations will use different orientations for their axes; a query point must be normalized in several new orientations. As we have seen in Section 3.1, however, applications that perform several point locations gain by performing the normalizations only once.

Let $\mathcal{P}$ be a set of c-oriented polygons with disjoint interiors and assume, without loss of generality, that one of the c possible orientations is parallel to the x-axis. To obtain a family of sparse rectangularizations, we decompose each polygon P of $\mathcal{P}$ into trapezoids (some of which can degenerate into triangles) by slicing through each vertex of P with the longest horizontal segment that is contained in P. Each trapezoid has (one or) two edges that are horizontal, which we call its top and bottom edges, and a left and a right edge. Since $\mathcal{P}$ is c-oriented, we can partition the resulting set of trapezoids into $c - 1$ subsets $\mathcal{P}_1, \ldots, \mathcal{P}_{c-1}$ according to the orientation of their left edge.

Consider one subset $\mathcal{P}_i$. Compute the horizontal trapezoidation of the left edges of trapezoids in $\mathcal{P}_i$, denoted $ADJ(\mathcal{P}_i)$, because the endpoints of each left edge are connected to the horizontally adjacent left edges. If we apply a skew transformation to make the left edges vertical, then the horizontal adjacency map is a rectangularization of $\mathcal{P}_i$.

Lemma 3.3 $ADJ(\mathcal{P}_i)$ *is a sparse rectangularization of $\mathcal{P}_i$ with linear size.*

Lemma 3.3 enables us to perform fast point location in $\mathcal{P}_i$: first we perform fast point location in $ADJ(\mathcal{P}_i)$ and then we test in constant time whether the query point is inside the trapezoid that intersects the rectangle of $ADJ(\mathcal{P}_i)$ that contains the query point. Observe that the query point needs to be normalized in both the y direction and in the direction perpendicular to the left edges of $\mathcal{P}_i$ in order to perform fast point location in $ADJ(\mathcal{P}_i)$. Applying this scheme to each $\mathcal{P}_i$ leads to the following theorem.

Theorem 3.4 *After c normalizations, one can perform 2-dimensional point location among disjoint c-oriented polygons with n vertices in $O((\log\log n)^2)$ query time using $O(n)$ space and $O(n \log n)$ expected preprocessing time.*

We can obtain the same result for fat triangles, where a triangle is *fat* if every internal angle contains at least one of a set of c fixed orientations.

Corollary 3.5 *After c normalizations, one can perform 2-dimensional point location among n disjoint fat triangles in $O((\log\log n)^2)$ query time using $O(n)$ space and $O(n \log n)$ expected preprocessing time.*

One way to extend these algorithms to three dimensions is to consider subdivisions into prisms whose bases are parallel c-oriented polygons or fat triangles. The next theorem is a generalization of Theorem 3.1.

Theorem 3.6 *In a subdivision of 3-space into prisms whose bases are parallel c-oriented polygons, with constant c, or parallel fat triangles we can perform point location in $O(\log n)$ time using linear space and $O(n \log n)$ expected preprocessing time.*

3.3 Vertical ray shooting queries

The problem of vertical ray shooting among horizontal objects in space can be seen as a generalization of 3-dimensional point location in a subdivision. In this section we apply the fast point location technique to speed up vertical ray shooting queries among horizontal rectangles or horizontal c-oriented or fat triangles. Because one normalization can serve for several point locations, we can improve query times from $O(\log^2 n)$ to $O(\log n(\log\log n)^2)$ for these problems. Similar improvements are possible for any structure that uses a (rectangular or c-oriented) point location structure as an associated structure.

The problem we study is this: Let S be a set of horizontal objects (parallel to the xy-plane) in 3-space. We want to preprocess S such that the first object hit by a ray directed vertically downward (parallel to the z-axis) can be determined efficiently. First we consider horizontal axis-parallel rectangles and later horizontal c-oriented polygons and fat triangles.

Theorem 3.7 *Vertical ray shooting queries in a set of n horizontal axis-parallel rectangles in 3-space can be answered in $O(\log n(\log\log n)^2)$ time using $O(n\log n)$ space. Preprocessing takes $O(n\log^2 n)$ expected time.*

Proof: Use a segment tree on the x-intervals of rectangles with point location as a secondary structure. ∎

For c-oriented triangles, we use a theorem of Alt et al. [1], which says that the boundary complexity of the union of a set of homothetic triangles is linear in their number.

We partition the triangles into $\binom{c}{3}$ sets, depending on the orientations of the edges; each set S consists of homothetic triangles. We can answer the ray shooting query on each set independently and choose the best result.

Theorem 3.8 *Vertical ray shooting queries among a set of n horizontal c-oriented triangles in 3-space can be answered in time $O(\log n(\log\log n)^2)$ with a structure that uses $O(n\log n)$ space. This structure can be built in $O(n\log^2 n)$ expected time.*

Proof: For homothetic triangles, build a binary tree whose leaves are the triangles ordered by z-coordinate. At internal nodes store the boundary of the union of triangles in its left subtree in a point location structure. ∎

For fat triangles we have a similar result with a slightly worse space bound because the union of fat triangles can have superlinear complexity.

Theorem 3.9 *Vertical ray shooting queries among a set of n horizontal fat triangles in 3-space can be answered in $O(\log n(\log\log n)^2)$ time and $O(n\log n\log\log n)$ space. Preprocessing takes $O(n\log^2 n\log\log n)$ expected time.*

Acknowledgements

We thank Kurt Mehlhorn for discussions on perfect hashing and Mark Overmars for discussions of fat objects.

References

[1] H. Alt, R. Fleischer, M. Kaufmann, K. Mehlhorn, S. Näher, S. Schirra, and C. Uhrig. Approximate motion planning and the complexity of the boundary of the union of simple geometric figures. In *Proc. 6th Ann. ACM Symp. Comp. Geom.*, pages 281–289, 1990.

[2] B. Chazelle. Filtering search: A new approach to query-answering. *SIAM J. Comp.*, 15(3):703–723, 1986.

[3] M. Dietzfelbinger, A. Karlin, K. Mehlhorn, F. Meyer auf der Heide, H. Rohnert, and R. E. Tarjan. Dynamic perfect hashing: Upper and lower bounds. In *Proc. 29th FOCS*, pages 524–531, 1988. Revised version: Bericht Nr. 77, Reihe Informatik, Paderborn, Januar 91.

[4] H. Edelsbrunner. *Algorithms in Combinatorial Geometry*. Springer-Verlag, Berlin, 1987.

[5] H. Edelsbrunner, G. Haring, and D. Hilbert. Rectangular point location in d dimensions with applications. *Computer Journal*, 29:76–82, 1986.

[6] H. Edelsbrunner and H. A. Maurer. A space-optimal solution of general region location. *Theoretical Comp. Sci.*, 16:329–336, 1981.

[7] M. L. Fredman, J. Komlós, and E. Szemerédi. Storing a sparse table with $O(1)$ worst case access time. *JACM*, 31(3):538–544, 1984.

[8] M. T. Goodrich and R. Tamassia. Dynamic trees and dynamic point location. In *Proc. 23rd Ann. ACM STOC*, pages 523–533, 1991.

[9] D. Johnson. A priority queue in which initialization and queue operations take $O(\log \log D)$ time. *Math. Systems Theory*, 15:295–309, 1982.

[10] R. J. Lipton and R. E. Tarjan. Applications of a planar separator theorem. In *Proc. 18th FOCS*, pages 162–170, 1977.

[11] K. Mehlhorn and S. Näher. Bounded ordered dictionaries in $O(\log \log N)$ time and $O(n)$ space. *Info. Proc. Let.*, 35:183–189, 1990.

[12] H. Müller. Rasterized point location. In H. Noltemeier, editor, *Proc. WG 85*, pages 281–294. Trauner Verlag, 1985.

[13] F. P. Preparata. Planar point location revisited. *Int. J. Found. Comp. Sci.*, 1(1):71–86, 1990.

[14] P. van Emde Boas. Preserving order in a forest in less than logarithmic time. In *Proc. 16th FOCS*, pages 75–84, 1976.

[15] P. van Emde Boas. Preserving order in a forest in less than logarithmic time and linear space. *Info. Proc. Let.*, 6:80–82, 1977.

[16] P. van Emde Boas, R. Kaas, and E. Zijlstra. Design and implementation of an efficient priority queue. *Math. Systems Theory*, 10:99–127, 1977.

[17] D. E. Willard. Log-logarithmic worst-case range queries are possible in space $\Theta(N)$. *Info. Proc. Let.*, 17:81–89, 1983.

Decomposing the Boundary of a Nonconvex Polyhedron

BERNARD CHAZELLE and LEONIDAS PALIOS

Department of Computer Science
Princeton University
Princeton, NJ 08544, USA

Abstract. We show that the boundary of a three-dimensional polyhedron with r reflex angles and arbitrary genus can be subdivided into $O(r)$ connected pieces, each of which lies on the boundary of its convex hull. A remarkable feature of this result is that the number of these convex-like pieces is independent of the number of vertices. Furthermore, it is linear in r, which contrasts with a quadratic worst-case lower bound on the number of convex pieces needed to decompose the polyhedron itself. The number of new vertices introduced in the process is $O(n)$. The decomposition can be computed in $O(n + r \log r)$ time.

1 Introduction

Because simple objects usually lead to simpler and faster algorithms, it is often useful to preprocess an arbitrary object and express it in terms of simpler components. In two dimensions, for example, polygon triangulation is a standard preprocessing step in many algorithms ([1], [5], [8], [11], [16], [17]). Similarly, in three dimensions, a polyhedron can be expressed as a collection of convex pieces or tetrahedra in particular (see [2], [4], [6], [15] for discussions on such decompositions). Of course, the size of the decomposition is critical for the application that uses it. Unfortunately, a convex partition of a polyhedron may be of size quadratic in the description size of the polyhedron in the worst case [4], which makes it unattractive from an efficiency point of view. It would be, therefore, of interest to have partitions into a guaranteed small number of simple components.

In this paper, we consider the problem of subdividing the boundary of a nonconvex polyhedron of arbitrary genus into a small number of connected *convex-like* pieces. By convex-like piece, we mean a polyhedral surface which lies entirely on the boundary of its convex hull. Our result is that the boundary of a nonconvex polyhedron that has r reflex angles can be subdivided into no more than $10r - 2$ such pieces. It is interesting to note that the number of pieces is independent of the number of vertices of the polyhedron, and it is linear in r. The algorithm proceeds in two phases. In the first phase, the boundary of the polyhedron is disassembled into at most $2r+2$ pieces whose internal edges are not reflex edges of the polyhedron, and whose intersections with planes normal to a fixed plane Π are collections of chains monotone with respect to a fixed direction. The partition is achieved by cutting first along the reflex edges of the polyhedron, and second along the edges that are incident upon facets whose projections onto the fixed plane Π overlap; figuratively, we cut along the "ridges" and "keels" of the polyhedron. The second phase further splits the pieces into convex-like ones by clipping them with planes normal to Π that pass through the endpoints of the polyhedron's reflex edges. The clipping is carried out in such a way that it introduces only $O(n)$ new vertices. The entire algorithm runs in linear time, provided that the boundary of the given polyhedron has been triangulated. Boundary triangulation takes $O(n + r \log r)$ time.

In two dimensions, the problem is very simple, and admits a linear-time solution that produces the minimum number of polygonal curves into which the boundary of a polygon (possibly with holes) can be cut so that each such curve stays on the boundary of its convex hull. The algorithm first disconnects the boundary of the polygon at its *cusps* (the vertices whose incident edges form an interior angle larger than π), and then breaks each resulting piece in a greedy fashion to enforce the convexity condition. In other words, we start at one end of such a piece and keep walking along it for as long as each encountered edge lies on the convex hull of the subpiece traversed so far, disconnecting it otherwise. The whole process takes linear time. A second algorithm can be obtained by taking the two-dimensional equivalent of the first phase of our algorithm as outlined above. The boundary of the polygon is disassembled at the cusps and at the local extrema with respect to some fixed direction, say, the vertical direction z (Figure 1). It can be proven by induction that a polygon of r cusps has at most $r + 2$ local extrema, which implies that the total number of pieces produced cannot exceed $2r + 2$. This is almost optimal in the worst case, since for any r there is a polygon of r cusps whose boundary cannot be disassembled into fewer than $2r + 1$ pieces each lying on the boundary of its convex hull (Figure 2).

The paper is structured as follows. In Section 2 we introduce our notation and prove a lemma to facilitate the analysis of our algorithm. The algorithm and its complexity analysis are presented in Section 3. Finally, in Section 4 we summarize our results, and discuss some open questions.

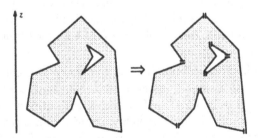

Figure 1

Figure 2

2 A Geometric Framework

A polyhedron in $\Re^3$ is a connected piecewise-linear 3-manifold with boundary. Its boundary is connected and it consists of a collection of relatively open sets, the *faces* of the polyhedron, which are called *vertices*, *edges*, or *facets*, if their affine closures have dimension 0, 1, or 2, respectively. By virtue of the definition of a polyhedron, no two of its faces can be self-intersecting, dangling or abutting. An edge e of a polyhedron is said to be *reflex* if the (interior) dihedral angle formed by its two incident facets exceeds π. By extension, we say that a vertex is *reflex* if it is incident upon at least one reflex edge.

A *patch* of a polyhedron P is a collection of facets or subsets of facets of P with their adjoining edges and vertices. The edges of a patch that do not lie on its relative boundary are called *internal*. Let us try to extend to patches some of the definitions pertaining to polygons. A patch is said to be *connected* if any two points of the patch can be connected by two disjoint (except for their endpoints) paths that lie on the patch. Under this definition, the two patches of Figure 3 are not considered connected. Unless it consists of a single facet, a connected patch has at least one internal edge. A connected patch is said to be *simple* if it is bounded by a single non-intersecting closed curve; otherwise we say that the patch contains holes. A patch is called *monotone* with respect to a plane if no two distinct points of the patch project normally to the same point of the plane. Finally, a patch σ is *convex-like* if it lies on the boundary of the convex hull H_σ of its vertices and the interiors of both P and H_σ lie on the same side with respect to each of the facets of σ. The latter condition implies that none of the internal edges of a convex-like patch is a reflex edge of the given polyhedron. The following lemma gives the necessary conditions for a patch to be convex-like.

Lemma 1. [7] *Let σ be a patch of a polyhedron P none of whose internal edges are reflex edges of P. If σ is simple and monotone with respect to a plane Π onto which it projects into a convex polygon, then σ is convex-like.*

Finally, we introduce the notion of extrema. A point p of a polygon (polyhedron resp.) is called an *extremum* with respect to an oriented line λ, if the intersection of the polygon (polyhedron) with a small enough disk (3-ball resp.) centered at p lies entirely in the one of the two closed half-spaces defined by the hyperplane normal to λ that passes through p. The extrema can be characterized as *negative* or *positive*

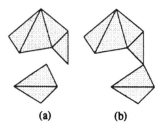

(a)   (b)

Figure 3

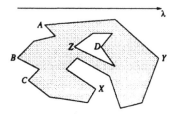

Figure 4

depending on whether the above intersection lies in the nonnegative or nonpositive half-space respectively. For the polygon of Figure 4, for instance, the vertices A, B, C and D are negative extrema with respect to λ, while X and Y are positive extrema. The vertex Z is not an extremum. Clearly, if no edge of the polygon or polyhedron is normal to λ, only vertices can be extrema.

Very often in the following, we consider the intersection of the polyhedron with a plane normal to the x-axis. For reasons of brevity, we call such an intersection as a *yz-cross-section* of the polyhedron.

3   The Decomposition Algorithm

Our goal is to subdivide the boundary of a nonconvex polyhedron P of n vertices and r reflex edges into $O(r)$ connected convex-like patches. The polyhedron is given in any one of the standard representations, e.g., winged-edge (Baumgart [3]), doubly-connected-edge-list (Muller and Preparata [13]), quad-edge (Guibas and Stolfi [9]), so that all the face incidences either are explicitly stored or can be found in linear time. To simplify the description of the algorithm, we assume that no facet of P is perpendicular to the z-axis, and no edge is normal to the x-axis. These assumptions are not restrictive; they can be checked in linear time, and, if necessary, enforced by rotating the system of reference. We also assume that the boundary of the polyhedron P is triangulated. Boundary triangulation can be achieved in $O(n + r \log r)$ time by employing the polygon triangulation algorithm of Hertel and Mehlhorn [10] on each nontriangular facet of P.

The algorithm consists of two phases, each of which splits already existing patches (the entire boundary of P is the initial patch to work on): the first phase guarantees that each of the resulting patches (i) does not have any internal edge that is a reflex edge of P, and (ii) its intersection with a plane normal to the x-axis is a collection of chains monotone with respect to the z-axis. Monotonicity with respect to the xz-plane is achieved in the second phase, where it is also ensured that the projections of the patches onto this plane are convex polygons. The resulting patches are therefore convex-like, by virtue of Lemma 1.

3.1   The First Phase

As mentioned earlier, no internal edge of a convex-like patch can be a reflex edge of P. We therefore need to cut along all the reflex edges. (Note, however, that the internal edges of a patch can all be nonreflex edges of P, and still the patch may not be convex-like. Think of a patch spiraling around several times.) We are now tempted to embark the second phase of the algorithm in which the patches are decomposed by means of cuts along the intersections of the boundary of the polyhedron with planes normal to the x-axis that pass through the reflex vertices of the polyhedron. This, however, will not necessarily produce the desired partition. Consider, for instance, the polyhedron of Figure 5, which is constructed by

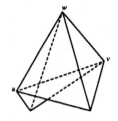

Figure 5

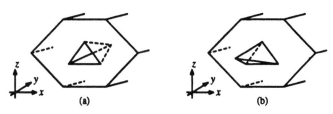

Figure 6

gluing two tetrahedra along a common facet uvw, where u and v are the vertices with the smallest and largest x-coordinates. The above cutting rules imply that the only cut introduced is along the single reflex edge uv. The boundary of the polyhedron will therefore still form a connected patch, which clearly is not convex-like. To rule out such cases, we sweep the polyhedron with a plane normal to the x-axis, and we further disassemble its boundary along the extrema (with respect to the z-axis) of its current cross-section. Recall that no edge of P is normal to the x-axis; as a result, the extrema of the polyhedron with respect to the x-axis are vertices of P. Moreover, no facet of P is normal to the z-axis, implying that the extrema in the cross-section are vertices, the intersections of edges of P with the sweep-plane. (Figure 8 depicts a typical cross-section of the polyhedron after all the cuts have been introduced.) Cutting along an edge has the effect that its incident facets are no longer considered adjacent, although they may still belong to the same patch (Figure 6(a)). Figure 6(b) in turn shows a patch that gets disconnected due to cuts along reflex edges.

Although a direct implementation of the plane-sweep approach requires sorting, we can in fact avoid it. The reason is that determining whether an edge is a reflex edge of P or whether it contributes extrema (with respect to the z-axis) in the corresponding cross-section of the polyhedron requires only information local to the edge, namely the relative position of its two incident facets. We therefore process the edges in any order, cutting along those that satisfy the above criterion. Clearly, the entire phase takes time linear in the number of edges, while no more than linear space is needed.

Figure 7 shows a typical patch at the end of the first phase; its boundary (the edges along cuts) is shown high-lighted. The patch may not be simple, and walking along its boundary may degenerate to a polygonal line traversed in both directions. Moreover, the way cuts are introduced implies that any point p of such a patch σ that projects on the boundary of σ's projection onto the xz-plane belongs to the closure of a reflex edge or an edge of the polyhedron

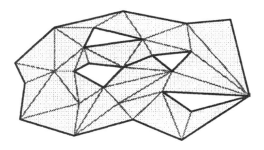

Figure 7

contributing extrema along the z-axis in a yz-cross-section that intersects the edge;

thus, in either case, it belongs to the boundary of the patch σ. The converse is not necessarily true, however. Finally, it can be proven that the number of patches produced at the end of the phase is no more than $2r + 2$ (see [7]).

3.2 The Second Phase

The previous phase produces patches that are not necessarily simple and may form spirals around the y-axis. On the positive side, however, the internal edges of each such patch are nonreflex edges of the polyhedron, and the intersection of such a patch with any plane normal to the x-axis consists of a number of disconnected polygonal lines, each of which is monotone with respect to the z-axis. The latter is not sufficient to ensure that the patch is monotone with respect to the xz-plane; patch monotonicity is guaranteed, if the patch is decomposed into subpatches so that the polygonal lines that form the above intersection belong to different subpatches. This will be our goal in this phase, i.e., to decompose each of the patches produced in the previous phase into subpatches, the intersection of each of which with any plane normal to the x-axis is a single connected polygonal line monotone with respect to the z-axis. Then, making sure that the projections of the subpatches on the xz-plane are convex polygons implies that the final patches are convex-like (Lemma 1).

The method that we are going to use parallels the way a nonconvex polygon (that may contain holes) is partitioned into convex pieces by using cuts parallel to a chosen direction to resolve the polygon's cusps. Namely, each patch is split by means of cuts along the intersections of the patch with planes normal to the x-axis that pass through the reflex vertices of the given polyhedron. Let us consider the intersection of such a plane passing through a reflex vertex v of P with the boundary of P (after the first phase). The intersection consists of several polygons (possibly with holes) whose boundaries are split into polygonal lines monotone with respect to the z-axis. Unless v is a point hole in one of these polygons or a polygon reduced to a point (in which case we do nothing), v is incident upon two of these polygonal lines (Figure 8). If v is the vertex of such a chain with the smallest z-coordinate, we refer to the chain as the *up-chain* of v; otherwise, it is the *down-chain* of v. A reflex vertex may be incident upon two up-chains, two down-chains, or one of each kind (see vertices w, u, v in Figure 8 respectively). It should be noted that these chains associated with a reflex vertex may or may not belong to the same patch; eventually they will not.

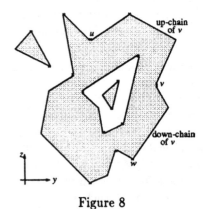

Figure 8

Cutting along the up- and down-chains of each reflex vertex will result in the desired decomposition. For simplicity, we cut along the up- and down-chains in separate passes, so that no cuts are advancing in opposite directions at the same time.

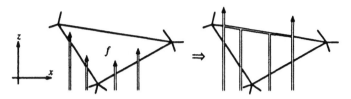

Figure 9

We describe below our method to generate cuts along the up-chains of the reflex vertices. Let us first remark that a brute-force approach may cut on the order of n facets every time we process a reflex vertex, which would produce a decomposition of an unacceptably large $\Omega(nr)$ size. Our plan is to discontinue some of the cuts that might slice too many facets. The key idea is illustrated in Figure 9. Assume that we advance four cuts upwards through facet f. The point is to extend only the leftmost and rightmost ones through adjacent facets upwards past f; we stop the remaining two cuts at the edge incident upon f with the largest x-extent, and we generate a cut along that very edge, which extends between the intersections of the edge with the rightmost and leftmost cuts. Thus, we maintain the following invariant:

> **Cut-invariant:** At most two cuts cross any edge propagating
> through a facet to an adjacent one.

We deal with each patch in turn; we sweep through it, moving upwards, advancing the cuts as we proceed. Each facet of the patch is processed after its adjacent facet(s) below it in the patch have been processed. Recall that if a cut has been generated along an edge during the first phase, the two facets incident upon the edge are no longer considered adjacent (although they may still belong to the same patch). Bearing this in mind, it is clear that some facets have no adjacent facets below them, and so these facets can be processed right away. Figure 10 depicts a patch with these facets highlighted.

We start by inserting these facets in a queue Q intended to store the facets of the

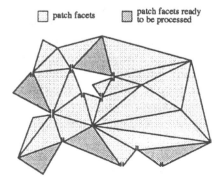

Figure 10

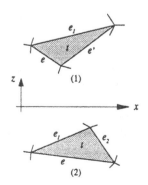

Figure 11

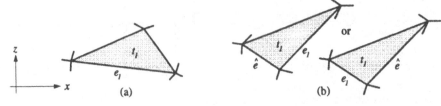

Figure 12

patch that are ready to be processed. Initially, our cut-invariant holds trivially. Then, for each element t of Q we iterate on the following procedure: depending on whether t is as shown in Figure 11(1) or 11(2), we execute step 1 or step 2 respectively.

1. The facet t is as shown in Figure 11(1), i.e., the edge e_1 incident upon t with the largest x-extent is "above" the other two edges e and e' incident upon t. Cuts may be propagating in t through the edges e and e', while an additional cut may emanate from the vertex incident to both e and e'. If the total number of cuts proceeding through t is no more than 2, then the cuts are simply extended all the way through t ready to proceed to adjacent facets, and they are associated with e_1. If, however, the number of cuts is larger than 2, we apply the idea illustrated in Figure 9. We extend only the leftmost and rightmost cuts all the way through t, and associate them with e_1. All remaining cuts are extended up to e_1 and are stopped there, while a cut is generated along e_1 between the intersection points of e_1 and the leftmost and rightmost cuts. So, our cut-invariant is maintained in this way. Finally, if no cut was made along e_1 during the first phase, we check whether the facet t_1, other than t, incident upon e_1 is a candidate for the queue Q. Namely, if t_1 is as in Figure 12(a), then it is inserted in Q; otherwise, t_1 is as in Figure 12(b) and is inserted in Q only if either a cut has been generated during the first phase along the edge $\hat{e}$, or the facet, other than t_1, incident upon $\hat{e}$ has already been processed.

2. The facet t is as shown in Figure 11(2), i.e., the edge e incident upon t with the largest x-extent is "below" the other two edges e_1 and e_2 incident upon t. If cuts are propagating through e, we extend them through t and we associate them with e_1 or e_2 depending on which edge they intersect. By induction, our cut-invariant is maintained. Next, we test the facets t_1 and t_2 (other than t), that are incident upon e_1 and e_2 respectively as candidates for the queue Q. (Note that either one or both t_1 and t_2 may not exist as cuts along e_1 or e_2 during the first phase may have disconnected the patch at these edges.) The procedure is the same as that involving t_1 in the previous case. Specifically, if t_1 is as shown in Figure 12(a), i.e., e_1 is the edge incident upon t_1 with the largest x-extent, we insert t_1 in Q; if not, in which case t_1 is as in Figure 12(b), we insert t_1 in Q only if either we have generated a cut along the edge $\hat{e}$ during the first phase, or the facet, other than t_1, incident upon $\hat{e}$ has already been processed. The same test is also applied to t_2, if it exists.

Then, t is removed from Q and we proceed with the next facet in Q.

Figure 13 shows a snapshot as the procedure is applied to the patch of Figure 10.

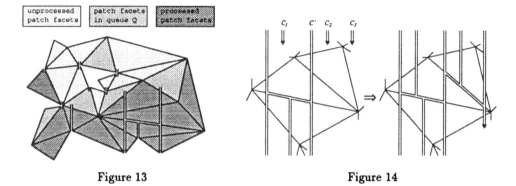

| Figure 13 | Figure 14 |

We can prove inductively that when the queue Q eventually becomes empty, all facets of the patch have been processed. Furthermore, our cut-invariant ensures that no more than two cuts are crossing an edge cutting both its incident facets.

This completes our work with the up-chains. Next we apply the same procedure with respect to the down-chains. Note that we take into consideration any cuts generated during the previous pass involving the up-chains. Consider, for instance, Figure 14. If a cut c_1 along a down-chain reaches an edge along which we generated a cut during the previous pass, then the cut is stopped there and its processing is considered completed. Moreover, if two cuts c_2 and c_3 along two down-chains reach an edge e that has been cut by an up-chain cut c' and c_2 is between c' and c_3 (with respect to the x-axis), then c_2 is stopped there, and a cut between the intersections of e with c' and c_3 is generated along e. This establishes our cut-invariant for cuts along down-chains.

At the end, pieces of a single facet of the polyhedron may belong to several different patches. Interestingly, however, due to our cut-invariant, no more than four cuts through any given facet proceed to adjacent facets both up and down. These would be the leftmost and rightmost cuts through the facet along both up- and down-chains. Any other cuts stop at the edge incident upon the facet with the largest x-extent. The total time spent in this phase is linear in the number of facets of the polyhedron plus a constant overhead per segment of each cut. As the total number of these segments is proven in Section 3.3 to be linear in the size of polyhedron, the total time required for this second phase is $O(n)$.

3.3 Description of the patches produced

Each of the patches produced by the algorithm consists of a portion of the boundary of the polyhedron that is clipped from left and right by two planes normal to the x-axis (cuts along up- or down-chains), and at the top and bottom by either an edge of the polyhedron (a reflex edge, for instance), or a polygonal line consisting of edges that contribute extrema in the cross-section of the polyhedron. Moreover, the patches are simple and monotone with respect to the xz-plane. To prove this, we need only show that the intersection of any patch produced by the second phase with a plane normal to the x-axis is a single polygonal line monotone with respect to the z-axis. Suppose, for contradiction, that this is not true. Then, since the patch is connected, there exist points p and q of the patch that have equal x-coordinates, and are such that a plane normal to the x-axis passing through them intersects the patch in a single polygonal line, whereas it intersects it into two disconnected polygonal lines if we move it slightly either to the left or to the right along the x-axis (to the right, in the case shown in Figure 15). (Note that the points p and q may coincide.) There exist then line segments e_1 and e_2 incident upon p and q respectively, along which we generated cuts (e_1 and e_2 may coincide in the case that p and q coincide). These cuts do not proceed along up- or down-chains of vertices, and thus e_1 and e_2 lie on edges of the polyhedron. Moreover, as the

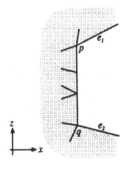

Figure 15

projections of patches onto the xz-plane exhibit internal angles at most equal to π at the projections of new vertices introduced during the second phase (see Figure 9), p and q must be vertices of the polyhedron. In fact, they must both be reflex ones, otherwise e_1 and e_2 would not have been cut. (Recall that e_1 and e_2 would have to contribute extrema along the z-axis at the corresponding yz-cross-sections, which is impossible if p and q are nonreflex vertices as shown in the figure.) This leads to contradiction, since, if this were the case, then cuts along the up-chain of p and the down-chain of q would have split the patch.

In light of Lemma 1, the above discussion and the following lemma establish that the produced patches are convex-like.

Lemma 2. [7] *The projection of any patch on the xz-plane is a convex polygon.*

Description size and total number of patches. Let us consider a polyhedron P of f facets and e edges, r of which are reflex. We compute the total number of edges of all the patches, where edges along the cuts are counted twice. The analysis proceeds in an incremental way by taking into account the new edges that each step of our algorithm introduces. The triangulation of the boundary of P does not affect the order of magnitude of the number of edges, so that this number is $O(e)$ before the beginning of the first phase. As the cuts of the first phase proceed along edges of P, the number of edges of all the patches at the end of the phase is at most twice their number after the boundary triangulation. During the second phase, several new edges are introduced in the following two ways: (i) an edge is split into two when a cut intersects it, or (ii) a new edge is introduced by cutting through a facet. Recall that for each reflex vertex we generate a cut along the intersection of the corresponding patch with a plane normal to the x-axis that passes through that vertex. This cut proceeds towards both directions away from the vertex, and may extend across several facets. Let us treat the two portions of the cut as separate cuts; since the number of reflex vertices does not exceed $2r$, the number of cuts generated during the second phase of our algorithm is at most $4r$.

Our goal is to charge each new edge to either a facet or a cut generated in the second phase. Specifically, each facet is charged with the number of edges that result from cuts traversing it and proceeding through adjacent facets, while each

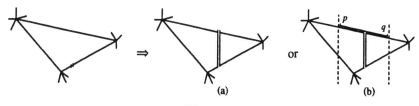

Figure 16

cut is charged with the number of additional edges that its traversal through the very last facet causes. In particular, each facet is charged with 12 units; note that a cut that traverses a facet proceeding to adjacent ones leads to the creation of three new edges (see Figure 16(a)), and that at most 2 such cuts are allowed per facet, for each of the two passes of the second phase (cut-invariant). In a similar fashion, each cut is charged with 3 units to pay for the new edges created in the facet that the cut traverses last. Additionally, a fourth unit is charged to each cut to pay for the extra edge that is created in the case that a cut is generated along a portion of the edge with the largest x-extent incident upon a facet (edge pq in Figure 16(b)). It should be obvious that this charging scheme pays for all the edges introduced during the second phase. Summing everything up, we end up with a total of at most $O(e) + 4 \times 4r + 12f$ edges, which is linear in the size of the input.

We now estimate the number of patches that are finally produced. As mentioned already, the first phase produces no more than $2r + 2$ patches. We partition the set of cuts generated during the second phase into two classes:

(i) those that extend all the way to the boundary of the corresponding patch produced during the first phase, and

(ii) the remaining ones, that is, those that were stopped at an edge that had been cut at least twice (Figure 16(b)).

Each cut in class (i) increases the number of patches by at most 1, whereas each cut in class (ii) increases the number of patches by at most 2 (see Figure 16). As the total number of cuts is no more than $4r$, and the number of cuts in class (ii) is bounded above by $4r - 4$, the total number of patches produced cannot exceed $2r + 2 + 4r + 4r - 4 = 10r - 2$.

4 Conclusion

Our results are summarized in the following theorem.

Theorem 3. *The boundary of a nonconvex polyhedron of n vertices and r reflex edges can be subdivided into 10r − 2 patches, each of which lies on the boundary of its convex hull. The decomposition can be carried out in $O(n + r \log r)$ time and $O(n)$ space.*

Unfortunately, the cuts performed may pass through facets of P; this has the disadvantage of introducing new vertices into the resulting decomposition. It would be of interest, instead, to achieve a boundary decomposition into a small number of convex-like pieces by means of cuts along edges of the given polyhedron only.

A different question is to find an algorithm that produces the minimum number of convex-like pieces. Is this problem NP-complete, as are many optimization questions in partitions and coverings ([12], [14]) ?

References

1. E. Arkin, R. Connelly, and J.S.B. Mitchell, "On Monotone Paths Among Obstacles, with Applications to Planning Assemblies," *Proc. 5th Annual ACM Symposium on Computational Geometry* (1989), 334–343.

2. C.L. Bajaj, and T.K. Dey, "Robust Decompositions of Polyhedra," *Technical Report* (1989), Dept. of Computer Science, Purdue University.

3. B.G. Baumgart, "A Polyhedron Representation for Computer Vision," *Proc. 1975 National Comput. Conference*, AFIPS Conference Proceedings **44** (1975), AFIPS Press, Montvale, NJ, 589–596.

4. B. Chazelle, "Convex Partitions of Polyhedra: A Lower Bound and Worst Case Optimal Algorithm," *SIAM Journal of Computing* **13** (1984), 488–507.

5. B. Chazelle, and L.J. Guibas, "Visibility and Intersection Problems in Plane Geometry," *Discrete and Computational Geometry* **4** (1989), 551–581.

6. B. Chazelle, and L. Palios, "Triangulating a Nonconvex Polytope," *Discrete and Computational Geometry* **5** (1990), 505–526.

7. B. Chazelle, and L. Palios, "Decomposing the Boundary of a Nonconvex Polyhedron," *Technical Report* (1992), Dept. of Computer Science, Princeton University.

8. L.J. Guibas, J. Hershberger, D. Leven, M. Sharir, and R.E. Tarjan, "Linear Time Algorithms for Visibility and Shortest Path Problems Inside Triangulated Simple Polygons," *Algorithmica* **2** (1987), 209–233.

9. L.J. Guibas, and J. Stolfi, "Primitives for the Manipulation of General Subdivisions and the Computation of Voronoi Diagrams," *ACM Transactions on Graphics* **4** (1985), 75–123.

10. S. Hertel, and K. Mehlhorn, "Fast Triangulation of a Simple Polygon," *Lecture Notes on Computer Science* **158** (1983), 207–218.

11. D.G. Kirkpatrick, "Optimal Search in Planar Subdivisions," *SIAM Journal on Computing* **12** (1983), 28–35.

12. A. Lingas, "The Power of Non-Rectilinear Holes," *Lecture Notes in Computer Science* **140** (1982), 369–383.

13. D.E. Muller, and F.P. Preparata, "Finding the Intersection of two Convex Polyhedra," *Theoretical Computer Science* **7** (1978), 217–236.

14. J. O'Rourke, *Art Gallery Theorems and Algorithms*, Oxford University Press (1987).

15. J. Ruppert, and R. Seidel, "On the Difficulty of Tetrahedralizing 3-Dimensional Non-Convex Polyhedra," *Proc. 5th Annual ACM Symposium on Computational Geometry* (1989), 380–392.

16. S. Suri, "A Linear Time Algorithm for Minimum Link Paths inside a Simple Polygon," *Computer Vision, Graphics, and Image Processing* **35** (1986), 99–110.

17. G.T. Toussaint, "On Separating two Simple Polygons by a Single Translation," *Discrete and Computational Geometry* **4** (1989), 265–278.

Convex Polygons Made from Few Lines and Convex Decompositions of Polyhedra

John Hershberger
DEC Systems Research Center

Jack Snoeyink
University of British Columbia

Abstract

We give a worst-case bound of $\Theta(m^{2/3}n^{2/3} + n)$ on the complexity of m convex polygons whose sides come from n lines. The same bound applies to the complexity of the horizon of a segment that intersects m faces in an incrementally-constructed erased arrangement of n lines. We also show that Chazelle's notch-cutting procedure, when applied to a polyhedron with n faces and r reflex dihedral angles, gives a convex decomposition with $\Theta(nr + r^{7/3})$ worst-case complexity.

1 Introduction

Some tasks of spatial solid modeling and computer-aided design (CAD) can be performed more easily on convex polyhedra than on non-convex ones. Examples include intersection, mesh generation, checking separation, and construction from primitives. Thus, a procedure that decomposes a polyhedron into convex pieces can be an important first step in these tasks.

Define the *complexity* of a set of polygons or polyhedra to be the total number of their vertices. By Euler's formula (discussed in any book that covers planar graphs) the number of vertices is linearly related to the numbers of edges and faces. Because the complexity of a decomposition into convex pieces has a direct impact on the difficulty of solving the convex subproblems, one would like a decomposition with small complexity. Finding the decomposition with minimum complexity is NP-hard [15], but in 1984 Chazelle [3] gave a procedure for which the complexity of the decomposition of a polyhedron depends on the number of its faces, n, and the number of its *reflex edges* or *notches*, r—an edge is reflex if its dihedral angle is greater than 180°.

Chazelle's procedure works by cutting notches until only convex pieces remain. We describe it in more detail in Section 3, where we consider the problem:

Problem 1.1 *Given a polyhedron P with n faces and r reflex edges, what is the complexity of Chazelle's decomposition into convex pieces?*

Chazelle's analysis showed that his procedure created $O(r^2)$ pieces with total complexity $O(nr^2)$ in $O(nr^3)$ time. Last year, Dey [8] showed an implementation with running time $O(nr^2 + r^3 \log r)$. He also claimed that the complexity of the decomposition was bounded by $O(nr + r^2 \alpha(r))$. Unfortunately, his complexity analysis is incorrect. Our results in Section 3.1

show that Chazelle's algorithm can produce decompositions with complexity $\Omega(nr + r^{7/3})$. Section 3.2 shows that this is also an upper bound on the complexity.

We prove our bounds by looking at two planar problems that are of independent interest.

Problem 1.2 *What is the total complexity of m convex polygons with disjoint interiors whose sides lie on n lines?*

Problem 1.3 *What is the horizon complexity of a segment that intersects m faces in an "arrangement" constructed in n stages, where stage i adds to the arrangement some of the segments given by intersecting line ℓ_i with the faces of the "arrangement" of stage $i - 1$?*

Problem 1.2 is more general than Problem 1.3; thus, the upper bound of $O(m^{2/3}n^{2/3} + n)$ that we prove for the first problem in Section 2.2 also applies to the second. We have recently learned that Halperin and Sharir [12] have proved a similar bound on the number of turns of m monotone, disjoint, convex chains defined on n lines. Our upper bounds on these problems can also be derived from their theorem. We give matching lower bounds for both problems in Section 2.1.

We can state both problems in terms of erased arrangements; define an *erased arrangement* of n lines to be a convex subdivision whose segments lie on n lines. The complexity of a face in an erased arrangement is the number of its corners. An *incrementally-constructed erased arrangement* is an erased arrangement formed by first ordering the lines $\ell_1, \ell_2, \ldots, \ell_n$ and then adding segments on line ℓ_i in stage i such that each stage is an erased arrangement. Figure 1 shows two erased arrange-

Figure 1: Erased arrangements

ments, of which only the second can be an incrementally-constructed erased arrangement. We also make one observation on erased arrangements.

Observation 1.1 *Let L be a set of lines in general position. A subdivision of the plane induced by segments of L is an erased arrangement of L iff whenever two segments meet they cross or one ends on the other.*

In the setting of erased arrangements, Problem 1.2 asks for complexity of m faces in an erased arrangement. Clarkson et al. [6] proved an $O(m^{2/3}n^{2/3} + n)$ upper bound on the complexity of many faces in an arrangement of lines; our upper bound proof in section 2.2 is based on their proof and our result includes theirs as a special case. Problem 1.3 asks for the complexity of the horizon of a segment in an incrementally-constructed erased arrangement. The complexity of $\Theta(m^{2/3}n^{2/3} + n)$ should be contrasted with the linear complexity of the horizon in a line arrangement [4, 10, 9]. The next section solves Problems 1.2 and 1.3 by analyzing the complexity of faces in an erased arrangement.

2 Erased arrangements

Because each face of an erased arrangement is convex, it is clear that each line of L can contribute at most one segment to its boundary, and so a single face can have maximum complexity n. In the next two subsections, we consider the maximum total complexity of multiple faces in an erased arrangement. It is important to remember that the complexity of a face in an erased arrangement is the number of corners of the face, which may be much smaller than the number of segments that end on the boundary of the face. This is appropriate for the many algorithms that navigate arrangements by walking from corner to corner.

2.1 Lower bounds for many faces

In this section we examine lower bounds on the complexity of many faces and on the horizon complexity of a line in an erased arrangement. A construction of Edelsbrunner and Welzl [11] establishes the the following theorem:

Theorem 2.1 (Edelsbrunner and Welzl [11]) *The complexity of m faces in an arrangement of n lines can be $\Omega(m^{2/3}n^{2/3} + n)$, for $1 \leq m \leq \Theta(n^2)$.*

This immediately implies the same bound for erased arrangements.

We modify their construction to show that for any $m < n$, there is an incrementally-constructed erased arrangement and a single line ℓ such that m consecutive faces along ℓ have complexity $\Omega(m^{2/3}n^{2/3} + n)$. This bound should be contrasted with the horizon complexity of a line in an arrangement of n lines, which is linear in n [4, 10]. The $\Omega(m^{2/3}n^{2/3} + n)$ bound finds application in Section 3.1 where we bound the complexity of a convex decomposition of a polyhedron from below.

In the construction of Edelsbrunner and Welzl, the m high-complexity faces are centrally symmetric and are centered on the points of a Cartesian grid; their diameters can be made less than ϵ for any positive ϵ. Define the *x-projection* of a face f to be the interval defined by the min and max x-coordinates of points inside f. We can rotate the arrangement and scale the high-complexity faces so that the x-projections of all these faces are disjoint and no line of the arrangement is vertical. Let $[x_1, x_1'], \ldots, [x_m, x_m']$ be the x-projections of the m high-complexity faces, and define $x_0 = -\infty$ and $x_{m+1} = \infty$. Translate the arrangement so that the high-complexity faces lie above the line $y = 1$. Choose points $p_1, p_2, \ldots, p_m$ such that p_i lies inside the ith high-complexity face.

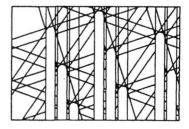

Figure 2: Channels to complex faces

We incrementally construct an erased arrangement on $n + 2m + 1$ lines. The first $2m$ lines that we draw are the vertical lines $x = x_i$ and $x = x_i'$, for $i = 1, \ldots, m$. This creates $2m + 1$ vertical slabs. Then we insert the horizontal line $y = 1$, drawing a segment from x_{i-1}' to x_i for every i, but leaving the slabs $[x_i, x_i']$ free. Now add the lines of the original arrangement in any order, following the rule that for any line ℓ:

1. no segment of ℓ is drawn below $y = 1$,

2. for all i, if ℓ passes below p_i, then no segment of ℓ is drawn in the interval $[x_i, x_i']$, and

3. all other segments are drawn.

Figure 2 shows a clipped window into the essential part of the construction. In Theorem 2.2 we prove that the horizontal line $y = 0$ intersects $2m + 1$ faces with total complexity $\Omega(m^{2/3}n^{2/3} + n)$.

Theorem 2.2 *For any $m \leq n$, there exists an erased arrangement on n lines such that some line intersects $2m + 1$ consecutive faces with total complexity $\Omega(m^{2/3}n^{2/3} + n)$.*

Proof: Incrementally construct the erased arrangement described above. Construction rule 1 guarantees that the line $y = 0$ intersects only $2m$ edges of the erased arrangement. Rules 2 and 3 ensure that the face that $y = 0$ intersects in $x_i < x < x_i'$ is bounded by the two vertical slab boundaries and by the upper edges of the ith high-complexity face of the arrangement of Theorem 2.1. Because the high-complexity faces are centrally symmetric, this includes half of the complexity of the ith high-complexity face. ∎

2.2 The upper bound for many faces

In this section we show that the complexity of m faces in an erased arrangement of n lines is $O(m^{2/3}n^{2/3} + n)$. This bound will also apply to the more restricted cases in which the erased arrangement is incrementally constructed or the m faces lie on a common line.

We begin with a weaker upper bound of $O(mn^{1/2})$ of the type proved by Canham [2, 9]. Then we use the machinery of random sampling to divide the arrangement into pieces that involve fewer lines and invoke the Canham bound on these pieces. This technique was pioneered by Clarkson et al. [6] in their analysis of the complexity of m faces in an arrangement. In our application, we must be careful to capture the face complexity that spills out of the pieces. We recently learned of Halperin and Sharir's theorem [12] bounding the number of turns of m monotone, disjoint, convex chains defined on n lines, which is proved by the same technique.

Lemma 2.3 *The complexity of m faces in an erased arrangement of n lines is $O(mn^{1/2} + n)$.*

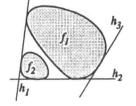

Figure 3: No $K_{2,3}$ subgraph

Proof: Form a bipartite graph G whose vertex sets are the set of m faces and the set of $2n$ halfplanes on the n lines. Join a face f to a halfplane h by an edge if h contains f and contributes a segment to the boundary of f. We bound the complexity of the m faces by bounding the number of edges of G.

Because the faces are convex, if three halfplanes h_1, h_2, and h_3 form part of the boundary of two faces f_1 and f_2, then f_1 and f_2 must intersect; see figure 3. Since all faces are disjoint, it follows that there is no $K_{2,3}$ in G. By a result of extremal graph theory [14], G has at most $O(mn^{1/2} + n)$ edges. ∎

Canham's original proof for line arrangements [2], which does not use extremal graph theory, can also be modified to apply here. It gives a smaller constant in the big-O.

By probabilistic arguments [6, 7], we know that we can choose a sample of $\Theta(r)$ of the lines and decompose its arrangement into r^2 trapezoids, which Clarkson et al. [6] call *funnels,* such that every trapezoid is cut by $O(n/r)$ lines. We will choose $r = O(m^{1/2})$. If we merge the trapezoidation with the erased arrangement, we find that the complexity of our original m faces may be spread over several trapezoids and that each trapezoid may intersect several faces. We will account for the complexity on a per-trapezoid basis.

We distinguish eight types of trapezoid/face interactions. We label each pair (t, f), where t is a trapezoid and f is a face, with one of the following:

type X: The trapezoid t does not intersect the interior of the face f (i.e., $f \cap t = \emptyset$).

type I: The trapezoid t is completely contained in the face f.

type C: One or more corners of trapezoid t lie inside the face f, but (t, f) is not type I.

type 0—type 4: The corners of the trapezoid t lie on or outside the face f, the intersection $f \cap t \neq \emptyset$, and 0, 1, 2, 3, or 4 edges of t intersect the interior of f.

In the next paragraphs, we account for the number of line segments contributed to the total by each type of interaction. Lemma 2.4 shows that the contributions are $O(nr)$ from all types except type 0 and type 1. Lemmas 2.5 and 2.6 show that the contributions of type 0 and type 1 are $O(m(n/r)^{1/2} + nr)$.

Lemma 2.4 *At most $O(nr)$ segments bound faces in the trapezoid/face pairs of types X, I, C, 2, 3, and 4.*

Proof: Let us look at the cases.

type X: If a trapezoid t does not intersect the interior of the face f, then t makes no contribution to f. Therefore, the total type X contribution is zero.

type I, C, 4, or 3: Because faces are disjoint and convex, any given trapezoid t can be involved in pairs of these types with only a constant number of faces. (One face of type I, four of type C, one of type 4 and/or two of type 3.) Furthermore, the contribution to a given face f in t lies on a convex curve involving only segments of the $O(n/r)$ lines that intersect t. Since there are r^2 trapezoids, the total contribution of these types is $O(r^2) \cdot O(n/r) = O(nr)$.

type 2: For any trapezoid t, we can partition the faces that intersect two edges of t into six classes, depending on which pair of edges they intersect. By the argument of Lemma 2.3, no two faces touching the same two edges of the trapezoid can touch the same side of a line ℓ. Thus, line ℓ can contribute to at most two faces of each class and the number of segments contributed by type 2 pairs involving t is $6 \cdot O(n/r)$. The total type 2 contribution is $O(nr)$.

Combining these results proves the lemma. ∎

Lemma 2.5 *The total type 0 contribution is $O(m(n/r)^{1/2} + nr)$.*

Proof: Consider a trapezoid t that is in m_i type 0 pairs, meaning that it completely contains m_i faces. We form an erased arrangement that involves only the $O(n/r)$ lines that cut t by extending the segments inside t that intersect the boundary of t to infinity. (This may cause more intersections, but won't change the arrangement inside t.)

The contribution of type 0 faces in t is $cm_i(n/r)^{1/2} + cn/r$, for some constant c, by Lemma 2.3. Summing the contributions over all trapezoids gives

$$\sum_{i \le r^2} cm_i \left(\frac{n}{r}\right)^{1/2} + c\frac{n}{r} = cnr + c\left(\frac{n}{r}\right)^{1/2} \sum_{i \le r^2} m_i = O(m(n/r)^{1/2} + nr)$$

and establishes the lemma. ∎

Lemma 2.6 *The total type 1 contribution is* $O(m(n/r)^{1/2} + nr)$.

Proof: We prove first that the number of type 1 pairs is $O(m)$, then that their total contribution is $O(m(n/r)^{1/2} + nr)$.

For each face f, form a graph G_f on trapezoids: join the trapezoids t and t' by an edge of G iff t and t' are neighboring trapezoids in the trapezoidation and the interior of f intersects the segment common to t and t'. Since the face f is connected, G_f is connected.

By a slight abuse of terminology, we will say that a trapezoid t is a type k node in G_f if it is in a type k pair (t, f). If t is a type 2 node in G_f, then t has neighbors on only two sides. It may have up to $\Omega(r)$ neighbors

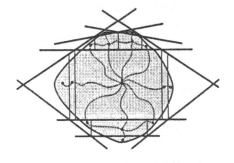

Figure 4: Graph G_f of the shaded face f

on either side, but if a neighbor is of type 1 or 2 in G_f, then it is the only neighbor on its side of the trapezoid t. Thus, we can group the type 2 nodes of G_f into maximal-length chains. A chain is incident to at most one type 1 node on each end.

The graph G_f can now have one of two forms. It can consist of a chain of zero or more type 2 nodes with type 1 nodes on each end. If all faces had such graphs there would be $2m$ type 1 pairs. Alternatively, G_f can contain a type 3, 4, C, or I node. Each node of these types has at most four neighbors of type 1 or 2, so the total number of type 1 nodes in all such graphs is bounded by four times the number of trapezoid/face pairs of types 3, 4, C, and I. Since each trapezoid can be in only a constant number of pairs of types 3, 4, C, and I (see Lemma 2.4), there are $O(r^2)$ such nodes. Combining these bounds shows that the total number of type 1 pairs is $O(m + r^2) = O(m)$.

Now we use the Canham bound, as in the proof of Lemma 2.5, to bound the type 1 contribution along each edge of a trapezoid. Again, consider a trapezoid t that is involved in m_i type 1 pairs, meaning that m_i faces cross one edge of t. We form an erased arrangement without decreasing the complexity of these m_i faces by extending the segments inside t that intersect the boundary of t to infinity. Then Lemma 2.3 states that the contribution from these faces is $O(m_i(n/r)^{1/2} + n/r)$ and summing the contributions over all trapezoids gives $O(m(n/r)^{1/2} + nr)$ as before. ∎

We have shown that the total complexity of m faces is $O(m(n/r)^{1/2} + nr)$. By choosing $r = (m^{2/3}n^{-1/3})$, we balance the two terms of the sum and establish the following upper bound theorem. Notice that $r^2 = m \cdot (m/n^2)^{1/3} = O(m)$ as promised, since $m = O(n^2)$.

Theorem 2.7 *In any erased arrangement of n lines, the total complexity of any set of m faces is $O(m^{2/3}n^{2/3} + n)$.*

3 Polyhedral decomposition

As mentioned in the introduction, Chazelle's procedure gives a decomposition of a polyhedron P whose complexity depends on n, the number of P's faces, and r, the number of P's reflex edges or notches. We begin this section with a brief description of this procedure and prove lower and upper bounds on the decomposition complexity in two subsections.

Chazelle's notch-cutting procedure works in two stages. First, it chooses a *notch plane* π_i for each notch $e_1, e_2, \dots, e_r$ of P, such that π_i contains e_i and forms dihedral angles of at most 180° with the two faces incident to e_i.

Then it considers each notch in some arbitrary order and resolves notch e_i by cuts in π_i. For the first notch e_1, the polyhedron P is cut along the component of $P \cap \pi_1$ that contains e_1. This may break notches into segments called *subnotches*, but no new notches are created. For each successive notch e_i, all subnotches of e_i are resolved as above by cutting simultaneously along π_i in the pieces of P that contain pieces of e_i. Thus, after the procedure considers each of the r notches, only convex pieces remain. We examine the two-dimensional subproblem of cutting along π_i in more detail in Section 3.2.

Notice that this procedure creates additional vertices. If no extra vertices are permitted, then convex decompositions may not exist. Ruppert and Seidel's result [17] implies that determining whether a decomposition using the original vertices exists is NP-complete.

Two other decompositions should be mentioned. Chazelle and Palios [5] gave a procedure that decomposes polyhedra that are homotopic to a ball and have n faces and r reflex edges into convex pieces with total complexity $O(n + r^2)$ in time $O((n + r^2)\log r)$. This complexity is best possible. Their decomposition depends strongly on the topology and does not use notch-cutting; in fact, it is not difficult to see that notch-cutting techniques cannot do better than $\Omega(nr)$.

Paterson and Yao [16] analyzed binary space partitions to show that the complement of n triangles in space could be decomposed into convex pieces with $O(n^2)$ total complexity in $O(n^3)$ time. This gives quadratic-size decompositions of polyhedra as a special case.

3.1 The lower bound on decomposition complexity

We convert the horizon lower bound of Section 2.1 to a polyhedron with n faces and r reflex edges, then give a sequence of cuts such that the complexity of the resulting convex decomposition is $\Omega(nr + r^{7/3})$. Our polyhedron is homotopic to a ball.

Sketch the erased arrangement of r lines described in Section 2.1, for which the line $y = 0$ intersects r faces and has horizon complexity $O(r^{4/3})$, on the xy-plane for use as a template. Imagine the x-axis extending to the right, the y-axis extending away and the z-axis extending upward.

Form a rectangular box $[x_{lo}, x_{hi}] \times [0, y_{max}] \times [0, 3]$, where x_{lo}, x_{hi}, and y_{max} are chosen so that the box covers all the high-complexity cells on the xy-plane. (That is, $x_{lo} < x_1$ and $x_{hi} > x'_r$.) We will cut this box so that the regions in the xy-plane that form the horizon of the line $y = 0$ become vertical prisms. Then, by cutting parallel to the xy-plane from r notches that see this horizon, we will get $\Omega(r^{7/3})$ complexity. Making sure that the r notch planes also intersect n faces adds the $\Omega(nr)$ term.

We create several gadgets, which figure 5 attempts to illustrate. Round off the left near edge of the box by n faces parallel to the z-axis that

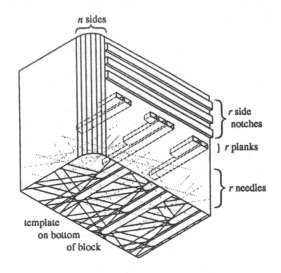

Figure 5: The lower bound construction

form no reflex edge nor intersect the erased arrangement in the projection on the xy-plane.

In the upper third of the box, $2 < z < 3$, place r *side notches* parallel to the x-axis and having rectangular cross-sections. These notches will be removed by horizontal cuts (parallel to the xy-plane).

In the middle third, we place r *planks*. Plank i is formed by subtracting from the box all points with z-coordinates in $[5/4, 7/4]$ whose xy-projection lies in the trapezoid bounded by $y = 0$, $x = x_i$, $x = x'_i$ and the line that joins the points of tangency of the latter two lines with the high-complexity face containing p_i. Each plank contributes 8 reflex edges.

Finally, in the lower third $0 < z < 1$, we place r *needles* above the lines of the template erased arrangement. The needle for the line ℓ is formed by subtracting a long, thin, rectangular box from the outer box. The subtracted box intersects one side of the outer box and comes near the other. The needle has a vertical face (parallel to the z-axis) whose supporting plane intersects the plane $z = 0$ in the line ℓ. The z-coordinate ranges of the needles are different, so no two needles intersect. Each needle contributes 8 reflex edges. (The intersection of two needles would form a hole in the box and would introduce more reflex edges.)

We now make three families of cuts. First we cut with the sides of the planks that are in the planes $x = x_i$ and $x = x'_i$. These cuts form $2m + 1$ vertical slabs. The side notches above the planks and the needles below are all cut, but do not interfere with the formation of the slabs.

Next, on each needle, we cut along the vertical face that is above a line of the template erased arrangement. By the placement of the planks, such a cut extends to the upper third of the ith slab only above the template segments drawn in the xy-plane. Notice that other needles and their cutting planes do not interfere with the propagation of the cut.

When we finish cutting vertically along each needle, the lower third of the box has been decomposed into vertical prisms—one for each face of the full arrangement of the lines of the template. The planks in the middle third stop some cuts, so that in the upper third, the far halves of the high-complexity faces (the side away from $y = 0$) can be seen from the side notches. A cut from a single side notch has $\Omega(n)$ complexity from the near left corner and $\Omega(r^{4/3})$ from the high-complexity faces. Summing over all side notches gives the $\Omega(nr + r^{7/3})$ lower bound. The remainder of the cuts can then be made arbitrarily—they can only increase the complexity.

Thus, we have proved

Theorem 3.1 *There is a polyhedron of genus zero with n faces and r reflex edges, and a sequence of cuts resolving the reflex edges, such that Chazelle's algorithm gives a decomposition into convex pieces with complexity $\Omega(nr + r^{7/3})$.*

3.2 The upper bound on decomposition complexity

In this section we bound the complexity of Chazelle's decomposition algorithm by analyzing the cuts along a notch plane. Let us take a closer look at how the kth notch e_k and its subnotches are resolved.

The set of notch planes induces an arrangement of lines on the notch plane π_k. When a notch e_j, with $j < k$, is resolved by cutting, some segment of $\pi_j \cap \pi_k$ may appear in π_k. These segments end on $P \cap \pi_k$ or on $\pi_i \cap \pi_k$ for some $i < j$, just as they would in an incrementally-constructed erased arrangement. The subdivision is not an erased arrangement, however, because the vertices of $P \cap \pi_k$ do not accord with Observation 1.1. Given a plane polygon Q, we define an *erased subdivision of Q based on l lines* to be a subdivision in which each segment not on the boundary $\partial(Q)$ ends by meeting another segment (possibly of $\partial(Q)$) in a T-junction. It is no longer the case that all faces of an erased subdivision of Q are convex, but the only reflex angles are those of Q.

We can now state and prove the main result of this section, Theorem 3.2.

Theorem 3.2 *Let P be a connected n-gon with r reflex vertices. The complexity of the horizon of a segment e in an erased subdivision of P based on r lines is $O(n + r^{4/3})$.*

In the proof of Theorem 3.2, we use a simple combination lemma:

Lemma 3.3 *A single connected component of the intersection of a connected j-gon and a connected k-gon has $O(j + k)$ sides.*

Proof: Let P and Q be connected, but not necessarily simply-connected, j-gons and k-gons, respectively. Let I be a connected component of $P \cap Q$. The vertices of I come from P, from Q, and from the intersection of edges of P with edges of Q. We call an edge of I *bad* if its endpoints are both vertices of the last type. Clearly, the number of non-bad edges is at most $2j + 2k$.

We count the bad edges that come from edges of Q by considering the face F of the arrangement of $P \cup \{$bad edges from $Q\}$ that contains I. Face F is a connected polygon with boundary cycles $\sigma_1, \sigma_2, \ldots, \sigma_m$ formed by edges of P and bad edges from Q. Since bad edges from Q cannot appear consecutively, we can bound them by the number of edges from P.

Notice, now, that no edge of P appears in more than one boundary cycle. For each cycle σ_i, list the edges of P that appear in sequence—no sequence can have an alternating $\ldots a \ldots b \ldots a \ldots b \ldots$ subsequence. By Davenport-Schinzel sequence bounds [13], the total sequence length is at most $2j$. The number of bad edges that come from edges of Q has the same bound.

A symmetric argument counts bad edges on P and establishes the $O(j + k)$ total bound. ∎

We prove Theorem 3.2 in three steps. First, we find an erased subdivision of P based on a set of $O(r)$ lines such that the horizon complexity of the segment e is contained in $s = O(r)$ convex faces, $F_1, F_2, \ldots, F_s$. Second, we find an erased arrangement of $O(r)$ lines whose intersection with the polygon P contains the faces $F_1, F_2, \ldots, F_s$. Third, we form a horizon polygon Q in this erased arrangement that has $O(r^{4/3})$ complexity by Theorem 2.7. Applying the combination lemma to P and Q obtains the desired $O(n + r^{4/3})$ bound.

To begin, let S be the set of segments of the erased subdivision of P that are not segments of the boundary $\partial(P)$ and let L be the set of r lines that contain S.

Step 1: Inside P, the reflex angles in the subdivision induced by P and S are those of P. To remove them, we process reflex vertices in some order; we extend one of the edges incident to the current vertex until it hits a segment of S or of the boundary $\partial(P)$. Since this cutting introduces no new reflex angles and forms only $O(r)$ new faces, the horizon complexity is now contained in $s = O(r)$ convex faces $F_1, F_2, \ldots, F_s$.

Before we proceed to step 2, forming an erased arrangement of $O(r)$ lines, let us explain why we need it. We currently have s convex faces and we could make them lie in an erased arrangement by extending edges of P through all convex corners into the exterior of P. This, however, would be an erased arrangement of $n + r$ lines, leading to an $O(n^{2/3}r^{2/3} + n)$ bound on the complexity of the horizon. By placing the faces $F_1, F_2, \ldots, F_s$ inside an erased arrangement of $O(r)$ lines and applying the combination lemma, we obtain a tight bound of $O(n + r^{4/3})$.

We should also caution the reader that $\Omega(r^{4/3})$ maximally-connected portions of P can contribute to the F_i's, even though P has only r reflex angles. Thus, an approach that cuts off the contributions of $\partial(P)$ to F_i by a segment in F_i can introduce more than $O(r)$ segments.

Step 2: We apply Wenger's construction [18], which encloses s convex polygons inside convex polygons with a total of $12s$ sides, to the faces $F_1, F_2, \ldots, F_s$. First, enclose the faces in a triangle and triangulate the complement of their interiors inside the triangle. Next, discard triangulation edges, which join pairs of faces, until every region is bounded by three triangulation edges. Triangulation edges are undirected, so the ordering of faces and subscripts in the next sentences does not matter. For each triangulation edge (F_i, F_j), choose a line ℓ_{ij} that separates the interiors of faces F_i and F_j. Then, for all i, form the polygon Q_i by intersecting the halfspace defined by ℓ_{ij} that contains F_i, for all edges (F_i, F_j)

incident to F_i. As Wenger proves, the polygons $Q_1, Q_2, \ldots, Q_s$ have disjoint interiors and have $12s$ edges in total [18].

To convert this into an erased arrangement of $O(r)$ lines, place the non-P edges of F_i inside Q_i. Extend those that ended on P to end on Q_i and extend all edges of Q_i until they end on the boundary of some other polygon Q_k.

Lemma 3.4 *The construction above gives an erased arrangement based on $O(r)$ lines whose intersection with P has the faces $F_1, F_2, \ldots, F_s$.*

Proof: If all lines used are in general position, then each of the segments stops on some other line segment. If they are not in general position, then they can be perturbed without changing the face complexity and points where three of more lines meet can be replaced by constructions like the "vortex" in Figure 1. Thus, the construction gives an erased arrangement.

Since the segments of $Q_1, Q_2, \ldots, Q_s$ and the non-P segments of $F_1, F_2, \ldots, F_s$ are the only ones used, the number of lines in this erased arrangement is $O(r)$.

To argue that the intersection with P contains the faces $F_1, F_2, \ldots, F_s$, we first show that no segment was extended into the interior of F_i. Because F_i is convex, none of its own extended segments intersected its interior; they all stop on Q_i. But this also implies that the extended segments of F_j stop on Q_j, which is disjoint from the interior of Q_i. Finally, the segments of $Q_1, Q_2, \ldots, Q_s$ were all extended in the exterior of $\bigcup_{k \leq s} Q_k$, so they also did not affect the interior of F_i.

All non-P boundary segments of F_i were included in the erased arrangement, however, and the rest are contributed by P in the intersection. ∎

Step 3: Applying Theorem 2.7 to the faces containing $F_1, F_2, \ldots, F_s$ in the erased arrangement constructed in step 2 gives us an $O(r^{4/3})$ bound on the complexity of those faces.

In the construction, moreover, the only segments added inside the original horizon were the cuts through the reflex vertices of P that formed $F_1, F_2, \ldots, F_s$. Removing these cuts from the erased arrangement gives a subdivision (not an erased arrangement) in which the horizon of segment e has $O(r^{4/3})$ complexity spread across the faces incident to e. If we now clip the at most r segments that intersect e just before they touch e and double them, we obtain a polygon that encloses the horizon and has the same complexity bound. Finally, applying the combination lemma gives the desired $O(n + r^{4/3})$ bound and establishes Theorem 3.2.

Because of the discussion at the beginning of this section, this theorem has an immediate corollary.

Corollary 3.5 *The complexity of Chazelle's decomposition is $O(nr + r^{7/3})$.*

Proof: Every vertex in the decomposition is a vertex of the polyhedron P or is introduced by some cut.

The kth cut, in the notch plane π_k, is the horizon of segment e_k in an erased subdivision of $P \cap \pi_k$ based on the set of $k-1$ lines $\pi_j \cap \pi_k$ with $j < k$. Since P has r reflex edges, $P \cap \pi_k$ has at most r reflex vertices and Theorem 3.2 gives a bound of $O(n + r^{4/3})$ for the complexity of the cut. Summing over all cuts gives $O(nr + r^{7/3})$ total complexity. ∎

References

[1] R. J. Canham. A theorem on arrangements of lines in the plane. *Israel J. Math.*, 7:393–397, 1969.

[2] B. Chazelle. Convex partitions of polyhedra: A lower bound and worst-case optimal algorithm. *SIAM J. Comp.*, 13(3):488–507, 1984.

[3] B. Chazelle, L. J. Guibas, and D. T. Lee. The power of geometric duality. *BIT*, 25:76–90, 1985.

[4] B. Chazelle and L. Palios. Triangulating a nonconvex polytope. *Disc. & Comp. Geom.*, 5:505–526, 1990.

[5] K. L. Clarkson, H. Edelsbrunner, L. Guibas, M. Sharir, and E. Welzl. Combinatorial complexity bounds for arrangements of curves and spheres. *Disc. & Comp. Geom.*, 5:99–160, 1990.

[6] K. L. Clarkson and P. W. Shor. Applications of random sampling in computational geometry, II. *Disc. & Comp. Geom.*, 4:387–421, 1989.

[7] T. K. Dey. Triangulation and CSG representation of polyhedra with arbitrary genus. In *Proc. 7th Ann. ACM Symp. Comp. Geom.*, pages 364–372, 1991.

[8] H. Edelsbrunner. *Algorithms in Combinatorial Geometry*. Springer-Verlag, Berlin, 1987.

[9] H. Edelsbrunner, J. O'Rourke, and R. Seidel. Constructing arrangements of lines and hyperplanes with applications. *SIAM J. Comp.*, 15:341–363, 1986.

[10] H. Edelsbrunner and E. Welzl. On the maximal number of edges of many faces in an arrangement. *J. Comb. Theory, A*, 41:159–166, 1986.

[11] D. Halperin and M. Sharir. On disjoint concave chains in arrangements of pseudolines. *Info. Proc. Let.*, 40:189–192, Nov. 1991.

[12] S. Hart and M. Sharir. Nonlinearity of Davenport-Schinzel sequences and of generalized path compression schemes. *Combinatorica*, 6:151–177, 1986.

[13] T. Kővári, V. T. Sós, and P. Turan. On a problem of K. Zarankeiwicz. *Colloq. Math.*, 3:50–57, 1954.

[14] A. Lingas. On the power of non-rectilinear holes. In *Proc. 9th ICALP*, volume 140 of *LNCS*, pages 369–383, Berlin, 1982. Springer-Verlag.

[15] M. S. Paterson and F. F. Yao. Efficient binary space partitions for hidden-surface removal and solid modelling. *Disc. & Comp. Geom.*, 5:485–503, 1990.

[16] J. Ruppert and R. Seidel. On the difficulty of tetrahedralizing 3-dimensional non-convex polyhedra. In *Proc. 5th Ann. ACM Symp. Comp. Geom.*, pages 380–392, 1989.

[17] R. Wenger. Upper bounds on geometric permutations for convex sets. *Disc. & Comp. Geom.*, 5:27–33, 1990.

Maintaining the Visibility Map of Spheres while Moving the Viewpoint on a Circle at Infinity

Hans-Peter Lenhof   Michiel Smid*

Max-Planck-Institut für Informatik

W-6600 Saarbrücken, Germany

Abstract

We investigate 3D visibility problems for scenes that consist of n non-intersecting spheres. The viewing point v moves on a flightpath that is part of a "circle at infinity" given by a plane P and a range of angles $\{\alpha(t)|t \in [0:1]\} \subset [0:2\pi]$. At "time" t, the lines of sight are parallel to the ray $r(t)$ in the plane P, which starts in the origin of P and represents the angle $\alpha(t)$ (orthographic views of the scene). We describe algorithms that compute the visibility graph at the start of the flight, all time parameters t at which the topology of the scene changes, and the corresponding topology changes. We present an algorithm with running time $O((n + k + p)\log n)$, where n is the number of spheres in the scene; p is the number of transparent topology changes (the number of different scene topologies visible along the flightpath, assuming that all spheres are transparent); and k denotes the number of vertices (conflicts) which are in the (transparent) visibility graph at the start and do not disappear during the flight.

1 Introduction

In this paper we investigate a dynamic 3D visibility problem, where the viewing position moves on a circular arc around the origin. We consider a scene, that consists of n non-intersecting spheres $s_1, \cdots, s_n$. The flightpath f is a part of a "circle at infinity" given by a plane P and a range of angles. The range of angles $\{\alpha(t)|t \in [0:1]\} \subset [0:2\pi]$ is parametrized by a "time" parameter $t \in [0:1]$. Here, $\alpha(t)$ is a monotonically increasing function. At time t, the lines of sight are parallel to the ray $r(t)$ in the plane P, which starts in the origin of P and represents the angle $\alpha(t)$ (see Figure 1).

We describe an algorithm which computes the visibility graph at the start of the flightpath and time parameters t, at which the topology of the scene changes,

---

*This author was supported by the ESPRIT II Basic Research Actions Program, under contract No. 3075 (project ALCOM).

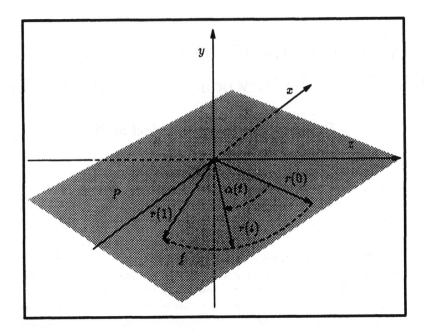

Figure 1: Flightpath f, the plane P and the range of angles $[\alpha(0) : \alpha(1)]$.

together with the corresponding topology changes. The algorithm has a running time of $O((n + k + p)\log n)$, where p is the number of transparent topology changes (the number of different scene topologies visible along the flightpath, assuming that all spheres are transparent); and k denotes the number of intersections (conflicts) which are in the transparent visibility graph at the start of the flightpath and do not disappear during the flight. Note that $0 \le k \le n^2$ and $0 \le p \le n^3$. In the transparent visibility graph we store not only the visible, but also the hidden edges and vertices.

Visibility problems of this kind arise in the field of molecular graphics, when, for example, molecules are rotated. In this application the atoms of the molecules are represented by spheres, the so called van der Waals hulls of the atoms.

A short overview about similar problems concerning polygonal scenes can be found in [1]. In that paper, Bern et al. present algorithms having running times $O((n^2 + p)\log n)$, $O(n^2 \log n + p)$ and $O(n^2 + p \log n)$, where n is the total number of edges of the polygonal objects in the scene. For terrains, they obtain a running time of $O((n + l)\lambda_3(n)\log n)$, where l is the number of opaque topology changes. For terrains and vertical flightpaths, they get a running time of $O(n \lambda_4(n)\log n)$. The latter result was also obtained by Cole and Sharir [2].

Other results and methods can be found in [6] and [4]. In [6], Plantinga and Dyer use "aspect graphs" to solve visibility problems. Mulmuley [4] gives algorithms for hidden surface removal for a polygonal scene with respect to a moving viewpoint.

His first algorithm preprocesses the set A of scene polygons in time $O(n_r(A) \log n + n(\log n)^2)$ and builds a cylindrical partition of $\mathbf{R}^3$ of size $O(n_r(A) + n)$. Here, $n_r(A)$ denotes the number of "regular crossings", which is smaller than n^2. Given any viewpoint $v \in \mathbf{R}^3$, the scene visible from v can be generated in $O((\log n)^2 + l \log n)$ time, where l is the size of the "fictitious" scene. Mulmuley's second algorithm computes all k_s "semi-opaque" topological changes between successive scenes in case the viewpoint is moving on a linear flightpath. The algorithm takes $O(k_s \log n + n^2 \alpha(n) \log n)$ time, where n is the number of edges of the polygons in the scene.

As far as we know, the present paper is the first one that considers these problems for spheres.

In Section 2 we show what kind of events change the topology of the scene. We describe in Section 3, how the spheres (resp. their centers) move in the projection plane during the flight. In Section 4 we give the algorithm for computing the events that cause changes in the topology of the scene and analyze its running time. Section 5 contains some concluding remarks.

2 Transparent topology changes

We consider a scene in 3-space, that consists of n non-intersecting spheres $s_1, \cdots, s_n$ with centers $M_1, \cdots, M_n$. The flightpath f is a "circle at infinity" given by a plane P and a range of angles $\{\alpha(t)|t \in [0:1]\} \subset [0:2\pi]$. The orbit f is parametrized by the time parameter $t \in [0:1]$. At time t, the scene is projected on a plane $\mathcal{P}_t$, which is orthogonal to the ray $r(t)$. This ray $r(t)$, which represents the angle $\alpha(t)$, is contained in the plane P and starts at the origin of P. We assume wlog that (1) the origin of the object space is always projected in the origin of $\mathcal{P}_t$ and (2) the intersection of P and $\mathcal{P}_t$ is the x-axis of plane $\mathcal{P}_t$. We investigate the topology changes in the planar graph G_t, which represents the projection of the scene in $\mathcal{P}_t$ at time t. The vertices of G_t represent all visible and non-visible intersection points of the circles in $\mathcal{P}_t$. These circles are the images of the spheres under the projection in $\mathcal{P}_t$. The image c_i^t of the sphere s_i in plane $\mathcal{P}_t$ is called the *circle of* s_i at time t. The edges of the graph G_t represent the circular arcs in $\mathcal{P}_t$. A transparent topology change occurs at time t, if the graph G_t changes, that is, for all sufficiently small $\epsilon > 0$, $G_{t-\epsilon}$ and $G_{t+\epsilon}$ are non-isomorphic graphs.

Lemma 1 *A transparent topology change occurs at time t if and only if*

(1) there are two circles c_i^t and c_j^t, which touch at time t and do not touch at times $t - \epsilon$ and $t + \epsilon$ for all sufficiently small $\epsilon > 0$, or

(2) there are three circles c_i^t, c_j^t and c_k^t, which intersect in one point at time t, or

(3) there are two circles c_i^t and c_j^t, which are identical at time t.

The proof of Lemma 1 is obvious. We refer to topology changes being of type (1), (2) or (3), respectively, according to their classification in Lemma 1.

Instead of moving the viewing point, we can also rotate the scene around a rotation axis and keep the viewing point fixed. We transform the whole scene in $O(n)$ time, such that (1) the viewing point is at $z = +\infty$, i.e., the projection plane $\wp$ is parallel to the xy-plane, (2) the rotation axis is the y-axis, (3) the origin of the object space is projected in the origin of the projection plane and (4) the xz-plane of the object space is projected in the x-axis of the projection plane $\wp$.

Let $A := \alpha(1) - \alpha(0)$, i.e., A is the total angle by which the spheres are rotated. Instead of considering the original problem, we investigate the transformed rotation problem. We compute the graph G_0 at the start of the rotation and all topology changes in the projection plane $\wp$ that occur when the spheres are rotated by the angle A around the y-axis.

3   Conflicts between circles

For each $i \in \{1, \cdots, n\}$ we denote the radius of s_i by r_i and the center of s_i at time t by $M_i(t) = (x_i(t), y_i(t), z_i(t))$. Then, $m_i(t) := (x_i(t), y_i(t))$ is the image of $M_i(t)$ under the projection at time t.

In order to compute all topology changes of type (1), we determine all conflicts between circles in the projection plane. To be more precise, we say that two spheres s_i and s_j *conflict* at time t, if $c_i^t \cap c_j^t \neq \emptyset$, or c_i^t is contained in the interior of c_j^t (or vice versa). Two spheres s_i and s_j have a conflict, if there is a time parameter $t \in [0:1]$, such that s_i and s_j conflict at time t.

We determine all pairs s_i, s_j of spheres that have a conflict. For each such pair, we determine all time parameters t at which they conflict and the corresponding parametrized curves

$$c_{ij}^l(t) := \begin{cases} undef & \text{if } c_i^t \cap c_j^t = \emptyset \text{ or } c_i^t \cap c_j^t = c_i^t \\ c_i^t \cap c_j^t & \text{if } c_i^t \cap c_j^t \text{ is one single point} \\ p_l & \text{if } c_i^t \cap c_j^t = \{p_l, p_r\} \text{ and} \\ & p_l \text{ lies to the left of } \overrightarrow{m_i(t)m_j(t)} \end{cases}$$

$$c_{ij}^r(t) := \begin{cases} undef & \text{if } c_i^t \cap c_j^t = \emptyset \text{ or } c_i^t \cap c_j^t = c_i^t \\ c_i^t \cap c_j^t & \text{if } c_i^t \cap c_j^t \text{ is one single point} \\ p_r & \text{if } c_i^t \cap c_j^t = \{p_l, p_r\} \text{ and} \\ & p_r \text{ lies to the right of } \overrightarrow{m_i(t)m_j(t)}. \end{cases}$$

The curves $c_{ij}^l(t)$ and $c_{ij}^r(t)$ describe the movements of the intersection points of c_i^t and c_j^t during the rotation.

If $A \leq 2\pi$, the set of all time parameters t at which s_i and s_j have a conflict consists of at most three subintervals of $[0:1]$. This set of parameters and the

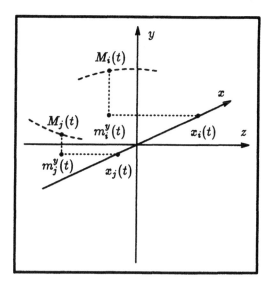

Figure 2: The orbits of $m_i^y(t)$ and $m_j^y(t)$ in the xz-plane.

corresponding curves can be computed in constant time for each pair s_i, s_j. (See below.)

How can we determine all pairs s_i, s_j of spheres having a conflict? The circles c_i^t and c_j^t have a conflict if and only if $|m_i(t) - m_j(t)| \leq r_i + r_j$. Thus, we have to consider the orbits of the centers in the projection plane. Since $y_i(t) = y_i$ and $y_j(t) = y_j$ for all t, the orbits $O_i = \cup_{0 \leq t \leq 1} m_i(t)$ and $O_j = \cup_{0 \leq t \leq 1} m_j(t)$ are line segments that are parallel to the x-axis in the projection plane. In order to determine if the corresponding circles have a conflict, we have to determine

$$\min_{0 \leq t \leq 1} |m_i(t) - m_j(t)|,$$

the minimal distance between $m_i(t)$ and $m_j(t)$ over all times. If this minimum is less than or equal to $r_i + r_j$, then c_i^t and c_j^t have a conflict.

Since $y_i(t) - y_j(t) = y_i - y_j$ is constant and

$$\min_{0 \leq t \leq 1} |m_i(t) - m_j(t)| = \sqrt{(y_i - y_j)^2 + (\min_{0 \leq t \leq 1} |x_i(t) - x_j(t)|)^2},$$

it suffices to consider

$$\min_{0 \leq t \leq 1} |x_i(t) - x_j(t)|,$$

the minimal distance of the x-coordinates of $m_i(t)$ and $m_j(t)$ at any time.

We consider the orbits of $m_i^y(t) := (x_i(t), z_i(t))$ and $m_j^y(t) := (x_j(t), z_j(t))$ of the centers $M_i(t)$ and $M_j(t)$ under the parallel projection in the xz-plane. (See Figure 2.)

Let $d_i = |m_i^y(0)|$, $d_j = |m_j^y(0)|$, $\phi_i = \arccos(x_i(0)/d_i)$ and $\phi_j = \arccos(x_j(0)/d_j)$. We consider the x-coordinates

$$x_i(t) = x_i(\alpha(t)) = d_i \cos(\alpha(t) + \phi_i) \text{ and } x_j(t) = x_j(\alpha(t)) = d_j \cos(\alpha(t) + \phi_j)$$

as a function of the angle $\alpha(t)$. Moreover, we define $d_{ij}(\alpha(t))$ as the difference of the x-coordinates of $m_i(t)$ and $m_j(t)$. We have

$$
\begin{aligned}
d_{ij}(\alpha(t)) = \\
= \quad & x_i(\alpha(t)) - x_j(\alpha(t)) = d_i \cos(\alpha(t) + \phi_i) - d_j \cos(\alpha(t) + \phi_j) \\
= \quad & d_i \cos(\alpha(t)) \cos(\phi_i) - d_i \sin(\alpha(t)) \sin(\phi_i) \\
& -d_j \cos(\alpha(t)) \cos(\phi_j) + d_j \sin(\alpha(t)) \sin(\phi_j) \\
= \quad & [d_i \cos(\phi_i) - d_j \cos(\phi_j)] \cos(\alpha(t)) - [d_i \sin(\phi_i) - d_j \sin(\phi_j)] \sin(\alpha(t)) \\
= \quad & \sqrt{(d_i \cos(\phi_i) - d_j \cos(\phi_j))^2 + (d_i \sin(\phi_i) - d_j \sin(\phi_j))^2} \cdot \sin(\alpha(t) + \varphi), \quad (1)
\end{aligned}
$$

where

$$\tan(\varphi) = -\frac{d_i \cos(\phi_i) - d_j \cos(\phi_j)}{d_i \sin(\phi_i) - d_j \sin(\phi_j)}.$$

We discuss $|d_{ij}(\alpha(t))|$ as a function of $\alpha(t)$, where $0 \leq \alpha(t) < 2\pi$. The function has its minimum (which is zero) for $\alpha(t) = -\varphi$ and $\alpha(t) = -\varphi + \pi$, and its maximum for $\alpha(t) = -\varphi + \pi/2$ and $\alpha(t) = -\varphi + 3\pi/2$. The function $|d_{ij}(\alpha(t))|$ increases from $\alpha(t) = -\varphi$ to $\alpha(t) = -\varphi + \pi/2$, and from $\alpha(t) = -\varphi + \pi$ to $\alpha(t) = -\varphi + 3\pi/2$. It decreases from $\alpha(t) = -\varphi + \pi/2$ to $\alpha(t) = -\varphi + \pi$, and from $\alpha(t) = -\varphi + 3\pi/2$ to $\alpha(t) = -\varphi + 2\pi = -\varphi$.

The following lemma is an immediate consequence of this monotonicity property.

Lemma 2 *The value of $\min_{0 \leq t \leq 1} |d_{ij}(\alpha(t))|$ is equal to $|d_{ij}(\alpha(0))|$, $|d_{ij}(\alpha(1))|$ or 0.*

It follows from Lemma 2, that we can find all conflicts between circles in the projection plane in the following way: First, we compute all conflicts in the graphs G_0 and G_1. Then we search all pairs s_i, s_j of spheres with

$$\min_{0 \leq t \leq 1} |d_{ij}(\alpha(t))| = 0 \text{ and } |y_i - y_j| \leq r_i + r_j.$$

Lemma 3 *Suppose the rotation angle $A = \alpha(1) - \alpha(0)$ satisfies $A \leq \pi$. Then, $\min_{0 \leq t \leq 1} |d_{ij}(\alpha(t))| = 0$, if and only if one of the two following conditions holds:*

(1) $x_i(0) \leq x_j(0)$ and $x_i(1) \geq x_j(1)$

(2) $x_i(0) \geq x_j(0)$ and $x_i(1) \leq x_j(1)$.

Proof: From Equation (1), we know that the relative order of the x-coordinates of $m_i(t)$ and $m_j(t)$ changes only at the (at most 2) moments when these x-coordinates become equal. $\square$

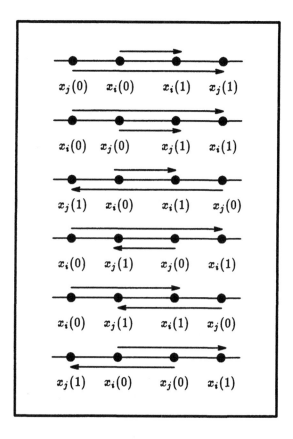

Figure 3: Situations where $\min_{0 \le t \le 1} |d_{ij}(\alpha(t))| = 0$.

In Figure 3, we characterize the different situations where $\min_{0 \le t \le 1} |d_{ij}(\alpha(t))| = 0$. We assume that the four numbers are different and consider the case that $x_i(0) < x_i(1)$. The case $x_i(0) > x_i(1)$ gives symmetric events.

Lemma 3 tells us that $\min_{0 \le t \le 1} |d_{ij}(\alpha(t))| = 0$ if and only if one of the following conditions holds

(a) $x_i(0) = x_i(1)$ and $x_i(0)$ is contained in the interval with endpoints $x_j(0), x_j(1)$,

(b) $x_i(0) < x_i(1)$ and
   $[(x_j(0) \le x_j(1)$ and $([x_i(0) : x_i(1)]$ is contained in, or contains $[x_j(0) : x_j(1)]))$
   or
   $(x_j(0) > x_j(1)$ and
   $(x_i(0) \in [x_j(1) : x_j(0)]$ or $x_i(1) \in [x_j(1) : x_j(0)]$ or $[x_j(1) : x_j(0)] \subset [x_i(0) : x_i(1)]))]$,

(c) $x_i(0) > x_i(1)$ and
$[(x_j(0) > x_j(1)$ and $([x_i(1) : x_i(0)]$ is contained in, or contains $[x_j(1) : x_j(0)]]))$
or
$(x_j(0) \leq x_j(1)$ and
$(x_i(0) \in [x_j(0) : x_j(1)]$ or $x_i(1) \in [x_j(0) : x_j(1)]$ or $[x_j(0) : x_j(1)] \subset [x_i(1) :$
$x_i(0)]]))]$.

(Note that these three cases are mutually exclusive.) Thus, the problem to compute for a given s_i all s_j with $\min_{0 \leq t \leq 1} |d_{ij}(\alpha(t))| = 0$ is reduced to a few simple interval queries (note that points are degenerate intervals). These interval queries can be answered efficiently using data structures that are based on priority search trees.

Theorem 1 ([3]) *Let S be a set of n intervals on the real line. There exists a dynamic data structure $T(S)$ storing S, such that we can insert intervals into S and delete intervals from S in time $O(\log n)$. The data structure has size $O(n)$. Given a query interval we can enumerate*

- *all s intervals $[a : b] \in S$ containing the query interval in time $O(\log n + s)$*

- *all s intervals $[a : b] \in S$ contained in the query interval in time $O(\log n + s)$.*

See McCreight [3] for a proof of Theorem 1. The structure $T(S)$ is called the *ic-structure* of S. Note that $T(S)$ is based on two priority search trees.

4 Determining all topology changes

In this section we give the algorithm that computes all topology changes in the scene, when the spheres in the object space are rotated by an angle $A \leq \pi$. The algorithm consists of three steps.

STEP 1: We compute all conflicts at $t = 0$ and $t = 1$. This can be done using a slightly modified version of the standard plane sweep algorithm. (See Nurmi [5].)

For every pair s_i and s_j of spheres, which have a conflict in G_0 or G_1, we test, whether and when a topology change of type (1) or (3) arises during the rotation. We store all these topology changes in an event queue Q_e sorted in order of increasing time parameter t.

STEP 2: We search for all pairs s_i, s_j such that

$$\min_{0 \leq t \leq 1} |d_{ij}(\alpha(t))| = 0 \text{ and } |y_i - y_j| \leq r_i + r_j.$$

Note that we can find pairs of spheres, which we have already detected in STEP 1. We consider the points $m_i^- := (x_i(0), y_i - r_i)$ and $m_i^+ := (x_i(0), y_i + r_i), 1 \leq i \leq n$. We sort these points in order of increasing y-coordinates. We do a sweep from $y = -\infty$ to $y = \infty$. Each time, we reach one of these points, the sweep line stops. We now explain what has to be done when we stop at point p.

Assume $p = m_i^-$. At this moment, the set S_{lr} of intervals $[x_j(0) : x_j(1)]$ (movement from left to right) with $y_j - r_j \le y_i - r_i \le y_j + r_j$ is stored in an ic-structure $T(S_{lr})$ and the set S_{rl} of intervals $[x_j(1) : x_j(0)]$ (movement from right to left) with $y_j - r_j \le y_i - r_i \le y_j + r_j$ is stored in an ic-structure $T(S_{rl})$.

(a) If $x_i(0) = x_i(1)$, we search for all intervals in $T(S_{lr})$ and $T(S_{rl})$ which contain $x_i(0)$. Then, we insert the interval $[x_i(0) : x_i(1)]$ in the tree $T(S_{lr})$.

(b) If $x_i(0) < x_i(1)$, we search for all intervals $[x_j(0) : x_j(1)]$ in $T(S_{lr})$ which are contained in, or contain the interval $[x_i(0) : x_i(1)]$, and for all intervals $[x_j(1) : x_j(0)]$ in $T(S_{rl})$ which contain $x_i(0)$ or $x_i(1)$, or are contained in $[x_i(0) : x_i(1)]$. Then, we insert the interval $[x_i(0) : x_i(1)]$ in $T(S_{lr})$.

(c) If $x_i(0) > x_i(1)$, we search for all intervals $[x_j(0) : x_j(1)]$ in $T(S_{lr})$ which contain $x_i(0)$ or $x_i(1)$, or are contained in $[x_i(1) : x_i(0)]$, and for all intervals $[x_j(1) : x_j(0)]$ in $T(S_{rl})$ which are contained in, or contain $[x_i(1) : x_i(0)]$. Then, we insert the interval $[x_i(1) : x_i(0)]$ in $T(S_{rl})$.

For every pair s_i, s_j we find, we test whether we have already detected it in STEP 1. If this is not the case, we compute when topology changes of type (1) and (3) take place for this pair. All these topology changes are stored in the event queue Q_e.

Assume $p = m_i^+$. Then, we delete the interval with endpoints $x_i(0), x_i(1)$ from the corresponding ic-structure. More precisely, if $x_i(0) \le x_i(1)$, we delete the interval $[x_i(0) : x_i(1)]$ from $T(S_{lr})$. If $x_i(1) < x_i(0)$, then we remove $[x_i(1) : x_i(0)]$ from $T(S_{rl})$.

STEP 3: Finally we determine all topology changes of type (2). For each sphere s_i which has at least one conflict, we do a sweep from $t = 0$ to $t = 1$. During this sweep we only consider the set C_i of spheres which have a conflict with s_i. This set C_i is obtained from Q_e.

During the sweep, we maintain a balanced binary search tree B, which contains the intersection curves $c_{ij}^*(t), * \in \{l, r\}$ defined at time t (i.e., $c_{ij}^*(t)$ is not equal to $undef$). At time t, the points $c_{ij}^*(t)$ in the current version of the tree B are sorted in order of increasing angle $\delta_{ij}^*(t) := \arccos((x_{coord}(c_{ij}^*(t)) - x_i(t))/r_i)$. In the event queue Q_s of the sweep we store all time parameters t when (1) a curve $c_{ij}^*(t)$ changes from undefined to defined or vice versa, (2) $c_i^t \cap c_j^t$ becomes equal to c_i^t for some sphere $s_j \in C_i$, or (3) $c_{ij}^a(t) = c_{ik}^b(t), a, b \in \{l, r\}$, for $j \ne k$.

When we stop during the sweep, in order to insert or delete a curve $c_{ij}^*(t)$ from the tree B or to swap two curves' positions, we have to compute all intersection points of the new neighbors in B. This can be done in constant time for every pair of new neighbors $c_{ij}^a(t), c_{ik}^b(t), a, b \in \{l, r\}$. The equations of the intersection points, which can be found in the appendix of the full version, imply that there are only a constant number of parameters t, where $c_{ij}^a(t) = c_{ik}^b(t)$. If $c_{ij}^a(t) = c_{ik}^b(t)$, then we have found a parameter t where three circles c_i^t, c_j^t, c_k^t intersect in one point. If $c_i^t = c_j^t$, then every point $c_{ik}^t(t)$ with $k \ne j$, which is in B at time t, is a point, where three circles c_i^t, c_j^t, c_k^t intersect. All these parameters t and the corresponding topology changes are stored in the event queue Q_e.

As mentioned above, we carry out this procedure for every sphere $s_i, i = 1, \cdots, n$, which has at least one conflict with another sphere.

We now analyze the running time of the algorithm. Let k be the number of conflicts between the circles $c_1^0, \cdots, c_n^0$ in the graph G_0, which do not disappear during the rotation. Let p_1 be the number of transparent topology changes which we find in STEP 1. The integer p_2 is defined as the number of transparent topology changes we find in STEP 2, which we have not found in STEP 1. Finally, p_3 is the number of transparent topology changes of type (2). Then, $p := p_1 + p_2 + p_3$ is the total number of topology changes during the rotation.

The conflicts in the graphs G_0 and G_1 can be computed in $O((n + k + p_1) \log n)$ time using the plane sweep algorithm of [5]. Note that we determine both visible and occluded intersections. The sweep in STEP 2 can be done in $O(n \log n + k + p_1 + p_2)$ time. Since intersections between two curves $c_{ij}^a(t), c_{ik}^b(t), a, b \in \{l, r\}$ can be computed in constant time, the entire STEP 3 can be done in $O((k + p_1 + p_2 + p_3) \log n)$ time. Thus, we get a running time of $O((n + k + p) \log n)$ for the entire algorithm.

Since an arbitrary angle $A \leq 2\pi$ can be partitioned into 2 angles that are at most π, we have obtained the following result:

Theorem 2 *Let S be a scene, which consists of n non-intersecting spheres. All transparent topology changes for a flightpath, which is a part of a "circle at infinity" given by a plane P and a range of angles, can be computed in time $O((n + k + p) \log n)$. Here, p is the number of transparent topology changes and k is the number of transparent conflicts at the start, which do not disappear during the flight.*

5   Concluding remarks

We have given an algorithm for maintaining the visibility map of a collection of non-intersecting spheres when the viewpoint moves on a circle at infinity. The algorithm has a running time of $O((n + k + p) \log n)$.

Of course, one open problem is to improve our time bound. First, it would be interesting to find an algorithm that spends only constant time for each topology change, instead of $O(\log n)$.

Second, it would be interesting to develop algorithms whose running times depend on the number of *opaque* topology changes, instead of the number of transparent changes.

Finally, it is not known whether our techniques can be applied to similar problems for other objects besides spheres.

References

[1] M. Bern, D. Dobkin, D. Eppstein and R. Grossmann. *Visibility with a Moving Point of View*. Proc. 1st Annual ACM-SIAM Symp. on Discrete Algorithms 1990, pp. 107-117.

[2] R. Cole and M. Sharir. *Visibility Problems for Polyhedral Terrains*. J. of Symbolic Computation, vol. 7 (1989), pp. 11-30.

[3] E. M. McCreight. *Priority Search Trees*. SIAM Journal of Computing, vol. 14 (1985), pp. 257-276.

[4] K. Mulmuley. *Hidden Surface Removal with Respect to a Moving Viewpoint*. Proc. 23th Annual ACM Symposium on Theory of Computing (1991), pp. 512-522.

[5] O. Nurmi. *A Fast Line-Sweep Algorithm For Hidden Line Elimination*. BIT, vol. 25 (1985), pp. 466-472.

[6] W. H. Plantinga and C. D. Dyer. *An Algorithm for Constructing the Aspect Graph*. Proc. 27th IEEE Foundations of Comp. Science (1986), pp. 123-131.

Voronoi Diagrams of Moving Points in Higher Dimensional Spaces*

Gerhard Albers
Lehrstuhl für Informatik 1
Universität Würzburg
D-8700 Würzburg, Germany

Thomas Roos[†]
Departement Informatik
ETH Zentrum
CH-8092 Zürich, Switzerland

Abstract

The modeling of realistic dynamic scenes often requires the maintenance of geometric data structures over time. This is the subject of a rising discipline called dynamic computational geometry. In the present work we investigate the behavior of spatial Voronoi diagrams under continuous motions of the underlying sites. Nevertheless, the methodology presented can be applied to many other geometric data structures in computational geometry, as well.

Now, consider a set of n points moving continuously along given trajectories in d-dimensional Euclidean space, $d \geq 3$. At each instant, the points define a Voronoi diagram which changes continuously except of certain critical instants, so-called topological events.

We classify the appearing events which cause a change in the topology of the Voronoi diagram and present an algorithm for maintaining the Voronoi diagram over time using only $O(\log n)$ time per event which is worst-case optimal. In addition, we give an $O(n^d \lambda_s(n))$ upper bound on the number of topological events. Thereby $\lambda_s(n)$ denotes the maximum length of an (n, s)-Davenport-Schinzel sequence, and s is a constant depending on the underlying trajectories of the moving sites.

Our work generalizes the most recent results by [FuLe 91], [GuMiRo 91] and [ImIm 90] to three and higher dimensions. Application areas include motion planning problems (such as air traffic control) as well as pattern matching problems in static and dynamic scenes.

1 Introduction

Voronoi diagrams are a fundamental tool expressing the proximity of geometric objects. So, it is not surprising that they appear in many variations in computational geometry as well as other scientific areas related (see [Au 90] for a survey on this topic).

A problem of recent interest has been of allowing the set of objects S to vary continuously over time. This "dynamic" version has been studied in the case of points in the Euclidean plane by [ImSuIm 89], [AoImTo 90] and [Ro 90]. Most recently, [Al 91] and [Ro 91] generalized these ideas with respect to the dimension ($d = 3$) and the order of the Voronoi diagram.

In this paper, we consider the following problem: We are given a set S of n points in d-dimensional Euclidean space, $d \geq 3$, each of which is continuously moving along a given

---

*Work on this paper was partially supported by the Deutsche Forschungsgemeinschaft (DFG) under contract (No 88/10 - 1, 2).

[†]This work was carried out while the second author was at the University of Würzburg.

trajectory. At each instant in time, the points define a Voronoi diagram. As the points move, the Voronoi diagram changes continuously, but at certain critical instants in time, topological *events* occur that cause a change in the dual graph, the Delaunay diagram. Our goal is to characterize the elementary topological events in order to *maintain* the Voronoi diagram over time in some useful data structure.

The main result is to prove a new $O(n^d \lambda_s(n))$ upper bound on the number of topological events, where $\lambda_s(n)$ denotes the maximum length of an (n, s)-Davenport-Schinzel sequence and s is a constant depending on the motions of the point sites. In the special case of points moving along polynomial curves of degree m (so-called polynomial m-motions), we get $s = (d+2) m$. Thus, our results are a linear-factor improvement over the naive $O(n^{d+2})$ upper bound.

In the case that only k of the n points of S are moving (while the remaining $n - k$ stay fixed), our bound on the number of events becomes $O(k n^{d-1} \lambda_s(n) + (n - k)^d \lambda_s(k))$, which is approximately $O(n^d)$ for fixed k. This generalizes the results by [GuMiRo 91] to $d \geq 3$.

We also present a numerically stable algorithm for the update over time of the topological structure of the Voronoi diagram, using only $O(\log n)$ time for each topological change. It is known (cf. [Ro 91]) that this update time is worst-case optimal (even in the planar case).

2 The Topological Structure of Higher Dimensional Voronoi Diagrams

This section briefly summarizes the elementary definitions and properties of higher dimensional Euclidean Voronoi diagrams of point sets. As usual, we let $d(.,.)$ denote Euclidean distance.

We are given a finite set $S := \{P_1, \ldots, P_n\}$ of $n \geq d + 2$ sites in d-dimensional Euclidean space $\mathbb{E}^d$, $d \geq 3$. (As usual, the dimension d is assumed to be a constant.) The perpendicular *bisector of P_i and P_j* is defined to be the hyperplane

$$B_{ij} := \{x \in \mathbb{R}^d \mid d(x, P_i) = d(x, P_j)\}$$

The (convex) *Voronoi polyhedron of P_i* is given by

$$v(P_i) := \{x \in \mathbb{R}^d \mid \forall_{j \neq i} \; d(x, P_i) \leq d(x, P_j)\}$$

The vertices of the Voronoi polyhedrons are called *Voronoi points* and the bisector portions on the boundary are called *Voronoi k-faces* (according to their affine dimension k). Finally the *Voronoi diagram of S* is defined by

$$V D(S) := \{v(P_i) \mid P_i \in S\}$$

The embedding of the Voronoi diagram into d-dimensional real space provides a graph that we call the *geometrical structure* of the underlying Voronoi diagram.

Now we turn our attention to the dual graph of the Voronoi diagram, the so-called *Delaunay graph $DT(S)$*. If S is in general position – i.e. no $d+2$ points of S lie on a common hypersphere and no $d + 1$ points of S lie on a common hyperplane – every Voronoi $(d - i)$-face in $V D(S)$ corresponds to an i-face in $DT(S)$, for $i = 0, \ldots, d$.

In the following, we use a *one-point-compactification* to simplify our discussion. We augment set S by adding the "point at infinity", yielding a new set of sites $S' := S \cup \{\infty\}$. The extended Delaunay graph is then given by

$$DT(S') = DT(S) \cup \{(P_i, \infty) \mid P_i \in S \cap \partial CH(S)\}$$

So, in addition to the Delaunay graph $DT(S)$, every point on the boundary of the convex hull $\partial CH(S)$ is connected to ∞. We call the underlying graph of the extended Delaunay graph $DT(S')$ the *topological structure* of the Voronoi diagram. In contrast with $DT(S)$, $DT(S')$ has the nice property that there are exactly $d+1$ $(d+1)$-tuples adjacent to each $(d+1)$-tuple in $DT(S')$. This will significantly simplify the description of the algorithm presented below.

Next, we adopt two functions[1] from [GuSt 85] providing a nice classification of the $(d+1)$-tuples of the extended Delaunay graph $DT(S')$. In particular, let $v(P_0, \ldots, P_d)$ denote the center of the hyperball $C(P_0, \ldots, P_d)$ of $d+1$ sites $P_0, \ldots, P_d \in S$, we have:

$$\{P_0, \ldots, P_d\} \in DT(S') \iff v(P_0, \ldots, P_d) \text{ is a Voronoi point in } VD(S).$$
$$\iff C(P_0, \ldots, P_d) \text{ contains no point of } S \text{ in its interior.}$$
$$\iff \forall_{P' \in S \setminus \{P_0, \ldots P_d\}} \text{ OUTSIDE}(P_0, \ldots, P_d, P') :=$$
$$\text{sign} [\text{VOL}(P_0, \ldots, P_d) * \text{INS}(P_0, \ldots, P_d, P')] = 1$$

Naturally, an analogous statement can be given for the extended $(d+1)$-tuples. If $\{P_0, \ldots, P_d\}$ and $\{P_0, \ldots, P_{d-1}, \infty\}$ are adjacent $(d+1)$-tuples in $DT(S')$ with $VOL(P_0, \ldots, P_d) > 0$, we have:

$$\{P_0, \ldots, P_{d-1}, \infty\} \in DT(S') \iff P_0, \ldots, P_{d-1} \text{ are the vertices of a } (d-1)\text{-face on}$$
$$\text{the boundary of the convex hull } \partial CH(S).$$
$$\iff \forall_{P' \in S \setminus \{P_0, \ldots P_{d-1}\}} \text{ OUTSIDE}(P_0, \ldots, P_{d-1}, \infty, P') :=$$
$$\text{sign} [\text{VOL}(P_0, \ldots, P_{d-1}, P')] = 1$$

The proof of these statements is straightforward. In the following, these classifications will be very useful to characterize the elementary topological events of higher dimensional dynamic Voronoi diagrams.

3 Higher Dimensional Dynamic Voronoi Diagrams

The contents of this section is to describe the changes in the topological structure of a set of continuously moving points in d-dimensional Euclidean space $\mathbb{E}^d$, $d \geq 3$.

For that, we are given a finite set of $n \geq d+2$ continuous trajectory curves in d-dimensional Euclidean space $\mathbb{E}^d$, $S := S(t) := \{P_1(t), \ldots, P_n(t)\}$. We make the following assumptions about the trajectories: First, we assume that the points move without collisions, or in other words: $\forall_{i \neq j} \forall_{t \in \mathbb{R}} P_i(t) \neq P_j(t)$. In addition, we demand the existence of an instant $t_0 \in \mathbb{R}$ when

---

[1] These functions VOL and INS (mnemonic for "volume" and "insphere") are defined as follows:

$$\text{VOL}(P_0, \ldots, P_d) := \begin{vmatrix} 1 & P_{01} & \cdots & P_{0d} \\ \vdots & \vdots & & \vdots \\ 1 & P_{d1} & \cdots & P_{dd} \end{vmatrix}, \quad \text{INS}(P_0, \ldots, P_{d+1}) := \begin{vmatrix} 1 & P_{01} & \cdots & P_{0d} & P_{01}^2 + \cdots + P_{0d}^2 \\ \vdots & \vdots & & \vdots & \vdots \\ 1 & P_{d1} & \cdots & P_{dd} & P_{d1}^2 + \cdots + P_{dd}^2 \\ 1 & P_{d+11} & \cdots & P_{d+1d} & P_{d+11}^2 + \cdots + P_{d+}^2 \end{vmatrix}$$

$S(t_0)$ is in general position which is necessary to obtain a definite topological structure at the starting position t_0.

Now, consider the situation at a moment $t \in \mathbb{R}$ when all points in $S(t)$ are in general position. On the one hand, by investigating the continuity of a suitable product of determinants, it is easy to see, that a sufficiently small continuous motion of the points does not change the fact that the points are in general position. On the other hand, the topological structure $DT(S')$ is completely determined by the *active* Voronoi points which currently appear in $VD(S)$ and by the d-tuples of sites forming the boundary of the convex hull $\partial CH(S)$. Therefore, the topological structure can only change in the following two different situations:

Case (1)   The appearance (disappearance) of an inactive (active) Voronoi point.

Case (2)   The appearance (disappearance) of a point on the boundary of the convex hull.

However, in both cases the loss of general position of the points $S(t)$ is necessary for changing the topological structure $DT(S'(t))$. This proves that the topological structure $DT(S')$ is *locally stable* as long as the points are in general position.

In order to address the question of sufficient conditions, we proceed with an investigation of the *elementary changes* of the topological structure of a Voronoi diagram. In the two-dimensional case, it is well-known (see, e.g., [Ro 90]) that such elementary changes can be described as "SWAP"s of adjacent triangles in $DT(S')$. However, in higher dimensions these transitions turn out to be more complex.

In our first case above, an inactive Voronoi point $v(P_0, \ldots, P_d)$ becomes activated, if the last point $P_{d+1} \in S$ leaves the variable circumsphere $C(P_0, \ldots, P_d)$. As well, an active Voronoi point $v(P_0, \ldots, P_d)$ becomes inactivated, if a point $P_{d+1} \in S$ enters this variable circumsphere. Additionally, we assume at that instant t' (when $d+2$ points lie on a common hypersphere) that no further point of S lies on the boundary of the circumsphere $C(P_0, \ldots, P_d)$. If we select $\varepsilon > 0$ sufficiently small, the entrance of the point P_{d+1} can be described as follows:

$$\text{OUTSIDE}(P_0, \ldots, P_d, P_{d+1})(t' - \varepsilon) = 1$$
$$\text{OUTSIDE}(P_0, \ldots, P_d, P_{d+1})(t' + \varepsilon) = -1$$
$$\forall_{P' \in S \setminus \{P_0, \ldots, P_{d+1}\}} \forall_{t \in [t'-\varepsilon, t'+\varepsilon]} \text{OUTSIDE}(P_0, \ldots, P_d, P')(t) = 1$$

In fact, there happens a real *zero-crossing* of the function $\text{OUTSIDE}(P_0, \ldots, P_{d+1})$ because the point P_{d+1} changes the side of the sphere $C(P_0, \ldots, P_d)$ at the instant t'.

How can we describe the resulting change of the topological structure? For that, we investigate the active $(d+1)$-tuples of $DT(S'(t'-\varepsilon))$ at an instant $t'-\varepsilon$, with $\varepsilon > 0$ sufficiently small. At first, it is apparent that the *local* topological structure in the neighborhood of $v(P_0, \ldots, P_d)$ is completely determined by the points $S_d := \{P_0, \ldots, P_{d+1}\}$. Thus, we only have to consider all $d+2$ subsets of points of S_d of size $d+1$. These subsets can be generated, for example, by eliminating the i-th element for $i = 0, \ldots, d+1$, respectively. So, let

$$\pi_i := \begin{cases} (P_1, \ldots, P_{d+1}) & i = 0 \\ (P_0, \ldots, P_{i-1}, P_{i+1}, \ldots, P_{d+1}) & \text{iff} \quad 1 \leq i \leq d \\ (P_0, \ldots, P_d) & i = d+1 \end{cases}$$

denote the sequence which has been obtained after eliminating the i-th element.

Using the fact, that the determinants considered are alternating forms (i.e. transposing two rows in any determinant changes its sign), we'll prove now that there exists a complete, disjoint *partition* of the π_i's into two subsets A and B, with $2 \le |A|, |B| \le d$, such that:

$$(*) \begin{cases} \forall_{\pi_i \in A} & \text{OUTSIDE}(\pi_i, P_i)(t' - \varepsilon) = 1 \quad \text{and} \quad \text{OUTSIDE}(\pi_i, P_i)(t' + \varepsilon) = -1 \\ \forall_{\pi_i \in B} & \text{OUTSIDE}(\pi_i, P_i)(t' - \varepsilon) = -1 \quad \text{and} \quad \text{OUTSIDE}(\pi_i, P_i)(t' + \varepsilon) = 1 \\ \forall_{P' \notin \pi_i \cup P_i} \forall_{t \in [t'-\varepsilon, t'+\varepsilon]} & \text{OUTSIDE}(\pi_i, P')(t) = 1 \end{cases}$$

These equations are obviously equivalent (due to the classification above) to the following so-called (i,j)-*transition*[2] of the local topological structure:

$$\boxed{\{\pi_i \in DT(S'(t' - \varepsilon)) \mid \pi_i \in A\} \quad \longleftrightarrow \quad \{\pi_i \in DT(S'(t' + \varepsilon)) \mid \pi_i \in B\}}$$

Next, we proceed by constructing the announced sets A and B.

$$A := \{\pi_i \mid \text{sign}\,[\text{VOL}(P_0, \ldots, P_d)] = -\text{sign}\,[\text{VOL}(P_0, \ldots, P_{i-1}, P_{d+1}, P_{i+1}, \ldots, P_d)]\}$$
$$B := \{\pi_i \mid \text{sign}\,[\text{VOL}(P_0, \ldots, P_d)] = \text{sign}\,[\text{VOL}(P_0, \ldots, P_{i-1}, P_{d+1}, P_{i+1}, \ldots, P_d)]\}$$

In other words, set A and set B include all π_i's where P_i and P_{d+1} lie on different or the same side of the hyperplane spanned by the sites $P_0, \ldots, P_{i-1}, P_{i+1}, \ldots, P_d$, respectively. Thereby, we assume that the sites of π_i do not change their orientation at the instant t'.

Now, if we use the fact that the sequence (π_i, P_i) can be obtained from the sequence $(P_0, \ldots, P_{d+1})$ by $d - i + 1$ transpositions, we have for any $\pi_i \in A$:

$\text{OUTSIDE}(\pi_i, P_i)$
$= \text{sign}\,[\text{VOL}(\pi_i)] * \text{sign}\,[\text{INS}(\pi_i, P_i)]$
$= (-1)^{d-i}\,\text{sign}\,[\text{VOL}(P_0, \ldots, P_{i-1}, P_{d+1}, P_{i+1}, \ldots, P_d)] * (-1)^{d-i+1}\,\text{sign}\,[\text{INS}(P_0, \ldots, P_{d+1})]$
$= (-1)^{d-i+1}\,\text{sign}\,[\text{VOL}(P_0, \ldots, P_{i-1}, P_i, P_{i+1}, \ldots, P_d)] * (-1)^{d-i+1}\,\text{sign}\,[\text{INS}(P_0, \ldots, P_{d+1})]$
$= \text{OUTSIDE}(P_0, \ldots, P_{d+1})$

Analogously, we obtain for any $\pi_i \in B$: $\text{OUTSIDE}(\pi_i, P_i) = -\text{OUTSIDE}(P_0, \ldots, P_{d+1})$. With that, the desired equations $(*)$ hold immediately. Notice that there always exists such a hyperplane spanned by the sites $P_0, \ldots, P_{i-1}, P_{i+1}, \ldots, P_d$ separating P_i and P_{d+1}, which proves $|A| \ge 2$. The proof of $|B| \ge 2$ is also straightforward. A three-dimensional example of a $(2,3)$-transition is depicted in figure 1 (from [Al 91]). The transition described is also equivalent to a *fusion* of the corresponding Voronoi points, which come together and disappear at the instant t', and the creation of new (dual) Voronoi points – according to the transition rule above.

Considering our second case, the appearance or disappearance of a point $P' \in S$ on the boundary of the convex hull $\partial CH(S)$ is equivalent to the activation or deactivation of the extended Delaunay edge (P', ∞).

At first, (P', ∞) becomes activated, if P' enters the boundary of the convex hull on a $(d - 1)$-face formed by the sites $(P_0, \ldots, P_{d-1})$. According to the classification above, the circumsphere $C(P_0, \ldots, P_{d-1}, P')$ contains no points in its interior already shortly before P' enters the boundary of the convex hull. If we assume at the instant of coplanarity, that no

---

[2]Thereby, i and j denote the cardinality of set A and B, respectively.

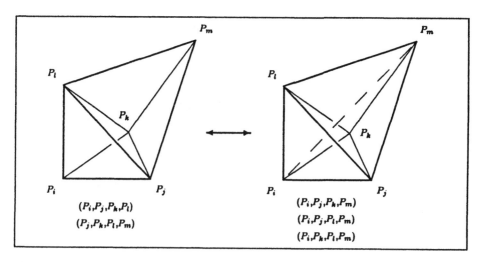

Figure 1: A reversible (2,3)-transition with the active Delaunay $(d+1)$-tuples in $\mathbb{R}^3$.

other point of S lies on the hyperplane H formed by these points and if we regard the interior of the infinite sphere through the points $P_0, \ldots, P_{d-1}$ and ∞ as the open halfspace that is bounded by the hyperplane H and lies outside the convex hull $CH(S)$, then we can apply the results of our first case.[3] The deactivation of the edge (P', ∞) can be dealt with analogously. To summarize our results, we present the following theorem.

Theorem 1   Elementary changes in the topological structure $DT(S')$ of the Voronoi diagram $VD(S)$ are characterized by (i,j)-transitions of adjacent $(d+1)$-tuples in $DT(S')$, except in degenerate cases. Thereby, the indices obey the conditions $i + j = d + 2$ and $2 \leq i, j \leq d$.

According to this, we have solved the question concerning *sufficient* conditions. Roughly speaking, topological events are characterized by non-degenerate loss of *local* general position, i.e. the loss of general position of adjacent $(d+1)$-tuples in the topological structure. Notice, that the same topological events are generated by several pairs of $(d+1)$-tuples representing the same sites. In this connection, the *original advantage* of the one-point compactification becomes apparent. It allows us the convenience of treating both cases similarly: as simple transitions in the extended dual graph $DT(S')$.

Up to now, we have been ignoring a technicality caused by degeneracies: it may be that more than $d + 2$ points in $S(t)$ are lying on a common hypersphere at the same instant or that more than $d + 1$ points in $S(t)$ are coplanar at the same instant. In both cases, we *recalculate* the local topological structure of the interior of the convex polygon described by the points at a moment $t + \varepsilon$. However, it is necessary to select $\varepsilon > 0$ in such a way, that the moment of recalculation precedes the next topological event.

---

[3] If we replace P_m by ∞, the activation of the extended dual edge (P_i, ∞) can be regarded as a left-to-right transition in figure 1.

4 New Upper Bounds

In this section, we generalize the recent results by [GuMiRo 91] to higher dimensions. As we have seen in the previous section, topological events are characterized by loss of general position. So, it is quite natural to assume that there exist at most $s \in O(1)$ zeros of the functions INS(...) and VOL(...) which are computable in constant time each. Indeed, this additional assumption can be regarded as a certain kind of *non-periodicity* condition, which is achieved, for example, in the case of polynomial curves of bounded degree. This assumption implies that each subset of S' of size $d + 2$ generates at most a constant number of topological events and, following that, a trivial $s \binom{n+1}{d+2} \in O(n^{d+2})$ upper bound on the number of topological events. By a Davenport-Schinzel argument, we improve this naive upper bound by (roughly) a linear factor.

First, it is clear that the maximum number of *extended* topological events is bounded by $s \binom{n}{d+1} \in O(n^{d+1})$, since this is the maximum number of instants at which $d + 1$ points of S can become coplanar. Therefore we only have to deal with such topological events when $d + 2$ points of S lie on a common hypersphere. The basic observation is that every topological event belongs to a local transition of altogether $d + 2$ Delaunay $(d + 1)$-tuples leaving the bounding Delaunay $(d - 1)$-faces unchanged. With that, we are able to determine the total number of topological events by adding for every imaginable Delaunay $(d - 1)$-face $(P_0, \ldots, P_{d-1})$ the number of adjacent topological events that do not destroy this $(d - 1)$-face.

With this intention, we consider an arbitrary d-tuple $(P_0, \ldots, P_{d-1})$ of different points and the line $B_{0,\ldots,d-1}(t, \mu)$ which is given by the formulation below:

$$B_{0,\ldots,d-1}(t,\mu) \quad := \quad m_{0,\ldots,d-1}(t) + \mu\, n_{0,\ldots,d-1}(t) \quad \text{where} \quad \mu \in \mathbb{R},$$

$$m_{0,\ldots,d-1}(t) \quad := \quad \frac{1}{d} \sum_{i=0}^{d-1} P_i(t) \quad \text{and} \quad n_{0,\ldots,d-1}(t) \perp H_{0,\ldots,d-1}(t)$$

In other words, $m_{0,\ldots,d-1}(t)$ denotes the *center of gravity* of the d sites and $n_{0,\ldots,d-1}(t)$ a *normal vector* to the affine hyperplane $H_{0,\ldots,d-1}(t)$ spanned by the d points. In addition, let $h_{0,\ldots,d-1}^{>}(t)$ and $h_{0,\ldots,d-1}^{<}(t)$ denote the two open halfspaces bounded by $H_{0,\ldots,d-1}(t)$.

Now, whenever the Delaunay $(d-1)$-face $(P_0, \ldots, P_{d-1})$ exists, there are exactly two $(d+1)$-tuples $\{P_0, \ldots, P_{d-1}, P'\}$ and $\{P_0, \ldots, P_{d-1}, P''\} \in DT(S')$ adjacent to this Delaunay face with

$$P' \in S'_{>} \ := \ \left(h_{0,\ldots,d-1}^{>} \cap S \right) \cup \{\infty\} \quad \text{and} \quad P'' \in S'_{<} \ := \ \left(h_{0,\ldots,d-1}^{<} \cap S \right) \cup \{\infty\}$$

If we look at the μ-values $\mu_x(t)$ of the circumcenters of the circumspheres $C(P_0, \ldots, P_{d-1}, P_x)$ on the bisector $B_{0,\ldots,d-1}(t, \mu)$, the upper $(d+1)$-tuple is obviously characterized by the minimum value:

$$\mu_{min}(t) \quad := \quad \min_{P_x \in S'_{>}} \mu_x(t)$$

This can be seen by imaging a point (circumcenter) starting from $m_{0,\ldots,d-1}(t)$ and moving along the line $B_{0,\ldots,d-1}(t, \mu)$ until a first point $P_x \in S'_{>}$ is captured by the variable circumsphere touching the sites $P_0, \ldots, P_{d-1}$. Naturally, an analogous construction can be done for the lower $(d + 1)$-tuple.

Now, if we investigate those moments when the upper $(d+1)$-tuple changes[4] we can restrict ourselves to those intervals in which $h^>_{0,\ldots,d-1} \cap S \neq \emptyset$. Next, we look closer at the functions $\mu_x(t)$ and their pairwise points of intersection:

Case (1) $\mu_x(t) = \mu_y(t) < \infty$
Both circumspheres $C(P_0,\ldots,P_{d-1},P_x)$ and $C(P_0,\ldots,P_{d-1},P_y)$ are identical, which implies that all $d+2$ points lie on a common hypersphere. By our non-periodicity assumption, this can happen only s times.

Case (2) $\mu_x(t) = \mu_y(t) = \infty$
These moments have no influence on the complexity of the minimum function $\mu_k(t)$, since we have restricted ourselves to intervals where $\mu_k(t) < \infty$.

Finally, we can summarize both cases with the statement that two different functions $\mu_x(t)$ and $\mu_y(t)$ have at most s relevant intersections. Thus, the theory of Davenport-Schinzel sequences ([Sh 88]) implies that the minimum function $\mu_k(t)$ has worst-case complexity $O(\lambda_s(n))$, where $\lambda_s(n)$ is the maximum length of a Davenport-Schinzel sequence of length n and order s. Summing over all $\binom{n}{d}$ tuples of points $(P_0,\ldots,P_{d-1})$, we obtain the following theorem.

Theorem 2   Given a finite set $S(t)$ of n continuous trajectories in d-dimensional space $\mathbb{R}^d$, the maximum number of topological events over time is $O(n^d \lambda_s(n))$.

                  If only $k \leq n$ points of S are moving (while the remaining $n-k$ stay fixed), this upper bound becomes $O(k\, n^{d-1} \lambda_s(n) + (n-k)^d \lambda_s(k))$, which is approximately $O(n^d)$ for fixed k.

The proof of the second part of this theorem is straightforward if we consider the $O(k\, n^{d-1})$ moving and $O((n-k)^d)$ fixed d-tuples separately. The crucial fact is that each fixed d-tuple generates only $O(\lambda_s(k))$ instead of $O(\lambda_s(n))$ topological events (compare [GuMiRo 91]).

5   Dynamic Scenes

The topological structure of a Voronoi diagram under continuous motions of the points in S can be maintained by the following algorithm:

Algorithm :   Preprocessing :

                  1. Compute the topological structure $DT(S'(t_0))$ of the starting position.

                  2. For every existing pair of $(d+1)$-tuples in $DT(S'(t_0))$ calculate the potential topological events.

                  3. For the set of the potential topological events create an event queue (priority queue).

---

[4]Notice, that P' can only be replaced by another point of $S'_>$, because the Delaunay $(d-1)$-face $(P_0,\ldots,P_{d-1})$ is not destroyed during the topological event.

Iteration :

1. Determine the next topological event and decide whether it is an (i, j)-transition or a recalculation.

2. Process the topological event and do an update of the event queue.

We look closer to the individual steps of the algorithm and their time and storage requirements. In the first preprocessing step, we compute the initial Delaunay triangulation $DT(S(t_0))$ and augment it with extended dual edges, obtaining $DT(S'(t_0))$ in $O(n^{\lceil \frac{d}{2} \rceil})$ time and and space (e.g., using the optimal algorithm by [Se 90]).

In the second preprocessing step, we continue with a flow of the $(d-1)$-faces in $DT(S'(t_0))$ computing the potential topological events. If m denotes the number of $(d+1)$-tuples which appear in the initial topological structure, this step can be done in $O(m)$ time.

In the third preprocessing step, we build up the event queue for the set of potential topological events. The topological events are stored in a priority queue according to their temporal appearance, with the corresponding $(d+1)$-tuples stored with each event. This step and therefore the entire preprocessing step requires $O(n^{\lceil \frac{d}{2} \rceil} + m \log m)$ time and $O(n^{\lceil \frac{d}{2} \rceil})$ space.

To determine the next topological event, we simply pop the event queue in time $O(\log n)$. Assuming that the degree of degeneracy remains constant, then one can decide in constant time if the event is an (i, j)-transition or a (local) recalculation.

Now, each topological event destroys only a constant number of adjacent $(d+1)$-tuples while creating also a constant number of new ones. Thus, in order to update the event queue, all we have to do is to delete the destroyed pairs of $(d+1)$-tuples and their corresponding topological events in the event queue and to insert the new ones. Thus, we spend time $O(\log n)$ per event (which [Ro 91] shows is worst-case optimal even under linear motions of the points in the plane). In summary, we have:

Theorem 3 Given a finite set $S(t)$ of n continuous trajectories in d-dimensional space $\mathbb{R}^d$. After preprocessing requiring $O(n^{\lceil \frac{d}{2} \rceil} + m \log m)$ time and $O(n^{\lceil \frac{d}{2} \rceil})$ space, we can maintain the topological structure in worst-case optimal $O(\log n)$ time per event. Thereby, m denotes the initial complexity of the Voronoi diagram at the starting position.

6 Concluding Remarks and Open Problems

We have presented an algorithm for maintaining Voronoi diagrams of moving points over time. The major open question remaining is to prove that the presented bounds on the number of events are tight.

The algorithm presented here has been implemented in the planar case ($d = 2$) on a SUN workstation, using [SuIr 89] methods for numerically stable evaluation of the functions involved. Extensive tests suggest that the number of topological events grows like $\Theta(n \sqrt{n})$ in the *average case* under linear motions chosen at random. We also expect in higher dimensions that the average number of topological events is significantly smaller than the derived worst-case bounds.

Dynamic Voronoi diagrams can be used for planning the motion of a disk in a dynamic scene of continuously moving points (compare [RoNo 91]). Additionally, there are many related geometric structures and problems in computational geometry which can be solved very efficiently if the Voronoi diagram is known in advance (for a survey see, e.g., [Au 90] or [Ro 91]).

A typical application of higher dimensional dynamic Voronoi diagrams which arises in the area of spatial path planning (such as air-traffic control) is the maintenance of the *closest pair* or the *all-nearest-neighbors* over time. It is also quite interesting to apply the pattern matching methods by [ImSuIm 89] and [AoImTo 90] to higher dimensions.

Acknowledgement

The authors wish to thank Rolf Klein, Joe Mitchell, Hartmut Noltemeier and Derick Wood for the invaluable discussions and suggestions.

References

[Al 91]   G. Albers, *Three-Dimensional Dynamic Voronoi Diagrams* (in German), Diploma thesis, University of Würzburg, July 1991

[Ag 87]   A. Aggarwal, L. Guibas, J. Saxe and P. Shor, *A Linear Time Algorithm for Computing the Voronoi Diagram of a Convex Polygon*, Proc. 19th Annual ACM Symposium on Theory of Computing, New York City, 1987, pp 39 – 45

[AgShSh 89] A. Aggarwal, M. Sharir and P. Shor, *Sharp Upper and Lower Bounds on the Length of General Davenport-Schinzel Sequences*, Journal of Combinatorial Theory, Series A, Vol. 52, 1989, pp 228 – 274

[At 85]   M.J. Atallah, *Some Dynamic Computational Geometry Problems*, Computers and Mathematics with Applications, Vol. 11, 1985, pp 1171 – 1181

[AoImTo 90] H. Aonuma, H. Imai, K. Imai and T. Tokuyama, *Maximin Locations of Convex Objects and Related Dynamic Voronoi Diagrams*, Proc. 6th ACM Symposium on Computational Geometry, Berkeley, 1990, pp 225 – 234

[Au 90]   F. Aurenhammer, *Voronoi Diagrams – A Survey of a Fundamental Data Structure*, Report B 90 - 09, FB Mathematik, Serie B Informatik, Nov. 1990

[Ed 87]   H. Edelsbrunner, *Algorithms in Combinatorial Geometry*, EATCS Monographs in Computer Science, Springer, Berlin - Heidelberg, 1987

[FuLe 91]   J-J. Fu and R.C.T. Lee, *Voronoi Diagrams of Moving Points in the Plane*, Int. Journal of Computational Geometry & Applications, Vol. 1, No. 1, 1991, pp 23 – 32

[GuKnSh 90] L. Guibas, D.E. Knuth and M. Sharir, *Randomized Incremental Construction of Delaunay and Voronoi Diagrams*, Proc. 17th Intern. Colloquium on Automata, Languages and Programming ICALP 90, LNCS 443, Springer, 1990, pp 414 – 431

[GuSt 85]   L. Guibas and J. Stolfi, *Primitives for the Manipulation of General Subdivisions and the Computation of Voronoi Diagrams*, ACM Transactions on Graphics, Vol. 4, No. 2, April 1985, pp 74 – 123

[GuMiRo 91]   L. Guibas, J.S.B. Mitchell and T. Roos, *Voronoi Diagrams of Moving Points in the Plane*, Proc. 17th International Workshop on Graph-Theoretic Concepts in Computer Science, LNCS 570, pp 113 – 125

[ImIm 90]   H. Imai and K. Imai, *Voronoi Diagrams of Moving Points*, Proc. Int. Computer Symp., Taiwan, 1990, pp 600 – 606

[ImSuIm 89]   K. Imai, S. Sumino and H. Imai, *Geometric Fitting of Two Corresponding Sets of Points*, Proc. 5th ACM Symp. on Computational Geometry, 1989, pp 266 – 275

[O'Ro 91]   J. O'Rourke, *Computational Geometry Column 12*, SIGACT News, Vol. 22, No. 2, Spring 1991, pp 26 – 29

[PrSh 85]   F.P. Preparata and M.I. Shamos, *Computational Geometry – An Introduction*, Springer - Verlag, New York, 1985

[Ro 90]   T. Roos, *Voronoi Diagrams over Dynamic Scenes (Extended Abstract)*, Proceedings 2nd Canadian Conference on Computational Geometry, Ottawa, 1990, pp 209 – 213

[Ro 91]   T. Roos, *Dynamic Voronoi Diagrams*, PhD Thesis, University of Würzburg, Sept. 1991

[RoNo 91]   T. Roos and H. Noltemeier, *Dynamic Voronoi Diagrams in Motion Planning: Combining Local and Global Strategies*, Proc. 15th IFIP Conference on System Modeling and Optimization, Zurich, 1991

[Se 90]   R. Seidel, *Linear Programming and Convex Hulls Made Easy*, Proc. 6th ACM Symposium on Computational Geometry, Berkeley, 1990, pp 212 – 215

[Sh 88]   M. Sharir, *Davenport-Schinzel Sequences and their Geometric Applications*, pp 253-278, NATO ASI Series, Vol. F40, Theoretical Foundations of Computer Graphics and CAD, R.A. Earnshaw (Ed.), Springer, Berlin Heidelberg, 1988

[SuIr 89]   K. Sugihara and M. Iri, *Construction of the Voronoi Diagram for One Million Generators in Single-Precision Arithmetic*, private communications, 1989

Sorting Multisets Stably in Minimum Space

Jyrki Katajainen[1,2] and Tomi Pasanen[2]

[1] Department of Computer Science, University of Copenhagen,
Universitetsparken 1, DK-2100 Copenhagen East, Denmark
[2] Department of Computer Science, University of Turku,
Lemminkäisenkatu 14 A, SF-20520 Turku, Finland

Abstract. In a decision tree model, $\Omega(n \log_2 n - \sum_{i=1}^{m} n_i \log_2 n_i + n)$ is known to be a lower bound for sorting a multiset of size n containing m distinct elements, where the ith distinct element appears n_i times. We present a *minimum space* algorithm that sorts *stably* a multiset in asymptotically *optimal worst-case time*. A Quicksort type approach is used, where at each recursive step the median is chosen as the partitioning element. To obtain a stable minimum space implementation, we develop linear-time in-place algorithms for the following problems, which have interest of their own:
Stable unpartitioning: Assume that an n-element array A is stably partitioned into two subarrays A_0 and A_1. The problem is to recover A from its constituents A_0 and A_1. The information available is the partitioning element used and a bit array of size n indicating whether an element of A_0 or A_1 was originally in the corresponding position of A.
Stable selection: The task is to find the kth smallest element in a multiset of n elements such that the relative order of identical elements is retained.

1 Introduction

The sorting problem is known to be easier for multisets, containing identical elements, than for sets, in which all elements are distinct. The complexity of an input instance depends on the multiplicities of the elements. When only three-way comparisons are allowed, $\Omega(n \log_2 n - \sum_{i=1}^{m} n_i \log_2 n_i + n)$ is known to be a lower bound for sorting a multiset with multiplicities $n_1, n_2, \ldots, n_m$ (where $n = \sum_{i=1}^{m} n_i$) [18]. Mergesort and Heapsort can be adapted to sort multisets in $O(n \log_2 n - \sum_{i=1}^{m} n_i \log_2 n_i + n)$ time [20] (without knowing the multiplicities beforehand). An optimal in-place implementation based on Heapsort also exists [18], but due to the nature of Heapsort this algorithm is not *stable*, i.e., the relative order of identical elements is not necessarily retained. The main concern of this paper is, how to ensure time-optimality, space-optimality, and stability at the same time, i.e., the problem left open by Munro and Raman [18].

In our accompanying paper [11] we proved that randomized Quicksort can be adapted to sort a multiset stably and in-place such that the running time will be optimal up to a constant factor with high probability. In the present paper we improve this result by showing that multisets can be sorted in optimal time also in the worst case. To adapt Quicksort for sorting multisets, one should perform a three-way partition at each recursive step [23]. For this purpose, we use the linear-time, in-place algorithm for stable partitioning presented in [11]. The standard way to make Quicksort worst-case optimal is to use the median as the partitioning element.

The basic problem encountered is how to select the kth smallest element in a multiset of n elements such that the relative order of elements with equal values is the same before and after the computation. This is called the *stable selection problem*. Actually, we shall also study the following variant of the selection problem, called the *restoring selection problem*: find the kth smallest of n elements such that after the computation the elements are in their original order. The latter problem has applications in other areas, e.g., in adaptive sorting (cf. [16]). An in-place solution for both of these problems is immediately obtained, if we scan through the elements and calculate for each the number of smaller elements. This will, however, require $O(n^2)$ time. On the other hand, if we allow $O(n)$ extra space, the linear-time selection algorithm [1] (or its in-place variant, see [13]) can be used to solve the problems simply by coupling with each element its original position. (After the selection, the elements can be permuted into their original positions as described for example in [19, Lemma 1]. Note that we need n extra bits to do this.)

To solve the restoring selection problem, we implement the prune-and-search algorithm of Blum et al. [1] more carefully. The algorithm is based on repeated partitioning. Therefore the fast, in-place algorithm for stable partitioning is used here. In order to reverse the computation we need a space-efficient solution for the *unpartitioning problem* defined as follows. Assume that an array A of size n undergoes a stable partition. Let the resulting subarrays be A_0 and A_1 with respective sizes n_0 and n_1. The problem is to recover A from its constituents A_0 and A_1. The information available is the partitioning element used and a bit array containing n_0 zeros and n_1 ones. The interpretation of the ith b-bit in position j is that the ith element of A_b is the jth element of A ($b \in \{0, 1\}$). In Section 3 we introduce an algorithm for stable unpartitioning that runs in linear time and requires only a constant amount of additional space.

In Section 4 we show how the restoring selection problem is solved in linear time using $O(n)$ extra bits. By means of this, we are able to develop an algorithm for stable selection that requires linear time and only $O(1)$ extra space. This algorithm presented in Section 5 is then used in the final sorting algorithm which we describe and analyse in Section 6.

Before proceeding we define precisely what we mean by a *minimum space* or *in-place algorithm*. In addition to the array containing the n elements of a multiset, we allow one storage location for storing an array element. This is needed, for example, when swapping two data elements. The elements are regarded to be atomic. They can only be moved and compared with the operations $\{<, =, >\}$ in constant time. Moreover, we assume that a constant number of extra storage locations, each capable for storing a word of $O(\log_2 n)$ bits, is available and that operations $\{<, =, >, +, -, \text{shift}\}$ take constant time for these words. An *unrestricted shift operation* takes two integer operands v and i and produces $\lfloor v \cdot 2^i \rfloor$.

2 Tools for Building Minimum Space Algorithms

In this section we briefly review the basic techniques for minimum space algorithms.

Blocking: The input array is divided into equal sized blocks. Often blocking with blocks of size $\sqrt{n}$ or $\log_2 n$ works well. (This requires that good estimates for the numbers $\sqrt{n}$ and $\log_2 n$ are available but these are easily computed from n in $O(n)$ time.) Most efficient in-place algorithms in the literature are based on the blocking technique (see e.g. [7, 8, 9, 19, 21, 22]).

Internal buffering: Usually some blocks are employed as an *internal buffer* to aid in rearranging or manipulating the other blocks in constant extra space. This idea dates back to Kronrod [12] (see also [21]) and is frequently used in minimum space algorithms. If the goal is a stable algorithm, the internal buffer should be manipulated carefully, since otherwise the stability might be lost.

Block interchanging: A block X can be reversed in-place in linear time by swapping the pair of end elements, then the pair next to the ends, etc. Let X^R be X reversed. The order of two consecutive blocks (not necessarily of the same size) X and Y may be interchanged by performing three block reversals, namely $YX = (X^R Y^R)^R$. This idea seems to be part of computer folklore.

Bit stealing: Let x and y be two elements, which are known to be distinct. Depending on the order we store the elements in the array, we obtain extra information. The order xy, $x < y$, may denote a 0-bit and the order yx a 1-bit. This technique has been used for example by Munro [17] in his implicit dictionary. With $\lceil \log_2(n+1) \rceil$ stolen bits it is possible to implement a counter taking values from the interval $[0..n]$, but the manipulation of this counter will take $O(\log_2 n)$ time.

Packing small integers: Let us assume that we have t small integers each represented by m bits. That is, the integers are from the domain $[0..2^m - 1]$. Further, assuming that $t \cdot m \leq \log_2 n$, the integers can be packed into one word w of $\lceil \log_2 n \rceil$ bits. Let us number the bits of w from right to left such that the righmost (least significant) bit has number 0 and the leftmost bit has number $\lceil \log_2 n \rceil - 1$. Now the integer i_j $(j = 1, 2, \ldots, t)$ is stored by using the bits $(j-1)m, \ldots, jm - 1$ of w. Each integer is easily recovered from w in constant time if multiplications and divisions by a power of two are constant time operations. The value v of i_j is obtained as follows:

$$v = \{w - [(w \text{ shift } -jm) \text{ shift } jm]\} \text{ shift } -(j-1)m.$$

(Observe that in our algorithms m can be chosen to be a power of two, so we do not need general multiplication.) With a code similar to this the value of i_j can be updated. Previously the packing technique has been used for example in [2, 14].

In some in-place algorithms also the modification of the input data is allowed (see e.g. [4, 6] or [22, Theorem 3.2]). However, we consider this as an illegal trick.

3   Stable minimum space unpartitioning

The heart of our selection and sorting algorithms will be the linear-time, minimum space algorithm for stable partitioning given in [11]. Another important subroutine is a fast, minimum space algorithm for stable unpartitioning which is the topic of this section. We show that the computation of the partitioning algorithm is reversible, even if the steps executed are not recorded.

In an abstract setting the *stable partitioning problem* can be defined as follows: Given an n-element array A and a function f mapping each element to the set $\{0, 1\}$, the task is to rearrange the elements such that all elements, whose f-value is zero, come before elements, whose f-value is one. Moreover, the relative order of elements with equal f-values should be retained. Let the resulting subarrays be A_0 and A_1. For the sake of simplicity, we call the elements of A_0 *zeros* and elements of A_1 *ones*. The *stable unpartitioning problem* is to recover A from its constituents A_0 and A_1. The information available is the f-function and a *placement array*, a bit array of size n indicating whether an element of A_0 or A_1 was originally in the corresponding position of A. Observe that in our formulation of the problem it is essential that f is known during unpartitioning.

Stable merging can be seen as a special case of stable unpartitioning, since the placement array is easily created by scanning the input of a merging problem with two cursors. By unpartitioning the original merging problem is solved. This indicates that it might be possible to generalize the algorithms for stable merging to solve the stable unpartitioning problem. Generally, this cannot be done since most algorithms for stable merging utilize the fact that the A_b-elements appear in sorted order. In unpartitioning this is not necessarily the case. However, there are great similarities between our unpartitioning algorithm and parallel merging algorithms given in [10].

The stable unpartitioning is easily done in linear time when $O(n)$ extra space is available. Algorithm A to be described next does this by scanning the placement array and storing the site together with each A_b-element. During the scan two cursors C_0 and C_1 are maintained, the former pointing to A_0 and the latter to A_1. Initially, C_b will point to the first element of A_b. If the jth position contain the bit b, then the C_bth element of A_b is coupled with its *site j* and the counter C_b is advanced. After computing the sites, the elements are permuted to their final positions (for details, see [19, Lemma 1]). Here we need one more bit for each element, telling whether the element is in its final position or not. Hence we have

Lemma 1. *Algorithm A solves a stable unpartitioning problem of size n in $O(n)$ time with n bits and $n + O(1)$ counters, each requiring at most $\lceil \log_2(n+1) \rceil$ bits.*

For the time being let us assume that n, the number of elements is a power of two. Lateron we show how to get rid of this assumption. In our improved algorithms we divide the input into blocks by using a *blocking factor* $\lg n$ or $2^{(\lg n)/2}$ ($\approx \sqrt{n}$), where $\lg n$ denotes the smallest power of two greater than or equal to $\log_2 n$. Since n is assumed to be a power of two, it is divisible by both of these blocking factors.

Before proceeding, we will introduce some terminology. Let us call a block containing only zeros as a *0-block* and a block containing only ones as a *1-block*. If a block is a 0-block or 1-block it is called a *0/1-block*. Further, let *0&1-block* denote a block consisting of two sequences, a sequence of zeros followed by a sequence of ones. The basic idea of our algorithms is simply to transform the original problem to n/t similar subproblems of size t. That is, by using the terminology introduced above, the goal is to transform one 0&1-block of size n to n/t 0&1-blocks of size t such that in each subblock the number of zeros and ones is equal to that of 0- and 1-bits in the corresponding part of the placement array. Hence, after this transformation the subproblems can be solved locally.

When the input array is divided into the blocks of size t, one complication is that one of the blocks might be a 0&1-block while the others are 0/1-blocks. The single 0&1-block is handled as follows. We first interchange the zeros of the block to the end of the input array and then move them gradually into the blocks they belong.

This can be done by repeated block interchanges. Each zero of the 0&1-block takes part in at most n/t interchanges, whereas each one of the input takes part in at most two block interchanges. Therefore the total work done here is $O(n)$. The blocks at the end of the array are called *finished* if they got all their zeros. The last *unfinished*, or *half-finished*, 1-block may have obtained only some of the zeros that should be there. The half-finished block is however seen as a 1-block, though the zeros at the end are kept untouched.

Let the *leader* of a 0-block be its *first* element and the *leader* of a 1-block its *last* element. Now the basic steps of the transformation, called *one-to-many transformation*, are the following (see also Fig. 1):

(a) $\quad 0_5 \; 0_7 \; 0_8 \; 0_9 \; 0_{12} \; 0_{14} \; 1_1 \; 1_2 \; 1_3 \; 1_4 \; 1_6 \; 1_{10} \; 1_{11} \; 1_{13} \; 1_{15} \; 1_{16}$

(b) $\quad 1 \quad 1 \quad 1 \quad 1 \quad 0 \quad 1 \quad 0 \quad 0 \quad 0 \quad 1 \quad 1 \quad 0 \quad 1 \quad 0 \quad 1 \quad 1$

(c) $\quad 0_5 \; 0_7 \; 0_8 \; 0_9 \; 1_1 \; 1_2 \; 1_3 \; 1_4 \; 1_6 \; 1_{10} \; 1_{11} \| 0_{12} \; 0_{14} \; 1_{13} \; 1_{15} \; 1_{16}$

(d) $\quad 1_1 \; 1_2 \; 1_3 \; 1_4 \; 0_5 \; 0_7 \; 0_8 \; 0_9 \; 1_6 \; 1_{10} \; 1_{11} \| 0_{12} \; 0_{14} \; 1_{13} \; 1_{15} \; 1_{16}$

(e) $\quad 1_1 \; 1_2 \; 1_3 \; 1_4 \; 0_5 \; 0_7 \; 0_8 \; 1_6 \; 0_9 \; 1_{10} \; 1_{11} \| 0_{12} \; 0_{14} \; 1_{13} \; 1_{15} \; 1_{16}$

(f) $\quad 1_1 \; 1_2 \; 1_3 \; 1_4 \; 0_5 \; 0_7 \; 0_8 \; 1_6 \; 0_9 \; 0_{12} \; 1_{10} \; 1_{11} \; 0_{14} \; 1_{13} \; 1_{15} \; 1_{16}$

Fig. 1. One-to-many transformation. (a) Example input with $n = 16$ and $\lg n = 4$. (b) Placement array. (c) Single 0&1-block is handled. (d) Block permutation is performed. (e) 0/1-blocks are transformed to 0&1-blocks. (f) Half-finished block is cleaned up.

1. Divide the input array A into blocks of size t.
2. If there exists a 0&1-block then move its zeros to their own blocks.
3. Merge the unfinished blocks such that the sites of their leaders are in sorted order. This way the elements will come closer to their final positions.
4. Transform the unfinished 0/1-blocks to 0&1-blocks such that each element is placed in its own block.
5. Move the zeros (if any) at the end of the half-finished block over the ones in the block.

To perform Step 1 only the value t has to be computed but this is easily done in linear time. Step 2 requires linear time as well (cf. the discussion above). Step 5 requires only $O(t)$ time. The most critical parts are the merging of the blocks (Step 3) and the transformation from 0/1-blocks to 0&1-blocks (Step 4). We show first that

Step 4 can be executed in linear time. The proof of the next lemma is similar to that given in [19, Lemma 2, Step 3] or [10, Section 3.2].

Lemma 2. *Step 4 of the one-to-many transformation can be done in linear time for any blocking factor t.*

Proof. Let $X_1, X_2, \ldots, X_{n/t}$ be the order of the blocks after the block permutation in Step 3. Consider any boundary between a 1-block and a 0-block in this sequence. Since the leader of a 1-block is its last element and the leader of a 0-block is its first element, no element has to be moved across the boundary. Let us therefore divide the sequence $X_1, X_2, \ldots, X_{n/t}$ into pieces, where each piece consists of two subsequences, a sequence of 0-blocks followed by a sequence of 1-blocks. Let us number the pieces from 1 to p.

Consider an arbitrary piece $X_{i_1}, X_{i_2}, \ldots, X_{i_j}$ and assume that this piece contains ℓ_i 0-blocks and m_i 1-blocks. Now only some of the zeros in the last 0-block should be moved to the left and some of the ones in the first 1-block should be moved to the right (see Fig. 2). The zeros to be moved in the last 0-block are obtained into their correct blocks by performing at most m_i block interchanges. Each one is involved in at most one block interchange. Therefore the work here is proportional to $m_i \cdot t + n$. In the same way, one can show that the work required when moving the ones (now at the end of the first 0&1-block) to the right is proportional to $\ell_i \cdot t + n$. Since $\sum_{i=1}^{p}(\ell_i + m_i) = n/t$, the claim follows. $\qquad\square$

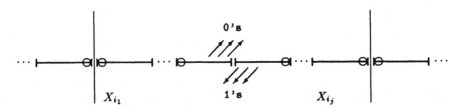

Fig. 2. Block sequence $X_{i_1}, X_{i_2}, \ldots, X_{i_j}$. A leader of each block is marked by a circle.

The question that remains to be answered is how the merging in Step 3 is implemented. First, assume that the blocking factor is $\lg n$. Now one possibility is to store the sites of the leaders explicitly. If these are available, Step 3 can be implemented by using any in-place merging algorithm. Since the blocks are of equal size they can be easily swapped in time proportional to their size. Hence, Step 3 can be done $O(n)$ time.

Algorithm B performs the one-to-many transformation as described above and solves the subproblems of size $\lg n$ by Algorithm A. Now Lemma 1 implies the result of the next lemma.

Lemma 3. *Algorithm B solves a stable unpartitioning problem of size n $(= 2^k)$ in $O(n)$ time with $O(n/\log_2 n)$ counters, each requiring $O(\log_2 n)$ bits.*

In Algorithm C the sites of the leaders are stored in a bit array. This means that we need $O(\log_2 n)$ time when manipulating a site. Hence the total time needed for the one-to-many transformation is $O(n \log_2 n)$. However, the number of element moves is only linear! The resulting 0&1-blocks are unpartitioned by Algorithm B. The critical observation is that we have to store only $O(\log_2 n / \log_2 \log_2 n)$ counters, each of $O(\log_2 \log_2 n)$ bits (and $O(1)$ indices, each of $O(\log_2 n)$ bits). The total number of bits required is only $O(\log_2 n)$. Therefore we can pack the integers into few words and manipulate them effeciently with shift operations. Thus each block is handled in $O(\log_2 n)$ time using $O(1)$ words of $O(\log_2 n)$ bits. The performance of Algorithm C is stated in the following lemma.

Lemma 4. *Algorithm C solves a stable unpartitioning problem of size n ($= 2^k$) in $O(n \log_2 n)$ time, using an array of $O(n)$ bits and a constant amount of words of $O(\log_2 n)$ bits, but makes only $O(n)$ moves.*

The space requirements can be further reduced by using bit stealing. (Note that in order to use bit stealing the f-function must be known.) Let us now divide the input into blocks of size $\sqrt{n}$ (or more precisely into blocks of size $2^{(\lg n)/2}$). The first $c\sqrt{n}$ zeros and $c\sqrt{n}$ ones are saved in an internal buffer, where c is a suitably chosen constant. Of course, it might happen that we do not have as many zeros or ones as needed. Such an input instance is however easily solved by moving the elements that we have shortage of to their proper places one-by-one. For example, if we had a shortage of zeros, each zero would be involved in at most $c\sqrt{n}$ block interchanges, whereas each one in at most one interchange. This totals $O(n)$ time. The same can be done if we had a shortage of ones. Hence assume that we have sufficiently many zeros and ones.

One can view the merging in the one-to-many transformation as an unpartitioning problem (cf. the discussion in the beginning of this section). Now this unpartitioning is implemented by Algorithm C and the elements of the internal buffer are used to steal the bits needed. The new placement array is computed in linear time by scanning through the original placement array. Since the size of the placement array created is about $\sqrt{n}$ it can be stored as a part of the internal buffer. Step 3 of the transformation requires $O(\sqrt{n} \log_2 n)$ time for comparisons and index calculations, and $O(\sqrt{n})$ block swaps; so $O(n)$ time in total. Hence, the whole transformation requires linear time. The subproblems of size $\sqrt{n}$ are also solved by Algorithm C and the bits required are stolen from the internal buffer. The post-processing step, where the elements of the internal buffer are moved to their proper places, is again done in linear time by repeated block interchanges.

We have thus obtained a new algorithm, call it Algorithm D, which is as fast as Algorithm C but requires only a constant amount of additional space.

Lemma 5. *Algorithm D solves a stable unpartitioning problem of size n ($= 2^k$) in $O(n \log_2 n)$ time and constant extra space, but makes only $O(n)$ moves.*

Our final algorithm, Algorithm E is again based on $\lg n$-blocking. The general structure of Algorithm E is similar to that of the previous algorithms. Now Algorithm D is employed for implementing Step 3 of the one-to-many transformation and Algorithm B for unpartitioning the blocks of size $\lg n$. As in Algorithm D the one-to-many transformation takes $O(n)$ time, but now only a constant amount of additional space is needed. As in Algorithm C, we use the technique of packing small integers to solve the subproblems in $O(\log_2 n)$ time with a constant number of words of $O(\log_2 n)$ bits. The total time for solving the subproblems is linear. Therefore Algorithm E requires $O(n)$ time and $O(1)$ extra space.

Up to now we have assumed that n, the number of elements is a power of two. If this is not the case, the following recursive method can be used to reduce the original problem to subproblems, whose size is a power of two. First, compute by repeated doubling the largest 2^k that is smaller than n. Second, scan through the first 2^k positions of the placement array and count the total number of 0-bits n_0 and 1-bits n_1 in there. Third, interchange the block of zeros (if any) lying after the first n_0 zeros with the block of the first n_1 ones. Fourth, unpartition the first 2^k elements with Algorithm E. Finally, use the same method recursively for unpartitioning the last $n - 2^k$ elements. Since Algorithm E runs in linear time, the running time of this method is proportional to $\sum_{k=\lfloor \log_2 n \rfloor}^{0} 2^k$, which totals $O(n)$ time. As comparared to Algorithm E, the space requirements are increased only by an additive contant. Hence, we have proved the following theorem.

Theorem 1. *A stable unpartitioning problem of size n can be solved in $O(n)$ time and $O(1)$ extra space.*

4   Restoring Selection

In this section we implement the (slow) linear-time selection algorithm of Blum et al. [1] to solve the restoring selection problem space-efficiently. In the next section this algorithm is then used to solve the stable selection problem in minimum space.

Let us recall the essence of the prune-and-search algorithm for selecting the kth smallest element in the multiset S of n elements (cf. the implementation given in [5, Algorithm 3.17] which requires $O(\log_2 n)$ extra space):

1. If n is "small" then determine the median p of S in a brute force manner and return p.
2. Divide S into $\lfloor n/5 \rfloor$ blocks of size 5, ignore excess elements.
3. Let M be the set of medians of these blocks. Compute the median p of M by applying the selection algorithm recursively.
4. Partition S stably into three parts $S_<, S_=,$ and $S_>$ such that each element of $S_<$ is less than p, each element of $S_=$ is equal to p, and each element of $S_>$ is greater than p.
5. If $|S_<| < k \le |S_<| + |S_=|$ then return p. Otherwise call the selection algorithm recursively to find the kth smallest element in $S_<$ if $k \le |S_<|$, or the $(k - |S_<| - |S_=|)$th smallest element in $S_>$ if $k > |S_<| + |S_=|$.

Next we describe the implementation details that will make it possible to restore the elements into their original positions. In Step 1 the median of small sets is computed by the quadratic algorithm that will not move the elements. (We do not specify, when to switch to the brute force algorithm, but refer to any textbook on algorithms, e.g. [5, Section 3.6].) In Step 3 the medians of the blocks are also found without moving the elements. To access a block median we store an offset indicating the place of the median inside the block. Here we need $3n/5 + O(1)$ bits in total. A convenient place to store the set M is at the front of the input array. In [5, Algorithm 3.17] it is shown how the elements of M are moved in-place. It is easy to reverse this computation.

In Step 4 the multiset S is partitioned stably by using the linear-time minimum space algorithm [11]. Now we use $2n$ bits to indicate whether before partitioning the corresponding position contained an element of $S_<, S_=$, or $S_>$. By using the stable unpartitioning algorithm developed in Section 3, we can reverse the computation done in Step 4. In Step 5 it is again convinient to move the multiset $S_\diamond (\diamond \in \{<, >\})$ that we shall work with to the front of the array. If the multisets are stored in order $S_<, S_> S_=$ or $S_>, S_< S_=$, the block interchanges performed can be easily reversed.

The sizes of the manipulated multisets are stored in unary form. At each "recursive call" we have to store also the type of the call telling whether the procedure was called in Step 3 or Step 5. This requires only $O(\log_2 n)$ bits. When these sizes and types are available the recursive calls can be handled iteratively. The overall organization of the storage is simply a "stack" of bit sequences. Of course, these sequences are stored in a bit array. From the standard analysis of the prune-and-search algorithm it follows that the total number of extra bits needed is linear.

We summarize the above discussion in the following theorem.

Theorem 2. *The restoring selection problem of size n can be solved in $O(n)$ time using an extra array of $O(n)$ bits and a constant amount of words of $O(\log_2 n)$ bits.*

In adaptive sorting it is extremely important not to destroy the existing order among the input data. Therefore our algorithm for restoring selection could be used to improve the space-efficiency of some adaptive sorting algorithms (e.g. that of Slabsort presented in [15]). We leave it as an open problem whether there exists a minimum space algorithm for restoring selection. Such an algorithm would make it possible to develop new in-place sorting algorithms that are also adaptive.

5 Stable Selection

Next we show that stable selection is possible in linear time in minimum space. Our construction is based on a minimum space algorithm for selecting an approximate median. When this is used as a subroutine in the standard prune-and-search algorithm (cf. Section 4), instead of the median-of-medians method, an in-place algorithm for stable selection is obtained.

Let $S = \{x_1, x_2, \ldots, x_n\}$ be a multiset. Further, let the *rank* of an element $x_j \in S$ be the cardinality of the multiset $\{x_i \in S \mid x_i < x_j \text{ or } (x_i = x_j \text{ and } i \le j)\}$. An element x is said to be an *approximate median* of S, if there exists an element $x_i \in S$ such that $x_i = x$ and that the rank of x_i is in the interval $[\alpha n..(1 - \alpha)n]$, for some fixed constant α, $0 < \alpha \le 1/2$. In the following, we do not try to determine any value for the constant α; the existence of such a constant is enough for our purposes.

To find an approximate median for a multiset S such that the relative order of the identical elements is not changed, we use $\sqrt{n}$-blocking. The median of blocks of size $\sqrt{n}$ is computed by the algorithm of Section 4. After computing the median of a block, the block is partitioned stably and in-place such that the elements equal to the median come to the front of the block. Then the first element of each block is used to find the median of medians. Here we use the trivial quadratic-time algorithm that do not move the elements. It is easy to see that the final output is an approximate median of S (cf. the analysis of the standard selection algorithm). The overall running time is linear and the number of extra bits needed $O(\sqrt{n})$.

This algorithm can be further improved by bit stealing. Next we show how the extra bits can be stolen from an internal buffer which is created as a preprocessing step. Our technique is similar to that used by Lai and Wood [13] in their selection algorithm, or Levcopoulos and Petersson [14] in their adaptive sorting algorithm. The contribution here is that the bits can be stolen without losing stability.

Assume that t bits are needed, $t \in O(\sqrt{n})$. Let S_0 denote the first $2t$ elements of the original input S. Now sort S_0 stably for example by the straight selection sort algorithm. This takes $O(t^2)$ time, that is in our case linear time. First, consider the case where none of the elements appears more than t times in S_0. By pairing the first element with the $(t + 1)$st element, the second element with the $(t + 2)$nd element, and so on, t pairs of different elements are obtained. These pairs are then used to represent the bits required. An approximate median is then searched for the elements in $S \setminus S_0$. Since the size of the buffer is proportional to $\sqrt{n}$ the result will still be an approximate median (under the assumption that n is large enough, but recall that small multisets are handled separatively).

Second, assume that some element x appears more than t times in S_0. Partition S (including S_0) stably into two parts: S_1 containing the elements equal to x, and S_2 containing the elements not equal to x. If the cardinality of S_2 is less than t the input instance is easy. The block S_2 is sorted by the stable, quadratic-time selection-sort algorithm and then S_1 is embedded into the result of this sort by a single block interchange. In this case, even the actual median can be returned in linear time. If the cardinality of S_2 is greater than t, the first t elements of S_1 (forming S_3) and the first t elements of S_2 (forming S_4) are used to create the internal buffer. To do this the blocks $S_1 \setminus S_3$ and S_4 are interchanged. Finally, an approximate median is searched for the elements belonging to $S \setminus (S_3 \cup S_4)$.

To summarize, an approximate median of n elements can be found in $O(n)$ time, using $O(1)$ extra space, such that the relative order of the identical elements is retained. The routine for finding an approximate median can be applied in the prune-and-search selection algorithm, instead of using the median-of-medians method. Hence, the result of the following theorem follows from the analysis of the prune-and-search algorithm.

Theorem 3. *The stable selection problem of size n can be solved in $O(n)$ time, using only $O(1)$ extra space.*

6   Stable sorting of multisets

In this section we describe and analyse a Quicksort type algorithm that sorts multisets stably in optimal time and minimum space. Let us assume that S, the multiset to be sorted is non-empty. The basic steps of the algorithm are:

1. Find the median p of S.
2. Partition S stably into three parts $S_<, S_=, S_>$ such that each element of $S_<$ is less than p, each element of $S_=$ is equal to p, and each element of $S_>$ is greater than p.
3. Sort the two multisets $S_<$ and $S_>$ recursively if they are not empty.

In Step 1 the median is determined stably and in-place by the algorithm of Section 5. Step 2 is implemented stably and in-place by using the algorithm given in [11]. To avoid the recursion stack in Step 3 we can use the implementation trick (based on stoppers) proposed in [3] or [24]. Hence, also this step can be implemented in-place (without losing stability).

Now we are ready to prove our main result.

Theorem 4. *Quicksort can be adapted to sort stably a multiset of size n with multiplicities $n_1, n_2, \ldots, n_m$ in $O(n \log_2 n - \sum_{i=1}^{m} n_i \log_2 n_i + n)$ time and $O(1)$ extra space.*

Proof. According to the previous discussion our implementation is stably and in-place. So let us concentrate on analysing the running time of the algorithm.

Let $S_1, S_2, \ldots, S_m$ be the minimum partition of the input into classes of equal elements. Without loss of generality, we can assume that the elements in S_i are smaller than those in S_j, for all $i < j$. Furthermore, let the cardinality of these subsets be $n_1, n_2, \ldots, n_m$, respectively. Now we denote by $T(i..k)$ the time it takes to sort the classes $S_i, S_{i+1}, \ldots, S_k$. Since median finding and partitioning are done in linear time, there exists a constant c such that the running time of the algorithm is bounded by the following recurrence

$$T(i..k) \leq \begin{cases} T(i..j-1) + T(j+1..k) + c(\sum_{h=i}^{k} n_h) & \text{for } i < k \text{ and} \\ cn_i & \text{for } i = k. \end{cases}$$

Let us use the following shorthand notations: $N_1 = \sum_{h=i}^{j-1} n_h$, $N_2 = \sum_{h=j+1}^{k} n_h$, $N = \sum_{h=i}^{k} n_h$. It is easy to establish by induction that $T(i..k) \leq c(N \log_2 N - \sum_{h=i}^{k} n_h \log_2 n_h + N)$ ($0 \log_2 0$ means 0). This is because $N_i \leq N/2$ ($i = 1, 2$) and therefore $N_1 \log_2 N_1 + N_2 \log_2 N_2 + n_j \log_2 n_j + N_1 + N_2 \leq N \log_2 N$. Hence we have proved that $T(1..m) \in O(n log_2 n - \sum_{i=1}^{m} n_i \log_2 n_i + n)$. $\qquad\square$

Acknowledgements

The discussions with Niels Christian Juul and Ola Petersson are gratefully acknowledged.

References

1. M. Blum, R.W. Floyd , V. Pratt, R.L. Rivest, R.E. Tarjan: Time bounds for selection. *Journal of Computer and System Sciences* **7** (1973) 448–461

2. S. Carlsson, J.I. Munro, P.V. Poblete: An implicit binomial queue with constant insertion time. *1st Scandinavian Workshop on Algorithm Theory*. Lecture Notes in Computer Science **318**. Springer-Verlag, 1988, pp. 1–13

3. B. Ďurian: Quicksort without a stack. *Mathematical Foundations of Computer Science 1986*. Lecture Notes in Computer Science **233**. Springer-Verlag, 1986, pp. 283–289

4. T.F. Gonzalez, D.B. Johnson: Sorting numbers in linear expected time and optimal extra space. *Information Processing Letters* **15** (1982) 119–124

5. E. Horowitz, S. Sahni: *Fundamentals of Computer Algorithms*. Computer Science Press, 1978

6. E.C. Horvath: Stable sorting in asymptotically optimal time and extra space. *Journal of the ACM* **25** (1978) 177–199

7. B.-C. Huang, M.A. Langston: Practical in-place merging. *Communications of the ACM* **31** (1988) 348–352

8. B.-C. Huang, M.A. Langston: Fast stable merging and sorting in constant extra space. *International Conference on Computing and Information*, 1989, pp. 71–80

9. B.-C. Huang, M.A. Langston: Stable dublicate-key extraction with optimal time and space bounds. *Acta Informatica* **26** (1989) 473–484

10. J. Katajainen, C. Levcopoulos, O. Petersson: Space-efficient parallel merging. *PARLE'92 - Parallel Architectures and Languages Europe* (to appear)

11. J. Katajainen, T. Pasanen: Stable minimum space partitioning in linear time. *BIT* (to appear)

12. M.A. Kronrod: Optimal ordering algorithm without operational field. *Soviet Mathematics* **10** (1969) 744–746

13. T.W. Lai, D. Wood: Implicit selection. *1st Scandinavian Workshop on Algorithm Theory*. Lecture Notes in Computer Science **318**. Springer-Verlag, 1988, pp. 14–23

14. C. Levcopoulos, O. Petersson: An optimal adaptive in-place sorting algorithm. *8th International Conference on Fundamentals of Computation Theory*. Lecture Notes in Computer Science **529**. Springer-Verlag, 1991, pp. 329–338

15. C. Levcopoulos, O. Petersson: Sorting shuffled monotone sequences. *Information & Computation* (to appear)

16. A.M. Moffat, O. Petersson: An overview of adaptive sorting. *Australian Computer Journal* (to appear)

17. J.I. Munro: An implicit data structure supporting insertion, deletion, and search in $O(\log^2 n)$ time. *Journal of Computer and System Sciences* **33** (1986) 66–74

18. J.I. Munro, V. Raman: Sorting multisets and vectors in-place, *2nd Workshop on Algorithms and Data Structures*. Lecture Notes in Computer Science **519**. Springer-Verlag, 1991, pp. 473–480

19. J.I. Munro, V. Raman, J.S. Salowe: Stable in situ sorting and minimum data movement. *BIT* **30** (1990) 220–234

20. J.I. Munro, P.M. Spira: Sorting and searching in multisets. *SIAM Journal on Computing* **5** (1976) 1–8

21. J.S. Salowe, W.L. Steiger: Simplified stable merging tasks. *Journal of Algorithms* **8** (1987) 557–571

22. J.S. Salowe, W.L. Steiger: Stable unmerging in linear time and constant space. *Information Processing Letters* **25** (1987) 285–294

23. L.M. Wegner: Quicksort for equal keys. *IEEE Transactions on Computers* **C34** (1985) 362–367

24. L.M. Wegner: A generalized, one-way, stackless Quicksort. *BIT* **27** (1987) 44–48

This article was processed using the LaTeX macro package with LLNCS style

A Framework for Adaptive Sorting

Ola Petersson[*]   Alistair Moffat[†]

Abstract: A sorting algorithm is adaptive if it sorts sequences that are close to sorted faster than it sorts random sequences, where the distance is determined by some measure of presortedness. Over the years several measures of presortedness have been proposed in the literature, but it has been far from clear how they relate to each other. We show that there exists a natural partial order on the set of measures, which makes it possible to say that some measures are superior to others. We insert all known measures of presortedness into the partial order, and thereby provide a powerful tool for evaluating both measures and adaptive sorting algorithms. We further present a new measure and show that it is a maximal known element in the partial order, and thus that any sorting algorithm that optimally adapts to the new measure also optimally adapts to all other known measures of presortedness.

1 Introduction

Despite the fact that $\Theta(n \log n)$ time is necessary to sort n elements in a comparison-based model of computation [12], some instances seem easier than others and can be sorted faster than indicated by the lower bound. This observation was first made by Burge [1], who identified this *instance easiness* with the amount of existing order (*presortedness*) in the sequence.

After being considered by, among others, Cook and Kim [3] and Mehlhorn [12], the concept of presortedness was formalized by Mannila [11], who studied several measures of presortedness. Mannila also studied how a sorting algorithm can profit from, and thereby adapt to, existing order. A sorting algorithm is *adaptive* with respect to a measure of presortedness if it sorts all sequences, but runs faster on those having a high degree of presortedness according to the measure.

Different measures of presortedness capture different sequences as presorted, and Mannila implicitly posed the question of whether one can somehow compare measures. In this paper we show that it is indeed possible. We introduce a natural relation, denoted $\supseteq$, which defines a partial order on the set of measures. For two measures M_1 and M_2, if $M_1 \supseteq M_2$ then every sorting algorithm that optimally adapts to M_1 optimally adapts to M_2 as well. Hence, from an algorithmic point of view M_1 is at least as good as M_2. As we insert all known measures into the partial order, our work has consequences for most of the previous results in the area. More importantly, however, the partial order provides a framework for evaluating both new measures of presortedness and new adaptive sorting algorithms.

We further present a new measure of presortedness that is a maximal element in the known partial order. Thus, every sorting algorithm that optimally adapts to our new measure is optimal with respect to all known measures.

---
[*]Dept. Comp. Sci., Lund Univ., Box 118, S-221 00 Lund, Sweden. Email: ola@dna.lth.se
[†]Dept. Comp. Sci., Univ. Melbourne, Parkville 3052, Australia. Email: alistair@cs.mu.oz.au

2 Measures of presortedness

Let $X = \langle x_1, \ldots, x_n \rangle$ be a *sequence* of n distinct elements x_i from some totally ordered set. A sequence obtained by deleting zero or more elements from X is called a *subsequence* of X. Further, let $|X|$ denote the *length* of X and $\|S\|$ the *cardinality* of a set S. S_n is the set of all permutations of $\{1, \ldots, n\}$, and $\log x = \log_2(\max\{2, x\})$.

Instance easiness is evaluated by a *measure of presortedness*, a non-negative integer function on a sequence X that reflects how much X differs from the sorted permutation of X. We follow Mannila [11] and present some well known measures of presortedness. The most common measure is the number of *inversions* in a sequence, that is, the number of pairs of elements that are out of order. More formally, $Inv(X) = \|\{(i,j) \mid 1 \leq i < j \leq n \text{ and } x_i > x_j\}\|$. Another well known measure of presortedness is the number of *runs* within the input, defined as $Runs(X) = \|\{i \mid 1 \leq i < n \text{ and } x_i > x_{i+1}\}\| + 1$. The third measure considered by Mannila was *Rem*, the minimum number of elements that need to be removed to leave a sorted sequence, i.e., $Rem(X) = n - \max\{k \mid X \text{ has an ascending subsequence of length } k\}$. The fourth measure we would like to mention is $Exc(X) = $ the minimum number of exchanges of arbitrary elements needed to sort X.

3 Adaptive sorting algorithms

An adaptive sorting algorithm is a sorting algorithm that adapts to the presortedness in the input. The more presorted a sequence is, the faster it is sorted. Moreover, the algorithm adapts without a priori knowledge of the amount of presortedness.

Besides Straight Insertion Sort, the most widely known adaptive sorting algorithm is Natural Mergesort [8]. Given a sequence, the algorithm first finds the runs by a linear scan. These runs are then repeatedly merged pairwise until there is just one run left, which is the sorted sequence. It is easy to see that the time consumed by Natural Mergesort on a sequence X of length n with $Runs(X) = k$ is $\Theta(n \log k)$.

Another simple adaptive sorting algorithm is by Cook and Kim [3]. The algorithm adapts to *Rem* and starts by spending linear time removing $\Theta(Rem(X))$ elements from the input X such that what remains in X is a sorted sequence. The removed elements are sorted separately by a worst-case optimal algorithm. Finally, this sorted sequence is merged with the sorted sequence that remained in X. The time consumed on a sequence X of length n with $Rem(X) = k$ is $\Theta(n + k \log k)$.

Is it possible to profit more from the presortedness than the given algorithms do? The concept of an optimal algorithm was given in a general form by Mannila [11]. We use the following equivalent definitions.

DEFINITION 1 *Let M be a measure of presortedness, and T_n the set of comparison trees for the set S_n. Then, for any $k \geq 0$ and $n \geq 1$,*

$$C_M(n, k) = \min_{T \in T_n} \max_{\pi \in below_M(n,k)} \{\text{the number of comparisons spent by } T \text{ to sort } \pi\},$$

where $below_M(n, k) = \{\pi \mid \pi \in S_n \text{ and } M(\pi) \leq k\}$.

DEFINITION 2 *Let M be a measure of presortedness, and S a comparison-based sorting algorithm that uses $T_S(X)$ steps on input X. We say that S is M-optimal, or optimal with respect to M, if $T_S(X) = O(C_M(|X|, M(X)))$.*

When proving optimality of an adaptive sorting algorithm we use

THEOREM 1 *Let M be a measure of presortedness. Then*

$$C_M(n,k) = \Theta(n + \log \|below_M(n,k)\|).$$

PROOF The lower bound follows from a combination of trivial and straightforward information theoretic argument; the upper bound is due to Fredman [6].   □

A lower bound for comparison-based sorting algorithms with respect to a measure of presortedness is thus obtained by bounding the size of the *below*-set from below.

Mannila [11] proved that $C_{Runs}(n,k) = \Omega(n \log k)$. As Natural Mergesort was shown to match this bound above, Definition 2 gives that Natural Mergesort is optimal with respect to *Runs*. For *Rem*, it is not hard to prove that $C_{Rem}(n,k) = \Omega(n + k \log k)$ [11]. Hence, Cook and Kim's algorithm is *Rem*-optimal.

In general, proving that a sorting algorithm is M-optimal is done in two parts. One part is a lower bound on $C_M(n,k)$, which is obtained by bounding the cardinality of the *below*-set from below. The other part is an upper bound on the time used by the algorithm, expressed in terms of M.

4   A partial order on the measures

Presented with a measure it is natural to ask is whether it is somehow related to some other measure(s). Is it even "superior" to other measures, whatever that means?

Let us examine the relationship between *Runs* and *Rem*. By definition, each run in a sequence X, except the last one, is defined by a consecutive pair of elements x_i and x_{i+1}, for which it holds that $x_i > x_{i+1}$. At most one of these elements can belong to a longest ascending subsequence in X. Hence, at least one of them contributes to $Rem(X)$. It follows that $Runs(X) \leq Rem(X) + 1$, for any sequence X.

It is now tempting to conclude that *Runs* is therefore "at least as good" as *Rem*, since any sequence with a low *Rem* value has also a low *Runs* value. To make such a statement we have to be very careful. Since we are concerned with sorting, what we would really like to prove is that any sequence can be sorted at least as fast by a *Runs*-optimal algorithm as by a *Rem*-optimal one. But this is not true. Consider

$$X = \langle 1, 2, \ldots, n - \sqrt{n} - 1, n - \sqrt{n}, n, n-1, n-2, \ldots, n - \sqrt{n} + 1 \rangle,$$

for which $Runs(X) = \sqrt{n}$ and $Rem(X) = \sqrt{n} - 1$. As $C_{Runs}(n, \sqrt{n}) = \Theta(n \log n)$, a *Runs*-optimal sorting algorithm may spend $\Theta(n \log n)$ time on X, while any *Rem*-optimal algorithm must complete in time $O(C_{Rem}(n, \sqrt{n} - 1))$ which is $\Theta(n)$.

Motivated by the preceding discussion, we make the following definition.

DEFINITION 3 *Let M_1 and M_2 be measures of presortedness. We say that*

- M_1 *is superior to* M_2, *denoted* $M_1 \supseteq M_2$, *if*

$$C_{M_1}(|X|, M_1(X)) = O(C_{M_2}(|X|, M_2(X)));$$

- M_1 *is strictly superior to* M_2, *denoted* $M_1 \supset M_2$, *if* $M_1 \supseteq M_2$ *and* $M_2 \not\supseteq M_1$;

- M_1 *and* M_2 *are equivalent, denoted* $M_1 \equiv M_2$, *if* $M_1 \supseteq M_2$ *and* $M_2 \supseteq M_1$.

The importance of Definition 3 becomes evident if we combine it with Definition 2:

THEOREM 2 *Let M_1 and M_2 be measures of presortedness.*

- *If $M_1 \supseteq M_2$, every M_1-optimal sorting algorithm is M_2-optimal.*

- *If $M_1 \supseteq M_2$ and S is a sorting algorithm that is not M_2-optimal then S is not M_1-optimal.*

To show that measure M_1 is superior to measure M_2 we need an upper bound on $C_{M_1}(n, k)$, which can be obtained from an upper bound on an algorithm [2]:

LEMMA 3 *Let M be a measure of presortedness, for which $C_M(n, k) = \Omega(f(n, k))$, where f is non-decreasing on n and k. If there exists a comparison-based sorting algorithm S that sorts a sequence X in $C_S(X) = O(f(|X|, M(X)))$ comparisons, then $C_M(n, k) = \Theta(f(n, k))$.*

Using Lemma 3 it follows that $C_{Runs}(n, k) = \Theta(n \log k)$, by the upper bound on Natural Mergesort, and that $C_{Rem}(n, k) = \Theta(n + k \log k)$, by the upper bound on Cook and Kim's algorithm.

Consider now the measures *Rem* and *Exc*. Carlsson et al. [2] proved that $C_{Exc}(n, k) = \Omega(n + k \log k)$ and $Rem(X) \leq 2 \cdot Exc(X)$, for any sequence X.

LEMMA 4 *Rem $\supseteq$ Exc.*

PROOF By the above upper bound on $C_{Rem}(n, k)$ and the results of Carlsson et al.,

$$
\begin{aligned}
C_{Rem}(|X|, Rem(X)) &= \Theta(|X| + Rem(X) \log Rem(X)) \\
&= O(|X| + Exc(X) \log Exc(X)),
\end{aligned}
$$

which is $O(C_{Exc}(|X|, Exc(X)))$. Definition 3 then gives the lemma. $\square$

To prove that *Rem* is *strictly* superior to *Exc* we need the following lemma.

LEMMA 5 *Let M_1 and M_2 be measures of presortedness. If, for all constants n_0 and $c > 0$, there exists a sequence X of length $n > n_0$ such that $C_{M_1}(n, M_1(X)) < c \cdot C_{M_2}(n, M_2(X)))$, then $M_2 \not\supseteq M_1$.*

THEOREM 6 *Rem $\supset$ Exc.*

PROOF It remains to prove that $Exc \not\supseteq Rem$. Consider the almost sorted sequence $X = \langle n, 1, 2, \ldots, n-1 \rangle$. We have $Rem(X) = 1$ and $Exc(X) = n-1$. As $C_{Rem}(n, 1) = \Theta(n)$ and $C_{Exc}(n, n-1) = \Omega(n \log n)$, Lemma 5 implies that $Exc \not\supseteq Rem$. $\square$

In general, proving that measure M_1 is superior to measure M_2 involves the steps we went through above. First, we need an upper bound on $C_{M_1}(n, k)$, which is given by the number of comparisons used by a sorting algorithm that adapts to M_1, and a lower bound on $C_{M_2}(n, k)$. The upper bound is then shown to match the lower bound by deriving an upper bound on M_1 in terms of M_2. To prove strict superiority we further give a sequence X of length n, which can be extended to arbitrary large n, for which $C_{M_2}(n, M_2(X))$ is asymptotically greater than $C_{M_1}(n, M_1(X))$.

Let us return to the comparison of *Runs* and *Rem*. We gave a sequence proving that $Runs \not\supseteq Rem$. To see that the two measures are unrelated consider the sequence $X = \langle n/2+1, n/2+2, \ldots, n, 1, 2, \ldots, n/2 \rangle$. It holds that $Runs(X) = 2$ and $Rem(X) = n/2$. From above, we have $C_{Runs}(n, 2) = \Theta(n)$ and $C_{Rem}(n, n/2) = \Omega(n \log n)$. Hence, $Rem \not\supseteq Runs$, by Lemma 5.

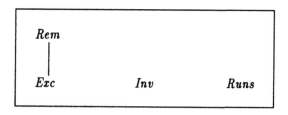

FIGURE 1: The partial order $\mathcal{P}(\mathcal{M}, \supseteq)$ obtained so far.

The relation $\supseteq$ defines a *partial order* on $\mathcal{M}$, the set of measures of presortedness, which we will denote by $\mathcal{P}(\mathcal{M}, \supseteq)$, and illustrate by a Hasse diagram. See Figure 1.

The transitivity of $\supseteq$ is a very helpful property when comparing measures of presortedness. For instance, as $Rem \supseteq Exc$ and $Rem \not\supseteq Runs$, it follows that $Exc \not\supseteq Runs$. Among the measures we have seen so far, Rem and Exc are the only ones that are related with respect to $\supseteq$. Hence, no arcs are missing in Figure 1.

5 Extending the partial order

We insert all known measures of presortedness into the partial order $\mathcal{P}(\mathcal{M}, \supseteq)$. A listing of these measures is given in Table 1; Figure 2 shows the resulting partial order. The proofs of the relationships shown are omitted in this extended abstract; however, they all appear in Petersson [14]. We can also prove that no arcs are missing in Figure 2, i.e., it shows all relations among the measures.

MEASURE	REF.
$Max(X) = \max_{1 \le i \le n} \lvert i - \pi(i) \rvert$	[2, 5]
$Block(X) = \lVert \{i \mid 1 \le i < n \text{ and } \pi(i) + 1 \neq \pi(i+1)\} \rVert + 1$	[2]
$Osc(X) = \sum_{i=1}^{n} \lVert \{j \mid 1 \le j < n \text{ and } \min\{x_j, x_{j+1}\} < x_i < \max\{x_j, x_{j+1}\}\} \rVert$	[9]
$SUS(X) = \min\{k \mid X \text{ is a shuffle of } k \text{ upsequences}\}$	[10]
$SMS(X) = \min\{k \mid X \text{ is a shuffle of } k \text{ monotone sequences}\}$	[10]
$Enc(X) = $ the number of encroaching sequences	[10, 15]
$Ham(X) = \lVert \{i \mid 1 \le i \le n \text{ and } i \neq \pi(i)\} \rVert$	[4, 5]
$DS(X) = \sum_{i=1}^{n} \lvert i - \pi(i) \rvert$	[3, 4]
$Par(X) = \max\{j - i \mid 1 \le i \le j \le n \text{ and } x_i \ge x_j\}$	[5]

TABLE 1: Other measures of presortedness that appear in the literature. Here, π is the permutation of $\{1, 2, \ldots, n\}$ for which $\pi(i) = rank(x_i, X) + 1$, for $1 \le i \le n$, where $rank(x_i, X) = \lVert \{x_j \mid 1 \le j \le n \text{ and } x_i > x_j\} \rVert$.

It is natural to ask if there is a single measure, superior to all measures in Figure 2, that is, a maximal element in $\mathcal{P}(\mathcal{M}, \supseteq)$. Indeed such a measure exists, namely

$$M(X) = \min\{C_{Block}(\lvert X \rvert, Block(X)), C_{Osc}(\lvert X \rvert, Osc(X)), C_{SMS}(\lvert X \rvert, SMS(X))\}.$$

However, this measure is somewhat artificial. We would prefer a measure that assumes the desired properties in a more natural manner. Starting from a general insertion sort algorithm, the next section is devoted to finding such a measure.

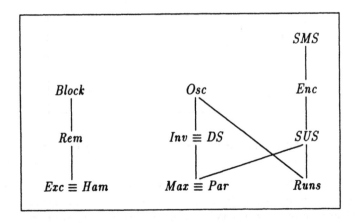

FIGURE 2: $\mathcal{P}(\mathcal{M}, \supseteq)$ after insertion of all measures in Table 1.

6 Insertion sort algorithms

The following describes a generic insertion sort algorithm [11].

> **procedure** *Insertion Sort* $(X: sequence)$
>> **for** $i := 1$ **to** n **do**
>>> Insert x_i in the sorted sequence formed by $x_1, \ldots, x_{i-1}$
>
> **end**

The adaptivity of the algorithm depends on how the sorted sequence is implemented. Using an array gives Straight Insertion Sort, running in $\Theta(n + Inv(X))$ time.

Mehlhorn [12] suggested applying an augmented (2-3)-tree, in which an insertion can be made in $O(\log \ell_i)$ (amortized) time, where ℓ_i is the distance from the end of the sequence to the position where x_i is inserted. That is, it supports exponential and binary search starting from the end of the sequence. The total time consumed by the algorithm, called A-Sort, is $O(n \log(Inv(X)/n))$, which is Inv-optimal.

Mannila [11] proposed that instead of being oblivious and starting the search from the last element in the sequence, one should exploit locality, and start from the previous element inserted. The algorithm, Local Insertion Sort, is implemented by storing the sorted sequence in a finger tree [12], letting the finger point at the latest inserted element. In this way x_i is inserted in $O(\log d_i)$ amortized time[1], where

$$d_i = \|\{k \mid 1 \leq k < i \text{ and } \min\{x_{i-1}, x_i\} < x_k < \max\{x_{i-1}, x_i\}\}\| + 1,$$

that is, the distance from the previous insertion point to the sought position. Thus, Local Insertion Sort runs in time $T_{LIS}(X) = O\left(\sum_{i=2}^{n} \log d_i\right)$. Consider the measure of presortedness obtained by multiplying[2] the d_i's:

$$Loc(X) = \prod_{i=2}^{n} d_i.$$

Then $T_{LIS}(X) = \Theta(n + \log Loc(X))$.

---

[1] The search for the insertion point takes $O(\log d_i)$ time in the worst case, and the actual insertion takes $O(1)$ amortized time.

[2] An equivalent measure, $logDist(X) = \sum \log d_i$, was studied by Katajainen et al. [7].

LEMMA 7 $C_{Loc}(n, k) = \Omega(n + \log k)$.

PROOF We bound the size of $below_{Loc}(n, k)$ from below. Divide the identity permutation in S_n into $n/k^{1/n}$ parts of equal length $k^{1/n}$, and permute the elements within each part. Then all elements are in their correct part, and thus, $d_i < k^{1/n}$. Hence, for every permutation π, constructed in this way, we have $Loc(\pi) < \prod k^{1/n} \leq k$, that is, $\pi \in below_{Loc}(n, k)$. The number of different π's constructed this way is

$$\|below_{Loc}(n, k)\| \geq \left(k^{1/n}!\right)^{\frac{n}{k^{1/n}}}.$$

Taking the logarithm and applying Theorem 1 completes the proof. □

As Local Insertion Sort matches this bound, we have by Definition 1 and Lemma 3

THEOREM 8 *Local Insertion Sort is Loc-optimal and* $C_{Loc}(n, k) = \Theta(n + \log k)$.

Using the techniques from Section 4, we can prove

THEOREM 9 *Loc* $\supset$ *Osc, Loc* $\supset$ *Block, and Loc* $\not\supset$ *SUS*.

Local Insertion Sort is based on the assumption that if the input is presorted, most insertions will occur close to the previous insertion, where the closeness is determined by the distance in *space*. Another natural way of measuring the distance is in terms of *time*. That is, the minimum number of insertions performed since one of the elements adjacent to the insertion position was inserted. To make a formal definition of the measure, we first extend the definition of d_i:

$$d_{i,j} = \|\{k \mid 1 \leq k < i \text{ and } \min\{x_i, x_j\} < x_k < \max\{x_i, x_j\}\}\| + 1.$$

Hence, $d_{i,j}$ tells the distance from x_j, $j < i$, to the insertion position of x_i. Note that $d_i = d_{i,i-1}$. Now, for $i > 1$ let

$$t_i = \min\{j \mid 1 \leq j < i \text{ and } d_{i,i-j} = 1\}.$$

As t_i tells the amount of history needed for inserting x_i, we call the corresponding measure of presortedness *Hist*, and define it analogously to *Loc*:

$$Hist(X) = \prod_{i=2}^{n} t_i.$$

Extending the analogy, we can define a *Historical Insertion Sort* that performs exponential and binary search in time rather than in space, and so

$$C_{HIS}(X) = O\left(\sum_{i=2}^{n} \log t_i\right) = O\left(n + \log \prod_{i=2}^{n} t_i\right) = O\left(n + \log Hist(X)\right) \quad (1)$$

comparisons are required. A data structure that supports the insertions within the stated resource bounds is the *Historical Search Tree* [13]. Basically, this structure is a family of $\log\log n$ sorted lists, implemented by finger search trees, where the i'th list, $0 \leq i \leq \log\log n$, contains the 2^{2^i} most recently inserted elements.

Replacing d_i by t_i in the proof of Lemma 7 gives that $C_{Hist}(n, k) = \Omega(n + \log k)$. Hence, $C_{Hist}(n, k) = \Theta(n + \log k)$, by eq. (1). The following theorem, given without proof, shows *Hist* to be a significant new measure of presortedness.

THEOREM 10 *Hist* $\supset$ *Block, Hist* $\supset$ *Inv, Hist* $\not\supset$ *Runs, and Loc* $\not\supset$ *Hist.*

It is easy to find sequences that are intuitively almost sorted but for which most insertions are costly regardless of whether the cost is a function of distance in time or in space. Katajainen, Levcopoulos, and Petersson [7] discussed a variant of Local Insertion Sort that exploits locality in time as well as in space, to some extent. The idea was simply to finger the k latest inserted elements, allowing more than just the most recent one to influence the insertions. The fingers were stored in a search tree, and the algorithm can be viewed as first searching backwards in history to refine the target interval in the finger tree, and then searching the interval in space, starting from both endpoints simultaneously. They also provided a technique to dynamize the algorithm, such that an appropriate number of fingers is allocated during the sorting process. The resulting algorithm is never asymptotically slower than Local Insertion Sort, and significantly faster on infinitely many sequences. However, a major drawback is that the historical search is not sufficiently adaptive; it is just binary, while the search in space is exponential and binary. Therefore, each insertion takes $\Omega(\log k)$ comparisons.

In order to improve upon the above variants of Insertion Sort, let us make an analogy in which their weaknesses and strengths become even more apparent. In the following, 'remembering' and 'watching' correspond to searching in history and space, respectively. Then, Straight Insertion Sort does not remember anything and is blind. A-Sort does not remember anything either, but it has good eyes. Local Insertion Sort has good eyes and one unit of memory, but this is not good enough. Historical Insertion Sort remembers everything but is blind. Finally, the multiple finger algorithm has a reasonably good memory and good eyes; however, it is unable to see till it has finished remembering. It is evident that a better algorithm should have both a good memory and good eyes, as well as being capable of combining these qualities efficiently.

This leads us to *Regional Insertion Sort*, where the assumption made is that most insertions will occur close to *some* element which was inserted recently, but neither necessarily close to the most recent one, nor immediately adjacent to a recent element. Hence, the new algorithm is more generous than its ancestors in what sequences are regarded as presorted.

Before presenting the algorithm, we describe an associated measure of presortedness. When inserting x_i, suppose we first traverse history to find an element from which to start searching in space. Then the shortest possible total distance covered is given by

$$r_i = \min_{1 \le t < i} \{t + d_{i,i-t}\},$$

where t and $d_{i,i-t}$ correspond to the time and space components, respectively. Note that choosing $t = 1$ gives $r_i \le 1 + d_{i,i-1} = 1 + d_i$, and choosing $d_{i,i-t} = 1$ gives $r_i \le t_i + 1$. Hence, r_i combines the distances in time and space, and can be viewed as two-dimensional. We define *Reg* as the product of the r_i's:

$$Reg(X) = \prod_{i=2}^{n} (r_i - 1),$$

where we subtract one to get the value 1 rather than 2^{n-1} when X is sorted. Replacing d_i by r_i in Lemma 7 gives $C_{Reg}(n,k) = \Omega(n + \log k)$, which is the bound that Regional Insertion Sort is to match.

To adapt to *Reg* Regional Insertion Sort performs an exponential and binary search in history and space simultaneously, interleaving the operations. The sorted sequence is stored in a finger tree, which supports the search in space. For the search in history, we maintain a data structure which essentially consists of fingers into the finger tree.

We first spend one comparison in history to find the closest two fingers among the two most recent ones. Second, we search distance two in space from the fingers. If we have not found the insertion position, we spend two comparisons in history to find the closest fingers among the four most recent ones, followed by searches in space within distance four from each of the found fingers. In this way we interleave the searches. In the ℓ'th round the 2^{2^ℓ} most recent elements influence the search in history, and the insertion position is found if it is within distance 2^{2^ℓ} from one of these elements, in which case the search stops, and we insert the element in the finger tree and a finger to the element in the historical search structure.

Suppose that the search stops after ℓ rounds and $O(2^\ell)$ time. This must mean that the search width $w = 2^{2^\ell}$ has become large enough that within the w most recently inserted elements there is at least one that is within w elements spatially of the insertion position. Let w_i be the minimum search width required in the i'th insertion to guarantee termination of the search:

$$w_i = \min_{1 \leq w < i} \{w \mid \exists t : 1 \leq t \leq w \text{ and } d_{i,i-t} \leq w\}$$

Then, by construction, the searching time during the i'th insertion is $O(\log w_i)$.

LEMMA 11 $w_i \leq r_i$.

PROOF The claim follows if we can show that

$$\exists t : 1 \leq t \leq r_i \text{ and } d_{i,i-t} \leq r_i,$$

since this is sufficient to establish r_i as a member of the set that w_i minimizes.

Consider the value t that minimizes r_i, that is, $r_i = t + d_{i,i-t}$. Since $t \geq 1$ and $d_{i,i-t} \geq 1$ we have both that $t \leq r_i$ and that $d_{i,i-t} \leq r_i$, and so t meets the specification. This gives the lemma. $\square$

Hence, the number of comparisons spent in the searches by Regional Insertion Sort is

$$O\left(\sum_{i=2}^{n} \log w_i\right) = O\left(\sum_{i=2}^{n} \log r_i\right) = O(n + \log Reg(X)).$$

Since the main purpose of this paper is demonstrating the relations between new and known measures of presortedness, we have concentrated on the number of comparisons, rather than the time, consumed by the presented algorithms, because these establish upper bounds on $C_M(n, k)$. Regional Insertion Sort needs some more details even for bounding the number of comparisons. We have not described how the searches and insertions in the historical data structure can be carried out. The searches are supported in $O(\log t)$ time by the Historical Search Tree, mentioned above. The insertions we are demanding seem much harder to implement efficiently. However, if we just count comparisons, we can rebuild the entire data structure prior to each insertion in no comparisons and linear time, by essentially copying the sorted sequence stored in the finger tree. We therefore conclude that $C_{RIS}(X) = O(n + \log Reg(X))$, and hence

THEOREM 12 $C_{Reg}(n,k) = \Theta(n + \log k)$.

As *Reg* is a generalization of *Loc* and *Hist*, the following theorem is expected.

THEOREM 13 *Reg* $\supset$ *Loc and Reg* $\supset$ *Hist*.

What might be more surprising is

THEOREM 14 *Reg* $\supset$ *SMS*.

PROOF Recall the definition of the measure *SMS* from Table 1. Levcopoulos and Petersson [10] proved that $C_{SMS}(n,k) = \Theta(n \log k)$. Let X be a sequence of length n with $SMS(X) = k$. We show that $Reg(X) = k^{O(n)}$, from which it follows that

$$C_{Reg}(|X|, Reg(X)) = O(|X| \log SMS(X)) = C_{SMS}(|X|, SMS(X)),$$

proving *Reg* $\supseteq$ *SMS*, by Definition 3.

Let X_j be any monotone subsequence in a decomposition of X into k monotone subsequences. Consider the sum of the r_i's taken over all elements in X_j:

$$\sum_{x_i \in X_j} r_i = \sum_{x_i \in X_j} \min_{1 \le t < i}\{t + d_{i,i-t}\} \le \sum_{x_i \in X_j} t(i) + d_{i,i-t(i)},$$

where $t(i)$ may take any value $1 \le t(i) < i$. We bound the sum from above by picking suitable values of $t(i)$. If x_i is the first element in X_j, let $t(i) = i - 1$. Otherwise, suppose x_ℓ is the element preceding x_i in X_j, in which case we let $t(i) = i - \ell$. Then the sum of the $t(i)$'s will telescope, and total to at most n. Similarly, the sum of the $d_{i,i-t(i)}$'s cannot exceed $2n$, since each of the elements *not* in X_j can be in at most one of the partitions created by the elements in X_j, and the first element in X_j contributes at most an additional n to the sum. We conclude that $\sum_{x_i \in X_j} r_i \le 3n$. As this applies for every monotone subsequence, we have $\sum_{i=2}^{n} r_i \le 3kn$. Since $Reg(X) = \prod_{i=2}^{n} r_i$ is maximized when the r_i's are about the same, we have

$$Reg(X) \le \prod_{i=2}^{n} 3k = k^{O(n)},$$

which proves the first part of the claim, as stated above.

To show that the superiority is strict, consider the sequence X obtained from the sorted sequence by permuting the first $\sqrt{n}$ elements, so that *SMS* is maximized for that part of the sequence. Then $SMS(X) = \Theta(\sqrt[4]{n})$, while $Reg(X) \le \sqrt{n}^{\sqrt{n}}$. As $C_{Reg}(n, \sqrt{n}^{\sqrt{n}}) = \Theta(n)$ and $C_{SMS}(n, \sqrt[4]{n}) = \Theta(n \log n)$, Lemma 5 implies that $SMS \not\supseteq Reg$, which concludes the proof. $\qquad\square$

We close this section with a theorem summarizing our knowledge of measures of presortedness with respect to $\supseteq$:

THEOREM 15 *Figure 9 shows all the relations among the measures.*

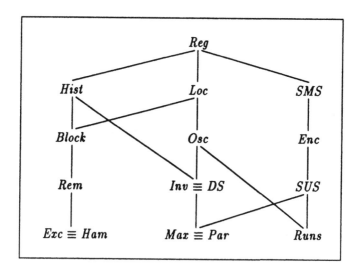

FIGURE 3: $\mathcal{P}(\mathcal{M}, \supseteq)$ after insertion of *Loc*, *Hist*, and *Reg*.

7   Conclusions

The partial order on measures of presortedness is important for two reasons. Firstly, the formalism we have introduced for comparing measures of presortedness and the partial order established by this formalism mean that any proposed new measure can have its usefulness evaluated using the partial order as a yardstick. Moreover, since every sorting algorithm defines a curve crossing the partial order, the order can also be used for evaluating the usefulness of any adaptive sorting algorithm. The transitivity of the relation $\supseteq$ helps this process, since it is not necessary for the adaptivity of the algorithm to be investigated for every measure.

At the top of the partial order we have described three new measures, *Loc*, *Hist*, and *Reg*. Local Insertion Sort is *Loc*-optimal, and we have sketched algorithms adapting to the other two measures. In this paper our primary interest has been in developing the partial order, so we have only needed to bound the number of comparisons required by these algorithms. The efficient implementation of Historical and Regional Insertion Sort remains an open problem. To date our best implementations require $\Omega(n \log \log n + \log k)$ time [13], and so are not optimal. Should it exist, a *Reg*-optimal implementation of Regional Insertion Sort would be optimal for all known measures of presortedness.

Acknowledgements

The first author is grateful to Christos Levcopoulos, with whom the results brought together in Section 5 have been reported in a series of joint papers. We would also like to thank Svante Carlsson, Jyrki Katajainen, Heikki Mannila, and Arne Andersson for stimulating discussions. This work was in part supported by The Australian Research Council.

References

[1] W.H. BURGE. Sorting, trees, and measures of order. *Inform. and Control*, 1:181–197, 1958.

[2] S. CARLSSON, C. LEVCOPOULOS, AND O. PETERSSON. Sublinear merging and Natural Mergesort. *Algorithmica*. To appear. Prel. version in *Proc. Internat. Symp. on Algorithms*, pages 251–260, LNCS 450, Springer-Verlag, 1990.

[3] C.R. COOK AND D.J. KIM. Best sorting algorithms for nearly sorted lists. *Comm. ACM*, 23:620–624, 1980.

[4] G.R. DROMEY. Exploiting partial order with Quicksort. *Software—Practice and Experience*, 14:509–518, 1984.

[5] V. ESTIVILL-CASTRO AND D. WOOD. A new measure of presortedness. *Inform. and Comput.*, 83:111–119, 1989.

[6] M.L. FREDMAN. How good is the information theory bound in sorting? *Theoret. Comput. Sci.*, 1:355–361, 1976.

[7] J. KATAJAINEN, C. LEVCOPOULOS, AND O. PETERSSON. Local Insertion Sort revisited. In *Proc. Internat. Symp. on Optimal Algorithms*, pages 239–253. LNCS 401, Springer-Verlag, 1989.

[8] D.E. KNUTH. *The Art of Computer Programming, Vol. 3: Sorting and Searching*. Addison-Wesley, Reading, MA, 1973.

[9] C. LEVCOPOULOS AND O. PETERSSON. Adaptive Heapsort. *J. Algorithms*. To appear. Prel. version in *Proc. 1989 Workshop on Algorithms and Data Structures*, pages 499–509, LNCS 382, Springer-Verlag, 1989.

[10] C. LEVCOPOULOS AND O. PETERSSON. Sorting shuffled monotone sequences. *Inform. and Comput.* To appear. Prel. version in *Proc. 2nd Scandinavian Workshop on Algorithm Theory*, pages 181–191, LNCS 447, Springer-Verlag, 1990.

[11] H. MANNILA. Measures of presortedness and optimal sorting algorithms. *IEEE Trans. Comput.*, C-34:318–325, 1985.

[12] K. MEHLHORN. *Data Structures and Algorithms, Vol. 1: Sorting and Searching*. Springer-Verlag, Berlin, Germany, 1984.

[13] A.M. MOFFAT AND O. PETERSSON. Historical searching and sorting. In *Proc. Internat. Symp. on Algorithms*, pages 263–272. LNCS 557, Springer-Verlag, 1991.

[14] O. PETERSSON. *Adaptive Sorting*. PhD thesis, Department of Computer Science, Lund Univ., Lund, Sweden, 1990.

[15] S.S. SKIENA. Encroaching lists as a measure of presortedness. *BIT*, 28:775–784, 1988.

Author Index

Lecture Notes in Computer Science

For information about Vols. 1–535
please contact your bookseller or Springer-Verlag